CW00749718

◎ BEING HUMAN ◎
IN A
BUDDHIST WORLD

◎ BEING HUMAN ◎
IN A
BUDDHIST WORLD

AN INTELLECTUAL HISTORY OF MEDICINE IN EARLY MODERN TIBET

JANET GYATSO

COLUMBIA UNIVERSITY PRESS *NEW YORK*

Columbia University Press would like to express
its appreciation for assistance given by

◎ ◎ ◎

A special thanks to the Shelley & Donald Rubin Foundation
for crucial financial support toward the publication of this book.

COLUMBIA UNIVERSITY PRESS
Publishers Since 1893
New York Chichester, West Sussex
cup.columbia.edu

Library of Congress Cataloging-in-Publication Data
Gyatso, Janet, author.
Being human in a Buddhist world : an intellectual history of medicine in early
modern Tibet / Janet Gyatso.
pages cm
Includes bibliographical references and index.
ISBN 978-0-231-16496-2 (cloth : alk. paper) — ISBN 978-0-231-53832-9
(electronic)
1. Buddhism—Tibet Region—History. 2. Medicine, Tibetan—History. 3.
Medicine—Religious aspects—Buddhism. I. Title.
BQ7584.G83 2015
294.3'366109515—dc23
2014005127

Columbia University Press books are printed on permanent
and durable acid-free paper.
This book is printed on paper with recycled content.
Printed in the United States of America
c 10 9 8 7 6 5 4 3 2 1

COVER AND FRONTISPIECE ART: *Tibetan Medical Paintings: Illustrations to
the* Blue Beryl *of Sangye Gyamtso (1653–1705).* Plate: 9, detail.
COVER AND BOOK DESIGN: Lisa Hamm

CONTENTS

ILLUSTRATIONS

ACKNOWLEDGMENTS

Working on this project has been immensely pleasurable and engrossing. It has also posed considerable challenges in the research it called for and especially in matters of interpretation. The challenges were very much ameliorated and the pleasure greatly enhanced by my fortune in having the occasion to discuss many of the book's issues with my learned colleague (and erstwhile student) Yang Ga, now associate professor at the Tibetan Medical College in Lhasa. This consultation took place during the period when he was a doctoral student in Inner Asian and Altaic Studies at Harvard, between 2003 and 2010. Not only was Yang Ga crucial to the project in pointing me to key resources and helping me to puzzle through difficult passages. He also has been a true partner in thought, an exemplary combination of his traditional education in medicine and his modern academic one, learning from both how to be critical of easy readings and assumptions. He is also very finely attuned to rhetorical ploy. This book would be greatly diminished without his generous collaboration, if it were even possible at all.

I have also had the benefit of illuminating discussions on the history and theory of medicine with Tupten Püntsok, now at the Southwest Minorities University in Chengdu. He was the first to introduce me to the riches of the *Four Treatises* and the thought of Zurkharwa Lodrö Gyelpo, during a course on medicine that he gave at the Shang Shung Institute in Conway, Massachussetts, more than a decade ago. I am indebted too to Sonam Chimé of the

Tibetan Medical College and other senior students of the Venerable Troru Tsenam in Lhasa for providing me with photocopies of manuscripts, and to Denpa Dargye at the China Tibetology Research Center and Dr. Tashi at the Tibetan Beijing Hospital for answering my questions on pulse diagnostics, as kindly facilitated by Tsering Thar of the Central Minorities University in Beijing.

No one has been more stimulating to my thinking on the conceptual issues of this book than Charles Hallisey. He read many versions of parts of this project and his probing comments, not to mention countless references in fields far beyond Asian Studies, prompted me to frame the questions of the book in terms of large foundational issues for the history of thought. Conversations with Steven Collins over the years have contributed critically to my ongoing interests in the nature of historical thinking and the grounds of truth claims. I am also grateful to Sheldon Pollock for inviting me to contribute an essay to a collection on knowledge in early modern South Asia, which led me to recognize central issues in this project around modernity. Responses to an essay on some of these same issues at a faculty workshop at Harvard Divinity School supported by the Richard T. Watson Chair in Science and Religion brought my comparative thinking on early modern science in Europe to greater precision, especially comments by Amy Hollywood and Mark Jordan. I am grateful too for incisive questions on a set of lectures that I gave at the Wellcome Trust Center for the History of Medicine in London from Harold Cook, Vivienne Lo, and Ronit Yoeli-Tlalim. A lively colloquium organized by Indrani Chatterjee at the Rutgers Center for Historical Analysis also inspired me to broaden the scope of the questions I was bringing to the project at a formative moment in its development.

The late E. Gene Smith generously gave me many references in Tibetan cultural history over the course of this project; indeed, the impact of his enormous knowledge, always bestowed unstintingly, on my entire career in Tibetan Studies cannot be overestimated. Conversations with Matthew Kapstein over the years have also had a great impact on my appreciation of the history of Tibetan religion and society; I am particularly grateful as well for the very detailed comments he made on the manuscript as a reader for Columbia University Press. Kurtis Schaeffer made extremely helpful comments as well in the same capacity, as did Vesna Wallace, and this book has benefited accordingly. Critical input from David Shulman vastly improved the quality of the translations in this book. I owe thanks to many other colleagues and students in Tibetan Studies for helpful references and illuminating conversations on aspects of this project, especially Jake Dalton, who conveyed comments from his seminar at the University of California at Berkeley on a penultimate draft of the book manuscript, as well as Vincanne Adams, Ben Bogin, Bryan Cuevas, Rae Dachille, Georges Dreyfus, Frances Garrett, David Germano, Jonathan Gold, Amy Heller, Theresia Hofer, Sarah Jacoby, Samten Karmay, Deborah Klimberg-Salter, Leonard van der Kuijp, Donald Lopez, Robert Mayer, Arthur McKeown, Willa Miller, Robert Pomplun, Andy Quintman, Joshua Schapiro, Michael Sheehy, Elliot Sperling, Heather Stoddard, Tashi Tsering, and Gray Tuttle. I am also grateful to colleagues in Buddhist Studies for their assistance and reflections, including Collett Cox, Michael Hahn, Paul Harrison, Maria Heim, Robert Kritzer, Ryan Overbey, and Robert Sharf; to Sokhyo Jo, Natalie Koehle, Philip Kuhn, Michael Puett, Michael Radich, James Robson, and Pierce Salguero for very helpful suggestions, references, and ideas regarding East Asian science and cultural history; and to Daud Ali, Shrikant Bahulkar, Donald Davis, Wendy Doniger, Paul Dundas, Robert Goldman, D. Christian Lammerts, Anne Monius, Narayana Rao, and John Taber for an array of comparative reflections on the history of science and related fields in South Asia.

This project was the beneficiary of a fellowship from the National Endowment for the Humanities and a grant from the American Council of Learned Societies, which allowed me leave from my teaching responsibilities to complete the book manuscript. I was also fortunate to be a visiting fellow at the Institute for Research in the Humanities at the University of Wisconsin at Madison at an early moment in the project, during which time I was able to participate in the wonderful and wide-ranging conversations there in cultural history and the history of science. Finally, a generous grant from the Robert H. N. Ho Family Foundation allowed the inclusion of color illustrations in this publication, for which I am very grateful.

Expert editorial assistance was provided by three doctoral students: William McGrath of the University of Virginia, and Elizabeth Angowski and Ian MacCormack in Religion at Harvard. I am grateful to all three for their care and precision in helping to prepare the manuscript for publication. I would also like to express my appreciation for Wendy Lochner's supportive guidance throughout the publication process at Columbia, and the excellent assistance of Christine Dunbar, Jordan Wannemacher, Lisa Hamm, and Leslie Kriesel.

Finally and most deeply, I am grateful to my dear husband, Charles, for his constant love and encouragement over the many, many moons of writing this book. May the fruits of this effort give something good to the future, that to which he ever points my attention.

A TECHNICAL NOTE

All Tibetan proper names are spelled phonetically in this book, with the exception of the bibliography and bibliographical references in the notes, where they are given in transliteration according to the Wylie system. Tibetan book titles that recur repeatedly are translated into English, with Tibetan bibliographical references in the notes; a few works that are only mentioned once or twice in the discussion are not translated and are given in Tibetan transliteration. The large majority of Tibetan technical terms are provided in transliteration and usually in parentheses or in the notes, but several terms that appear frequently in the discussion or already have widely established usage in Western-language contexts are spelled phonetically. The transliteration for all phonetically spelled Tibetan terms and names may be found in the index.

The gender of pronouns referring to Tibetan doctors are masculine in this book, since from everything the texts indicate, the practitioners of Sowa Rikpa were virtually, if not entirely male until well beyond the period under discussion, and the lines of medical inheritance explicitly patrilineal.

ABBREVIATIONS

Aṣṭāṅga[hṛdaya]	Murthy 1991; Das and Emmerick 1998
Abhidharmakośa[bhāṣya]	Shastri 1970–73
Caraka	Ram and Dash 1976
Ci.	Cikitsāsthāna
Kaneko	Kaneko 1982
PT	Pelliot tibétaine
TBRC	Tibetan Buddhist Resource Center; www.TBRC.org
Toh.	Ui 1934
Śā.	Śārīrasthāna
Sū.	Sūtrasthāna
Suśruta[saṃhitā]	Bhishagratna 1998–99
Ut.	Uttarasthāna
Vi.	Vimānasthāna

Plate and vignette numbers (e.g., pl. 2.3) refer to Parfionovitch 1992, vol. 2.

◎ BEING HUMAN ◎
IN A
BUDDHIST WORLD

INTRODUCTION

This book studies how knowledge changes. What enables epistemic shift, and what constrains it? How do historians recognize such a shift? If it is not a full rupture with the past, how should we weigh the relative import of continuity and difference? What impact do conceptual loyalties have on the possibility of new thought, or the articulation of unique and particular experience? Do norms function differently when they are enshrined in writing and when they are construed by the senses? Under what circumstances may norms be jettisoned? What role do deference, prestige, and rhetoric have in the formation of knowledge? And what work is done by the cachet of alterity and resistance?

In plumbing the annals of academic medicine in Tibet, this book seeks to account for double movements. It studies how medical learning, a mix of the main Asian health-care systems of its day, fostered a probative attitude to religious authority, even as it grew to maturity within the great institutions of Tibetan Buddhism. Yet medicine remained deeply informed by religious cosmologies, in ways that both facilitated and inhibited scientific thinking. Tibetan medical scholars both drew on and held at bay their religious heritage. In so doing, they sought how best to attend to the everyday ills of their patients, on principles legible and credible in a Buddhist world.

This project ponders key issues for the history of science, including the disjunctions—and conjunctions—between scientific approaches to knowledge and religious ones. It studies moments when learned physicians set aside revealed scripture in favor of what they observed in the natural world. It finds medical theorists resisting ideal system of any pedigree, and endeavoring instead to account for idiosyncrasy and unpredictability. They relied most of all on traditions of medical knowledge from the larger world around them. But they adapted a range of methods and concepts developed in Buddhist contexts as well—from epistemology to ethical discourse to meditative visualization—in novel ways that could inform scientific aims. They also did so in order to account for aspects of human embodiment that are not determined by material conditions alone. This book studies the methodological self-consciousness that allowed certain leading medical theorists to intentionally mix disparate streams of thought and practice. That meant confronting, in unprecedented ways, the possibility that the Buddha's dispensation did not encompass everything that needed to be known for human well-being.

◎ ◎ ◎

Tibetan religion, particularly the philosophies and practices of Buddhism, has received much scholarly attention in the last century. But there has been little recognition of knowledge traditions in Tibet that were not primarily religious in nature. This book seeks to correct the imbalance by studying an outstanding example of alterity in Tibetan intellectual and cultural history. Medicine in Tibet underwent a major transformation at the same time that Buddhism itself was entering and becoming domesticated there, from the seventh through the twelfth century, coming into its own in the early modern period, and culminating during the rule of the Ganden Podrang government in the seventeenth to eighteenth century.[1] But medicine had different roots. Although early Indian Buddhism had fostered palliative care and reflected knowledge of human physiology, the Tibetan tradition of Sowa Rikpa drew most centrally on the full-service medical treatises of Indian Ayurveda, Galenic and Islamic medical conceptions from western Asia, Chinese medicine, and other old strands of knowledge on the Tibetan plateau. Medicine in Tibet did indeed grow, like Buddhism, into a highly scholastic tradition, producing an estimable body of historical and commentarial literature of its own. It was also influenced by concepts and ideals in the Buddhist literature being translated from Indic languages and then composed in Tibetan during the same period. The classical medical text in Tibet, the *Four Treatises*, even takes the exceptional step of framing itself as a teaching originally preached by the Buddha in his form as Bhaiṣajyaguru, the "Medicine Buddha." But Tibetan medicine continued to operate out of an explicitly worldly ethos and a distinctive sense of the empirical grounds for knowledge, and often adapted what it was taking from Buddhist heritages in novel ways suited for medical science. This makes for a fascinating and instructive history. Given the overwhelming hegemony of Buddhist knowledge systems in Tibet—and in spite of many moves to keep medicine under the purview of the Buddha's teachings—the gestures that medicine managed in the direction of autonomy are no less than astonishing. The vignettes explored in this book thus have a lot to tell us about what it takes for knowledge to recast its foundations, on conceptual and rhetorical registers alike.

Among the many sites of both engagement and disjuncture between Tibetan medicine and Buddhist formations is the esoteric branch of Buddhism often called *tantra*. Its contribution to everyday medicine is among the several interests of this book. In general, tantra represents a medieval development across Indian religions that produced some of the most outlandish practices, cosmologies, and mythologies in Asian history. This was so for tantra's Buddhist forms too. Yet tantric literature also describes the human body in considerable detail. While modern scholars have long puzzled over its transgressive sexual and violent practices or tried to penetrate the religious import of its theories of embodiment, this project studies another dimension of tantra's significance, that which was of interest to physicians.[2] Several of the ensuing chapters follow medical thinkers as they work through tantric anatomical and physiological categories, arguing about whether they should be taken literally or figuratively and finding ways that the insights of an arcane spiritual tradition might sometimes be useful for their purposes.

The project takes as its starting point the period of the "Great Fifth" Dalai Lama, who consolidated the Tibetan state in the seventeenth century. The Dalai Lama's new government was itself deeply grounded in a tantric Buddhist worldview. That already tells us a lot about the political stakes in medicine's assessment of traditional Buddhist knowledge. The book begins with the brilliant regent Desi Sangyé Gyatso, protégé of the Great Fifth, who endeavored to lift up medicine with state patronage. Amid a host of other innovations in knowledge and the instruments of governance in the capital, medicine stands out for its self-consciousness and for its voluminous scholarly output.

Seeking to understand medicine's status at the height of the Tibetan Buddhist state, the book works back through leading theorists of previous centuries, reaching finally into the roots of academic Tibetan medicine in the *Four Treatises,* composed by the mastermind Yutok Yönten Gönpo in the twelfth century. At this seminal moment in Tibetan civilization, Yutok and his disciples created an amalgam of medical knowledge on the basis of existing traditions in Tibet, along with works translated into Tibetan at the court of the seventh- to ninth-century kings and then again in small medical circles up through his own period. The study of medicine came to be housed in special academies at Buddhist monasteries. And yet the sociology of medicine remained distinct from monasticism. By the twelfth century, medicine had already begun to separate itself by virtue of its quotidian professional ethics, as well its own historiography and specialized textual corpus. By the late seventeenth century a freestanding institution for medical learning flourished in the Tibetan capital, atop a craggy mountain across from the Potala Palace.

Alas, the Chakpori Medical College that the Desi built in Lhasa no longer stands. Nor did Tibetan medicine produce case-study records, as are so plentifully available for Chinese medicine since the Song.[3] We know little of the everyday reality of medical practice in Tibet prior to the twentieth century. This book works largely through texts, along with one extraordinary visual record. It leaves aside the many oral medical traditions in Tibet, the medical knowledge of the Bönpos (practitioners of Bön, Buddhism's alter ego on the Tibetan religious scene), and especially the very huge array of ritual therapies—another chapter in the larger history of Tibetan healing practices, to be sure.[4] The focus is upon a select set of issues in Yutok's *Four Treatises* and certain commentaries and related literature, which I argue were formative in the development of the area of inquiry known as Sowa Rikpa, "the science of healing," or what I will usually just call "medicine."[5] I am most interested in the intellectual and cultural issues around medical knowledge, rather than the exigencies of clinical practice per se, although matters relating to the latter will have considerable significance as well.

◎ ◎ ◎

In the chapters that follow I deploy the categories of science and religion as heuristics, to highlight the processes by which medicine forged a certain distance between itself and the ways of knowing associated with ideals of human perfection and supernatural realms. Both categories have been the subject of critical interrogation in recent decades, and both are invoked here in relatively loose ways. The looseness not only speaks to the fact that the pair does not exactly map onto a parallel dyad in Tibetan or Sanskrit, but also acknowledges that whatever purchase the two rubrics do have in referencing groupings consolidating on the ground does not yield a ready ironclad distinction between them. Often in this book issues of the religious will be framed in terms of things Buddhist, and occasionally things Bön, as well as resonances in Ayurveda from other Indian religions. I will usually be referring to those aspects of the Buddhist dispensation that were trained upon timeless metaphysics, ideals of ascetic and ritual forms of self-cultivation, otherworldly cosmologies, and appeals to omniscient epistemic authority. But the entire historiographical rubric of Buddhism is fraught, a matter to which I will return below.

Problems attend the category of science too. The terms *rig pa,* as it occurs in Sowa Rikpa (i.e., *gso ba rig pa*), and again *dpyad pa,* as in *gso dpyad* or *sman dpyad*—which I render as "medical science" or some variant thereof—do name kinds of investigation that aim to foster critical and often empirically based ways of investigating and knowing about the world. But certainly neither term includes everything that the modern notion of science denotes, in either kind or degree. What is more, both *rig pa* and *dpyad pa* are used frequently in Buddhist contexts to denote critical ways of knowing with transcendent or spiritual aims. In fact, a common Buddhist grouping of five kinds of *rig pa,* or Sanskrit *vidyā,* often translated in modern scholarly literature as "the five sciences," counts medicine and salvific realization as two of its members. But I am less interested in terminological parity than in the epistemic orientations—conceptual and practice-based centers of gravity, if you will—that defined Tibetan medicine. I argue that these were reaching toward a scientific sensibility.

Although it did take into account the impact of moral factors on health, medicine focused on the physical treatment of illness.[6] This makes for a notable contrast with the many ritual means in Tibet for addressing illness, which may more readily be conceived as religious healing. This is not to say that everything associated with Sowa Rikpa eschewed ceremony or meditative practice. The *Four Treatises* does allude, albeit only occasionally and briefly, to ritual means to enhance the efficacy of medical practice, particularly for the treatment of demon-caused disease, but also to enhance the physician's self-conception as he makes medicine; it even provides a magical means, adapted from Indian medicine, to change the sex of a fetus.[7] The cycle of ritual teachings, called *Heart Sphere of Yutok,* that grew up around the figure of Yutok Yönten Gönpo, as well as the continued presence of ritual assemblies at medical colleges like Chakpori and certainly the Fifth Dalai Lama's own investment in the ritual transmission of medical knowledge are all signs of the complex intersection of religion, medicine, and matters of state, at least by the sixteenth century.[8] But the vast majority of the medical theory and therapy in the *Four Treatises* and its commentarial legacy betrays a prevailing and, I argue, defining concern with material etiologies and material remedies. The following chapters will focus on several outstanding instances of such concerns and their implications for medicine vis-à-vis the formations of religion in Tibet up through the seventeenth century.

A key piece of medicine's defining epistemic horizon can be detected in the attention the *Four Treatises* and related literature pay to the vicissitudes of ordinary human life. Coming from a Buddhist Studies background in twentieth-century American academia, a training focused on doctrinal systems and religious ideals, I have been struck by the nonjudgmental and matter-of-fact tone, in Tibetan medical discourse, regarding things like human sexuality and competitive ambition. Indeed, the medical treatises address any number of human foibles and bodily imperfections, all laid out with no final solution in sight save keeping oneself comfortable until one's eventual and certain demise. Of course, to describe illness, not perfection, is exactly what medicine is supposed to do. What's more, everyday foibles and ambitions are portrayed throughout the history of Buddhist literature too, and not always in derogatory terms, a point that influences the analysis of the *Four Treatises*' medical ethics in the final chapter of this book. One locus for such portrayal close to the time and place in which the Tibetan medical corpus was produced is Tibetan Buddhist autobiographical writing, which sometimes addresses mundane matters in quite quotidian terms. Still, even there it is hard to find an instance of the kind of thoroughgoing materialist and pragmatic sensibilities encountered so frequently in medical works.[9] And at the other end of the spectrum, the complex agenda of medical theory also distinguishes it from other kinds of Tibetan writing on practical matters. There were systematic compositions on grammar, astrology, and law, not to mention extensive governmental and military documents and local administrative records. But the Tibetan medical corpus may well prove exceptional in both scope and nature as a highly cultivated, scholarly production that is self-conscious of its epistemic and ethical presumptions and the implications of its relationship to religious values, and yet trained on the material conditions of ordinary life.[10] This book argues that the specificities of this mix made for a distinctive medical "mentality," a particular set of sensibilities and orientations whose nature and significance I will be considering throughout the ensuing chapters.[11]

And yet by the seventeenth century, at the very moment that medicine was really coming into its own, it bumped up against another key piece of its identity in Tibet: its place in the apotheosis of the Dalai Lama and the Ganden Podrang Buddhist state. Patronage of medicine contributed to the state's prestige and power, but the growing independence of medical learning was at odds with claims to the ultimate authority of Buddhist revelation in the person of the Dalai Lama. Encroaching powers of Qosot and other Mongol tribes, not to mention the ensuing Qing dynasty, threatened the autonomy of the Tibetan government. This pressure resulted in, among many other things, an increased urgency to showcase the transcendent truths undergirding enlightened incarnation and its hegemony on the Tibetan political stage. That may go a long way in accounting for why we are by no means seeing a scientific revolution here; by no means did Tibetan medicine blossom into full-throttle modern science in the way that similar rumblings in other parts of the globe did.[12] But rumblings were there nonetheless.

◎ ◎ ◎

One of the challenges for the study of Tibetan medicine is to appreciate the significance of its trajectory without undue influence from the historiography of early modern science in Europe. But that does not mean we should fail to notice the potential comparisons. Academic Tibetan medicine from the twelfth through the eighteenth century does display movements analogous to those in other parts of the world, if in much more modest form.[13] In contrast to what Dieter Schuh starkly claims about Tibetan scholars of astronomy never actually looking at the sky, there was indeed a turn to the evidence of the empirical in some quarters of Sowa Rikpa learning and practice.[14] We find this in Tibetan physicians' use of observation and dissection to augment and question the systems of human anatomy they had inherited from both medical and Buddhist scripture. We see it as well in their growing naturalistic knowledge of medicinal botany and an accompanying effort to describe it, catalogue it, and render it visually in encyclopedic form. Such activities were sometimes accompanied by second-order reflection on what it means for the physical world to be at odds with what is written in a text.

Some Tibetan medical theorists also recognized the contingency of truth in light of regional and climatic differences, and in the shifting circumstances of history. This is not merely evidence of documentary knowledge. It also bespeaks a different paradigm for truth, a point not lost on the actors in question. Certain medical debates depart from the primacy frequently accorded, in Tibetan religious writing, to a golden age of the past. Their presumptions also depart from a metaphysical proclivity to cast the highest reality of all as timeless and universal. Instead they evince a concern for historical specificity as the grounds of scientific truth. What goes along with this is a sense that the masters of old did not necessarily fathom what has recently been realized. That opens the door for a valorization of new knowledge. Most challenging of all, attention to the local specificities of the natural world and the realities of temporal sequence gave some medical commentators the ammunition to critically reject the Buddha's authorship of the root medical text. These turns constitute momentous departures from the center of the Tibetan Buddhist hegemon.

Still, none of these shifts entailed a wholesale dismissal of all Buddhist inheritance. Nor did medicine by any means always or consistently favor the empirical over other sources of knowledge. Double movements in Tibetan medicine's intellectual history will be seen in manifold ways throughout this book. So even though reading about the trials and tribulations of naturalists and surgeons in Europe helped me to recognize certain trends gathering momentum in Tibetan medicine, the question remains of what that actually tells us. For one thing, the movements in Tibet do not appear to have been connected to or influenced by what was happening in Europe. Certainly there were Jesuits and other Europeans at the Manchu court, exerting much influence on scientific inquiry in China since at least the illustrious career of Matteo Ricci.[15] But although Tibetans had significant relations with the Manchu court themselves, it is hard to show that they were impressed by—or even registered—the scientific winds from abroad. A telling case in point is the coincidence of the Fifth Dalai Lama's mission to the Manchu court with that of the Jesuit Father Johann Adam Schall von Bell (1591–1666), a prominent astronomer. There is no evidence that the two even met. As Gray Tuttle notes, the Dalai Lama only mentions in passing von Bell's apparent ability to predict the weather, using the old Tibetan term for prophecy (*lung bstan*) and referring to the Jesuit as an Indian heterodox astrologer.[16] There were even European missionaries in central Tibet by the seventeenth century (although the most important of these, Ippolito Desideri, was in Lhasa after the period with which this book is concerned). And yet there are very detailed diaries of the Dalai Lama's daily activities and those of his retainers—not only for the

Great Fifth but also for the Sixth, whose brief reign was even more closely entwined with the Qing—in which there is no inkling, as far as I know, of Tibetan interest in European science.[17] Even on Desideri's sojourn in Lhasa, current scholarship has detected mostly the impact of Tibetan theology on the priest rather than much effect in the opposite direction.[18] So far, Tibetanists have found European influence only upon Tibetan geographical knowledge, starting in the nineteenth century, with perhaps one case of exposure to European astronomy a century before.[19]

The developments in Tibetan medicine proceeded on different grounds and out of different impulses. We might be tempted to invoke recent discussions on "alternative" or "multiple" modernities, but much of that literature is grounded in the European origins of the scientific revolution and other formations of modernity.[20] Sanjay Subrahmanyam's capacious notion of "connected histories" nonetheless remains useful for conceptualizing the complex set of factors that fed the genealogy of Tibetan medicine. Subrahmanyam's angle of vision is trained on gradual developments that resist easy distinction between tradition and modernity. To adopt that perspective means taking into account not only the cosmopolitan moment of the Fifth Dalai Lama's patronage but also the earlier patronage fostered by the seventh- to ninth-century Tibetan kings.[21] Indeed, the array of sources that the historian of Tibetan medicine should consider is daunting, stretching back at least to the turn of the Common Era, and including all the knowledge on human anatomy, *materia medica,* diagnostics, and therapeutics coming out of India, western Asia, and China, not to mention the agonistic debate culture of academic Indian Buddhism and the long engagement of Buddhist thought with theories of perception, among other things.

Scholars like Sheldon Pollock, Sudipta Kaviraj, Dominick Wujastyk, and Yigal Bronner have each recognized, at certain moments in Indian intellectual history prior to colonial contact, a historicist approach to the past, an increasing sense of regional differences in knowledge, and a willingness to criticize tradition and to valorize the new. Their historiography is exemplary for ways to analyze comparable movements in Tibetan medicine.[22] The same might be said of Benjamin Elman's study of Tibet's other great neighbor, which shows how the classical Confucian notion of "the investigation of things" and wide-ranging interests in natural history in the Ming dynasty fed the trend toward "evidentiary scholarship" (*kaozheng*) in many Chinese intellectual fields. A particularly germane comparison for the purposes of this book might be with the empirically oriented career of Kajiwara Shozen (ca. 1265–1337), a Japanese

Buddhist physician who came to reject karmic etiologies of illness in favor of physicalistic medical knowledge coming from China—although in this case innovation seems to have been unburdened by the meta-issues of authority and epistemology that were at work in academic Tibetan medicine.[23]

The movements toward evidentiary scholarship in medieval China ended up facilitating dialogue with Jesuit-brought objectivist science, particularly in mathematics, cartography, calendrical system, and astronomy, beginning in the seventeenth century (Western medicine, on the other hand, only started to be taken seriously by Chinese physicians in the nineteenth century).[24] Indian "indigenous modernities" also gained momentum through encounters with Western missionaries and colonialists, becoming "practices that modernized," as Daud Ali's rich analysis of Indian court culture suggests, and providing resources that could be relocated to newly emerging realms of civil society in the colonial period.[25] But no such use was made of the Tibetan medical movements studied in this volume, at least not until the second half of the twentieth century, when they served political ends in contemporary China.[26] Nor can that endpoint define a teleology for these movements earlier, when they were bidirectional at least, and certainly not representing any single-minded march toward modern science. To be sure, the findings of this book do raise questions about how features normally associated with modernity can also emerge in very different forms and times and with very different upshots. But the following chapters are primarily concerned with the ways and means of Tibetan medical thinking on its own terms, and how it developed on its own steam, as it were, in its particular cultural hothouse. Luckily a robust archive of medical writing over a period of six centuries allows us to track these developments closely.

◎ ◎ ◎

The book begins with the height of Sowa Rikpa's fortunes in the late seventeenth century. Chapter 1 explores one of its most impressive productions, a set of seventy-nine medical paintings whose creation was directed by the regent Desi Sangyé Gyatso. At this point medical learning had garnered unprecedented symbolic capital that directly fed state aspirations for power. The chapter studies how the Desi's medical paintings depict everyday life, anatomy, *material medica* and the practice of medicine; how they sideline Buddhist imagery; and how that all conveys broad cultural messages about the reach of the state.

Chapter 2 pulls back to the larger context and the ambitions of the medical establishment, reviewing the history of medicine in Tibet from the twelfth to seventeenth centuries and its heritage from South Asia, then focusing on its patronage by the Fifth Dalai Lama and the Desi. A leading light for medicine's growing empirical interests, the Desi was also the mastermind behind the image and power of the new Tibetan Buddhist state, and deeply invested in defending the ultimate authority of a Buddhist episteme. The chapter studies how this complex figure could defy the normal ethics of deference, how he broached the question of innovation, and how he and the rest of the Dalai Lama's court worked to preserve and advance medical knowledge.

The rest of the book goes backward in time, exploring the values and streams of thought that issued into the medical culture of the Desi's day. Chapters 3 through 5 track a series of disputes around textual authority and religious practice that came to articulate other ways of knowing based on history, material conditions, and the observable body. Each is brought to a head by Zurkharwa Lodrö Gyelpo, probably Tibet's most brilliant medical writer, who flourished a century before the Desi. Chapter 3 explores the long-raging debate about the authorship of the root text of Tibetan medicine, the *Four Treatises,* which had been attributed to the Buddha himself. But some scholars pointed to signs that betrayed the *Treatises'* more quotidian Tibetan origins. The chapter studies their cautious but ultimately subversive arguments for a historical human author and for a critical approach to mythological language, even while preserving the virtues of enlightened authorship.

Chapter 4 examines how discrepancies between ordinary sense perception, authoritative text, and idealized maps for meditative practice were adjudicated for medical anatomy. Opening with a physician from the Fifth Dalai Lama's court who publicly dissected some human corpses in order to count their bones, the chapter then turns back again to previous centuries. Apparently a question had arisen as to why the channels of the body described by tantric yogic manuals are not seen in the human corpse. Medical theorists tried to solve the problem by identifying parts of the empirical body with the tantric anatomy. Zurkharwa refused this. He proposed another solution that betrays a commitment to empiricist standards and at the same time a renewed recognition that yogic knowledge could help elucidate the connection between bodily materiality and subjective experience.

Chapter 5 studies a case where the commentators were exercised to correct an empirically untenable gender difference in the heart's position, suggested by the medical text itself. They did so by subtly changing the text's anatomical specifications, but also by recourse to tantric gender associations, which Zurkarwa would later suggest were but conventions. A bold physician from the eighteenth century went even further to reject the tantric explanation altogether, still preserving a subtle gender difference but proclaiming his own copious clinical experience to prove that the heart leans to the left in both sexes.

A Coda to chapters 3 through 5 reviews the Desi's reception of these several debates and highlights their agonistic and often self-serving rhetoric—habits in knowledge formation that characterized medical culture in Tibet throughout its history. The Desi's own rhetoric tips toward the authority of Buddhist revelation despite his contributions to the autonomy of medical learning on other fronts. This suggests the conservative side of medicine's state-level significance, echoes of which are still discernible in contemporary Tibetan medicine.

Chapter 6 turns to further rhetorical dimensions, now back at the formative moment of the late twelfth century, asking what impact medicine's attention to the empirical body might have had on the representation of women and gender. The *Four Treatises* was distinctive, in comparison to Indian Ayurveda, in noting that the male body should not be normative for general medicine. In some ways medicine displayed an open view on sexual identity and recognized a third sex, even valorizing its symbolic implications—in stark contrast with the exclusionary restrictions on the third sex in Buddhist monasticism. Some centuries later Zurkharwa came to articulate a category that specifically names gender as distinct from genital anatomy. Yet the medical treatise also has strident misogynist passages that echo similar language in Buddhist scriptures, here marshaled in a bald defense of patriliny and patriarchal privilege. The chapter reminds us that medical writing could serve as a site of social negotiation.

Chapter 7 looks at the early culture of Tibetan medicine's professional ethics and its special values and ways of learning. A twelfth- or thirteenth-century commentary to the *Four Treatises*' chapter on medical ethics provides an intricate description of key virtues based on clinical experience, including intimacy with the teacher, dexterity, artistry, nimble communication skills, and a keen appreciation of the absoluteness of death. It defines a "way of humans" that is trained upon compassion for patients, but is comfortable telling young medical students how to get ahead at the expense of their colleagues. The commentary speaks as well of the illustrious status of medicine in the royal period, offering a glimpse of the culture of prestige that fed the drive for medical excellence. It also illustrates how Buddhist scholastic categories could be adapted to articulate medical values. The human way of practicing medicine shaped the defining features of what I am calling the medical mentality, which held sway up to the height of Sowa Rikpa's achievements in seventeenth- and eighteenth-century Tibet.

The conclusion to the book reflects on the conceptual challenges Tibetan medicine faced and the strategies it marshaled to further medical knowledge and attend to patients in locally credible ways. It also turns for a final look at the complex interaction of medicine with Buddhist formations, and proposes ways to account for both influence and difference.

◎ ◎ ◎

I have moved backward in time rather than forward in order to start with the heyday of Tibetan medicine under the patronage of the Great Fifth and the Desi. The Desi's career perfectly encapsulates the uneasy fit of Buddhist revelation with scientific investigation in medicine. But the hermeneutical acrobatics in his medical writings are amply anticipated in the scholars who preceded him. So much of the intellectual history of Sowa Rikpa has been occupied with its position vis-à-vis Buddhist structures of knowledge and authority. That relationship makes for one of the central problematics of this book.

Pointing to ways that medicine in Tibet tested the cultural hegemony of Buddhism of course begs the question of what *that* category denotes, as already suggested. This is so especially for the noun but also for the adjective. In using such labels we risk complicity with long-standing habits of reification in modern scholarship on Buddhism, whose own historical conditions and agendas have received well-deserved scrutiny.[27] Some of these habits are changing as we begin to take better stock of the diversity to be discovered in vernacular sources and the ethnographic record, not to mention the normative scholastic documents themselves.[28] That diversity tells us that we need to be cautious in deploying totalizing constructs. Even on the limited set of issues raised in this book there are a wide range of stances and sensibilities, even if we confine ourselves to something like "Tibetan Buddhism," or Gelukpa Buddhism, or even the Buddhism of the Dalai Lama, for that matter. And that is not to mention the heterogeneity of whatever goes by the name of Indian Buddhism; although medical theorists drew on notions of karma or yogic anatomies or technologies of contemplation that they took from

Buddhist texts and practice communities in Tibet, from the larger South Asian perspective it is not always easy to specify what is distinctively "Buddhist" about those things either.

But if what has gone under the heading of Buddhism is far from univocal and the rubric cannot actually point to any bounded referent on the ground, it can still have heuristic value. In this book the category works in two key ways, one coming out of our own historiographical assessment, the other having to do with the rhetorical use to which Tibetan medical theorists put similar categories themselves. In the case of the former, naming macro-aggregative formations like the Buddhist canon, Buddhist monasteries, or Buddhist meditative tradition helps specify the primary loci for those habits of mind and practice—such as produced notions of omniscient epistemic authority, exalted states of consciousness, ideal paradises, or parts of the body that cannot be confirmed by the evidence of the senses—that came to be at odds with medical theory as it developed in Tibet. Such "religious" notions and orientations do not subsume all aspects of the received heritage of Buddhist teachings in Tibet; but they were one large part of that heritage. Attention to the contexts in which they were embedded and the practices to which they were connected helps us to better understand the implications of their dissonance with the distinctive medical mentalities considered in this book.

The other reason a rubric like Buddhism is important is the fact that Tibetan medical writers invoked cognate rubrics of their own, some of which were already broadly in play in Tibetan historiography and/or South Asian Buddhist literature. The way these writers construed the Buddha's legacy may not sit well with our own conception, but it behooves us to attend to their varying rhetorical agendas in doing so. For example, chapter 7 shows how the medical invocation of the Buddha's "True Dharma" (Skt. *saddharma*) served to mark a different approach to healing from the quotidian "human way" of medical ethics defined by the *Four Treatises.*

But as a sign of how fungible a category's rhetorical import could be, in another context the same True Dharma was deployed instead to attack those who considered medicine separable from the Buddha's teachings.[29] A similar impulse to make medicine Buddhist was behind the *Four Treatises*' insertion of itself into the literary universe of the "Word of the Buddha." And yet this move in turn was later resisted by medicine's more historicist wing, part of a growing effort by some critically minded commentators to distinguish disparate systems and traditions. The *Four Treatises* already uses another well-known category, the "insider" (often glossed elsewhere as "insider Buddhist," *nang pa sangs rgyas pa*), to distinguish one part of the work's audience from other kinds of medical practitioners, thereby suggesting the hybridity of medical knowledge and the separability of various threads therein. By the sixteenth century there were pointed distinctions between the "Secret Mantra" teachings of the Buddha and Sowa Rikpa, or again between the disparate senses that a given technical term could have in monasticism and in pulse diagnostics. There was also a growing sense that medicine should have a separate historiography, and an accompanying question on how much of the story of the Buddha's life and teachings really belongs in that. To be sure, each of these cases trades in a different construct; they do not issue into one overarching entity on the order of one of the "isms" of the world's religions. And yet, I suggest that the matters at stake overlapped in the minds of the medical writers and in some respects congealed. This was so whether the Buddha's teaching was being invoked to mark its difference from medical tradition or all of medicine was being swept under its purview. It is precisely that double position—both within and without the Buddhist dispensation—that makes Tibetan medicine's history so rich.

It is no accident that my own historiography goes in both directions as well. While using categories like "Buddhist scholastic practice" or "Buddhist scriptural conventions" or "Buddhist doctrinal terms" helps us pinpoint the loci for certain tensions with the developing mentalities of medical learning, the same domains also produced ideas and practices that were usefully adopted by the *Four Treatises* and its commentaries, elements not to be found in Ayurvedic or other Asian medical literature from which Tibetan medicine otherwise borrowed copiously. Unlike medicine's rhetorical invocation of broad rubrics for the Buddha's dispensation in moments of self-positioning, its adaptation of particular pieces of that received heritage sometimes served no rhetorical purpose at all, nor even took note of the mixed pedigree; it merely helped to articulate concepts relevant to medical practice. By the end of this book I will even argue that Buddhism conceived more broadly, beyond its soteriological dimensions and now intentionally defined for our own historiography, was the dominant civilizational force in the world in which Tibetan medicine flourished, and which contributed in foundational ways to medicine's distinctive profile there. But this is where the distinction between our own historiography and the construal of a Buddhist heritage by Tibetan medical writers becomes important, for it means that the foregoing proposition can have historical merit even if certain actors within that civilization took issue, on certain occasions, with the Buddha's teachings so named. To recognize a Buddhist imprint on virtually all aspects of Tibetan life after the eleventh century is not to preclude the possibility of alterity within that. Nor is the proposition of a Buddhist civilizational force on the same order as the efforts by certain other historical actors to make Sowa Rikpa Buddha Word. Recognizing the ways that Buddhist habits of thought and practice influenced Tibetan medicine is not the same as claiming that Tibetan medicine is "Buddhist medicine." By my own estimation at least, there are too many conceptual, practical, and historical disparities—sometimes explicit and with fundamental implications—for our characterization of Tibetan medicine to be that simple.

In identifying enduring tensions, I have not entirely abandoned a "conflictual" model in my reading of key moments in Tibetan medical thought, although I heartily agree with Frances Garrett that the relationship between medicine and Buddhism in Tibet was nothing if not complex.[30] If the issues at stake were not always parsed in overtly oppositional terms, that is in part because of a real political risk in questioning one of the most powerful foundations of authority in the land. Perhaps more important, there was also real ambivalence, even among the most empirically minded medical theorists, toward anything that would undermine the evident value of meditation and the high aspirations of Buddhist ethics. Yet there remained in medicine a commitment to the evidence of the material world, and that sometimes meant thinking on a different register than certain hallowed traditions of Buddhist learning. I argue in the final chapter that the figure of death on the horizon worked differently in the clinic than when framed by the possibility of transcendent salvation, and that this difference defines the heart of the alternate epistemic space toward which Tibetan medicine was reaching. Other kinds of shifts away from religious formations and toward the center of gravity that I am calling the medical mentality are also suggested throughout the book.

◎ ◎ ◎

While this project has uncovered non-Western parallels to tensions between religion and science in the West, that dynamic became evident in different ways, and certainly had a vastly different outcome historically. The Tibetan debates studied herein are also not readily mappable onto recent discussions of religion and science with respect to Buddhism. Those instances where

Tibetan theorists argued that medicine falls squarely within the Buddha's dispensation are not to be equated with nineteenth- and twentieth-century apologetics that maintain that Buddhism has always been scientific, or that Buddhism goes further than modern science and has much to teach it. Arguments of the latter sort have been advanced by Buddhist leaders from Ledi Sayadaw, Dharmapāla, and Taixu in the face of Western colonialism to recent participants in the neuroscientific dialogue with accomplished meditators.[31] They include interesting propositions worth considering on their own terms, but these are new ideas tailored specifically to the contemporary and global hegemony of biomedicine. They represent different kinds of impulses and respond to different kinds of challenges than those in the following chapters.

Again, the texts studied herein were all written prior to any significant contact with Western science or European colonialism. More to the point, Tibetan medical theory's display of allegiance to traditional authority structures often served to circumvent them. There was no clear mechanism to alter traditional cosmologies or anatomies in light of empirical evidence. The Buddhism that the thinkers in this book were engaging, be that positively or oppositionally, was not the same as the one envisioned by the current and very modern Fourteenth Dalai Lama when he says that Buddhism should change its doctrine wherever it is contravened by scientific data.[32]

In short, this book is less about a Buddhist science than it is about an evolving scientific tradition that flourished in a Buddhist world. If I nonetheless conclude with a suggestion to intentionally stretch our notion of Buddhism to include its civilizational reach, it is only in service of a better historical account of how this development unfolded. The material considered here *does* show an edge of a Buddhist world that was moving toward a scientific and empiricist mentality. I am most interested in how that effort fared, in the face of other sides of that same Buddhist world that would hold it in check.

◉ ◉ ◉

In the end there can be no absolute choice between understanding Tibetan medicine as inculcated with Buddhist ways of knowing and being, and as operating apart from Buddhist revelation with its own ways of knowing and being. Instead, we are attending to an intricate intellectual history. These medical theorists proceeded with caution and considerable rhetorical finesse. This book is thus very much a study of self-positioning, fine distinctions, innuendo, and multiple levels of irony, not to mention fuzziness (intentional or not), sleight of hand, split differences, and patent self-contradiction. Much of it is taken up with reading passages closely and trying to fathom some quite convoluted exchanges. But only in the thick texture of rhetoric—or to flip the metaphor, in the fine lines of the details—can we discern the contested spaces of Tibetan medicine. Such a reading defines a primary methodological orientation of this project.

Coupled with the need to read closely line by line is a commitment to tolerating, even appreciating, inconsistency. The stakes of the medical debates sometimes meant that scholars would mute their claims by blurring their implications. This alone poses a considerable hermeneutical challenge. But just as much, the writers in question were often not completely consistent because of the complexity of the issues themselves. It is thus usually more important to get a sense of the arena that constitutes an issue than to look for whatever particular solution is being floated. It is more important to recognize that certain medical theorists were *reaching toward* empirical accountability than to judge, from the vantage of the twenty-first century, whether they actually attained it or not. In this I propose a somewhat different upshot than the enduring distinction Georges Dreyfus finds between modern and traditional forms of rationality with respect to Tibetan monastic debate.[33] The difference may be partly a function of the material considered. It is certainly the case for Tibetan medicine too that freedom of thought was limited by allegiance to constitutive texts, as Dreyfus puts it. But the medical impulse to heed direct perception over scriptural authority nonetheless had the potential to reshape the entire epistemic matrix in ways that Dreyfus judges to be missing in scholastic debate culture. That impulse may not have realized full fruition, but the very fact of its existence is all the more significant in light of the competing claim to the authority of religious revelation that was also in play. Not entirely unlike the Western "scientific revolution" that never fully happened, or the ideals of racial or gender equality that have not yet been fully achieved, the annals of medical history explored here are most important for what they are reaching *for.* They tell us just this, that in certain particular conditions, scholars saw an opening to push knowledge further, tempting them out onto thin ice—until they looked around to see where they had come to, and realized that something might crack.

I am convinced that to read for processes (reaches, retreats, experiments, questions, worries) rather than positions requires a humanistic eye. Quite apart from whatever issue is at stake, we are best poised to appreciate the fact that a significant process is in progress when we remember how opaque our own ideas can be. Keeping sight of our own history allows the scholars we are reading to be as human as we are, to be stretching beyond themselves, to be still in the process of thought. So, if contemporary biomedicine is home to untold numbers of unsettled questions, so was Tibetan medicine. We read to find other human beings taken up with questions in ways that are meaningful to us, sometimes less to find out what answers they proposed than to appreciate their negotiation of complexities along the way. We find familiar urges even as we realize the disparity in resources to pursue them. To keep what we are reading at arm's length, always reminding ourselves that what we are seeing comes from "another culture," always resisting our own "presentist" concerns, as we are repeatedly urged to do, is to risk losing sight of the real importance that this segment of the history of science and religion has for us now.[34] This history helps us to understand what religion and science and Buddhism have been and can be, even if we have not yet settled the definition of any of those terms. And that is not to mention the pleasure in standing next to our colleagues from a Buddhist world, if for a moment, and watching with empathy and care as they attempt to carve out a space for medical learning.

◦ PART I ◦
IN THE CAPITAL

1

READING PAINTINGS, PAINTING THE MEDICAL, MEDICALIZING THE STATE

An extraordinary set of seventy-nine paintings, executed at the height of academic medicine in Tibet, serves well to launch this study.[1] These exquisite *tangka* scrolls portray in meticulous detail the anatomy, *materia medica*, diagnostics, therapeutics, pathologies, and healthy and deleterious forces that determine the human condition. Their encyclopedic reach and generic depictions participate in trends, seen much more fully in other parts of the world during the same period, toward producing scientific illustrations from life. They are also stunning testimony to the artistic sophistication that could be mustered in the seventeenth-century Tibetan capital, and to the state's investment in medical learning.

Beyond these paintings' pedagogical value for medicine itself, they point to medicine's import in other cultural arenas, especially for the new government of the Dalai Lamas. Their extraordinary range in content illustrates how medicine's focus upon the quotidian realities of human life made it capable of rendering the scope of control to which the state itself aspired. This can also be discerned in the ways that the medical paintings represent religion, in turns exalting and critiquing it, but most notably subordinating it to larger conceptions of human flourishing, of which it represented but one part.

Endeavoring to "read" this artifact raises a host of methodological issues. Not the least is the problem of how to discern cultural significance in a field where knowledge about everyday life is still scanty. The challenge becomes all the more daunting when trying to read visual images, where broad cultural expertise is required to appreciate their manifold implicit messages. From art historian Francis Haskell to a semiologist of the likes of Roland Barthes, scholars have long recognized the difficulties and the risks of such an endeavor.[2] In the present case, we are both helped and distracted by the fact that the images in question are closely tied to a textual corpus that they specifically represent: the massive four-volume commentary to the twelfth-century medical root text *Four Treatises*, written by Desi Sangyé Gyatso (1653–1705), chief minister and then regent of the Great Fifth Dalai Lama, Ngawang Losang Gyatso (1617–82). The Desi also oversaw the production of the paintings. Indeed, a key question for this chapter concerns what was achieved by translating the Desi's already comprehensive written commentary into visual form.

The puzzle of this redundancy also presents an opportunity. The ways that images say and do things that texts cannot—whether intentionally or not (as may be more faithfully the case)—has been a rich site of reflection in the semiology of art.[3] What becomes visible when words are replaced by images? Much of my method in answering this question is based in comparison and attention to difference: between what the text says and the extra data added by the images; between the medical paintings and dominant modes of other Tibetan art, particularly religious icons and narrative illustration; and between the Desi's stated aims for the set and the way its images point beyond them. These exercises help us to recognize the images' distinctive modes of representation, as well as the cultural and even political connotations of the set taken as a whole.

It is still rare in the study either of Buddhism or Tibet to explore modes of representation for their larger cultural and historical implications. Sheldon Pollock's magisterial reading of literary style in early modern South Asia is a recent exemplar of what is possible in a closely aligned field.[4] There has also been more cross-disciplinary thinking in the art history of Buddhism, moving far beyond questions of dating and provenance. Patricia Berger's study of the visual culture produced under the Qianlong emperor and its reflections of Qing imperial power is one excellent example.[5] The following reflections have also benefited from the history of science for both Asia and Europe, which has taken anatomical and botanical illustration as a key site of signal trends in the conception of knowledge. For models further afield, one might look to the exceptional powers of imagination and interpretation in sociologist Norbert Elias' reading of the drawings of the *Medieval House-Book* and what they reveal of the values, experiences, and atmosphere of the world of a medieval knight.[6] Such exemplary work underlines the rich potential of imagery for historians. The Tibetan medical paintings certainly provide a fine case in point.

The study that follows is illustrated by images from a copy of the painting set that was made under the auspices of the Thirteenth Dalai Lama in the early twentieth century. In both content and aesthetics, this copy very closely replicates the original set, some of which is still preserved in Lhasa.[7] The copy thus reliably mirrors the artistic and representational styles with which the paintings were first executed, and serves well as a basis for cultural and historical reflections.

PRECISION, PLAY, AND THE EVERYDAY

There is an immediate and palpable delight in looking at the medical paintings. One is struck at once by the beauty and vivid color of the large anatomical figures, as well as the pleasing order and neat rows of smaller vignettes. Each of the plates in the set evinces a sense of serene control and comprehensiveness. There is an exquisite precision and often intricate detail, even in the vignettes in their rows. Their individual distinctiveness is all the more striking for its contrast with these images' commonality of position within the ordered parade of registers. The dynamic is well illustrated by an example from the medical botany (fig. 1.1).

The artistry of the medical illustrations comes especially to the fore when their delightful details exceed their taxonomical import. This is determined by comparing the images with their corresponding description in the root text and the Desi's *Blue Beryl,* summarized as well in captions on the plates themselves.[8] A good example is a lively depiction of cooks in a series on types of prepared food (fig. 1.2). One cook samples the stew, one chops ingredients while the broth simmers, one checks under a lid, one bends down to adjust the fire, one keeps a spoon in the pot and another tool in the fire at the same time, one pours in more water, one looks away, seemingly distracted by something else. Note too the array of outfits and gender diversity. But none of these charming details has anything to do with the actual specifications of what is being denoted. The associated texts say nothing about the means to prepare the listed foods and certainly nothing about the people who do so.[9]

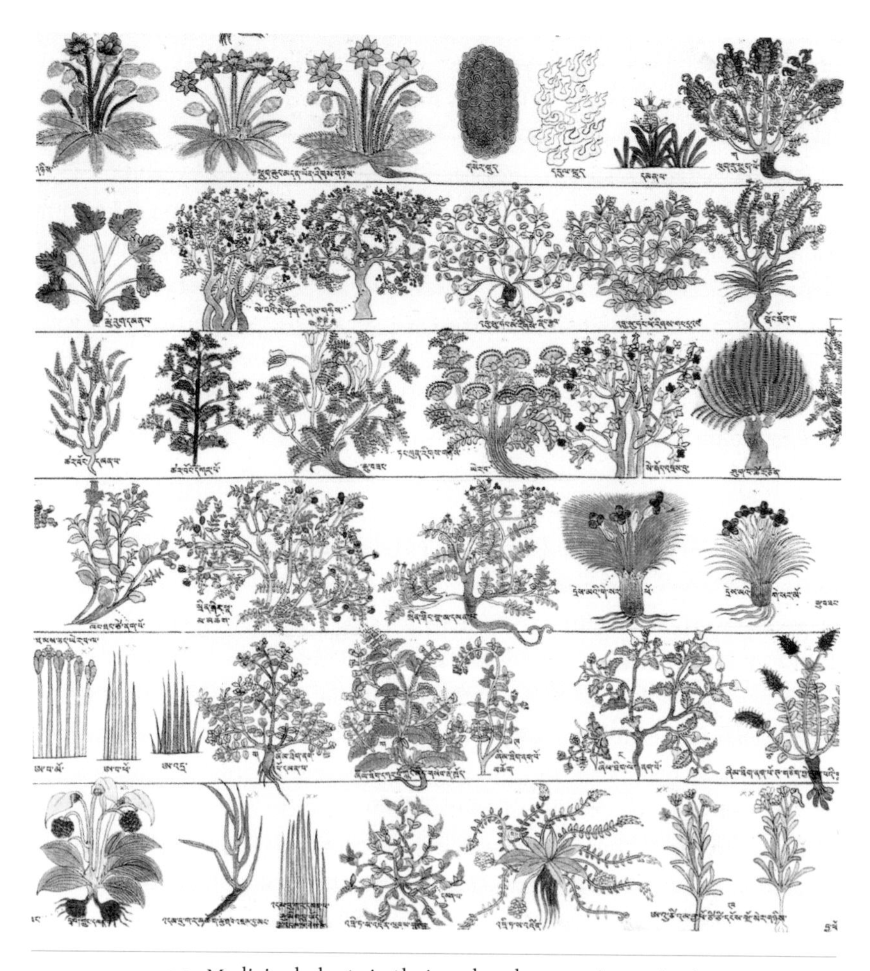

1.1 Medicinal plants in their ordered rows. *Plate 26, detail*

It appears rather that in rendering the taxonomical information of the texts, the artist also used the opportunity to depict an array of human types and familiar everyday scenes.

Imaginative extra detail can be found throughout the set. One good place to recognize it is in the section on categories of animals. They are not always pictured as mere passive creatures posing inertly for the zoology lesson. Rather, a number are shown doing something to amuse themselves or make themselves comfortable: a yak is licking his mate; a snow leopard is scratching itself under its neck; a cat is dozing (fig. 1.3).[10]

These animals are not taking their role in medical classification very seriously; in fact, they couldn't care less. Yes, their shapes and coats serve to illustrate the features of the species they represent. But even if their postures of sleeping or expressing affection also show typical behavior for these species, these details have nothing to do with the medical text's grounds for classification and are not mentioned in the text (nor is it terribly unique to leopards to scratch themselves or to yaks to lick their partners).[11] Rather, these gestures make the animals seem real. We might say that representing *both* the taxonomical information *and* the larger life of animals serves to enhance the set's credibility as an illustration of real objects in the world.[12]

In fact, much becomes visible when the details of depiction exceed overt didactic content. For now, note that the excess charm and delight have to do with everyday experience. This would be true for the artists, who would have drawn on such experience, as well as for the set's viewers, larger parts of whose lives would inform their engagement with these images, beyond the business of medical learning itself.

Aside from extraneous detail, there is also a more fundamental respect in which everyday realities come to the fore. The set's didactic content is *itself* about everyday human life. This fact is immediately striking to anyone with even the barest exposure to Tibetan art. The medical topics of these paintings set them apart from most illustrated scrolls produced in Tibet, predominantly religious in content. In contrast, the medical plates focus upon the ordinary material and social world and the ordinary people within it. In the course of portraying human anatomy, physiology, pathology, and pharmacology, the set pictures many, many scenes of quotidian life: people washing their hair, bathing in streams, giving each other massages, eating food, getting married. The predominance of such scenes and the relative infrequency of religious iconography in the set, with but a few exceptions, is remarkable.

A salient mark of the set's unusuality are the many images of couples having sex, on practically every plate. We do of course see tantric *yab yum* deities in coitus and occasionally yogis doing sexual yoga in other Tibetan art, but very rarely everyday sex. The one notable exception is from the old Buddhist tradition of rendering the "wheel of life" (Skt. *bhavacakra*). Warning against the ills of samsara, such illustrations provide imagistic renderings of the "twelve interdependent links": a person in the state of ignorance, a person being struck with the arrow of sensory perception, and so on, all shown to be in the clutches of the angry god of death, Yama (figure 1.4).[13] Sometimes one or two links are represented as a couple having sex, symbolizing the sexual

1.2 People cooking various medicinal foods and brews. *Plate 22, detail*

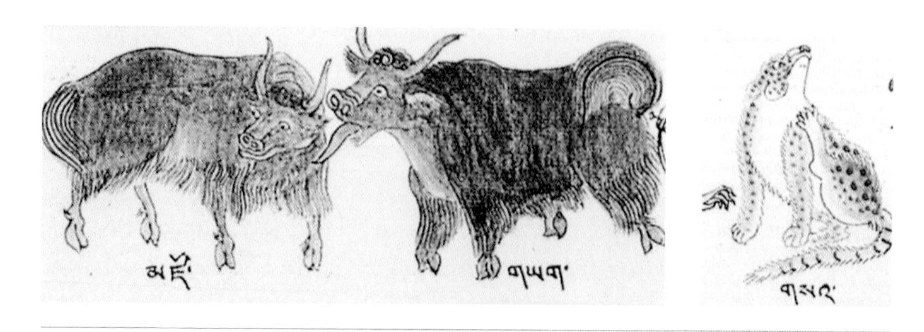

1.3 Animals in the *materia medica*. *Plate 21, details*

contact between man and woman that leads directly into ceaseless birth and death. In this eighteenth-century example, sex is portrayed twice in the outer rim, once elliptically at four o'clock, where the embracing couple represents "sensory contact," and again at nine o'clock, where it is the clincher that makes for samsaric "existence" (figure 1.5).

But otherwise such images are very rare in Tibetan art.[14] In contrast, the medical text speaks often of sex as one of the many kinds of human behavior that impact our health, and every time, it is duly represented on the painted plates as well. Figure 1.6 shows a few of the many examples in the set.

1.4 The Wheel of Life.

Eastern Tibet; 18th century. Pigments on cloth. Rubin Museum of Art, F1997.40.10 (HAR 591)

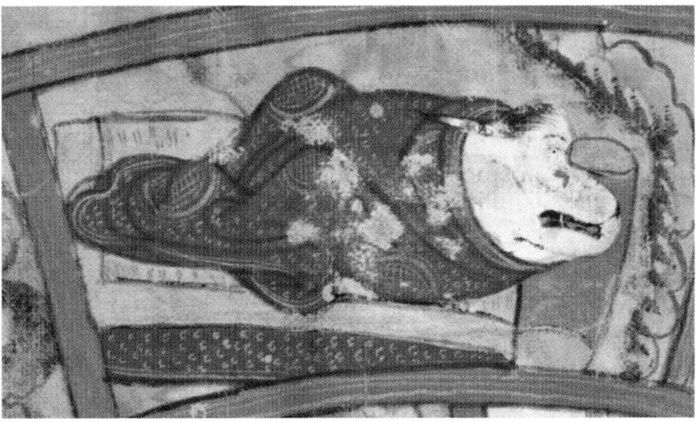

1.5 Sexual relations as sensory contact and samsaric existence. *Figure 1.4, details*

I dare say that the set's viewers in its time would have been mildly surprised to see so many and such matter-of-fact renderings of sex. Their lively variation would probably have been amusing, if anyone were to compare them. Once again we have a case of artistic license. The textual lists only mention sex as such, saying nothing that would necessitate the differences in rendering whereby some of the lovers are dressed, some not, some are outside with a landscape behind them, and so on. As do the cooks and the animals, these images gesture to a larger human world beyond their place in a taxonomical or etiological scheme.

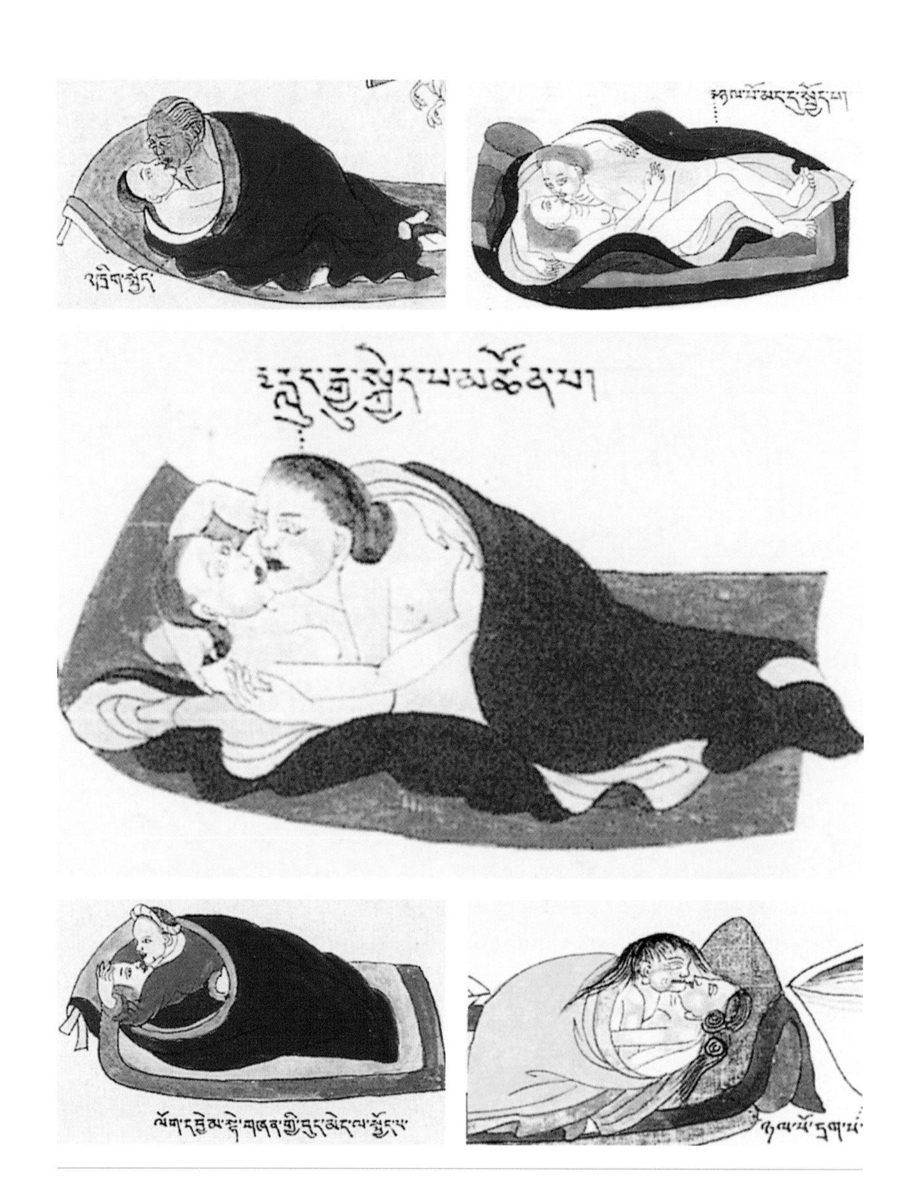

1.6 Medical images of sexual intercourse. *Details from various plates*

1.7 Newlyweds having excessive sex. *Plate 36, detail*

It is not the case, by the way, that there is no judgment on sex at all in these plates. In medical terms, excessive sex is often seen to be a problem, as is implied in this depiction of a newlywed couple. Too much sex can cause exhaustion as well as more serious physical conditions.[15]

Still, such a message is a far cry from the widespread negative characterization of sex in Buddhist teachings due to the ignorance and craving that bring us to it and leave us entangled.[16] Indeed, the medical texts often characterize sex in moderation as good for health, especially for males.[17]

Most of all, sex is just something that people do. And so, given its frequent appearance in medical knowledge, the artists took the opportunity to portray it in many guises. The set even has a depiction of homosexuality. It is one of several cases where an image is positioned alongside another that is tied to a caption, but lacks a caption itself. In this instance it is not clear whether the same-sex pair goes with the couple labeled as "copulating" on the left, or with the masturbating man to the right who is "depleting his vital fluids" (the text seems to indicate the latter). Both of those activities are listed in the Desi's *Blue Beryl* as things one should not do on the night before having a urinalysis.[18] For some reason, the painters decided to add an extra image

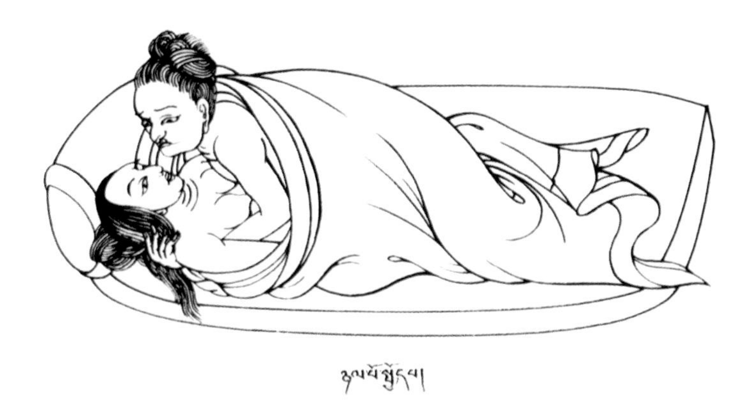

1.8 Ways of depleting one's sexual fluids. *Plate 62, detail*

showing another way that human beings deplete their sexual fluids. Homoeroticism is not portrayed elsewhere in Tibetan art, as far as I know. Quite matter-of-factly, albeit only by suggestion, the medical set pictures it as about to be performed by what looks like an adult monk with a smaller, probably younger one.

In the medical paintings, sex is simply a part of life, just like the plethora of other quotidian activities pictured, of people fighting, talking, farming, doing religious rituals, urinating, being born, and, of course, being sick. This focus upon everyday realities alone constitutes a major facet of the distinctively medical orientation toward knowledge and its representation.

. . . AND RELIGION

I just mentioned religious ritual as one of the many things that the painting set shows people doing. How indeed are Buddhist and other religious figures portrayed in these plates? Let us broach this key question by turning first again to visual impressions.

The very first plate features the Medicine Buddha in his medical Pure Land of Tanaduk, teaching the root medical text *Four Treatises* (fig. 1.9). There are also smaller cartouches of this preaching scene at the points in the set where each of the four treatises begins and ends, including a grand finale when the preaching is complete, taking up about a third of the last plate. But while these scenes might fulfill the expectations of viewers accustomed to religious iconography, they can actually be read two ways. The Buddha preaching scenes can be said to govern all that occurs between them. Or, the frame story of the Buddha's original teaching of the *Four Treatises* can be read as a set of bookends, readily bracketed in light of the near absence of religious icons elsewhere in the set. In fact, the historical veracity of the preaching scene was contested, and some scholars considered it a superficial gloss. They claimed that the attribution of the *Four Treatises* to the Buddha's original teaching was a pious fiction, added merely to give legitimacy to what really is a compilation of the knowledge of historical physicians.

We will study this contested matter in chapter 3. For now, what remains are the occasional Buddhist and other religious figures and practices in the rest of the set, when people are shown making offerings or hearing the Dharma as ways to improve their health, or priests are doing rituals to handle

1.9 The Medicine Buddha in the medical city of Tanaduk. *Plate 1*

demons. There are also two tableaux illustrating how a physician can use a Buddhist-style visualization while producing rejuvenation elixirs (plates 52 and 53).[19] But such cases are few and far between, mixed in with the hundreds and hundreds of other human activities. What is most striking is the degree to which we see, in Japanese historian Barbara Ruch's words, "sacred and profane in wholly comfortable continuity."[20] Even with the contested bookends, it would seem that religion is far from the heart of the set, which is rather about a larger world in which Buddhist and other religious practices are an integral—but just one—part.

This point takes us straight to central questions of this book. How is that larger world construed, and how are—or aren't—religious and medical formations distinguished therein? In this chapter I am looking in particular at how all this is conveyed visually. But the decentering of religious values and practices was long at work textually. On occasion the *Four Treatises* mentions the practice of the Dharma as something that contributes to physical well-being, but it is only one among many such factors. Equally, transgression of religious values or proscriptions becomes one of several causes of illness and serves further as a sign in dreams and other auguries, a bad omen for someone's chances of recovery from illness. None of that is surprising for people steeped in Buddhist culture. What *is* remarkable is the way all of the factors that contribute to illness, religious and not, are placed alongside each other, seemingly on an even playing field. Largely absent in these images, both textual and visual, is the kind of hierarchy so regularly seen in Buddhist writing, between the "worldly" and the "world-transcending," for example, or between conventional and absolute truth. If anything, the "way of humans" that so centrally occupies medicine takes precedence over the transcendent, as will be explored in chapter 7.

Consider the illustration of a set of patients a doctor should decline to treat (figure 1.10). We know that is what they are from their captions; the group is

1.10 Kinds of people not to take on as patients. *Plate 35, detail*

taken straight out of the text. They include people who have aversion to their religious mentor (the painters portray this as a person with his back to his teachers, one monastic and one tantric); people who are aggressive toward sentient beings (here illustrated as a man slaughtering an animal); people who don't listen to the advice of their doctor (a person looking away from the physician reading his pulse); people who are very busy; and people who are poor.[21] No distinction in kind is made between the fault of turning one's back on the Dharma and the fault of turning one's back on someone reading one's pulse.[22] More unsettling yet is to see the fault of turning one's back on the Dharma portrayed in the same size and visual register as the fault of being busy (which presumably means the patient will not have time to follow the doctor's advice), not to mention the fault of having little wealth, which renders such a patient also worthy of rejection by the doctor!

This leveling of the field between religious and mundane values and practices, already at work in the text, can make for jarring visual juxtapositions. For example, figure 1.11 illustrates human pubic hair next to the nectar produced in a ritual (*sman sgrub*). Both are part of a longer list of substances used in medicinal compounds, which also includes the flesh of a flying squirrel, fangs of a rabid dog, and urine of an eight-year old. The ritually produced substances come up seemingly randomly in this list of 175 items, the only identifiably "religious" one in the lot.

But both of the foregoing examples merely represent visual translations of the text. The subordination of religious figures within the larger field of medical representation is even more notable when we consider again some of the extraneous detail added by the artists. For example, there are many

1.11 Some of the substances used in medicinal compounds. *Plate 31, detail*

1.12 Five people who can be cured by a *dbang ril* pill. *Plate 31, detail*

instances where doctors are portrayed as monks; they are also frequently portrayed as laymen. Their status as monks reflects a social fact, rather than having anything substantive to do with their practice of medicine. I see no sign that a principled choice determines when these images show the doctor as a cleric or as a layman; the variation seems to be random. I can make the point even more clearly for those cases where patients are figured as monks. Actually, most patients in the painting set are in various states of undress, making their status undetermined. But consider the group of patients pictured in figure 1.12, which includes a monk on the far left. The group illustrates the power of a particular kind of medical substance (*dbang ril*) to heal five people, part of a larger category of medicines that have varying powers, indicated by the numbers of people each can help.[23]

While the text does not describe the five people at all—it is merely concerned with their number, not their kind—the painters illustrated five rather disparately dressed individuals. Now this is hardly a category to which monks are normally said to belong. This is not a group of people who are disciplined, or virtuous, or spiritually advanced. Indeed, nothing specific to being a monk qualified this one's inclusion. He is only there as an example of the general state of being human. In this context being human is of greater significance than being a monk. The inclusion of a monk in the group thus appears to be random: any number of kinds of people would have served equally well to represent the five patients.[24] It is almost as if his monkhood points to what about him is not a monk; being a monk here seems irrelevant, or sidelined at best.

THE GENESIS OF THE MEDICAL PAINTINGS

We have already noticed several respects in which the medical illustrations point to something bigger, beyond the specifics of medical learning. Their imaginative details reflect a wide range of human life beyond their ostensible pedagogical content. And religious figures and values, of such high status in Tibetan society otherwise, are sometimes positioned as merely one part of that larger domain of life. Both features suggest the broad reach of medical knowledge into Tibetan life.

The historical and discursive background, the set's genesis and the stated intentions of its creator, is germane to these paintings' distinctive modes of representation and cultural import. Indeed, the condition of their production constitutes one of the most significant dimensions of the set's larger horizons: the fact that it was sponsored by the Ganden Podrang, the government of the Dalai Lama, which had recently consolidated rule over central Tibet and beyond, into some of the eastern reaches of the Tibetan plateau. That also speaks to the primary audience for the set, which certainly included not only medical teachers and students but also the elite aristocratic and monastic hierarchs serving in the Dalai Lama's government.

The medical paintings were masterminded and commissioned by Desi Sangyé Gyatso, the famous regent after the Great Fifth Dalai Lama.[25] Their creation was part of a larger renaissance in the study and practice of medicine during the period. Patronage and advancement of medicine was a key ingredient in the Tibetan Buddhist state established under the Great Fifth in 1642.[26] I will provide a broader account of the Dalai Lama's and the Desi's lives and their roles in fostering medical learning in the next chapter. But the painting set stands as one of the best illustrations of the significance of the state's sponsorship of medicine in the larger cultural transformation of Tibetan society and governance under Ganden Phodrang rule.

The paintings were begun in 1687—several years after the death of the Fifth Dalai Lama—under the direction of the Desi.[27] Several artists and scholars were involved in their design and production.[28] More plates continued to be added to an initial set of sixty over a period lasting more than a decade. The final set of seventy-nine plates was completed at some point after 1697.[29]

We don't know for sure how much of the original set survives.[30] The version that illustrates this book was probably created during the reign of the Thirteenth Dalai Lama, Tupten Gyatso (1876–1933; r. 1895–1933), from an older set at the Mentsikhang Medical School in Lhasa.[31] This copy was then brought to Buryatia. Most recently it has been kept in the History Museum of Buryatia in Ulan Ude. I have chosen this version because of the clarity of its reproduction and the ready and kind permission of its modern publisher.

The original set was copied numerous times after its creation, although records are scanty.[32] The original set seems to have been deposited at Chakpori, the medical college established by the Desi in Lhasa in 1696.[33] Copies were in existence at least by the time of the Seventh Dalai Lama (1708–57) or Eighth Dalai Lama (1758–1804).[34] There were still parts of a copy at the Yonghe Gong Tibetan Buddhist temple in Peking in the first half of the twentieth century,[35] and there are reports that a good set was kept at Labrang Monastery in eastern Tibet.[36] Copies of pieces from the set are known elsewhere too.[37] We also know that single plates, primarily from the anatomical and pharmacological sections, were reproduced as individual black-and-white xylographs and distributed independently as teaching tools.[38] There was also a proliferation of pharmacological writing from the early eighteenth century onward containing illustrations of the *materia medica* embedded in the pages of the text, often based on the images in the Desi's painting set.[39]

By the time of the Thirteenth Dalai Lama, apparently many paintings from several versions of the set were to be found both in the Mentsikhang Teaching Hospital and in the Norbulingkha, the Dalai Lamas' summer residence in Lhasa. Numerous plates were also missing from the set at Chakpori. In 1923 the Thirteenth Dalai Lama, as part of an effort, in the words of contemporary medical historian Jampa Trinlé, "to preserve culture, particularly medicine and astrology," underwrote a project to recompile the full set under the direction of the outstanding physician of the time, Khyenrap Norbu (1883–1962).[40] This would have been when the copy illustrating this book was made. There is now at least one complete set at the Mentsikhang, which may well be the one compiled by Khyenrap Norbu. The Mentsikhang's holdings may contain some of the original paintings made under the direction of the Desi, transferred there at some point before Chakpori was destroyed in 1959.[41] There are also a number of medical paintings at the Norbulingkha that seem to have a very old pedigree.[42] In any event, comparison of the plates now at Mentsikhang with the Ulan Ude set and others in collections around the world shows how stable the Desi's iconography and aesthetics have remained. We also can compare the existing sets with the Desi's own summaries of the content of each plate that he recorded separately.[43] In fact, the captions and colophons on each plate are reproduced exactly in all of the currently known copies. Finally, a stylistic comparison of the existing versions shows very close reproduction, not only in the topical composition but also in virtually all of the visual details, including the examples studied in this chapter.[44] I have observed some very slight differences, such as in the rendering of breasts, hair, and genitalia in a few of the small vignette figures, between plates at the Mentsikhang and the one from Ulan Ude; these minor exceptions only prove the rule of how close the copies otherwise are.[45]

The Desi stands as the "author" of the painting set, both because he directed its execution and because the plates were formulated in concert with the organization and content of his *Blue Beryl* commentary on the medical root text, the *Four Treatises.* The Desi characterized the paintings as "following the meaning of the commentary" when he discussed the completed project with a group of doctors and scholars some years later.[46] The precise timing of the set's creation with respect to the writing of the *Blue Beryl* is not entirely clear. It may well be that the process of rendering medical knowledge in visual form contributed to the content of the written commentary. In fact, it sometimes appears that the Desi regarded the paintings as a more perfect creation than the text, as when he suddenly shifts from a discussion of the genesis of the *Blue Beryl* to list the contents of each plate, where we would have expected a chapter list of the written work.[47]

Without doubt, the two productions were of a piece. The execution of the paintings is illustrated in a mural in the Potala, itself probably painted in the early 1690s. The mural shows the Desi beginning to write the *Blue Beryl*, according to the caption. It also shows several assistants, including some of the artists who worked on the painting set. It appears from the larger illustration that the both the writing and painting took place in a room at the Jokhang Temple.[48]

By his own account, the Desi devoted much effort and creativity to the paintings' blueprint. He portrays himself as selective about which elements of the root medical text would actually be represented. In this he is aware of the discrepancy between text and image, recognizing that despite his intention to render the medical treatise in its entirety, some points of medical knowledge do not lend themselves to visual translation or there would be no benefit in so translating them, and so he would leave these out.[49] Indeed, many of the later plates in the set either omit entire chunks of the information in the *Four Treatises* and *Blue Beryl* or aver in their colophons that these topics are represented in abbreviated, synoptic form (*mtshar sdus*).

1.13 The Desi writes the *Blue Beryl* and instructs artists to render its medical teachings as paintings. Detail of mural at the Potala.
From Lha sdings Byams pa skal bzang, *ed. 2000, 137.*

The Desi speaks most about issues regarding the rendering of the anatomy and the *materia medica*, which were indeed appropriate for visual depiction and were at the heart of the set's pedagogical value. He mentions prominently the lack of models and clearly sees himself as inventing a visual tradition *de novo*. He recounts how he studiously "compared" a variety of textual sources to overcome the lack of direct instruction on how to draw such things.[50] He also collaborated with a number of his contemporaries, such as Lhünding Namgyel Dorjé, whose oral instruction determined the anatomy represented in two of the plates.[51] He enlisted the skills of artists Lhodrak Norbu Gyatso, who drew many of the outlines, and Lhepa Genyen, who applied the color.[52] It is not clear how many other artists or assistants were involved; the Potala mural labels the two artists at work as Tendzin Norbu and Lumshak Genyen.[53] The Desi also consulted experts in local botany.

Yet although the paintings are clearly the product of a team, the Desi is the one who oversaw their production, sent workers into the field for data, drew on his own education to fill out that data, and expressed his own opinion and experience in adjudicating what was appropriate.[54] In all of this he remained the director and indeed the impresario, weighing the value of varying texts and informants, adjudicating what was right and what was not, and above all engineering the circumstances to create a new form of medical knowledge.

AUTHORIAL INTENTION AND THE NEW: MEDICAL ILLUSTRATION IN TIBET

In a few passages the Desi speaks directly of his intentions in creating the painting set. Among other things, his comments address the question of what is to be gained by reiterating a medical text in visual format. What might images convey that text does not?[55]

In the colophon to his *Blue Beryl,* the Desi accounts for the painting project this way: "Basically, as a special aid to understanding, I thought to introduce on the basis of illustrations. If we could render the whole *Four Treatises* as illustrations—from the introduction to the *Root Treatise* to the end of the *Final Treatise*—it would aid understanding."[56] A few folia down, he asserts, "There has been no tradition of making a manual like this that compiles the treatise and commentary as illustrations that allow one to introduce by pointing a finger."[57] And then once again, at the end of his career, he writes that the paintings are "unprecedented illustrations, codified so that the meaning of the entire [*Four Treatises* and *Blue Beryl*] could be understood easily by anyone from a scholar to a child, just like one [can see] the shape of a myrobalan fruit in the palm of one's hand."[58]

The Desi's statements boil down to two main claims about the set of paintings *qua* illustrations. Illustrations of the root text and commentary are something new, and he intends them to provide clear and easily perceptible information. The second point makes an implicit comparison between pictures and words, and the greater immediacy of the former.

Let us scrutinize these two claims in turn. Not only will that help isolate the special nature of the Desi's paintings. Considering the degree to which the paintings actually do—or don't do—what the Desi says they are doing will also lead to some of the larger, if unarticulated import of both the paintings and medicine itself in the time period when the set was produced.

First, whether or not the set truly was unprecedented, it is telling that the Desi *claimed* that it was: "there has been no tradition of making a manual like this." To speak of newness is not something to which Tibetan scholars regularly lay claim. And while sometimes the label "unprecedented" can be applied lyrically to something deemed extraordinary, the Desi clearly uses the term in this context to indicate a new and groundbreaking means of representing knowledge. Such a claim is itself significant in a society where the authority of tradition and lineage, and the inspiration of the masters of the past, determine the value of virtually everything, from academic institutions and elite literature to folk culture.[59] Of course, Tibetan writers and artists did innovate, but they usually disguised their originality.[60] The Desi himself is often quite diffident about his own creativity, even in passages directly contiguous with those just cited, where he proffers that he has asserted nothing new in the *Blue Beryl* and has depended entirely on authoritative tradition.[61] As we will see in the next chapter, the recognition of the value of new knowledge—and an accompanying ambivalence about that—is a defining feature of the medical episteme in the Desi's day.[62] For now, something special is afoot in his unabashed touting of a new invention.

As for the fact of the matter, when the Desi speaks of the novelty of presenting "the treatise and commentary as illustrations," it is not entirely clear whether he is referring to the translation of text into images in Tibetan art more generally, or just in medicine. Certainly narrative texts had long been illustrated in Tibetan art in the centuries prior, both on *tangka* scrolls and as murals on walls of monasteries, palaces, and temples. Usually only the main events from the narratives are illustrated. In exceptional cases—paintings of the life of the Buddha, or his former lives, or of the Indian master Padmasambhava, or the Tibetan saint Milarepa—many individual episodes are portrayed in delightful detail.[63] But it is hard to think of an example that achieves anything like the degree of detail with which the medical paintings render the content of the *Four Treatises*, where sometimes every sentence, even every noun, of certain passages gets its own illustration.

In any case, the Desi is probably rather referring to the novelty in his world of illustrating the *Four Treatises* per se. He is right about that. In fact, there were but few efforts to draw medical information of any kind in Tibet before the Desi's time. The modern Tibetan historian Jampa Trinlé points to a number of titles of early Tibetan texts, some mentioned by the Desi himself in his history of medicine, that would seem to have contained illustrations of some kind, or at least described measurements and grids.[64] And while the Desi bemoans the paucity of illustration models and instructions, he does cite a number of guides that he recognizes from earlier medical tradition. He attributes medical illustration models (*dpe ris*) to Drangti Pelden Tsoché (thirteenth century) and to Zurkhar Nyamnyi Dorjé (fifteenth century).[65] Neither of these seems to survive now, although an apparently early manuscript with a few plant illustrations embedded in its pages has recently surfaced.[66] The Desi also mentions conventions for portraying mnemonic diagram trees (*sdong 'grems*) and for anatomical measurements (*yul thig*) that he learned from Lhünding Namgyel Dorjé and his predecessor, Lhünding Dütsi Gyurmé.[67] He further cites illustration traditions that he picked up from the lineage of Terdak Lingpa,[68] as well as one Mentangpa tradition of iconometry.[69]

But such models were rare, even in the Desi's time, and it is likely that whatever models did exist were limited in scope. Fernand Meyer points to a single diagram of moxibustion points with Tibetan captions from Dunhuang[70] and to grids for acupuncture points that were used in China, but notes the uniqueness of the Tibetan anatomical grids. He also notices the possible influence from Greco-Arab medical tradition in a few of the paintings, such as in the use of a certain crouching position that is otherwise unknown in Tibetan or Chinese drawing. The appearance of Greek medical traditions in Tibet since the eighth century has long been known to modern scholars. The existence of an early Arabic illustrated anatomy from the fourteenth century, quite disparate from the Tibetan examples and not known to have been available inside Tibet, might possibly have planted the seed, if indirectly, for doing something similar in the Desi's medical circles.[71]

In any event, there was no influence on the Desi from Indian medicine, which does not seem to have produced comparable illustrations of Ayurvedic knowledge at all up to that time. China, on the other hand, did have substantial medical illustration, part and parcel of a larger history of producing encyclopedias and *collectanea* in consonance with the Confucian interest in "the investigation of things," and in full swing by the late Ming.[72] There are illustrations of pharmacopoeia since the Tang dynasty. An outstanding example from the Ming period is a major catalogue of *materia medica* by the great naturalist Li Shizhen (1518–93). And yet their dependence upon a large archive of earlier, more general illustration manuals that were mostly reproduced in rote fashion in later versions has led historians of Chinese medicine to question whether the medical catalogues were really meant to provide realistic, didactically useful portrayals.[73] The spectacular production of the *Imperially Commissioned Golden Mirror of the Orthodox Lineage of Medicine* (*Yuzuan yizongjin-jian*) in 1742, during the heyday of the Qianlong emperor, shares in many of the aspirations of the Desi's medical paintings to rectify medical knowledge and present it in easy-to-digest format at a moment of great imperial power, but it too draws on stock illustration traditions.[74] In any case, *The Golden Mirror* reflects a very disparate medical tradition from the Tibetan example.[75] It also postdates the Desi's set of paintings by roughly fifty years.

Certainly farther afield, medical and naturalistic illustration was in the air in the Desi's day. The German botanist Johannes Kentmann (1518–77) had made great strides in the *Codex Kentmanus*. Another pioneer in botanical painting was his Dutch contemporary Carolus Clusius (1526–1609).[76] Major advances had also been made in anatomical illustration in Europe by the sixteenth century with the work of Leonardo da Vinci, and then again with the magnificent anatomical volumes produced by Andreas Vesalius. Further heights in comprehensiveness and virtuosity were achieved by the seventeenth century, not only in Europe but also across the cultural worlds adjacent to the Desi, from Persia to the Punjab to East Asia. An exceptional example from the west of Tibet, a seventeenth-century illustration of natural history and medicine, is, like the Tibetan medical paintings, tied to an older textual tradition beginning in the thirteenth-century cosmological and medical treatise by al-Qazwīnī.[77] But again, there is no sign of the Desi's familiarity with any such illustration traditions from abroad, Chinese, Persian, or otherwise.[78] Nor did he have an indigenous sheaf of stock illustrations for natural history upon which to draw. While his achievement used some rudimentary drawing protocols in Tibetan medicine, his claim to newness stands. No such collection of medical paintings, with plate after plate of hundreds of plants, minerals, anatomical detail, and vignettes of everyday life, had ever been seen before in Tibet, as far as we know. Indeed, it is hard to find an illustrated medical encyclopedia of anatomy, pharmacopeia, botany, etiology, diagnostics, therapeutics, and lifestyle that compares in scope, beauty, and systematic detail to the Desi's set from *any* place, at any point in time.

REPRESENTING THE EVERYDAY

There are also other dimensions of what might have been "unprecedented" in the Tibetan medical paintings, not directly mentioned by the Desi, but having everything to do with the novelty of his project to render medical knowledge visually. I already pointed to the sheer number of everyday subjects in the set, which would be readily evident to any viewer with even rudimentary exposure to Tibetan art. Now it is certainly not the case that there had been no representation at all of the everyday before the Desi's moment. But a brief consideration of some of those other cases, although not providing a systematic survey of Tibetan art (and yielding the occasional exception to practically every generalization that might be suggested), will still bring to the fore what was indeed unprecedented—even if not precisely new—in the Desi's project in terms of relative emphasis and proportion, semiology, and above all the cultural significance of representing scientific knowledge.

A few instances of secular scenes in art may be found in the earliest examples of painting from Tibetan cultural spheres, such as on a set of coffins found in western China, as well as in some fine murals from a cave near Dunhuang in the period of its Tibetan occupation.[79] More pertinent for the background of the style and content of the medical paintings, however, are the traditions of religious painting, in progress by the eleventh century, that developed in concert with the emerging institutions of Tibetan Buddhism.[80] In particular, from an early moment there were portraits of lay donors and historical personages, in murals at Tabo in western Tibet, for example, and at the bottom of painted scrolls and illustrated manuscripts.[81] Laypeople and ordinary activities continued to be portrayed in Tibetan painting throughout its history. But unlike in the medical set, the majority of such images are on the side, representing scenes pertinent to a central religious icon. In other words, their depiction is almost always in service of a religious didactic program, rather than for their own sake. The same can be said of plants used as decorations or offerings, animals as throne bearers, and the occasional carnivorous or gentle animals in an idealized landscape. Such animals stand as figures in a parable, or to represent an imagined holy place, or to illustrate a story. They rarely are there just to represent themselves, out of an interest in their own nature on its own terms.

That is not to say that there are never scenes that focus on everyday activities of laypeople. One could cite a detailed mural of artisans and tradesmen at work in the construction of the great temple at Sakya (thirteenth century).[82] There are also fine fifteenth-century depictions of building, dancing, and other such scenes in the murals at Tsaparang from the old kingdom of Gugé.[83] Most of the scenes of laypeople and ordinary objects and activities in Tibetan art occur in narrative paintings that illustrate the life of the Buddha or another master. Sometimes such scenes are given delightful and detailed attention and do seem to have been painted for their own sake, although the larger didactic program is still in view.[84] The same can be said for the tradition of painting an array of offerings to deities, which as it developed shows animals that are sometimes very frisky and engaged in various realistic behaviors of their own, not unlike those observed in the zoological images by the Desi's artists.[85] Such beautiful and sometimes exceedingly imaginative renderings would have certainly provided some models for the medical paintings. By the mid-seventeenth century, paintings of historical scenes could be very detailed and accomplished, especially in the exquisite series of murals of scenes from Tibetan history at the Potala itself, which surely would have been known to the artists rendering the Desi's medical set only a few decades later.[86] But it remains the case that none of these examples provided anywhere near a sufficient store of models for the medical topics that the Desi's painters needed to depict. The artists had to invent and improvise vastly in their task of rendering the numerous medical substances, species of animals, plants, foods, kinds of excrement, pathological symptoms, and therapeutic procedures, not to mention all the other many idiosyncratic everyday activities that figure into the medical treatises, such as being prosecuted for tax evasion, or sleeping exposed to midsummer heat, or being struck by a stone as compared to being struck by a cudgel, or reading when you can't sleep, or crying to the point of exhaustion, and so on and so on—far beyond the repertoire provided by existing Tibetan painting up to the Desi's time, even if the artists pored through everything in every collection on the plateau. Perhaps the very act of painting the huge variety of everyday scenes and topics of medical knowledge opened the floodgates more generally for the visual depiction of a full range of life, at least in the capital. A good example is found in the Potala, where there is a truly exceptionally detailed set of murals of recent events, including of the Desi's own life, executed just a few years after the medical painting set was undertaken.[87]

There are also further grounds to distinguish the medical set from other kinds of Tibetan painting, beyond subject matter per se. For one thing, a good deal of the quotidian realia encountered in most Tibetan painting depicts mythological or exemplary subjects rather than the people and objects that would be encountered in ordinary life, as do the medical images. Frequently, as in the illustrated lives of the Buddha, such images also endeavor to depict South Asian people and places rather than local examples of clothing and buildings and so on. And so while painting conventions borrowed from South Asian and Chinese traditions certainly did influence the medical illustrations in many ways, one of the distinctive features of the set is the degree to which it endeavored to portray the zoology and sociology that were directly observable to the artists and part of their own everyday lives. As I will discuss later, it was a point of pride that pains were taken to check the accuracy of the portrayal of the anatomy and minerals and plants, and some were drawn from real-life models.

Now there is indeed representation of local costume or natural life in other examples of Tibetan art. Donor portraits, while often quite small, do refer to real people and their regional attire and other cultural conventions. There is also quite wonderful and strikingly realistic portraiture of lamas, from as early as the eleventh to twelfth centuries, in both painting and sculpture. The exceptionally moving and idiosyncratic portrait of Butön Rinchen Drup (1290–1364) made in the fourteenth century is a fine example.[88] Portraiture in particular developed rapidly in the period when the medical set was produced, as seen in some of the striking portrayals of the Fifth Dalai Lama himself, as well as the seemingly realistic images of people at his court.[89] But now these cases highlight another very important difference to be discerned. They all portray historical religious leaders, or specific characters in a story, or politically powerful people and the events around them. In contrast, the medical paintings, while also purporting to portray real people and activities and things, do not reference *actual or particular* real people, events, or objects.

This pinpoints one of the central departures of the medical paintings from virtually all other Tibetan art. With the exception of a few cases considered below, along with the occasional tangential decoration or embellishment, the large majority of Tibetan visual art portrays *specific* figures and things and scenes. This would include deities and places that are not Tibetan or historical in any sense, but rather mythological or iconographical: they are still *particular* deities or *maṇḍalas*, with names and other specifications. This is certainly so in the wonderfully detailed scenes of historical events in the Potala murals executed in the early 1690s; or again the episodes from the life of the Ngor Abbot Rinchen Mingyur Gyeltsen (b. 1717), including the making of books and paintings;[90] or the portrait of Situ Panchen (1700–74) overseeing the creation of a set of narrative paintings, both from the eighteenth century.[91] All of these scenes' details contribute to the representation of real, actual events. Perhaps not every last ordinary person portrayed represents a specific individual, but they all still serve to fill out the scene as an approximation of the workers and others who were in fact there, to provide a realistic-looking picture of a particular historical scene.

Let us note, then, a crucial specification. Quite apart from the question of what the medical images represent, the Desi's set has but few precedents in featuring a different *kind* of representation. The medical paintings, although realistically rendered and focused on everyday life, do not purport to portray actual or specific people or events or particular deities or heavenly landscapes. With very few exceptions—several registers of the lineages of buddhas, deities, and Indian and Tibetan teachers of medical knowledge atop the first plates of the set; a few specific deities named as omens or part of the rejuvenation pantheon; and of course the preaching Medicine Buddha himself—the images in the medical paintings portray *generic* everyday things, and scenes, and deities, and people. Although always individualized, and representing a wide range of socioeconomic classes, occupations, stations in life, body type, and so on, each of the people here is just *some* man, *some* woman—not any particular one. Both the large anatomical figures and the smaller vignettes are meant to portray a *sample* spine, or hyena, or marriage ceremony, not an actual one. They portray typical examples of the categories they represent.

The generic quality of the medical images is reflected in the frequent reiteration of terms for "class" or "chart" (*gras; sde*; *kha byang*) in the captions and the colophons. These terms call attention to the act of classification that the set's plates "arrange" (*bkod pa*), as in plates 21–35. Arrangement of images along rows of registers was not new; it had long been known in early Nepalese-style icons in Tibet that display series of buddhas or deities. But there again the images portray particular deities in a *maṇḍala* or other iconographical set, not classificatory types per se. In contrast, the generic nature of the *materia medica* in particular is signaled by the term *dpe*, "model" or "example." This term is drawn from an old Tibetan medical tradition that standardized classification for medical plants (*'khrungs dpe*), written studies of which survive from as early as the eighth century, and whose meticulous illustration forms the central part of the Desi's seventeenth-century medical paintings.[92] Generic illustration is especially appropriate for scientific classification and marks the affinity of the Desi's set with similar works in other parts of the world. Kentmann, for example, was keenly aware of the fact that although the plants depicted in his *Codex Kentamanus* were painted from life, he was establishing types, not illustrating actual, individual plants.[93]

Now once again, there was other generic illustration in Tibetan painting. But examples are not easy to find, especially where they are the center of attention and pictured for their own sake. The cautionary images of the wheel of life mentioned above would be one such instance. Bryan Cuevas points to another instance of generic diagrams in the folia of a manuscript that serve as models (*dpe ris*) for drawing effigies to be used in an old tantric ritual cycle.[94] Closer to the medical paintings would be cases where generic items are presented as a kind of catalogue. There are a few, all much more limited in scope than the Desi's set and with much less intentional systems of classification. I already mentioned at least one possibly old manuscript that has generic line drawings of medicinal plants; this is of the standardized plant classification (*'khrungs dpe*) genre, and the images are embedded in the pages of the text.[95] There are also occasional cases where a classificatory mural was painted on the wall of a monastery to depict the requisites for monastic life, such as kinds of robes, begging bowls, and other allowed accoutrements, as well as some scenes on points of behavior.[96] A few other examples of generic illustration may be recognized from the same period and provenance—the same episteme, to invoke Foucault's critical term—as the Desi's painting set, when illustrated catalogues seem to have been on the rise. There is the famous set of paintings, after a series of visions by the Fifth Dalai Lama, that illustrate the implements to be used in magic rituals.[97] Like the paintings of monks' robes, the small drawings of effigies serve as a visual guide to what might be used in practice, but they still represent sample instruments or effigies, not the particular ones that the viewer would actually use. A related project, of cataloguing *techne*, is to be found in the generic depictions of meditative positions in the Lukhang Temple in Lhasa, painted for the Sixth Dalai Lama sometime after the Desi's medical paintings were completed.[98]

We further find a few generic images at the edges of otherwise specific icons. Tibetan *maṇḍalas* focus on particular deities or symbols that are iconographically determined, but they do frequently show at the periphery horrific scenes around cemeteries, such as jackals tearing out the victuals of a corpse; these too would be generic images, not referring to any specific incidents or characters.[99] There are also *maṇḍalas* connected to Kālacakra tradition that illustrate a few generic houses and elements at various locations in the cosmos.[100] And sometimes, but not often, generic figures seem to have been added at the whim of the artist to make an otherwise historically or locally specific scene more realistic. The murals at the Potala and Tashilhünpo depicting other monasteries in Tibet provide a few examples; some of the people are indeed labeled and refer to historical persons, but the tiny figures walking around the grounds or listening to a lecture at Drepung, for example, or making prostrations in front of Tashilhünpo do not seem to reference anyone or any event in particular.[101]

We need not go further with this survey of other Tibetan art that might have contributed to the distinctive appearance and semiotics of the Desi's medical paintings. None of the elements constituting this character is absolutely new, but taken together and in light of their unmatched scope, they bring to the fore what was indeed unprecedented about the set. Not only are these plates virtually unique for painting medical topics, they far surpass in number and variety all other illustration of everyday, local subjects, and also the everyday message and import of those subjects, where even the occasional allusions to religious practice or deities have only to do with their impact on human health.[102] As for semiology as such, the Desi's medical images are most closely aligned with a handful of other efforts to display generic things. These include an older tradition of portraying medicinal plants and a few other attempts to catalogue ritual objects and practices, some postdating the Desi's project. In the latter, generic depiction of the very wide range of items mentioned and classified in the medical treatise is at the very heart of its impulse. This is what the human body looks like, this is what a particular medical plant looks like, this is what the symptoms of an illness look like, this is a typical scene of a person plowing his field, here is what a group of people fighting might look like.

Let me just add a brief caveat at this juncture. For modern scholars of Tibet, the features now identified would seem a windfall. With their riches of description, the medical paintings can serve as a resource—so rare otherwise—for a picture of ordinary life, at least from the perspective of a few late seventeenth-century central Tibetan artists. Here we might learn something of birthing practices (pl. 5) or child care and child rearing (pl. 45); what kinds of tools were used in woodwork (pl. 56); what kinds of weaving styles and cloths and household items were produced and used (pls. 56–58); how houses were built (pl. 57); what a shoemaker might have looked like (pl. 57); what kinds of lamps were used for nighttime study (pl. 77); what weddings looked like (pl. 17); how people performed ransom rites and other rituals (for example, pl. 18), not to mention the wide array of clothing styles and headgear and hairstyles throughout the collection.

But any effort to extract sociological information has to proceed with caution. The paintings often do reflect sociological facts faithfully—such as, among many other things, the relative mix of lay and monastic physicians and the dearth of female physicians (until perhaps the late nineteenth century).[103] And yet the information in the set is often biased. For one example, the medical images are deeply androcentric. So while it seems to have been true that there were few, if any female practitioners of Sowa Rikpa, it is not the case that there were few, if any female patients! But virtually all patients are portrayed either overtly as male or with their gender not visible, except for women with specifically female maladies. I have studied this aspect of the paintings separately and will not go through the details here, but never is a female figure used to represent a general medical condition or anatomical feature common to all human beings, whereas male figures very often are.[104]

I will add other ways below that the set's images convey any number of social agendas. All this proves that we cannot uncritically take its depictions as a fully reliable mirror image of Tibetan society in the Desi's day. There are important ways that the set does present itself as such, but as historians we need to look at the significance of that, rather than taking it at face value. Generic depiction always involves a choice in terms of what is deemed typical, how that is framed, and what is included and what is left out. That certainly applies to the set's representations of religious figures and practices, as well as the ways that the set almost seems to present medical knowledge as isomorphic with the reach of the Tibetan state. Continuing now with the Desi's other stated aim for the medical paintings and what we can discern of the larger import of Tibetan medicine in all of this, we will start to approach the images' connotations and "cultural messages," in Roland Barthes' sense: the larger frame of reference to which the literal elements gesture, and which renders them intelligible.[105]

ILLUSTRATING THE REAL

Generic depiction is endemic to the illustrated scientific encyclopedia, but that also entails that the images are realistic. In order to portray a typical white rhodiola, say, or a typical human torso so that it is informative for medical purposes, it has to look like a real one, or at least show its pertinent distinguishing features in a way that matches real cases.

To consider the ways that these illustrations were realistic—or at least aimed to be—gets to the Desi's second main stated intention, to present medical learning in a manner that can be perceived and understood clearly. That meant, for the Desi, painting from life, at least in some cases. And while I cannot say if this was the first such effort in the history of Tibetan art, I know of no other instance where it is actually discussed, indeed valued as the optimal way to achieve a desired goal—in this case, the depiction of medical knowledge.

To be sure, we do not see here the full use of the dissected corpse as cornucopia for anatomical knowledge, as in European medicine at a similar point in time. But the corpse did serve as a model for the Desi's artists. One of them, Lhodrak Tendzin Norbu, seems to have based his rendering of some of the images on plate 49 upon his direct observation of dead bodies, in contrast with other anatomical illustrations on the same plate that are labeled as being in accord with the *Four Treatises* text instead.[106] But the Desi also evinces concern about the discrepancies between codified anatomical measurements from the text and what one sees in corpses "in reality" (*dngos su*). He notes that such differences would be due to the fact that the dead body under observation might have had a malady that caused variations from the norm. He notes further that humoral and other disturbances usually ensue after death. And he also remarks that the mere incidence of an empty stomach can make a big difference in how an arrow enters the body. Thus does he underline the idiosyncrasies of any particular moment or disposition of the body, and warns that the normative system of iconometric measurements (*yul thig*) should not be expected to describe precisely what one sees in actuality, given the variety of factors that distinguish a given case.[107] In such comments the Desi was addressing a classic tension for scientific illustration, whereby naturalistic depiction and the desire to systematically classify the material so depicted are actually at odds: it is hard to portray a general type realistically.[108]

The Desi also speaks of commissioning local experts to bring him plant specimens so that they could be accurately rendered. Even though he consulted "reliable commentaries," he remained "worried there would be a mistake" and called upon a variety of doctors, whose names he provides, to collect plants from several Himalayan areas for his inspection. He also asked them to confirm the accuracy of his artists' renderings of medicinal plants that were from their homeland and with which they would have had first-hand experience. He appointed a colleague, Jagowa Lozang Wangchuk, "to interview and investigate those who are not really doctors, of Lhomön ethnicity and from Latö, et cetera, according to their own explanation of how medicine grows, and so on, in their own area, until each said that the drawn renderings were right."[109] He reports in another case that he sent a colleague to collect a plant from a distant region in order to check whether its leaf was indeed notched in the way that it had been described in texts. "There are a few sections [of his writing] where Mendrongpa says he has seen the way the medicines grow in Sharmön, and that from among the kinds of [medicinal plants] that grow there, the leaf of *chu chung ba* is crooked, and so on. But since I had some doubts about that, I sent a person to collect the real thing (*ngo bo*) and compared."[110] The Desi's efforts start to remind us of the process under way in Europe a century or so earlier—albeit more comprehensively than in Tibet—as when Kentmann felt the need to consult widely on the exact appearance of the plants that he went on to illustrate in his *Code*.[111]

In discussing the genesis of the medical paintings, the Desi shows himself to be aware of the newness not only of using images to convey medical knowledge in Tibet but also of consulting the natural world as a way of adding to, and even sometimes correcting, the often vague information in texts from the past. The two methods are intimately related in the case of medical knowledge. The same would not be true, for example, of illustrating a novel, wherein one might draw the various people in the story based on live models, but the images would not represent the novel's characters as such—since there is no "as such" in the case of imagined people. In contrast, it is precisely by pointing to the real world that the illustration of medical knowledge is of value. Only by offering a faithful rendering of a plant or body part mentioned by a text—abstracted as a token of the typical in order to serve the classificatory aim, but still based on empirical observation of real samples—can visual images be of use in the clinical setting. Only realistic depiction could serve the Desi's second didactic aim to provide clear and easily understandable information, to introduce medical learning "by pointing a finger," and for that to be as lucid as a "fruit in the palm of one's hand."

While there can be little doubt that the Desi and his team obtained some real examples from which to draw, we are not sure how much of the set was actually executed with live models in sight. The Desi was also aware of the difficulty in drawing all the minute details of a particular subject, and he sometimes adds the seemingly redundant word *mtshon*, "show" or "indicate," to the captions. He does so especially, it seems, when a rendering is perforce only impressionistic or indicative—for example, at plate 7.7, illustrating the 7.5 million pores on the right leg; or at plate 8.9, meant to portray the 21,000 hairs on the head. Some of the smaller drawings that I am calling "vignettes" are also so labeled, like the set of images of people suffering from excessive bleeding on plate 40, or the figure at plate 48.40 showing a number of arrow and knife wounds to the body. But such caveats about the limits of illustration—supplied only occasionally and inconsistently—serve to show how seriously the Desi was taking the project to represent realistically. Most of the images in the set display the same schematic and largely flat quality of much other Tibetan painting (although some shading is attempted, as also found in other works), but the plates of medical botany, pharmacology, and anatomy appear very precise and would serve well to illustrate didactically the "meaning" of the medical text through a veritable pointing of the finger. This would also be true of the plethora of zoological depictions, showing what various bird and animal species look like (e.g., pl. 21, pl. 28); the various kinds of milk products in their appropriate containers (pl. 21); the head shapes that indicate the patient's disposition of humors (pl. 48); the points on the limbs suitable for bleeding (pl. 38); where exactly to press for pulse taking (e.g., pl. 60); and the illustrations of medical instruments (pl. 34). All would indeed show, in ways that words never could, what something looks like, and would indeed be invaluable tools to enhance the pedagogical effectiveness of a text.

ILLUSTRATIONS THAT DON'T POINT

But not all of the images in the medical painting set illustrate didactically. Noticing cases where certain images' pedagogical import is indirect at best—and there are many of them—will lead us to other registers of significance beyond the Desi's stated motivations, as well as the scientific value of the paintings per se.

A variety of matters mentioned in the *Four Treatises* do not easily lend themselves to visual rendering, since they are not readily available to visual perception at all, let alone a pointing of the finger. Yet despite the Desi's declaration that he would exclude such topics, the set frequently tries to illustrate things that are hard to show.[112] It often does so quite ingeniously and successfully. Many such examples appear not in the anatomy or the

1.14 Types of pulse qualities. *Plate 60, detail*

pharmacopeia, but in other small vignettes. The listless figure of a man who has a deficiency of vital substances (pl. 19.100) shows more than the text's litany of symptoms about this person's fear, emaciation, unhappiness, and fading complexion.[113] To see the image would give the viewer the gestalt of the entire condition, which presumably would make it easier to recognize in the clinic. Something similar could be said of the delightful attempts to convey—through color, texture, thickness, tension, and rhythm—the spirit and rhythm of the various pulses used in diagnosis. These are not exactly realistic representations, but they do manage to portray the pulse varieties well, via a kind of visual analogy.

Certainly the many vignettes of doctors engaged in professional activities—making medicine and gathering herbs (pl. 67); administering eye medicines and enemas (pl. 68); performing moxibustion, and so on—do not portray step-by-step instructions. They are too small and general to teach the viewer anything specific about procedure. Still, they might provide medical students with the general scene, giving an idea of the situation in which something is done. Consider for example the wonderful set of physicians on plate 61 taking pulse while talking, smiling, looking worried, or leaning toward the patient while looking away, as if concentrating intently. Such figures suggest the body language of the ethics of medical attention. It is the kind of suggestion that is enhanced immeasurably by the picture that is better than a thousand words.

1.15 Physicians taking patients' pulse. *Plate 61, details*

But many images in the set do not do much didactically at all. For example, many of these depict the content of their captions not iconically, to use Peircean terminology, but rather by means of symbolic or indexical signs. A man surrounded by flames illustrates a condition of excessive heat in the body (pl. 35.84). A spindle stands for a female child and an arrow for a male child (pl. 54). A son with a good overall prognosis is depicted as a boy with auspicious symbols above him in the air (pl. 59.53). To decode such signs requires cultural knowledge—for example, that the spindle is a sign for the female. Such culturally coded figures would *not* illustrate directly. Nor would they provide more easily accessible information than the words of the text. All they do is substitute visual signs for what the words say. In fact, most words (that is, except for onomatopoeia) are, in Peircean terms, symbols of their semantic contents, in that they are arbitrary signs of meaning. Symbolic or otherwise coded visual images just repeat—albeit in another language, i.e., in pictorial form—a similarly general and conceptual message as is conveyed by words. They don't lucidly illustrate medical knowledge, as the Desi claimed they would.

Similar semiological questions can be asked about other images in the medical set that are in fact iconic but still add no new information to what is already known from reading the text. The picture of a doctor sitting seemingly helplessly near various bowls of substances is used to represent the kind of doctor that the text criticizes as not knowing how to make medicine (pl. 37). Again, the image provides nothing that the words of the text already indicate; it is really just a visual placeholder. Another kind of doctor that the text criticizes does not know how to give a prognosis. The image, also from plate 37, portrays such a figure as just sitting there with his bag and book all wrapped up. What does this tell the viewer that is different from the discussion already provided in the text about the dangers of practicing medicine without receiving appropriate instruction?[114]

Many of the images of pathological symptoms are redundant too, especially for symptoms that are well known, such as diarrhea, which is illustrated by an image of someone excreting watery stool (pl. 44.65). Here we might say that the image only serves as a visual repetition of the text. Even

1.16 A doctor who does not know how to give a prognosis. *Plate 37, detail*

illustrations that do seem intended to give substantive visual information are frequently very general and fail to provide that. The illustrations of various skin diseases, pictured as people with nonspecific marks all over their body, are good examples. As Meyer well observed, these would not have been very informative. The Desi and his artists failed to make use of the opportunity to give close-up pictures of what particular skin lesions would look like. Rather, these images must be doing something else. We could say they are fillers, as Meyer suggests.[115] Or perhaps there were other ways that the paintings' images served pedagogically.

A comparable set of Chinese medical illustrations, Li Shizhen's *Bencao gangmu,* is understood to have little didactic use at all, given the highly conventional quality of the images, and scholars believe that it served more as a ritual device for healing.[116] No such use is known to have been made of the Tibetan set, whose practical function largely had to with teaching—as the Desi's intentions suggest. One of the prime functions of the painting set in the centuries after the Desi is seen in the use of individual plates, primarily those whose direct didactic value we have already noted. According to E. H. Walsh, writing in 1910, at least one of the plates, on the iconometry of the torso, was used for teaching at Chakpori, and each student possessed a copy.[117] But again, that says little about what purpose the rest of the plates served, especially the many individual vignettes whose value in a teaching context is not evident.

All of the set's beautiful images, whatever their semiotic function, also simply serve to please the eye and captivate the attention. Yang Ga, a contemporary teacher of Tibetan medicine in Lhasa, opined that visual images provide excitement and interest. They enliven a class that otherwise consists in a boring litany of categories.[118] We have already noted the sense of play and humanness in some of the images. If those features served the classroom purpose that Yang Ga suggests, it might have provided the motivation to paint all the chapters and main topics of the medical treatise, not just those usefully enhanced by visual elaboration.

There may have been other pedagogical uses for the set that would have motivated the aspiration to render visually the medical treatise in its entirety. One might have been to serve as a mnemonic device, a kind of visual "memory palace" such as is known in other cultural contexts.[119] Indeed, the Tibetan set's close counterpart in China, the Qianlong emperor's *Golden Mirror,* had as one of its main intentions to aid memorization.[120] Such a function—to provide an abbreviated collection of icons that would summarize the content of the *Four Treatises* and facilitate memorization via a quick glance—would obviate the question about those images that do nothing but repeat the textual content in visual form. But we have little information on how the set was used pedagogically, other than what has already been suggested. So far there is no evidence that it was either intended or later used as a mnemonic aid. It is also not the case that every last point of medical knowledge from the *Four Treatises* is portrayed in the painting set, as the Desi acknowledged. In particular, some of the diagnostic and therapeutic detail in the lengthy third section—the *Key Instructional Treatise*—is either elided or represented in very abbreviated form in the paintings. And yet the set covers every chapter and topic of the medical treatise, and often in very intricate detail, much of it very precise and pedagogically useful, even if other sections seem less so. The set gives a strong visual impression of completeness, just as the Desi himself characterized its scope in his statements of his overall intention.

DETECTING THE FRAME

Perhaps there were other motivations for endeavoring to paint the entire medical treatise—or better, effects that this accomplished—beyond pedagogy altogether, even if the Desi did not express or even intend them explicitly. A modern Tibetan historian reports that the set was exhibited each year at the Chakpori Medical College during a brief summer break, in the last seven days of the sixth month.[121] That already speaks to the set's value *qua* set, as an object of pride, or perhaps as a kind of self-promotion of medical learning. And although a few of the plates were reproduced and distributed individually, the entire set of paintings was also copied many times. That too indicates an interest in owning a complete set, quite beyond the pedagogical value of certain individual plates.

The medical painting set clearly had value for its interest to power. Desi Sangyé Gyatso writes that when the Sixth Dalai Lama, Tsangyang Gyatso, came to the throne, in the fire-ox year of 1697, the Desi offered the set into his hands.[122] Exactly what "offered into his hands" actually entailed is unclear. The set was not yet finished, only numbering sixty-two plates. It is in any case unlikely that the Dalai Lama kept it as a gift. It is more plausible that the Desi showed the set in its current form to the new ruler as an accomplishment to be proud of, something that would please and impress him. Then the Desi continued production until it was complete in seventy-nine plates, finally installing it in the new medical college he had built at Chakpori.

The offering to the ruler shows the set's status as an object of prestige that would have brought praise to the Desi—much as did naturalist illustration further afield, as when Kentmann dedicated a pictorial plant book to the Duke of Saxony in hopes of obtaining the post of court physician, or when Vesalius dedicated his *De humani corporis fabrica librorum epitome* to Phillip II of Spain.[123] But the accomplishment would also have been of value to the ruler himself. The Desi writes on another occasion that the completion of the medical paintings would mean that the Dalai Lama would have longevity, and that his government would last long as well. As he puts it, "The Omniscient Lotus-holder's pair of feet will be stable like a *vajra*, and the dominion of the heavenly appointed Ganden Podrang that integrates religion and politics will stay a long time."[124]

We can understand well enough why it would be thought that such a major contribution to medical knowledge would benefit the ruler's overall health and well-being. But how would medical paintings contribute to the longevity of a government bureaucracy? Note again the striking parallel with the aspirations around the completion of the Chinese *Golden Mirro*r collection; it too served in crucial ways to enhance the Qianlong emperor's power and reign.[125] But exactly how illustrated catalogues would accomplish that needs further thought.

The Desi's medical paintings could have had political import in several ways. One follows on my earlier point that these paintings do not provide a transparent picture of society for the purposes of the historian but rather reflect certain agendas at the time of their creation. But that does not mean that they don't *appear* to provide such a picture. Many of the images that "document" something about medical knowledge also "work" (to borrow a useful distinction from Dominic La Capra) to both mirror and create cultural ideals.[126]

Many of the images in the Desi's set model a larger social and even political ethos, quite beyond whatever they say about medical theory and practice. The picture of Tibetan society that the medical paintings conjure, partial and prejudiced to be sure, aids and abets the prejudices and aspirations of their creators and their intended audience, especially the elite officials and scholars in the circles of the Dalai Lama and the Desi. For these latter viewers in particular, the paintings offered a vivid picture of their social world. They show how beautiful women are imagined (pl. 53); they display disapproval and a touch of parody in the depiction of stages of drunkenness (pl. 22). In the many vignettes of couples there is a general gestalt of marital relations—endearing and attractive at that—in how these figures are shown to carry themselves in each other's presence. One set of examples may be found in the renderings of the impact of a couple's pulses on the sex of their offspring from plate 54 (fig. 1.17).

Also fashioned in the mix are a range of attitudes about religion. The Dharma is often positioned as a *summum bonum*, for example at the pinnacle of one of the diagram trees that schematize the medical realities of human life (pl. 2.41). The positive import of religious figures or practices is communicated frequently: to see a monk or a layperson with rosary beads (such people being of fine quality [*skyes bu dam pa*]) or a person with fame (this too is a monk) constitute auspicious signs for one's longevity (pl. 18.7–8). Practicing religion itself is good for longevity (e.g., pl. 18.85–91).

There are also critiques of religion. Plate 42, illustrating the chapter on the causes of contagious diseases, make a very loud point about the harmful practices carried out in the name of religion. The text listed intersectarian disputes propagated by *tāntrika*s, fights between groups within

1.17 Kinds of pulse combinations in couples. *Plate 54, detail*

1.18 Religious rituals that bring about infectious diseases in people. *Plate 42, detail*

the monastic community, and the heaving of destructive magical devices by non-Buddhist Indic *tīrthikas*, Buddhist monks, and Bönpos alike, along with other things that cause disease, like taking a vow to harm others, and physical factors like overly strenuous activity and unbalanced eating.[127] But once again the visual translation provides further detail and specification, starkly illustrating the travesty that religion participates in spreading infectious disease. It divides intersectarian tantric disputes into two scenes, one showing a variety of monks who practice tantra and the other rendering lay *tāntrikas* engaged in various related acts. It also separates the hurling of black magical devices into four individual and detailed scenes in the visual translation, one with yogis who are akin to *tīrthikas*, one with monks, one representing those who wear the garb of *tāntrika*s, and one with Bönpos, in sum representing the participation of many kinds of culprits in this deplorable practice. These two registers constitute one of the most extensive representations of Buddhist and related religious practices in the entire painting set. Their framing and visual vividness make for a harrowing recognition on the part of the viewer: this is indeed what religious people do in our world, and it is appalling.

I am not suggesting that each of the culture-shaping implications of so many of the images in the medical set necessarily reflected particular agendas of the Dalai Lama per se. My point is rather more basic, having to do with the very nature of the set. Given that the project to visually portray medical knowledge entails that the images must fill out the picture, if you will, of what the text says, it is well nigh inevitable that those pictures will mirror values and send messages that intersect, in some way or other, with the interests of their creators. And that again speaks to the many images that don't directly or usefully denote medical knowledge, but were nonetheless participating in other registers of significance that the painting set embodied.

A BIRD'S-EYE VIEW

The more one pores over the plates of the Desi's medical paintings, the more its range of topics astonishes. Even something as inscrutable as the subjective state of a person engaged in "big mental activity" is pictured. Thinking engrossing thoughts or doing a lot of mental work is another item from the list of what not to do the night before a urinalysis. It provides the artistic

1.19 A man thinking hard. *Plate 62, detail*

occasion for a study of intense reflection that remarkably has no reference to Buddhist terminology or practices of mental cultivation whatever (pl. 62). This is an entirely ordinary moment of thinking, just something that people might do—but shouldn't—on the eve of a physical exam.

There had long been indications in the texts that medicine understood its purview to be the world overall, and on different grounds than the soteriological orientation of Buddhist metaphysics, which also endeavors to account for the world and its contents.[128] For medicine, it is a matter of cataloguing the material foundations of human life for pragmatic purposes: being able to track the connection of humor imbalances with social conditions; reading pulse and urine as prognostications for human relations; realizing that everything one does has impact upon one's health. Such an ambitious reach of medical knowledge is already clearly suggested in the *Four Treatises*.[129] The Desi's medical paintings only enhance that impression, with their orderly rows of register after register, suggesting that even though the set illustrates one particular text and its commentary, it really contains the entire world. Skimming through, one gets the sense that this is a visual lexicon of *all* of the kinds of minerals, *all* of the kinds of humor imbalances, and again, *all* of the people who are likely to get sick and be treated.

In fact, the set's eminent skimmability itself speaks to both its pedagogical value and its import for the ruler. The fact that it provides relatively comprehensive coverage of the main points of the medical treatise makes it an easily scanned overview of the text's contents. For example, if one were trying to find all the different places where animals are discussed in the *Four Treatises,* it would be easier to scan the plates, which are labeled by chapter, than to skim the text itself for animal names. Or again, if one were looking for all the places where the *Blue Beryl* or the *Four Treatises* mention sex, one could easily flip through the plates to find the many images of couples in bed, and go from there to the appropriate chapters.

Scannability once again depends on completeness. At least in terms of intention, the set was envisioned as providing comprehensive coverage of everything, "from the introduction to the *Root Treatise* to the end of the *Final Treatise*," as the Desi put it.[130] In short, the paintings would be isomorphic in their overall coverage with the overall chapter structure of the *Four Treatises.* They would provide "control" of the material in the text.

The analogy between control of scientific learning and control of the state was not lost on the Tibetans' counterparts ruling China.[131] It might not have been lost on the Tibetan rulers either. Completeness and control had been accomplished many times in Tibetan scholastic domains. The compilation of the Buddhist canon, for example, has repeatedly enhanced the prestige and power of rulers.[132] Tibetans had also been producing textual encyclopedias for centuries, mostly as compendia of religious learning, but the scope of their contents increasingly included, by the seventeenth century, the material and social worlds.[133] The Desi's medical paintings represent a similar movement, yet the ways their subject matter is construed—and would have made the set a particularly germane gift on the occasion of the enthronement of the ruler—get back to the unprecedented sidelining of Buddhist elements in favor of a larger picture of the world in which religion is but one part. It is a different picture from the religiously condemned concept of samsara, and not what we might mean by the "secular." Nor would it be synonymous with the idea of "all sentient beings" that is the classical purview of the enlightened Buddhist sage. No matter how great the magisterial scope of religious metaphysics or philosophy, medicine could go a step further by viewing even that as a sociohistorical fact that can be perceived from the outside. Ironically, by having to focus on the vicissitudes of everyday illness, medicine has the potential to relativize even Buddhism—if for no other reason, perhaps, than the fact that even the most enlightened lama, even the Great Fifth himself, eventually gets sick and dies.

LOOKING DOWN FROM ABOVE

The way the world was envisioned in the Desi's medical paintings was a model of, and for, the values of the Dalai Lama's state. I would like to close this chapter with a few individual images that display the significances we have teased out for the painting set as a whole. In particular, some of the images visibly imply the very stance or position from which medical knowledge takes in its purview of the world—or from which the ruler might scan the subjects under his control.

We might think right away of the Potala and Chakpori: the two large edifices atop two adjacent mountains in the city of Lhasa. The first is the one from which the Dalai Lamas and the Desi ruled, and the second is the medical college that the Desi built during the same period (see chapter 2, figure 2.7). I don't mean to say that these towering structures are pictured, literally, at the edges of the medical paintings. But there is still, perhaps at the level of an *imaginaire,* a register to be detected that suggests the lofty position where the set's creators—and viewers—are located.

Consider this example, from plate 18. It is an image of a person doing religious practice—or more specifically, "accomplishing," i.e., performing the chanting, ritual offerings, and visualization *sādhana* meditations that help one identify with a tantric deity. It comes up as part of a list of ways to avert signs of impending death. These amount to a group of practices that produce

1.20 Man performing *sādhana* practices. *Plate 18, detail*

1.21 Painting of a deity. *Figure 1.20, detail*

merit and wisdom; others pictured are giving charity, reading the Dharma, performing ransom rituals, and meditation. *Sādhana* is a classic practice for both Buddhists and Bönpos, described in hundreds of Tibetan works.

But what is striking about the medical image of this practice is the generality of that prescription. The man pictured in front of his altar is doing a generic visualization, not of any deity in particular. This can be seen by a close look at the *tangka* of the deity pictured within the vignette, the one that the man is propitiating.

Although from afar it looks right—the standard fire halo, the fierce dancing posture, the animal skin skirt, the lotus base, the prone figure beneath the deity's feet, possibly the back of a consort—up close we see it is just an impressionistic sketch. The image bespeaks a lack of care about iconographical details. It is in fact no particular deity at all. Rather, this is a generic deity, and is different in important respects from a few overviews of visualization practice produced by Tibetan scholars that carefully discern the basic features of deity yoga and theorize each kind's functions and benefits.[134] Here, in contrast, the generic image is just a rough sketch. It does not stand in any considered or deliberate way for the specific image or ritual that will be inserted in its slot.[135] Medicine is merely telling people to do *some sādhana; any sādhana.* The image gestures to a "whatever" in which there need be no specific options in mind. *It is a depiction of a whatever deity as such.*

1.22 The Fifth Dalai Lama, flanked by small images of Padmasambhava and Tangtong Gyelpo. *Figure 2.4, detail*

I may be making a lot of a very small detail, but it is hard to find a vague rendering of a buddha or sacred deity elsewhere in Tibetan painting. The deity pictured on the man's altar cannot be equated with the occasionally less-than-fully detailed images sometimes found in biographical paintings of a master having a vision or depicting his teacher or tutelary buddha in the heavens above his head. In such instances the details are omitted because it is a side illustration, small in size, and therefore perforce tiny and rough. The figure being denoted, albeit imprecisely, *is* nonetheless the very deity that the main figure saw in his vision or regards as a master or divine guide. But in fact, as in this painting of the Fifth Dalai Lama, such small figures are often very carefully rendered and clearly refer to particular deities or persons, represented according to their specific iconography. (In this case they are easily

1.23 Monk reading a Dharma book. *Plate 18, detail*

recognizable as Padmasambhava and Tangtong Gyelpo, two masters of the past who were inspirational to the Great Fifth; at about one inch high, they are the same size as the deity on the medical plate.)

In the medical vignette of the layman meditating, something else entirely is being denoted. Details would not be relevant or even appropriate, even if there were all the room in the world to portray them.

What we see here is of course a function of the generic nature of the medical images explored earlier. Now we are prepared to detect an aspect of that feature's significance, in light of the larger circumstances of the paintings' production. One more example will help make the point. Right next to the meditating layman is a monk reading from a Dharma book.[136] It is another thing to do that will fend off impending death. The book is open and the page is showing, but written on it are only some squiggly lines.

Again, comparison with religious portrayal is illuminating. Sometimes we find a master holding a book with meticulously rendered Sanskrit or Tibetan letters on its open folios,[137] but more frequently books are unmarked, or like in the medical image, show only vaguely suggested letters. But these will refer to events in a master's life when he studied a particular text, or taught it, or had it printed; or to a particular collection held in a monastery; or to teachings associated with a specific bodhisattva, such as the *Perfection of Wisdom* sūtra that Mañjuśrī holds in his hand. It may not show a particular page, chapter, or work, but the rendering still refers to a specific event, practice in a particular lineage, or element of a particular iconography or tradition. It certainly cannot refer to a text with which the figure in question never had contact. In contrast, the medical rendering of the book has no such limitations. It seems almost to be saying, here's some Dharma stuff—any, it doesn't matter. It certainly doesn't matter, say, what school this book is from. No sectarian polemics or nuanced distinctions; it's just Dharma, any Dharma, and good for your health and longevity.

Religion would not disagree with that thesis. But its depiction in medicine is disconcerting; the Dharma may be a therapeutic trump card, the *summum bonum* of human activity, and yet pictured from a distance. The magisterial transcendence of the specificities of religion that these rough images display is not the same as in a modern still-life such as a Bible portrayed by van Gogh, its pages open but its letters obscured.[138] There the holy book is even more recontextualized than in the medical image, now in terms of its purely visual qualities of color and form. But the Dutch master still acknowledges that he is portraying the Bible, is still aware of the sacrilege he is knowingly committing. The perspective of the medical paintings, in contrast, shows no hesitation in its fully abstracted portrayal of the Dharma, despite the acknowledgment of its value. The Dharma of the people performing beneficial religious acts in the medical series is only Dharma from a bird's-eye view: seen from afar, performed by others, the others down there.

And so not only has religion been decentered and recontextualized in the medical paintings; it has also been made into a "whatever." We might go from this striking liberty that is visible, at least by suggestion, in several of the individual images to the rest of the painting set. The fact that it consists virtually entirely in generic images, rather than specific historical, mythological, or imagined subjects, entails for all of them a vantage point from which someone could even know what is typical in the first place. Couple that stance with the sense of control that the scope of medical knowledge suggests, its imputed magisterial sweep of everything about human life. Both features depend upon the apparatus that stands behind the set's creation as its condition of possibility, that can account for everything under the aegis of one overarching system. In depicting the medical images in ways that actually suggest that apparatus, the Desi had produced a vision of imperial power and its cultural implications that anticipated the Qianlong emperor's own patronage and systematization of medicine in the decades following. Marta Hanson writes that the Chinese scholars who created the *Golden Mirror* "employed evidential methods and imperial canonization as effective tools to support the Qing court's claim to its position as standard-bearer of Chinese civilization. . . . The *Golden Mirror* can thus be read as a manifestation of a distinct stage in the process of imperial documentation through which the Qing court simultaneously defined Chinese culture while assuming authority over it." She could just as well be talking about the Tibetan case.[139]

I will explore the relationship between medicine and the Dalai Lama's state in the following chapter. For now, the Desi's painting set provides a lucid example of how medical learning came to be pictured as isomorphic with the aspirations of the state. Fernand Meyer saw something important when he said of the Desi's gift to the young Sixth Dalai Lama that "beyond its technical purpose this iconography was also conceived as an aesthetic courtly work of prestige, and as a grand political gesture."[140] Perhaps the paintings' bird's-eye perspective, alluding to control and completeness, helps explain why they had the capacity to make such a gesture.

THE INDIVIDUAL IN THE UNIVERSAL

But the view from above that enables society to be grasped in a glance is only half the equation. The other side of what makes the Desi's medical painting set akin to a simulacrum of the Dalai Lama's rule is its close scrutiny of details. This too can be detected at the micro level in certain individual images, where the delightful humanity and idiosyncrasy found repeatedly in the set come especially to the fore. Such details actually reinforce the seeming distance of the bird's-eye view, although the two would seem to be in contradiction. How does the close-up square with an epistemic foundation in the overarching and generic? Scrutiny of a few more examples of Barthes' "superfluous detail," now in light of the paintings' larger connotations, will suffice to show how the details gesture to the vantage on high that made them possible.[141]

Consider this series of illustrations of head shapes from plate 47. Again, the most charming detail is extraneous to the explicitly denoted content.

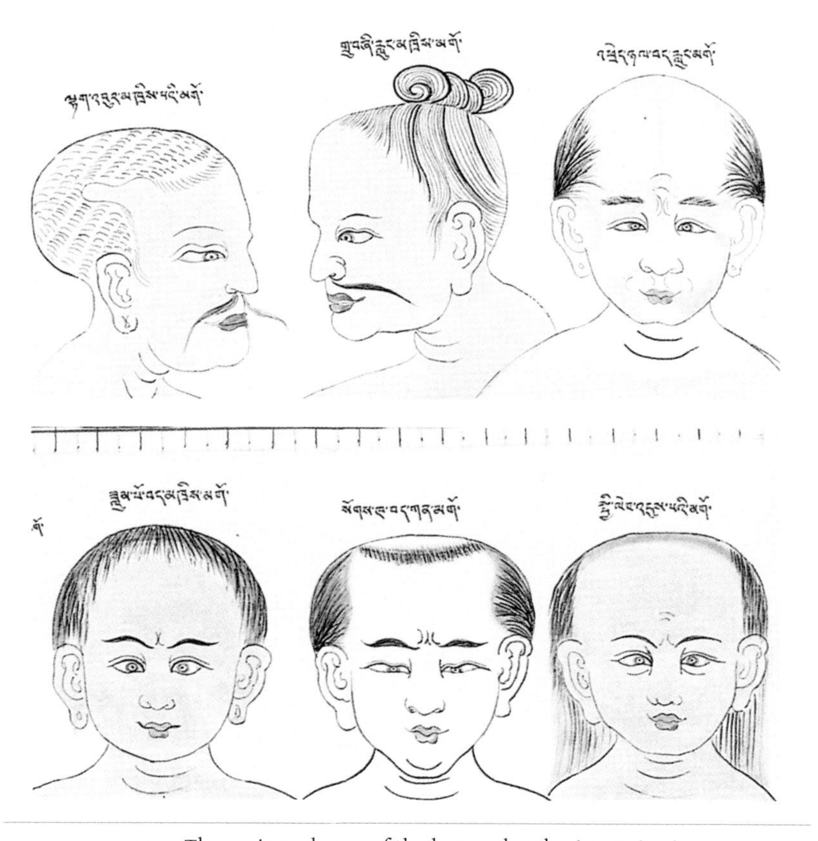

1.24 The various shapes of the human head. *Plate 47, details*

But we are still drawn to ask why: why portray models with such noticeably different hairdos, eyebrow thickness, eye shape, and expression, if what is meant to be conveyed is a lesson on head shape? It is not being said that these particular features have anything to do with these particular head shapes.[142] Why even show hair at all, or at least, why not provide schematic faces and hairstyles in order to focus attention on the matter at hand?

Or witness the idiosyncratic differences between two male figures who have been enlisted on the same plate to display the organs inside the thoracic cavity (fig. 1.25). The images betray an interest in, even an appreciation for, these men's human specificity, even the idiosyncrasies of their personalities, as they submit to the indignity of the display. Yet clearly the two men are also the same in being human. Either would have served to display any of the internal organs.

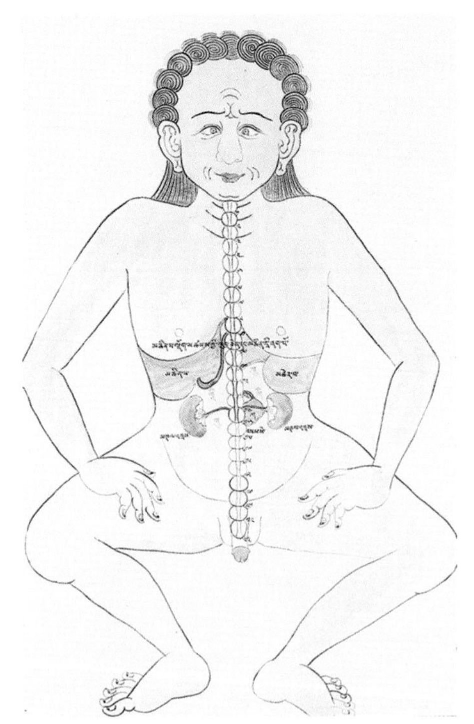

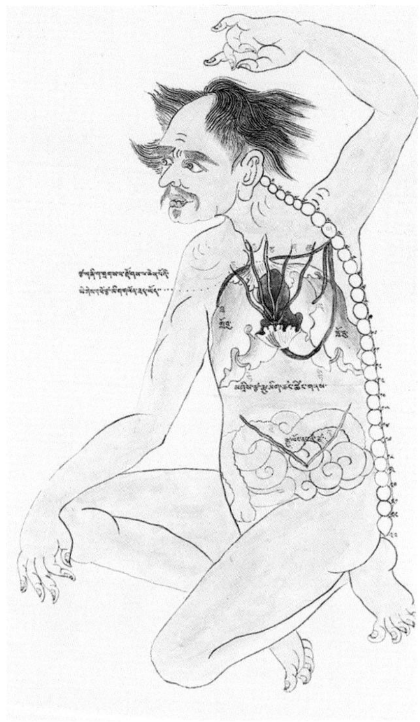

1.25 The organs inside the thoracic cavity. *Plate 47, details*

In this example, the two opposite features that we have been exploring are at work in the same visual message. On the one hand, the figures are unique individuals; on the other hand, both serve equally well as generic illustrations of human anatomy. In fact, if there were anything about their inner anatomy that was idiosyncratic, they would not be suitable as models for the lesson. And to be able to know that the two bodies are typical and not aberrant depends precisely upon taking a position high or transcendent enough to compare and to discern sameness among the many tokens of the principle at stake.

Paintings such as these put the viewer up very close and very far away at once. Perhaps the very distance that enables a depiction of the generic ironically is what also allows the idiosyncrasy. Perhaps it is only when one has some distance that one has the magnanimity to allow anything and everything to be portrayed. It is only then that one can contextualize, and tolerate, difference. And it might well be that medicine provides a better chance of achieving

1.26 Actions in the care of infants. *Plate 45, details*

that distance than does religion, with its decidedly judgmental gaze upon ordinary society. Medicine must simultaneously know the general patterns of illness and health and know how to discern those patterns in unique individuals. Medicine has far less ambition than Buddhism to transform everyone into an ideal. It rather earns its keep by focusing on the aberrations.

One final example from the medical paintings will illustrate how the distinctive way they combine generality with individuality might make them an excellent medium to display and fulfill the aspirations of the emerging state. A register of women on plate 45 illustrates various actions that mothers do that can cause illness in the child. But if we shift our focus, we can also notice the delightful—and once more extraneous—detail that each mother is dressed very differently.[143] In fact, their range of costume reflects various kinds of regional dress. Could it be that in one stroke, this vignette celebrates the array of different styles of attire on the Tibetan plateau, and conveys an isomorphism between the single banner of medical learning—the topic of child rearing—under which they are all gathered and the fact that the newly centralized state had assumed oversight over all the regions that their costumes represent? In other words, might the implication that one medical topic can subsume all of these different mothers bolster, or at least mirror, the proposition that one bureaucracy can govern all of the regions from which they hail—especially since these illustrations were produced by the ruler of that state?

I am not arguing that we can actually ascribe an intentional nationalist agenda to this set of mothers, or to the many other instances throughout the painting set where women and men are portrayed in garb and headdresses from many different regions of Tibet.[144] Short of a systematic study—which would surely occupy a lengthy chapter of its own—it appears at first glance that the variation in regional dress is rather random. But random variation still reflects a general assumption on the part of the paintings' creators that there is a wide range of local styles among the people whom medicine would serve. At least the very nonjudgmental purview that we have discerned—the willingness to portray all kinds of people, of all shapes and sizes, from all regions, and in all manner of compromising poses and medical conditions—would signal that medicine is applicable to everyone. Medicine is here pictured as being meant for everyone and anyone. And that reflects the ambitions of the Dalai Lama's reign too; even if this point is not referenced explicitly or systematically, it is there aesthetically.

"Everyone and anyone" in fact sums up the visual message of the medical painting set quite well, enabled precisely by the juxtaposition of specificity and generality. Medicine is portrayed as meant for every individual person, no matter how aberrant, funny, quotidian, or bad. But at the same time its knowledge is general enough to apply to anyone. Whether you have bushy locks and eyebrows or are clean-shaven, the doctors' anatomical knowledge will apply to you. The same applies to every and any animate thing. Look at this rendering of, say, a white rhodiola plant, and you can go out in the field and recognize its thousands of iterations. By virtue of the medical artists' exuberant license in portraying the ordinary world, taking the opportunity to stretch their wings, so to speak, in this unusually nonstandard iconographic field, *the very conditions of such license*, the power and position to know and control, are close to the surface and even visible.

If the dialectic between sweeping vision and delight in the idiosyncratic is a mirror of imperial aspiration, our "reading" of the medical *tangkas* has stumbled upon a familiar but fundamental dynamic, noted many times and in many registers for the same centuries in contexts across the globe. The distinctive tension between the valorization of the local and the rule of the general, between the valuation of particular practices and the universality of the right to pursue those practices, describes patterns of development on humanistic, political, and scientific fronts alike in the early modern period. Stephen Toulmin's lucid analysis of the clash between pluralism and the Cartesian totalizing vision, or again between humanism and rational science in the constitution of the modern cosmopolis, is an exemplary case in point.[145]

The study of aesthetics and semiotics in the medical painting set has revealed some parallels to these developments. Among them are a magisterial bird's-eye view, a pride and fascination with the panoply of differences within that sweep, and an impulse to classify. The latter remains in tension with a companion urge to represent real objects, seen in similar projects of scientific illustration in both early modern East and West Asia and farther afield. The Tibetan medical interest in natural and social worlds also marks a critical turn in epistemology, whereby the way that tokens can illustrate a larger principle has shifted in kind from, for example, theories of the Buddha and its many theophanies. In at least a few critical cases noted in the following chapters, it appears that rather than looking for system or manifestations of an ideal, the empirical observation of the natural world in medicine was trained on idiosyncrasy, asymmetry, and unpredictability.

The decentering of Buddhist figures and ideals in the Desi's medical paintings, along with the abundance of energy that they devote to ordinary life, reflects the turn toward the secular across the early modern world. But if we were able to discern in a few cases the outrageous implication that medicine could actually encompass a larger episteme that trumped the authority of religion, an episteme of which Buddhist traditions were only a part—even a "whatever"—it must quickly be added that such a suggestion was quickly muted, and subject to condemnation from many quarters. In the end, any idea that Tibet was participating in a scientific modernity such as was unfolding in Europe cannot be sustained. One very big reason is that despite these gestures, religion was far from stripped of sovereign authority in the Dalai Lama's Ganden Podrang, as Christianity came to be in Europe. Quite the contrary.

One of the central aims of this book is to track just these contradictions, not only on the aesthetic register but also, and even more so, in the historiography and theory of medicine that came to a head in the Desi's time. The Desi will continue to serve as the lynchpin in the next chapter, which unfolds more fully the history of medicine in Tibet and the particular mentality it was fostering—right in tandem, it seems, with his vision for the Tibetan Buddhist state.

2

ANATOMY OF AN ATTITUDE

MEDICINE COMES OF AGE

Not only did the Desi mastermind, from the heights of the Potala, an exceptional set of medical paintings. He also was a consummate historian who authored voluminous biographies of both the Fifth and Sixth Dalai Lamas and a detailed history of the Gandenpa school, in addition to his influential history of medicine, among other massive writings. The last and most powerful chief minister appointed by the Fifth Dalai Lama three years before his death, and then regent of Tibet until the enthronement of the Sixth, the Desi is an outstanding example of the lay intellectuals emerging in this period, particularly in circles of power. He was an influential collaborator with the Fifth Dalai Lama on many matters of culture and especially on the vision of the newly unified Tibetan Buddhist state; all of his scholarship fed or reflected his politics and conception of governance. A difficult but rewarding writer, highly subtle and often ironic, he wrote copiously on astrology and calendrical calculations and the principles of administrative law, as well as much about the Tibetan capital, its ritual observances, and its symbolic and geomantical implications.

Scholars have begun to study the rich range of the Desi's writings for what it tells about his times. But as much as noting *what* he says, we are also rewarded by appreciating *how* he says it. Such an exercise is valuable for the rich sense it gives not only of his own character but also of what was possible—how knowledge could be shaped and shifted—in the intellectual and political institutions of his day. Often his innovations consisted in

2.1 Desi Sangyé Gyatso. *Painting held in the Potala.*
After Bod kyi thang ga *1985, pl. 78. By permission of Cultural Relics Press*

a fine-tuning of long-standing tradition, as is the story for much of South Asian intellectual history in the early modern period.[1] But there were also moments when the Desi and some of his predecessors, especially in the field of medicine, were on the brink of disrupting the very foundations of epistemic authority in their world.

One is certainly struck with the Desi's panache, his confidence, and his vast learning—a point he often wants to impress upon his readers. As we look more closely, we see a restless mind: complex, critical, and yet simultaneously unsure, and often self-contradictory. At one point, in what must be one of the most curious moves in all of Tibetan historiography, the Desi bitingly criticizes the character of one of his most important medical predecessors of the previous century. Most surprisingly, he does this in the very course of relating this figure's life. A close look at this passage will be a way to introduce a number of questions about the reigning attitudes to tradition, innovation, and the newly centralized Tibetan Buddhist state during the Desi's day. I would like to ask what the culture around the study and practice of medicine might have contributed to those dispositions.

AN ESPECIALLY CRITICAL MENTALITY

It is an entirely standard thing for Tibetan historians to relate, in a few lines or pages—and invariably in highly complimentary terms—the lives of the eminent figures in the movement whose history they are recounting. Thus does the Desi's rendition of the life of Zurkharwa Lodrö Gyelpo, sixteenth-century author of the ground-breaking *Four Treatises* commentary called *Ancestors' Advice* and principal holder of the Zur medical lineage after its founder, Nyamnyi Dorjé (1395–1475), come right up in the section on the Zur school in his history of medicine. So the reader is hardly prepared, upon arriving at the six-page passage on Zurkharwa, for anything other than the standard laudatory overview of birthplace, parents, education, and accomplishments, along with the predictable hagiographical flourishes.[2]

Indeed, this eminent descendant of the Zur medical line had an excellent pedigree. The Desi's biography goes through in standard honorific fashion Zurkharwa's birth in 1509, his ordination as a Buddhist monk by Karma Trinlé (a prominent disciple of the seventh Karmapa), and his education at the Dharma center called Lekshé Ling in logic, poetry, and the sciences, including the classics of medicine.[3] His first teacher of medicine was Lingbu Chöjé, a student of Tsomé Khenchen.[4] Although he was squarely in the Zur tradition, he studied Jangpa works extensively, as well as the Drangti lineage at Sakya Mendrong, a major medical center at the time. The Desi does not fail to give Zurkharwa credit for his original research and scholarship. He graces Zurkharwa's career by mentioning the rumor that he found precious hand-smudged "golden notes" penned by Yutok Yönten Gönpo himself at Nyangmé.[5] Then Zurkharwa obtained the patronage of the local king Wangyal Drakpa, and wrote the *Ancestors' Advice* commentary in four years.[6]

Then he went to Lhasa, and in the manner of other great minds of the time, posted a set of questions on medical practice and theory on the ancient treaty pillar. The questions were posted three times, and yet no response came back. But his disciples encouraged him, so he wrote his own autocommentary.[7] This, the Desi relates, seemingly in admiration, was not understood by most people. (Let us not fail to note that this picture—the brilliant commentator whose work no one else understands—would come to portray the Desi's sense of his own career too.)

Apparently after some time a few luminaries did offer responses, but the Desi notes that Zurkharwa was not in accord with their ideas and therefore wrote his own response to the questions.[8] The Desi also goes on to mention Zurkharwa's work in editing the root medical text *Four Treatises* based on his own research. He recounts a story that Zurkharwa had requested patronage to print the newly edited root text from an important Rinpung leader, who first agreed but then did not deliver the funding. Later, the Desi reports, in the colophon to the Dratang edition of the *Four Treatises* that Zurkharwa finally got carved and printed at Lhokha Dratang, Zurkharwa praises his patron Yargyapa, who did underwrite the edition, but adds that "previously [the Rinpung leader] said 'I'll do this,' and then many people worked very hard, but just like the crow who came to die, nothing great became of it." "Thus did [Zurkharwa] utter satirical, rebuking words," the Desi comments, still using the normal honorifics.[9] The Desi finds it remarkable that Zurkharwa would criticize the Rinpung leader in this way.[10]

The Desi goes on to sketch the other works that Zurkharwa wrote. When he gets to a short essay on whether the *Four Treatises* were originally taught by the Buddha or not, the Desi summarizes Zurkharwa's position—that from the outer perspective, the work was taught by the Buddha, from the inner view it is a composition of a pandit, and from the esoteric vantage it is a composition authored by a Tibetan (we will see in chapter 3 why the Desi was so exercised about this).[11] Then he comments that in all of Tibetan medical writing, including that of Zurkharwa's own lineage master, no such point had ever been seen before. Here the tone begins to get sarcastic.[12] For Zurkharwa in this way to sound a great roar and demolish all of the previous authorities, including his own teachers, the Desi notes, is, well, as the great master Sakya Paṇḍita said: "Setting aside their shame for their own deep degeneracy, they hold aloft their own bad talk; some of the royalty in Kāñcīpura beat the drum of victory when they kill their father."[13]

The Desi then fumes about the one who hangs himself with his own colloquialized and idiosyncratic version of medical science.[14] He broods that while Zurkharwa's venerable predecessor, Nyamnyi Dorjé, had painstakingly drilled holes through diamonds, Zurkharwa just came along and added the thread.[15] The Desi then goes back to listing more of Zurkharwa's medical writings, and also a few works on poetic theory and other topics. But soon he returns to his critique. Other than luminaries from the past, Zurkharwa never praised the actual teachers with whom he studied medicine and poetry. The Desi quotes the *Final Treatise*'s critical description of one who "keeps one's teacher secret and puts oneself forward as great," to make the point that Zurkharwa was really just like that.[16]

The Desi concludes the story with the sad fact that Zurkharwa never completed his *Ancestors' Advice* or his history of medicine, due to a stroke.[17] In an acid sum-up, he remarks cattily that Zurkharwa had no prophecy or special memories to confirm that he was a reincarnation; his claim of being the incarnation of Nyamnyi Dorjé was only recognized by the Red Hat Karmapa. (Again the Desi is positioning himself, this time in contrast, for he was very proud of being an incarnation himself; and the Desi had a special dislike for the Karmapas and everyone associated with them, given their battles with the Dalai Lama's forces in his own day.) In brief, the Desi tells us, Zurkharwa just had superficial knowledge of topics like poetry, logic, and Mahāmudrā, while in medicine he didn't reach the level of Kyempa Tsewang (Zurkharwa's peer, discussed in the following chapters). And yet he still demolished teachers in his own lineage and looked down upon many other textual traditions. As for the Rinpungpa lord whom Zurkharwa criticized for never coming through with support to print the *Four Treatises*—the Desi now supplies his name, Ngawang Jikten Drakpa—well, *he* could tell the difference between gold and copper. He came to realize that Zurkharwa's edition of the *Four Treatises* changed its meaning and had hidden errors. And so what Zurkharwa's colophon said about the Rinpungpa—that "just like the crow who came to die, nothing great became of it"—was really Zurkharwa's own lot.[18]

In contemplating the severe breach of writerly etiquette that this critical biography represents, we at least have to set aside the silly claim that nothing ever came of Zurkharwa's work. If nothing else, we know that the Desi himself copied word for word much of Zurkharwa's *Ancestors' Advice* in huge swaths of his own *Four Treatises* commentary, the *Blue Beryl*, not to mention Zurkharwa's medical history, long passages of which show up in the Desi's own medical history. The reasons for his resentment likely stem from a combination of factors, ranging from Zurkharwa's association with the Karma Kagyü lords of Tsang who came to oppose the Dalai Lama to a more personal sense of competition that the Desi might have felt with a brilliant predecessor.[19] It doesn't take rocket science to know that the bad blood was related to the Desi's severe unhappiness over Zurkharwa's views on the original author of the *Four Treatises;* he already makes that clear in the passage itself. Disagreements and competitiveness were common in Tibet's polemical intellectual climate, and so were considered critiques of the views or practices of other scholars, even those from the same lineage, not to mention those of rival schools.[20] But I have yet to discover another case of an author abusing someone on *ad hominem* grounds in the very course of recounting his life story and achievements. That constitutes a fundamental transgression of the most basic presumptions of the Tibetan genre of life-telling. So we are left with a puzzle.

One line of speculation that I would like to pursue takes off from the simple observation that the Desi's main explicit reason for abusing Zurkharwa—that he lacked the requisite deference—sounds a lot like the Desi's own career. This is not to suggest that the Desi as biographer is modeling himself on his subject per se. Rather, perhaps it is just that medicine occasioned a growing sense of license for great minds to be, as part of the job description, short on patience and manners. Certainly if there were ever someone who had the nerve to sound the great roar to demolish previous authorities, it was the Desi himself, and not only on intellectual grounds but in light of an overall assessment of their intelligence and general competence.

In fact, far more frequently than personally attacking a respected figure of the past, the Desi even critically assesses the virtues and capacities of his still-living masters.[21] On several occasions he confides to his readers that Lhünding Namgyel Dorjé, whom we met in the last chapter as one of the Desi's valued resources for the medical paintings, is exemplary, yet in the end a disappointment. While the Desi reserves his most scathing invective for Zurkharwa, this more measured critique is damning nonetheless.

> Lhünding Namgyel Dorjé: His qualities and expertise are greatly famed as exemplary. At one point he put a critique of the chapter on recognizing medical plants in *Ancestors' Advice* on a stone pillar. But since no rebuttal of his main point appeared, I checked into the kinds of questions he raised about plant recognition and the quality of the plants and so on, and with the exception of fine distinctions regarding certain plants and one or two other points, my investigation of the *Four Treatises* in light of authoritative sources and reasoning [revealed his ideas] pretty much to accord with my own thinking. Indeed, he was in a lineage of noble physicians (*lha rje*) that had not been broken since the time of Tri Songdetsen. Even now he has made a vow to memorize a verse every day and he keeps to that. He can rattle off 10,000 pages with no problem. Accordingly, he is very sure on the words of the *Four Treatises*. What's more, the amount of words of the root *Kālacakratantra*, its great commentary, and with respect to the other sciences (*rig gnas*) that are in his head is vast. But regarding fine detailed issues and medical science, his medicine would seem to be limited. When he does make medications, they are flawed by his conceptual confusion, and there is little care for others in his way of teaching. Other than that, his mind is vast, and he is proud with manifest celebrity. [22]

A number of things are striking here: the apparently common practice for scholars to post questions on a pillar as a challenge to others; the role of noble lineage for Tibetan physicians; conceptions about education; the significance of memorization; the relative emphasis on practice over theory. But let us stick for now with the puzzle of the Desi's attitude—his gall—in being so candid. Not only did Lhünding have an impeccable medical pedigree, he was also the Desi's *own* teacher. In this passage, the Desi was actually in the course of discussing how Lhünding contributed two of the *tangka*s for the medical illustration set, and also how he taught the Desi human anatomy on the basis of Lhünding's own father's teachings.[23] The Desi later also asked him to teach the measurements of the thoracic cavity and the mnemonic diagrams to groups of local doctors.[24] And he clearly admired his work enough to be moved to respond to the set of questions that Lhünding had pasted to a pillar, which no one else was able to address. So this was a respected teacher whom the Desi trusted as an authority . . . at least on certain topics.

Lhünding's problem, for the Desi, is more nuanced than the socially unacceptable conceit for which he skewered Zurkharwa. The Desi is very particular. He withholds praise and bestows it only on that which he actually respects, and he is willing to criticize the parts of the picture that he finds wanting. He speaks of how he interviewed Lhünding, along with another colleague, Namling Panchen Könchok Chödrak, regarding certain issues about the anatomy of the channels. The Desi shows himself as wavering, at first dismayed at the difficult style of Namling Panchen's way of speaking, but later in agreement with his ideas and finding now Lhünding's opinions unconvincing.[25] Years later he would return to report on this interview again, adding that Lhünding was very proud and conceited and evinced little regard for others in the conversation. The Desi even says now that he had engaged Lhünding in questioning deliberately, in order to break his pride. And yet with regard to explaining the medical texts, recognizing medical plants, and the details of anatomy and the mnemonic diagrams, the Desi affirms that Lhünding was without peer during this period. So the Desi received many of his medical teachings in private instruction.[26]

The Desi holds back and retains a superior vantage from which he feels entitled to judge everyone. As he declares a few pages down in the earlier passage,

> If you look at the flow of their conversation, when I search among the current noble physicians, the one who is almost definite is Lhünding Namdor and also Nyemo Tseden, about whom I have heard talk that he is good at [memorizing] words and understanding the meaning. Other than those two, I think that [Namling Panchen] is better at explaining the *Four Treatises*.[27]

Few of his would-be respected colleagues are spared the Desi's ever-ready disparagement of their character and scholarly virtues. Even about Jango Nangso Dargyé—a highly respected physician of the Fifth Dalai Lama with whom the Desi worked closely and whom he praises as being stable even in old age in his knowledge of the *Four Treatises*—he adds wryly that otherwise this same venerable physician "either had not studied the commentaries very well or it was like the allegory of one's father's bowl." The expression refers to a nonprobative attitude toward what has been passed down from one's ancestors, and its critique of that approach is telling of the creeping modern sentiments of the time.[28] And this is not to mention all the other doctors whom the Desi asked, say, about the channels of the body, and who, let alone about the growth and life channels (which are subject to considerable debate), had "short tongues" even about the connecting channels (the evident arteries, veins, and nerves that doctors regularly use for bleeding and moxa). This mumness renders their ignorance as easy to read, the Desi avers, as the Nepalese *se'u* (a fruit whose insides can be known by looking at the color on the outside).[29]

Desi Sangyé Gyatso is intent on scrutiny. He is skeptical of those who are trying to protect their own reputation and look like they know something, and he is often not sure he believes what people say. About one of his local colleagues, Mengongpa, he quips

> even though he did not pass the oral exam on the *Four Treatises,* he says he passed it before; maybe it is so. He can memorize some bits but he doesn't have much answer to hard questions, and yet he can answer the middling ones, and has come up with a few more answers than the others.[30]

The Desi is tough, and he looks closely. He shows himself casting a critical eye even upon his closest friends, in his twenties when he lived with the renowned physicians Sumga Lhaksam and Darmo Menrampa.[31] Speaking of a time when the three were residing together and "extremely familiar with each other," he addresses his reader with remarkable intimacy, reporting how he could "occasionally [watch them] closely, and roughly see how well they could recognize medicine and were understanding the texts."[32] He reveals that Sumga had only roughly memorized three of the *Four Treatises,* and mostly based his practice on a few texts and oral tradition. Yet although his knowledge of medical science was as not vast as others', he had much compassion and made exceptional effort on behalf of patients. The Desi notes jocularly that Sumga's style matched his name, Lhaksam (lit., "Exceptional Considerateness"): he had virtuous thoughts for all. But even here he adds that his colleague also had some bits of anger in the heat of the moment.[33]

Apparently the Desi used the occasion of collaborating with other physicians as a way to spy on them. He feels free in his writings to divulge how one doctor held "great memory of the *Four Treatises* in his mind, but paid little attention to the practical instructions and supplements, so I could see that he did not practice"; or how another colleague, with whom he was also familiar, fashioned himself as a tantric master, but otherwise had not seen much of medical science.[34] As for the principal heir of Lhünding, a physician named Ganden Menla who would have had a prominent position and whom the Desi thanks elsewhere for helping him both to research his commentary to the *Instructional Treatise,* the most difficult part of the *Four Treatises*, and to write it down,[35] the Desi critiques him quite unabashedly. "He knew the *Four Treatises* well and could also recognize plants, but he was lazy and indolent and his effort was small."[36] And on another doctor, well known, among other things, for his reply to Zurkharwa's questions, the Desi is particularly harsh, if also jocular:

> Tibetan Learned One Mipam Gelek's composition of poetry is solid. Otherwise, regarding his general knowledge, it doesn't look like he really lives up to the meaning of his name. But he knows how to read and write, so he's got a few compositions.[37]

In this latter case, the Desi blurts out his barbs with no discernible relevance to what he has just been discussing, namely his own medical activities.

Again and again, the Desi scrutinizes and points out the weaknesses of the very colleagues who are his main teachers and collaborators. Given that his critical observations would have been there for all to see, right in his major, very well-known publications, we have to wonder. How was the Desi imagining his audience? Surely his primary readers included the very persons just named. Did he not care what they thought? Did he think that somehow his irreverence was acceptable, or even exemplary for a medical scholar? Or was he using these reflective passages as a kind of modernist autobiographical outlet, a chance to vent and to express himself as "who he truly is"?[38] If the latter, was the Desi even thinking about the message his attitude was sending? I am most interested in the possibility that although he may not necessarily have been conscious of it, his candidness and his criticalness were the products of—and models for—medical professional ethics.

Or maybe we should just conclude that the Desi was a very persnickety person, or perhaps better, chalk up his arrogant display to an exaggerated sense of entitlement, in light of his high government position and power. His arrogance likely helped, tragically, to get him executed in the end by the Lhazang Khan.[39] Tucci dubs the Desi "crafty";[40] there must be much in the foregoing that is a product of the Desi's own idiosyncratic character. But we need to add in other considerations beyond this personal assessment. We must note at least the growing trend in Tibetan writing toward realistic depiction, which reached new heights during the reign of the Fifth Dalai Lama.

In literature, one major site where individualistic and candid description is evident is in diaries and autobiographies.[41] While the Desi himself had much to do with the growth of detailed record-keeping in the period, he left no substantial account of his own life save a few short passages—along with a series of small autobiographical murals of himself in the Potala.[42] Perhaps the Desi sought other means of self-expression, as I speculated above regarding his handling of Zurkharwa's life story. We also saw how his medical paintings, with their commitment to realistic anatomy and botanical depiction, participated in (or in fact served to foster) this same trend. There can be no question that the Desi was a major player in developing the very practices that made for more modern and candid kinds of depiction in his day.[43]

I want to entertain the possibility that medical culture was one of the currents that fed these shifts toward realistic depiction. It is worth asking if the increased confidence in the independent witness of empirical evidence, and especially the idea that knowledge can be improved, made for a shift in attitudes toward tradition and authority that were felt in realms beyond academic medicine itself. In medicine such attitudes congealed as a stance or self-conception that might be characterized as a "mentality," using this historiographical term in a different but related way to its famous Annales School sense. As Michel Vouvelle has described it, the notion of mentality is deliberately vague in order to contrast with the more intentional nature of ideology.[44] Although the present study is based on analysis of elite texts rather than the nonsystematic sources and practices of the larger population to which the term has usually been applied, what Vouvelle dubs the "artistic" nature of the category helps capture for me the wide-reaching disposition, or epistemic orientation, that can be perceived in medical writing even though it was never formulated as such. This orientation would issue out of the demands of clinical practice and its attendant investment in empirically demonstrable fact. Concerns of this kind were not limited to medical circles and had been expressed by master Buddhist scholars like Sakya Paṇḍita (1182–1251) in entirely different contexts, but they seem to have come up remarkably readily in the orbit of discussions around medicine.[45] Actually some of the most radical expressions of skepticism regarding received traditions about the history of medicine, the articulation of the body's functioning, and the pursuit of scientific knowledge were voiced by critics whose specialty was not medicine per se.[46] But no matter who voiced them, it seems the stakes in getting things right in medicine sometimes meant that the usual circumspect style was jettisoned. Such a disposition looked back to the robustly competitive and self-promoting ethos already in operation in Tibetan medicine by the twelfth century, to be studied in chapter 7. But it also continued to evolve down into the Desi's day, part and parcel of a growing conviction about the rightness of the aims of medicine as it became an increasingly independent site of knowledge formation.

The Desi's transgression of protocol in his discussion of Lhünding illustrates the impact of the medical mentality well. The meticulous assessment of personal qualities and even impertinence with respect to his teacher have everything to do with what the passage is actually about. It is part of the long concluding section of the *Blue Beryl* commentary. Significantly, the Desi begins this closing section with a review of the entire history of medicine, first in India and then in Tibet. He frames his efforts to complete his commentary and the painting set—as we saw in the last chapter, the two projects were of a piece—in rhetoric familiar from Buddhist writings about the "degenerate age."[47] But both the commentarial and painting projects did not depend only, in standard scholastic fashion, on earlier commentaries and other written sources. The advice of contemporary colleagues and the direct observation of bodies and plant life were also critical to the entire endeavor. Again, fierce disputation was certainly rife across Tibetan Buddhist scholasticism, and self-assertion was also frequently to be found in autobiographical writing.[48] In addition, creative thought, sometimes based on an author's personal experience, was not unknown as a claim to epistemic authority, especially in works on meditation.[49] But the Desi's self-entitlement and adversarial excess in these medical passages are extreme. I am wondering if the particular challenges in the quest for knowledge of the physical world made a difference in the kinds of rigor and strict precision that were conceivable. Disputation was essential to the growth of science in Europe.[50] This also is true of ancient Indian medicine, and in fact the principles of logical debate and proof in India more generally seem to have emerged first in the classic Ayurvedic work *Carakasaṃhitā*.[51] Perhaps it is the combination of the special demands of medical learning with the felt urgency in the Desi's political sphere, if not the larger *zeitgeist* of the period, for medicine to improve that can best explain the Desi's unrelenting drive toward accuracy above all else.

In the passage above, Lhünding, like his other colleagues, is being assessed because he was one of the major sources on which the Desi's medical work depended. Lhünding's failings, like those of pretty much everyone else, serve as a sign of the degenerate age, in which accurate medical knowledge is hard to come by. It is an essential feature of the times, in the Desi's view, that the reign of the Fifth Dalai Lama was marked by obstacles and demons. Most doctors didn't study hard, many could not even read, people died young, doctors' reputations were thereby ruined, and new diseases caused by karma were rife.[52] This dismal set of conditions also becomes an apology for the Desi's own present commentary, both its achievements and its shortcomings. As we will see later in this chapter, the Desi was extremely anxious about the vulnerabilities of his own medical knowledge. But the current state of degeneration was equally the justification for his practice of quizzing everyone, to see who really knew what. The candid and discerning review of Lhünding's merits and shortcomings displays how hard the Desi himself tried to get things right, and how critical he was with his sources. The same probative attitude can be observed in his attitude to written sources. Although he is sure to mention a few reliable works of old, he also shows himself to be furiously "throwing out previous detailed commentaries and summaries, be they printed or not, whose explanations consist in dead words, confusions, and distortions" in the course of his own medical education.[53]

All this leads to a striking thought. Perhaps trying to get scientific matters right, trying to get the natural world right, made for a particular kind of license to transgress the traditional protocol of unquestioning deference to mentors. Perhaps what went along with this was a shift in attitudes to time and change. In fact, as will emerge in the following chapters, the projects to expand the scope of medical knowledge in which the Desi and other outstanding members of the Dalai Lama's court engaged had already been anticipated in a previous generation of medical scholarship, and its growing sense that knowledge from the past, generically, could be improved.

WHENCE THE STAKE IN MEDICINE? THE DESI AND THE DALAI LAMA

If the rising fortunes of Sowa Rikpa had estimable impact on the culture of the seventeenth-century Tibetan capital, we still need to ask why medicine had become so critical to a figure like the Desi in the first place. From everything we can gather, medicine had a very special significance among his many projects as architect and administrator of the new Tibetan Buddhist state. In the few autobiographical passages that he did leave, it appears that in his own eyes, his promulgation of Tibetan medicine was one of his crowning achievements.[54]

Even though he complains that "I passed the time working on politics and other traditions of knowledge, but did not have time to help sentient beings" (i.e., to work on medicine),[55] the Desi later avers that during this period

> many kinds of teachings came easily to me, about Sanskrit, arts and crafts, medicine, and the inner science. My own writing on the five great topics [of the five sciences] filled about twenty volumes, which I completed during my spare time while I was administering the government.[56]

Early on in his career the Desi had already produced the definitive work on Tibetan astronomical tradition and astrological calculation, *White Beryl*.[57] He also managed to complete his monumental *Blue Beryl* medical commentary in 1688, nine years after he took power as chief minister of Tibet in 1679.[58] This was a period taken up with the aftermath of the Fifth Dalai Lama's death, looking for the new incarnation, completing the Potala Red Palace construction, and many other duties.[59] He directed the medical paintings at the same time. He wrote several other voluminous works thereafter, concerning largely issues of governance and life in the Tibetan capital, but medicine was also the topic of his last major composition before his death. His influential history of Tibetan medicine, completed in 1703, when he was fifty-two, appears to be the only work he published after he had retired from his government post.[60]

In telling his own story again in the history, the Desi mentions his early interest in medicine, prior to the development of any propensities toward Buddhist realization. As a child he "would collect all kinds of plants and pretend they were medicine."[61] He grants that he did receive some education in medical theory and practice from Lhünding Namgyel Dorjé, but only regarding the mnemonic diagrams and the anatomical measurements. In a recurring refrain, the Desi complains that expert training in medicine was hard to get, and that he had to fill out what his teacher could not provide by studying texts on his own, especially the problematic *Instructional Treatise* containing the all-important specifics on therapeutics and remedies.[62] He emphasizes how he resorted to his own research to teach himself.

> I was able, with one or two exceptions, to figure out the meaning of most of [the *Four Treatises*]. On the few issues left over, I figured out the majority from the *Aṣṭāṅga* and its commentary, the translation from the Sanskrit that is located in the *Tengyur*, and also Tibetan-composed commentaries.

> I did deep research and therefore, even though I never received a word-by-word explanation, I got it into my mind. Mostly I asked questions about the word-by-word meanings of the *Instructional Treatise,* and then, because I looked at the texts, I solved my problems.[63]

The Desi understands the value of medicine in ethical terms. Medicine helps beings. As he says, "This science can pacify all the illnesses of this body of six elements, which is so hard to get a chance to have, and it gives the essence of happiness for body, speech and mind, all three. It is an unsurpassed gem for the general [populace]."[64]

He repeatedly invokes the "benefit of sentient beings" as the bottom-line motivation for studying and protecting medicine.[65] There can be no mistaking that this motivation echoes Mahāyāna Buddhist ethical discourse. Given the benefits of "the science that takes care of" people, the Desi can be selfless in his desire to protect it.[66] He does so "out of hopes to be able to benefit beings, and being unable to tolerate it if medicine were to decline. Other than that I have no need for this to be a route to further my own reputation, much less a desire to be known as masterful or knowledgeable."[67] That explains his frustration at finding so few trustworthy doctors and medical scholars. Sentient beings need to be helped by medicine; therefore he must be untiring in finding good doctors and creating ways to train new ones. It has nothing to do with his own reputation or esteem.

Although the Desi's characterization of his impulse to foster medicine was couched in classic Buddhist terms for selflessness, on a more personal—but also political—level, his drive to raise up medicine had much to do with his relationship with his master and mentor, the "Great Fifth" Dalai Lama, Ngawang Lozang Gyatso (1617–82), head of the Tibetan state since 1642. Indeed, the Desi and the Fifth Dalai Lama cannot be thought of separately on many grounds, and their collusion on behalf of medicine in the capital is surely one of them. On the Desi's own account, the Dalai Lama worked ceaselessly to cultivate good doctors and to acquire badly needed but rare and hard-to-get medicines for the benefit of Tibetans.[68] His keen interest in supporting and renewing medicine in Tibet was already in full swing before the Desi was born. He explicitly commissioned the Desi to spur the rejuvenation further, an order that the Desi conscientiously endeavors to fulfill.[69]

The Desi was the Dalai Lama's trusted administrative aide for years before his appointment as chief minister in the last years of the Great Fifth's reign.[70] His statements about his own life emphasize how far their relationship went back. As early as when the Desi was eight years old, he entered service at the Potala and then worked as a servant while the Dalai Lama trained him, "in an easy manner," in the "two ways" of politics and religion, "however much could fit into my mind at that age."[71] There was even an earlier moment—the Desi alludes to it more than once—when the Dalai Lama protected him and performed the traditional rite to guard a newborn from demons. In one passage the Desi reports that this occurred right after he was born, when the Dalai Lama had returned from his trip to China.[72] One wonders why the Dalai Lama was present at his birth. This question is behind the oral tradition, still circulating today, that the Desi was actually the Dalai Lama's biological son, despite the fact that at least one of the Desi's writings names his father and refers directly to his passing when he was four.[73] In any event, he also calls the Dalai Lama his father, or calls himself the Dalai Lama's eldest son, in the course of exulting their close relationship, but such terms are easily read as common euphemisms for teacher and close disciple.[74] If nothing else, the Fifth Dalai Lama and the Desi were certainly that.

The key moment to which the Desi ascribes his commitment to medicine was the Dalai Lama's ritual transmission of the *Heart Sphere of Yutok*, a set of teachings in Treasure format that had been associated with the author of the *Four Treatises,* Yutok Yönten Gönpo.[75] At one juncture in the rite, the Dalai Lama put the *Four Treatises* and a set of auxiliary medical writings from the same period, *Eighteen Pieces from Yutok,* into the Desi's hand. As the Dalai Lama played a special drum and recited a spell for a long time, the Desi reports that he had sudden clarity about all the medical teachings he had long labored to memorize. When the Dalai Lama then exhorted the Desi to protect and keep the medical teachings, he felt a spontaneous and strong impulse to carry out the order at once. He also realized that everything he had already done for medicine was a matter of his participation in the Dalai Lama's own body, speech, and mind of blessing and compassion.[76]

MEDICINE AND BUDDHISM

If it is hard to disentangle the Desi's personal bonds with the Dalai Lama from his commitment to fostering medicine, that is because both had everything to do with the new Tibetan Buddhist state. For both men there was a large overlap among medicine, rulership, and their conception of central Buddhist

2.2 The Fifth Dalai Lama, Ngawang Losang Gyatso.
Ngagwang Lobzang Gyatso, 5th Dalai Lama. Tibet; 17th century. Gilt copper alloy. Rubin Museum of Art, C2006.38.2 (HAR 65647). Photograph by Bruce M. White

ethical principles. There can be no gainsaying the many ways medicine, more perhaps than any other inroad into new forms of knowledge and governance in the same period, could be said to stand for the aims of the Ganden Podrang government.[77] This has as much to do with medicine's responsibility for the welfare of the people as with its power to protect the well-being of the Dalai Lama himself. The previous chapter considered the Desi's characterization of his production of medical paintings as causing the Dalai Lama to have a long life and the government to survive for a long time.[78] This led us to discern not only the paintings didactic value but also their higher order of significance. To have medical learning in place signifies stability and the health of the state and its ruler. We should understand the Desi's report on his star pupil, Chakpa Chömpel, in the same way, when he recited the entire *Four Treatises* in one session in front of the Fifth Dalai Lama, as an offering.[79] Much like the offering of paintings to the Sixth, significance exceeds overt content. The pupil is certainly not reciting the work to give the lama an empowerment, as Tibetan teachers often do, or to convey its information, or even to prove that he has it memorized. More fundamentally, such a performance of mnemic virtuosity in a work of great value would have simply been music to the ruler's ears.

The semiotics of these episodes' message is complex, overdetermined by at least two major genealogies that fed the confluence of medicine and the Dalai Lama's Buddhist state. One has to do with the long association between Buddhism and medicine in India. The other issues out of the close connection between the patronage of Buddhism and medicine in the story of the early Tibetan kings. The upshot, in the particular circumstances of Tibetan history, is that Buddhist symbols, discourse, and institutions directly facilitated the state's patronage of medicine.

Certainly the traditional overlap of Buddhist soteriology with care for the well-being of people eased the way for the Desi and the Dalai Lama to position themselves as simultaneously taking care of their subjects' material welfare and their ultimate enlightenment. The Buddha Śākyamuni's enlightenment had long been associated with his capacities as a healer, "the king of physicians," and his teachings were frequently characterized as a remedy for both the physical and spiritual ailments of sentient beings. Such an assumption is widely in evidence in the sūtras, both in the Pāli canon and in Mahāyāna works. It has everything to do with the fundamental problematic of suffering in the Four Noble Truths, which is very much about illness, death, and the painful cycle of human life. By the early Mahāyāna there was an entire sūtra devoted to the worship of the Buddha and his ritual healing powers in his specific form as "the Medicine Buddha," Bhaiṣajyaguru.[80]

The complex relationship between Indian medicine and Buddhism received careful attention in a classic study by Paul Demieveille.[81] While most canonical Buddhist writings on healing trade largely in the powers of meditation, ritual, and faith, a few scriptures display some interest in physical medicine, largely drawing on Ayurvedic knowledge. Most of these were well known to Tibetan scholars.[82] The *Suvarṇaprabhāsottamasūtra* provides a brief chapter on the physician Jaṭiṃdhara's teachings on the role of the humors, seasons, elements, and kinds of food.[83] The *Nandagarbhāvakrāntisūtra* contains some information on human embryology.[84] There are also signs of Ayurvedic knowledge in Vinaya texts in both Pāli and Sanskrit, and in the mention of medical theories, procedures, and diagnostics in the *Milindapañha*.[85] Kenneth Zysk has argued that Indian Ayurveda was in large part developed in the first place by the heterodox *śramaṇas*—rather than in brahmanical circles—since the *śramaṇas'* relative lack of taboos regarding purity and pollution allowed individuals to touch and examine bodies freely. Early Buddhist monks took part in this movement. They ran clinics for the monastic community, and some monastic codes required them to serve in caretaking and healing roles (Skt. *glānopasthāna,* lit., "standing before a patient") for fellow monks who were ill.[86] Serious meditators were also exposed to anatomical information in the course of the widespread Buddhist practice of observing decomposing corpses in charnel grounds, as a way to contemplate impermanence. The detailed knowledge of the body so engendered can be seen in the list of body parts in the mindfulness exercises described in *Satipaṭṭhānasutta*, for example, or in the *Visuddhimagga* of Buddhaghosa.[87] The interaction between conceptions of the body for the purposes of Buddhist meditation and those developed in Ayurveda continued into the tantric period, and some traces of the channel systems from late Indian Buddhist tantra even emerge in the Tibetan *Four Treatises*.[88]

By at least the beginning of the Common Era, Buddhist monastic establishments seem to have had hospitals on their grounds, such as at Nāgārjunakoṇḍa.[89] The place of medicine and the other "sciences" among Indian Buddhists and even at the major Buddhist monastic college of Nālandā gets outside confirmation in the seventh-century observations of the Chinese pilgrim Xuanzang.[90] There has even been speculation that some of the great minds of classical Ayurveda in India had Buddhist affiliations. A Chinese version of the *Sūtrālaṅkāra* suggests that Caraka was doctor to the famous Central Asian Buddhist king Kaniṣka, although the evidence is far from convincing.[91] Some have also wondered if Vāgbhaṭa had Buddhist training, since he had a master with the seemingly Buddhist name Avalokita, but that too is inconclusive. It seems rather that such Ayurvedic writers had fluid and pluralistic religious affiliations.[92]

In any event, there seems to have been no necessary dissonance between the practice of Ayurvedic medicine and the deployment of Buddhist healing rituals in India, as evidenced for example by the Bower manuscript, found in a *stūpa* that contained a Buddhist charm against snake bites together with several medical tracts.[93] This dual approach to healing continued as Buddhism traveled into East and Central Asia.[94] Buddhist Tibet, which certainly maintained its own battery of ritual healing traditions, is but one example of this pattern. Even the *Four Treatises* briefly mentions ritual and magical healing devices on occasion, despite its primary concern with physical medicine; a few of these employ Buddhist deities and basic visualization techniques, while others represent local exorcistic tradition, along with a few cases of ritual lore from Ayurveda.[95]

In this book I am nonetheless pointing to key moments in the case of Sowa Rikpa where distinctive emphases and orientations came into view, suggesting ways that medicine still constituted a separable knowledge system not reducible to formations of Buddhism. And at least in Tibet, despite a plethora of mixed signals, medicine sometimes even characterized itself that way. The entire question of how human malady is conceived is a prime case in point. Ayurveda and the *Four Treatises* primarily see it as a result of imbalances in the three humors (*doṣa,* i.e., wind, bile, and phlegm). Buddhist scriptures, on the other hand, pervasively understand the primary source of human pain to be the three passions (*kleśa,* i.e., ignorance, hatred, and greed), the occasional appearance of Ayurvedic conceptions of illness notwithstanding.[96] That is not to say that these two paradigms, one physical and one moral and mental, could not be bridged, as indeed we might expect, given the long interaction between Ayurveda and Buddhism in India. As one example, the important Buddhist scripture *Mahāparinirvāṇasūtra* argues for a close interdependence between the passions and the humors, although it still distinguishes between physical and spiritual malady.[97] The Ayurvedic treatises, for their part, do not fail to consider spiritual grounds for illness. From Caraka onward, Ayurveda notes continuities between psychic states and the physical experience of the body. Ayurvedic sources are also quite ambivalent about the status of karma in the etiology of physical states and disease,[98] while classic Buddhist Abhidharma doctrine acknowledges the role of the material elements in the moral ripening of karmic effects, and passages can even be found that countenance causes of human experience entirely independent of the working of karma.[99] As Wilhelm Halbfass has demonstrated, natural processes were not always considered to be entirely governed by retributive causality in Indian religious thought.[100] In short, the fact that the question remained open and was at times controversial does suggest that there is no iron-clad boundary between science and religion in India. It says something as well about the *Four Treatises*' own ambivalence that it alternates between attributing the condition of an embryo to the condition of the parents' semen and blood, to the karma of the reincarnating being, and even to the "low merit" of the mother; or again, that it too mentions a relationship between the three humors and the three passions.[101] The *Four Treatises* also on occasion makes general recommendations about religious practice and merit-making, assuming that these are good for the health, or create auspicious conditions, or counteract the noxious influence of local spirits. Recall the medical paintings' illustration of a man doing a generic meditation or a monk reading a nonspecific Dharma book.[102] In such moments Tibetan Sowa Rikpa is of a piece with religious healing in Buddhist and related contexts across Asia.

Yet overlap and fuzzy borders do not mean that systems cannot be distinguished, especially in terms of primary concerns and conceptual centers of gravity. As Halbfass rightly notes for Buddhism, orthodox doctrine continued to reject or discount materialist accounts of causality and was uneasy with postulating a buddha vulnerable to ordinary conditions outside of his moral control.[103] And as for medicine, it remains the case that both the Ayurvedic masterworks and the *Four Treatises* focus first and foremost on material, bodily conditions and material, bodily treatments—even in the case of illnesses caused by demons—and with but few exceptions, their therapies consist in a kind of care that is quite distinct from the ritual approaches to healing emphasized in Buddhist sources.[104]

As a sign of how fraught in Tibet was any such suggestion that the epistemic universe of medicine might indeed be separable from that of Buddhism, the *Four Treatises* took the extraordinary step, as we will study in the following chapter, of presenting itself as the "Word of the Buddha." In this it far outstrips in scope and comprehensiveness the brief medical passages ascribed to the Buddha to be found in a few canonical scriptures. The *Four Treatises* makes this move against all historical evidence to the contrary—evidence that members of the Tibetan intelligentsia later cited in protest. It is hard to understand this attribution of a full-service medical textbook to the Buddha without seeing an evident desire to *make* physical medicine part of the Buddhist dispensation, precisely out of a concern that it might be perceived otherwise.

There were already signs in South Asian Buddhist contexts that the place of medicine was not entirely unequivocal. Important works like *Brahmajālasutta* and *Theragāthā* include proscriptions against monks practicing medicine, listing a variety of medical procedures and functions that are prohibited and reflecting concerns about monks being distracted by business and profit.[105] Similar concerns are evident in the Chinese version of the *Smṛtyupasthāna,* which counts medicine as one of thirteen kinds of activity that are harmful for monks since it is a source of greed and envy. The latter even suggests that physicians often wish that there may be many sick people; in short, *śramaṇas* should heal sentient beings of the three passions rather than the three humors.[106] So when Buddhist scholastic treatises in the early centuries C.E. began to claim that a broad education was essential to the omniscience of the enlightened master, it may have been to counter these long-standing concerns. In the list of the five "sciences" (*vidyāsthānas*), codified in the *Mahāyānasūtrālaṅkāra* and other works, medicine (*cikitsāvidyā*) is positioned along with linguistics (*śabdavidyā*), logic (*hetuvidyā*), arts and crafts (*śilpakarmasthānavidyā*), and "that which is most one's own" (*adhyātmavidyā*), i.e., soteriology. These were said to constitute the five areas of expertise needed in order to achieve omniscience.[107] In this curious list in which transcendence, or ultimate truth, is placed alongside topics that can well be conceived apart from the religious, we can see two impulses at once. On the one hand, things like medicine and the plastic arts are listed as separate items, along with—but different from—the *summum bonum*. That would seem to indicate a certain distinction between what modern scholars have read as what is most "Buddhist" and the other four disciplines, which were explicitly defined as "outer" and "worldly" (*laukika*).[108] And yet by virtue of this same act of listing, in terms of both academic curriculum and the larger argument that such knowledge is necessary for the enlightened sage, a critical merger has been achieved. Although a major Tibetan scholar like Sakya Paṇḍita still had to argue strenuously for it (his own interest was largely in poetics and grammar),[109] by the time of the Fifth Dalai Lama the prestigious institutions of academic Buddhist learning in Tibet had long been convinced that medicine fell within the purview of a proper education.[110]

Of course, just because medicine is included in an educational curriculum trained ultimately on Buddhist enlightenment does not necessarily establish that medical theory and its means of implementation are of a piece with the rest of that education in substantive, epistemic terms. In both India and Tibet there were voices that queried that compatibility, and we will study several of these in the following chapters. Thus the larger issues at stake remained open questions into the Fifth Dalai Lama's day. But for his part, the Dalai Lama, and in important respects the Desi too, was deeply invested in the side fostering close identity between medicine and the teachings of the Buddha, an identity well suited to the religious and political aspirations of each.

LOOKING TO THE PAST: THE GOLDEN AGE OF TIBETAN MEDICINE

Indian Buddhism offered robust precedents to link the Buddha's compassion with the expectation that the enlightened teacher would master medical science. But there was more yet to the Dalai Lama's interest in medicine. The apotheosis of the Dalai Lama position as the bodhisattva Avalokiteśvara was intrinsic to the very foundation of the Ganden Podrang government. Buddhist deities had been deeply implicated in the narratives of Tibetan history for centuries, but the connection was only heightened by the works that the Desi wrote about the nature of the Dalai Lama, the Tibetan state, and the rituals of the capital.[111] The political dimension of the Dalai Lama's identity in turn strengthened the equation between medicine and Buddhism.

Some of the imperial ambition in this equation comes through in the Desi's invocation of duty cited above. The suggestion that medicine is as much the victim of the degenerate times, demons, and black magic as is Buddhism takes on particular significance from the perspective of the ascendancy of the Tibetan state. Just as the Buddha's dispensation is subject to successively worsening conditions, so is medicine, and both, therefore, require special measures and support from the ruler. Likewise, medicine (again mirroring Buddhism) is well suited to help protect people from the various sufferings that those degenerate times will bring on, and thus an appropriate avenue for the state to carry out its responsibilities.[112]

Most significant of all in the homology between the welfare of the Buddha's teachings, medicine, and the state may be the Desi's quotation of prophecies from the Tibetan Treasure literature. Those apocryphal but highly valued narratives of Tibetan history purport to represent the voices and intentions of the glory days of the early empire.[113] Treasure prophecies are usually invoked to account for the subsequent decline in religious morals, discipline, and learning in Tibet. But the Desi uses them to describe the state of medical learning during his lifetime as well. When he goes through pronouncements from the ancient ordinances of the kings or the visions of the Northern Treasures about the noxious demons who will plague Tibet, he adduces the result, rather than the usual point about the decline of the Dharma, that "other than having general knowledge, people's minds won't get into Ayurveda."[114] Characterizing both his own and the state's connection to medicine in such terms powerfully brings medicine into the heart of the new Buddhist nationalism.

The Treasure prophecies do more than just align medicine with the Buddha Dharma as something precious to be renewed and cultivated. They also tie medicine to the halcyon days of the specifically Tibetan past: the era of the great emperors of the Yarlung dynasty of the seventh to ninth centuries. The invocation of ancient royal glory had already enhanced the legitimation of the government of the Dalai Lama's own nemesis, King Tseten

Dorjé of Tsang, as well as the previous reign of Pakmodrupa.[115] The Treasure medium bolstered that constructed past, and had long served as a means to affirm sociopolitical needs in the present.[116] Treasure prophecies could confirm that a ruler had been predicted and even commissioned by the heroes of the past to do whatever he was doing in the present to advance the welfare of the Tibetan people. They also could provide a literary space to voice dissent. But most of all, the Treasure narratives and prophecies were key tools in the thematization of Tibet's exceptional destiny as a Buddhist land. They pointed repeatedly to Tibet's enemies, featuring warnings about invading hordes and rallying calls to fend them off. Their legitimating power helps explain why the Fifth Dalai Lama—and the Desi—took so much interest in Old School teachers like Terdak Lingpa who preserved Treasure lore and ritual.[117] Since medicine was also featured in the stories of the early kings' beneficent activities, the Desi's invocation of such scenes now ineluctably tied the Dalai Lama's and his own efforts to foster medicine to the dawn of Tibetan Buddhist civilization and its exalted destiny.

When reading the Treasure narratives, it is not easy to separate exuberant imagination and rhetorical flourish from what actually happened. Instances of the "invention of tradition" to be sure, the Treasure works nonetheless serve as veritable treasure troves of historical references and nuggets, even as they color their stories with the agendas of the time of their own composition. But the same can often be said of other Tibetan historiographical genres, such as the accounts of the coming of Buddhism to Tibet (*chos 'byung*), whose vision of the imperial period is often influenced by the Treasure *imaginaire* in any case. This is true of the specialized histories of medicine in Tibet as well, about which I will say more below. For now the pertinent point is that both the historical fact of the matter and rhetoric were at play throughout the historiography of medicine in Tibet, and especially for the Desi and the Dalai Lama in the seventeenth century. In short, medicine did indeed figure in the events of the Yarlung court. And then on top of that, the second-order *fact that* Tibetan medicine had its inception in the glory days of the empire conferred prestige on medicine later. This occurred in a manner quite akin to the way that certain strands of Tibetan Buddhism constructed a glorious past for themselves in Treasure texts.

One obvious place to see both the actual presence of medicine in the imperial past and its symbolic capital in the postdynastic period is the high status accorded to certain medical families. The class of doctors whose position was said to have been bestowed by the Yarlung rulers (*gnang rigs*) was distinguished from newly established and less honored medical lineages.[118] There are also reports of a title of "superior physician" (*bla sman*) conferred upon court physicians by the time of Tri Desongtsen (late eighth century); it remained in use through the twentieth century. The same stories also tell of the granting of land and inherited rights to medical clans, including release from military duty.[119] The medical titles of "high learning in medicine" (*sman rams pa*)[120] and "divine lord" (*lha rje*) are also said to have been granted by the early kings.[121] We have seen one indication of the lasting status of such titles in the Desi's evaluation of Lhünding Namgyel Dorjé.

The centrality of medicine in the projects of the Tibetan emperors—at the same moment that these rulers were courting Buddhist teachers and also mounting what, at least in retrospect, counted as the first iteration of a centralized Tibetan Buddhist government—is portrayed in most detail in the medical histories.[122] The story of medicine in the imperial period is recounted repeatedly in an old set of narratives that the Desi copied from previous sources, although the story varies and we are only beginning to sort out what was added by whom.[123] But across the board, these narratives present a surprising amount of detail on the initiative and vision of the Yarlung emperors to improve local folk knowledge of medicine, characterized as oral tradition, regarding diet, exercise, and physical therapy.[124] The means of improvement was to introduce superior foreign learning and expertise. Again, this is part and parcel of the larger story of the kings' effort to "tame" and benefit the Tibetans, in which *both* foreign Buddhism and foreign medicine served well. The emperors' interest in medicine is also tied in these stories to concerns about their own health and attempts to save a ruler's life when he is ill.[125]

The move to invite physicians from abroad to teach medicine is extended back to the descendants of the semimythical king Lha Totori Nyentsen, who are said to have invited physicians from the Central Asian land of Azha and from China. But it started in earnest with Songtsen Gampo, the warrior king who conquered significant portions of the Tibetan plateau and founded a centralized state in the seventh century A.D.[126] This ruler's Chinese consort is credited with bringing Chinese medical works, which were then translated into Tibetan. The same king is said to have convened an international council made up of an Indian doctor, a Chinese doctor, and one from Takzik, an Arabic or Persian land to the west.[127] Their names are transliterated into Tibetan script to correspond to Bharadhvāja, the mythical founder of Indic medicine; Huangdi, or the "Yellow Emperor," the mythical founder of Chinese medicine; and Galenos, the famous Greek physician. Clearly these three luminaries did not manage to visit the seventh- or eighth-century Tibetan kingdom, but there is reason to believe that physicians representing the three major medical traditions of the ancient world did teach and translate in the Yarlung court.[128] Several old medical writings in Tibetan that were preserved at Dunhuang speak of regional differences in medical traditions and even distinguish physical medicine from religious healing.[129] We also have the Tibetan translation of the Indian *Siddhasāra* of Ravigupta, which seems to date from around the end of the imperial period.[130] But the Tibetan medical histories list, page after page, the titles of scores of works on Greco-Arabic, Indian, and Chinese medicine that were brought to Tibet at this time, although there is little evidence of their existence today. These histories also list titles of works that were newly composed in Tibetan that synthesized or gathered together the various medical traditions circulating in this scene, some of which are familiar, such as the tradition of Caraka. The succeeding Tibetan kings, especially Mé Aktsom, also invited other physicans from Trom (Tib. Khrom, possibly the Greek "Eastern Roman" or Byzantine empire), whom some sources identify as the originators of the important Biji and ensuing Drangti lineages of doctors in Tibet.[131] There were also physicians from Himalayan regions such as the Kathmandu valley, Drugu, Dölpo, and the Kashmir area (Tib. Kha che), as well as parts of Central Asia.[132] The other great figure of the dynasty, Tri Songdetsen, expanded the project in the eighth century, when a medical college is already supposed to have been in existence. The important physician of the Biji line called Tsenpashilaha also flourished during his reign.[133]

The capacity to bring a wide-ranging array of medical experts to the court itself connotes the power of the imperium. The kings were not just patrons but also impresarios. They are shown cajoling their guests to stay longer, plying them with gifts. The Fifth Dalai Lama would repeat this gesture. The kings are also shown testing the skill of the physicians, and this reminds us a bit of the Desi. In one story the ailing king lies in another room, assessing his physicians' pulse-reading abilities by means of a rope tied in turn to a chicken, a cat, a millstone, and finally the king himself.[134] The sources recount the translation of a host of works by several Chinese teachers, often titled "Hwashang," and sometimes involving the well-known Tibetan translator Vairocana. An important work called *Somarāja*, said to combine many medical traditions, appears in this period.[135] Some medical historiography also locates the appearance of the main root text of Tibetan medicine, the *Four Treatises,* in this period, although that is disputed.

2.3 Yutok Yönten Gönpo. *Plates 5 and 8, details*

After the fall of the dynasty, patronage for medical learning was taken up by the emerging Buddhist monastic centers and scholars in central and western Tibet. Rinchen Zangpo (958–1055) translated the last great Indic medical work, Vāgbhaṭa's *Aṣṭāṅgahṛdayasaṃhitā,* along with the commentary by Candranandana known in Tibet as *Moon Ray.*[136] The *Aṣṭāṅga's* presentation of Ayurvedic medicine was influential in Tibet and was reworked into portions of the *Four Treatises.* The seeds of the main Tibetan medical lineages are in evidence by the following century. One luminary was the famous Buddhist scholar/systematizer Gampopa (1079–1153). Another was Jetsün Drakpa Gyeltsen (1147–1216), who was connected with the establishment of a medical college at Sakya Mendrong during the twelfth century, a school that preserved some of the important Biji medical streams coming from the west of Tibet.[137] Part of that same generation would have been the twelfth-century medical master Yutok Yönten Gönpo, also from southwestern Tibet, to whom, along with his disciple Sumtön Yeshé Zung, should actually be attributed the composition of the *Four Treatises.*[138]

Yutok and his disciple also authored other summaries of medical theory and practice.[139] As testimony to the symbolic capital of having origins in Indian Buddhism, Yutok or his followers cast his main masterwork as a scripture originally preached by the Buddha. A related story started to be told that the *Four Treatises* had been translated from Sanskrit into Tibetan by Vairocana during the reign of Tri Songdetsen, then hidden as Treasure, to be rediscovered by Drapa Ngönshé in the eleventh century and then eventually passed to Yutok. This invocation of Treasure transmission is one of several signs of the Old School affiliations of many of the early medical writers. So is the creation of an entire Treasure cycle of Buddhist practices associated with Yutok Yönten Gönpo, the *Heart Sphere of Yutok,* closely connected to other works on ritual medicine propagated in the Old School lineages.[140] Most of all, the Treasure myth facilitates the claim of the Buddha's authorship of the *Four Treatises*, and was therefore enthusiastically taken up and elaborated by the Desi in his own time.[141] Treasure lore also served the burgeoning medical tradition by providing a storyline that could enhance the status of a medical work through association—again, with both the Tibetan dynasty and ultimately the Indian Buddha.

The story of the patronage of medicine by the early kings served the program of the Dalai Lama and his minister well. The royal agency in inviting medical scholars from numerous Asian countries to the Tibetan court and commissioning the translation of medical treatises would be commemorated and renewed in the acts of the Fifth Dalai Lama. Medicine is cast as one of the great gifts of the Tibetan kings to their subjects. It would be a great gift of their seventeenth-century descendant too.

IMPERIAL REDUX: THE GREAT FIFTH AS MEDICAL IMPRESARIO

By inviting international physicians to his court in the seventeenth century, the Dalai Lama was shrewdly providing the occasion for scientific innovation and in the same stroke reinscribing the glorious, benevolent initiatives of old. The fact that medicine became one of his pet projects is a central reason it had such state-level significance during this period. It was a key component of the cosmopolitan spirit in the capital more broadly, and one of the Dalai Lama's signature contributions to a heady period that in other ways he might not have entirely controlled.

Although he did not initiate the momentous upheaval in central Tibet in the middle of the seventeenth century, there can be no question that Lozang Gyatso, the Fifth Dalai Lama, capitalized on the opportunity it offered. In the wake of the bloody battles when the powerful Zhelngo Sönam Chömpel (1595–1657) in the Dalai Lama's administrative retinue induced the Khosot leader Gushri Khan to defeat the King of Tsang, central Tibet was unified for

2.4 The Fifth Dalai Lama. *Scenes from The Life of The Fifth Dalai Lama (1617–1682). Tibet; 18th century. Pigments on cloth. Rubin Museum of Art, C2003.9.2 (HAR 65275). Photograph by Bruce. M. White*

2.5 The Fifth Dalai Lama in Beijing in 1653, with the Shunzhi Emperor.
Mural in Potala, detail. Photograph by Samten Karmay

the first time since the fourteenth century. The "Great Fifth" was anointed king of Tibet in 1642.[142] He proceeded to move his Ganden Podrang government (the "Tuṣita Palace") to the White Palace of the Potala, built to house the new Tibetan state government, in 1649. After a symbolic journey to meet the Manchu emperor Shunzhi in Beijing in 1652–53, he returned to rule until handing over power to the Desi in 1679, and died in 1682.

The period is documented in detail in the Dalai Lama's own diaries and then continued in the three-volume journalistic biography authored by the Desi. It marks a watershed in the establishment of bureaucratic structures that would remain in place until the middle of the twentieth century. We are only beginning to think about the relationship between the changes in the organization of political power under the Ganden Podrang and corresponding shifts in the intellectual and religious climate of the Tibetan cultural sphere.[143] The government oversaw in a newly systematic way the levying of taxes, estate governance and finances, the granting of serf households and land to worthy individuals both secular and religious, distribution of grain and other provisions in times of need, repair of bridges and ferries, construction of canals, setting of workers' wages, standardization of systems of corvée services, censuses, supervision of the judiciary networks and the penal system (on which the Desi wrote at length[144]), codification of calendrical calculation (on which again the Desi's writings were key), establishment of hunting-free zones, building of Buddhist images and installing of portraits, and the supervision and conscription of various monasteries and other religious groups to do particular rituals at particular intervals on behalf of the long life of the Dalai Lama and the welfare of the Ganden Podrang government (yet again, laid out in detail in the Desi's writings).[145] Another important part of the new state structure included the oversight and control of monastic property and authority. In the Desi's record of his administrative work after the death of the Dalai Lama there are detailed accounts of monasteries in central, eastern, and western Tibet, their assets, what rituals they were conscripted to do for the state, what offerings they were obliged to give to the government, and the appropriate conduct of monks in various categories.[146]

A major part of what was new and had significant cultural implications were the very practices of record keeping. A prime example is the assiduousness and comprehensiveness in the diaries of the Dalai Lama, with reams of details on the cost of all these governmental activities, the kind and amounts of offerings received, grants given back, prayers done, rites performed.[147] The contemporary scholarly community does not yet have free access to the archives of the Potala and the various monastic administrative units in central Tibet, but such documents do survive in massive numbers, and some in scattered smaller collections and archives have begun to be studied, especially by Dieter Schuh and his students.[148] We can expect that one product of the new documentary habits, beyond their pragmatic value, was a certain reflexive pride, a self-consciousness and an accompanying sense of control, accomplishment, completeness. I argued in chapter 1 that this sense is visually represented—and celebrated—in the beautiful systematicity of the medical paintings. It was celebrated and reiterated in state rituals and other cultural domains as well.

The expanded sense of the ability to rule and control was coupled with the glory of the past at the foundation of the new state. All converged in the Dalai Lama's person and how he was portrayed.[149] Not only was he the emanation of the bodhisattva Avalokiteśvara, he was also an incarnation of Songtsen Gampo, the first of the conquering emperors from the Yarlung period.[150] This once again ties him to the *imaginaire* of the Treasure literature. The Desi also had his own identity tied to the royal period when the Dalai Lama recognized him as the incarnation of Muné Tsenpo, or sometimes Mutik Tsenpo, both sons of Tri Songdetsen.[151] Military power, administrative reach, exalted Buddhist attainment, recuperation of early royal charisma: it was a potent mix. That the Dalai Lama inhabited these virtues well and channeled the mythical early kings' penchant for culture-building magnanimous acts is why he is known as the Great Fifth.

Although famously hegemonic in dealing with sectarian rivals such as the Jonangpas and the Karmapas, converting monasteries into Gelukpa establishments, and attempting to expand Tibetan control into Bhutan and Ladakh, the Dalai Lama's government also pursued a diplomatic course to strengthen his circle of allies. He negotiated with Mongol tribes and the new Qing court, and in South Asian arenas as well.[152] His government even allowed European missionaries at court.[153] The eclectic atmosphere in the capital served not only instrumental and political ends but also cultural advances such as were fostered by his patronage of medical science. This was well under way before the Desi came into office.

According to the Desi, the Great Fifth did not have much education in medicine himself, although we have to remember the Desi's own rhetoric regarding the scarce availability of medical education and the virtues of learning on one's own.[154] The Dalai Lama seems to have studied parts of the *Four Treatises* and various medical traditions current in central Tibet by the time he was twenty-three. Several of the early Tibetan medical lineages had by his day been established for centuries, even if some were in decline or had diminishing institutional bases.[155] There had been flourishing medical academies, at least through the fourteenth century, not only at the Buddhist monastic center at Sakya but also at Zhalu, which specialized in the *Aṣṭāṅgahṛdaya* system, and at Tsurpu, which fostered eclectic medical scholarship. The same was true for É Chödra at Bodong in the early fifteenth century. By the time the Jang line of physicians was consolidating at Ngamring, the capital of the Jang myriarchy, in the fifteenth century, there were oral medical examinations and regimes of memorization. There was also much medical learning at Latok Zurkhar, which became the home of the other major line of Tibetan medicine, the Zur. The Drigung Kagyü developed a further medical lineage, branching off from the Zur. The Dalai Lama's mentors were products of these major lines, including the eminent aristocratic physicians Jango Nangso Dargyé and Darmo Menrampa, as well as many other doctors and teachers at the court.[156]

Despite the Dalai Lama's education at the hands of these teachers, the Desi attributes his commitment to medicine to his assumption of the identity of the Medicine Buddha, particularly in the ritual context of conveying initiations.[157] There might be truth in that, for it seems that while the Dalai Lama entertained a growing conviction about the importance of the academic traditions stemming from *Four Treatises* study, he was also deeply invested in tantric healing traditions, several major transmissions for which he had conveyed by 1672.[158] It may well be that his confidence in conceiving his role in the preservation of Sowa Rikpa was buttressed through his ritual offices.

As early as 1645, even before his move to the Potala, the Dalai Lama had already undertaken major initiatives to build academic medical institutions. He established a medical school, Sorik Dropen Ling, first in the western wing of his own Ganden Podrang at Drepung and appointed Nyitang Drungchen Lozang Gyatso as the main instructor. The Dalai Lama also granted the school the right to "wage teas" and other provisions. Grants to support the students' living would be a recurring component of his patronage. At one point the Desi makes a key point explicit: under the Dalai Lama's instructions, he arranged for a grant of grain to Lhünding Namgyel Dorjé at his medical institution Tsoché Lhünding, in order to create an exemplary situation for the teaching of medicine and to support the students.[159]

The emphasis upon institution building continued throughout the period. Soon after starting the school at Drepung, the Dalai Lama established another medical college called Drangsong Düpé Ling at Samdruptsé, the fortress at Zhikatsé that became a prototype for the Potala architecture, and appointed Pöntsang Tsarongpa Tsewang Namgyel as its main teacher. There too he supplied the requisite provisions.[160] The Desi remarks that the number of scholars at these schools who had memorized and could recite the entire *Four Treatises* was more than in the past.[161]

After the Dalai Lama moved to the Potala, he established yet another new medical school, called Lhawang Chok or Sharchen Chok, which was put under the direction first of Jangopa and later of Darmo Menrampa. Again he gave supplies and scholarships; at least two names of particular students, Lhaksam, whom we have also already met, and Naza Lingpa are recorded.[162] The administration of these institutions participated in larger new trends in bureaucracy. So, for example, when Jangopa took over management of Dropen Ling at Drepung in 1646, he saw the need for more standardization and requested the Dalai Lama to write a charter (*bca' yig*)—a regulation manual for behavior and operations—for the medical school, to which the hierarch assented.[163] At the urging of Pöntsang Tseten Dorjé, the Dalai Lama composed a charter for medical practice at Samdruptsé as well.[164]

At some point the school at Drepung was apparently pronounced a failure. The Dalai Lama looked to other venues. This eventuated in the establishment of what was to become the central institution for medical learning in Tibet into the twentieth century. The new medical school on Chakpori, the "Iron Mountain," was actually built by the Desi, who reports that the project first dawned upon him as a vision. It came to him just after he had completed a set of 100,000 circumambulations around the central temple of Lhasa, a reminder of how deeply imbricated were religious ritual and imagery with the state-level fortunes of medicine in Tibet. The Desi continues that he had been practicing the austerity of circumambulating the Jowo Śakyamuni statue not for his own merit but to eliminate any impurities collected by the Dalai Lama.[165] On the afternoon of its completion, he climbed Iron Mountain, and when he reached the top he saw something different from how previous texts had characterized its shape. He saw it as the very city-on-the-mountain Tanaduk: the site of the Buddha's imputed preaching of the *Four Treatises.*[166] The Desi notes that this vision was at odds with Iron Mountain's empirical shape of three sides (Tanaduk would have had four). He sent his trusted student, Lhünding Ganden Menla, to reconnoiter. He was delighted at the

2.6 Chakpori circa 1920. *Photo by Charles Alfred Bell or Rabden Lepcha. Pitt Rivers Museum, University of Oxford 1998.286.146*

2.7 Long view of Chakpori on the Lhasa plain, taken from the roof of the Potala. *Photo by Frederick Spencer Chapman. Pitt Rivers Museum, University of Oxford PRM 1998.131.291*

auspicious sign that there were many medicinal herbs growing on the hill, contrary to what everyone had thought.[167]

The new school that came to be built on this hill was underwritten by the Ganden Phodrang government.[168] Completed in 1696, it was called in full Chakpori Rikché Dropenling (Iron Mountain Science Island That Benefits Beings). It quickly became a magnet for students in central Tibet. It was seen as a monastery, but may have allowed lay students of medicine as well.[169] While embedded in Buddhist ritual cycles and in much of the academic culture cultivated over the previous millennium in Tibetan Buddhist monasteries, the institution was exceptional as a freestanding school dedicated to the study of medicine. The Desi made his own works on medicine, the *Blue Beryl* and the *Practical Manual,* the center of the curriculum.[170] The student body quickly grew from an initial group of thirty older scholars to an assembly of seventy monks. The Desi had a strong hand in determining the nature of the oral examination and could boast by the end of his life that the student body included more than a few doctors at the *sman rams* level.[171] Contemporary

Tibetan historiography understands the school as initiating significant changes in educational practice. A modern-day historian of Tibetan medicine adds that the basis for evaluation introduced by the Desi was distinctive for its method of praising and promoting the good and demoting the bad as a stimulus for improvement.[172]

The creation and support of new medical schools initiated by the Dalai Lama marked a new emphasis on medical learning. The actors in these developments saw themselves as revitalizing what had already been fostered in earlier centers, but also as producing new resources, methods, and practices. There was a new urgency for the production of specific medical compounds, and that in turn meant assembling the requisite ingredients, efforts at which the Desi details at some length. One of the key medical accomplishments of the Fifth Dalai Lama that the Desi celebrates was a major production of medicines.[173]

It is significant that some of the most important substances were only obtained on the Dalai Lama's tour to China. That medicine could be renewed and advanced through foreign sources hearkened back, again, to moments in Tibetan history when Buddhist lineages and forms of knowledge were renewed and reconstituted by influxes of teachers from afar. But it is also emblematic of the unprecedented cosmopolitan atmosphere in Lhasa in the seventeenth century and the growing international interaction under the Dalai Lama's reign. Some of his impetus to search further afield for new medical knowledge was also his own personal needs, such as his problems with cataracts, but it quickly snowballed into a general zeal to expand medical sophistication in his court and for the benefit of the state. A good illustration is the Dalai Lama's discovery, around 1675, of an Indian doctor, Manaho, who was skilled in eye treatment and who was in Yarwo. He invited the doctor to Lhasa, gave him all the material support he needed for as long as he was there, and issued him travel documents. The Dalai Lama then placed his own physician, Darmo Menrampa, in Manaho's training.[174]

He also searched out medical experts from far-flung areas of Tibet itself who might have special techniques or medical traditions unknown in the capital. One such expert was Neluk of the Tö lineage from Drachi, who was brought to court to teach the methods of eye operation according to Mitrayogin and certain older Tibetan traditions. The Dalai Lama reports on the cataract treatment that he eventually received from Darmo Menrampa, which ameliorated the problems he was having with his vision.[175] Another Tibetan doctor brought from afar was Nyanang Namkha Lha, summoned to teach the method of making, boiling, and washing mercury, originally attributed to the Indian master Śabarīśvara, who passed it to the Tibetan siddha Orgyenpa Rinchen Pel (1229/1230–1309). These methods were conveyed to Darmo and Lhaksam and some local doctors, again for the purpose of benefiting the people, as the Desi puts it.[176]

In addition to soliciting new techniques, the Dalai Lama supported foreign medical scholars to translate new medical texts from India. Manaho was one of these. I am following the reading of contemporary Tibetan historians in observing that by commissioning the translation of foreign medical works, the Dalai Lama renewed and carried forward the benevolent medical activities of the kings. As one contemporary historian of medicine writes, many medical texts from other countries were translated during the reigns of the kings, but in later times no one was looking to do more new translation work, and Tibet had fallen behind in the medical knowledge of the day.[177] A striking moment in the court's efforts to correct that situation is captured in a colophon to one of the new translations, and it gives a sense both of the desire for more knowledge and of the urbane atmosphere at the time.[178] The colophon recounts that a group of scholars from India were standing in the courtyard of the Potala in 1664. Darwa Lotsawa, a scholar of grammar, was sent to search for someone in the group whom the Dalai Lama had heard was knowledgeable about medicine. When he found Godararañja, he asked Darmo Menrampa to study with him. Godararañja was later sent to bring back medicines from South Asia and to translate his own summary of Ayurveda into Tibetan with the help of Darmo and Darwa—this is the work in which the colophon appears. There were numerous other such figures. The Desi was aware of the significance of this enterprise and also listed the translation work of the Indian physician Dānadāsa, who rendered a circle diagram of seventy-two kinds of medical formulae, and an Ayurvedic practitioner named Raghunātha, who translated other works.[179]

Unlike in the past, the medical scholars commissioned by the Dalai Lama and his court were mostly translating not classics, but rather new Indic works. Although not everything old from India on medicine had been translated into Tibetan—the *Aṣṭāṅgahṛdaya* and its commentaries were the main classic Ayurvedic sources available to Tibetan readers[180]—there seems to have been more interest, in the Dalai Lama's day, in new techniques and knowledge, and a desire to get up to speed. To what degree this urge was participating in the larger movement afoot in the development of science during the period is hard to specify exactly. We can say at least that it was chronologically of a piece with other moments of medical innovation, including in India on the verge of its own, more abrupt encounter with modernity and colonialism.[181]

THE COURT OF THE GREAT FIFTH PRESERVES/ CREATES MEDICAL TRADITION

The Dalai Lama and his circle were concerned to recuperate medical knowledge from the past as well. But the backward-looking gaze seems to have centered upon Tibetan medicine rather than foreign resources. Active efforts were made to preserve what was perceived to be ancient knowledge and especially to seek out lineage holders of rare traditions and save them before they were lost (despite the ruler's concerted suppression on other fronts of the literary heritage of rival religious sects).[182] The Desi recalls with pride how the Dalai Lama asked Gelong Zhenpen Nyingpo to teach an old Tibetan commentary on the *Aṣṭāṅgahṛdaya* entitled *Snying po bsdus pa* and also the *Zin tig bces bsdus,* the *Gser bre* from the Drangti tradition, and the *Explanatory Treatise* commentary by Jangpa Namgyel Drakzang.[183] The Dalai Lama is also reported to have been looking for old teachings on the medical use of mercury. He searched at monasteries known for their old collections such as at Shelkar Dzong and inquired among scholars at the famous libraries of Sakya, but with no success. After much effort he located the material at Nyanang.

Although most of these efforts were trained on academic medicine of Ayurvedic ilk or the legacy of Yutok Yontan Gonpo, the period also saw the recuperation of old healing rituals. One example is the massive ritual reinstituted by the Dalai Lama to produce a kind of pill based on an arcane system from Zur and Drigung, with many rare ingredients collected from many places.[184]

Antiquarian projects to save, compile, edit, and revitalize the old must always be Janus-faced: even if such an effort is trained on the past, it is for reasons that speak to the current situation, that is, to something new. Thus the Dalai Lama's accomplishments in recuperating old Tibetan medical literature also served to standardize, canonize, create orthodoxies, foster the allegiances of the medical intelligentsia, and extend the ruler's reach over contemporary culture. They are comparable to the larger-scale but similar efforts to consolidate and control scientific scholarship by the Qianlong emperor a half-century later.[185]

One of the major accomplishments of the Dalai Lama's circle was a new compilation and blockprint publication of a set of writings, called *Eighteen Pieces from Yutok*, attributed to Yutok Yönten Gönpo and his disciples.[186] This is an important historical artifact whose riches have yet to be fully plumbed by scholars today. It includes, among other things, earlier drafts of the *Four Treatises* by Yutok and valuable works on techniques of treatment, which the Desi reports were already rare in his time.[187] With the encouragement of the Dalai Lama, the collection was pulled together by Darmo Menrampa after seeking out whatever prototypes (*ma dpe*) he was able to recover from a variety of sources.[188] The blockprint publication was sponsored by the Dalai Lama at the Ganden Püntsok Ling printery.[189] It appears that it was the first blockprint edition of the set, and which works belonged in the collection was debatable.[190] The authenticity of some as attributable to Yutok had already been questioned by previous scholars,[191] and it is unclear when it first came into existence as a set.[192] The Desi himself was quite disdainful of several works in the collection on the questionable reliability of their attribution to Yutok, among other grounds.[193] His objections may have also had to do with their content. One of the works, an early commentary on parts of the *Four Treatises* (the basis of the final chapter of this book), lays out a picture of medical ethics that, while fascinating for its instrumentalist ethos, may well have been embarrassing to later generations. And yet it was published and preserved regardless, in recognition of the value of old documents.

Preservation efforts can also require new creations. The very place of Yutok Yönten Gönpo as the founder of Tibetan medicine was only being fully codified in this period, as was the fixation on the *Four Treatises* at the expense of other medical works that had been circulating on the Tibetan plateau for centuries.[194] But there was considerable debate about whether Yutok was really the author of the *Four Treatises*, given that the work is framed as an original sermon by the Buddha. At some point around 1680 the Dalai Lama asked the medical scholars in his court to produce a full biography of the master. This alone would count as another of his great accomplishments for medicine, but it had a special twist. His scholars actually produced two biographies.[195] In effect they split the figure of Yutok into two, a Younger and an Elder. The Younger would be the twelfth-century historical person,[196] but the so-called Elder, who would have lived in the eighth and ninth centuries, appears to have been either entirely an invention or based on lore connected with a physician with that name who actually lived in the imperial period. Yang Ga suggests that such a figure did exist, albeit with no connection to the *Four Treatises.*[197] We do not know if the biography of the Elder Yutok produced in the late seventeenth century was based on any older prototype[198] or was created *de novo* out of certain auto/biographical fragments connected to the twelfth-century Younger Yutok, which also fed the latter's newly codified biography.[199] The Desi's own summary of the lives of the two Yutoks includes a few telling comments, to the effect that there has been some confusion between their careers, and that during the life of the Elder a goddess gave a prediction about a distinction between a Younger and Elder Yutok in the future.[200] Indeed, both figures have the exact same name; both of their fathers are named Yutok Khyungpo Dorjé; both travel to India multiple times to receive medical teachings. Still, the Desi himself strongly supported the thesis that there were two Yutoks.[201] And it is testimony to the influence of the seventeenth-century reworking that ever since, Tibetan medicine has thought in terms of two Yutoks.[202] But in fact, as Yang Ga has shown, an Elder Yutok from the imperial period who was involved with the *Four Treatises* does not seem to be known in Tibetan medical historiography prior to this moment.[203]

It is not entirely clear what was gained by creating a double to the twelfth-century Yutok who lived four hundred years earlier. We can say at least that the long biography of the Elder provided an opportunity to recount stories about medical ethics and the enlightened nature of medicine. But the rationale for locating the patriarch of Tibetan medicine during the imperial period is less obvious. The Treasure narrative that had been in place since the thirteenth century—whereby the *Four Treatises* was translated from Sanskrit into Tibetan at the imperial court and then hidden and later passed to the twelfth-century (Younger) Yutok—had already deflected authorship from the latter and located it instead with the Buddha. Perhaps the placement of an Elder Yutok in the imperial period served to acknowledge that someone with the name Yutok had a strong role in the formulation of the *Four Treatises,* but associated him with the royal dynasty several hundred years earlier. Having the father of Tibetan medicine live in the same glory days as the kings and the introduction of Buddhism to Tibet tied medicine all the more securely to that mythic topos. As the Desi himself says, the Dalai Lama commissioned new biographies to be created out of "a collection of old documents, to accord with students' minds."[204] What he probably was implying was that composing biographies of the Yutok patriarchs was meant to offer an inspirational story for medical practitioners of his day. I would add that codifying the lives of the Yutok masters in light of readers' appetite for the imperial mythos was a crucial piece in the consolidation of the *Four Treatises*' canonical status.

Yet a third key literary accomplishment for the advancement of Tibetan medicine and the legacy of Yutok Yönten Gönpo was the recognition of the phenomenal contribution of the scion of the Zur school, Lodrö Gyelpo—the very figure the Desi disparaged some decades later. At some point in the 1670s the Dalai Lama worked with the scholars in his court to publish the brilliant commentary on the *Four Treatises* that Zurkharwa had written in the previous century, *Ancestors' Advice*. Zurkharwa had only finished his comments on the *Root and Explanatory Treatises* plus two chapters of the *Final Treatise* when he suffered a stroke. According to the Desi's account, the original idea to bring this valuable work to light came from Darmo Menrampa and Namling Panchen. It was Namling who edited the existing *Ancestors' Advice*, and again the Dalai Lama provided material support for the printing. An odd comment by the Desi—that the Dalai Lama thought that to edit and print this incomplete work was "better than nothing"—further suggests that the mere resurrection of a classic was not enough.[205] The Dalai Lama and his associates did go on to complete the work. This resulted in the writing, by Darmo Menrampa, Namling Panchen, and others, of a new commentary to the rest of the *Four Treatises* under the same name of *Ancestors' Advice*.[206] Apparently they were able to base this partly on notes left by Zurkharwa himself. It was fortunate that those notes survived, especially with regard to the *Instructional Treatise*, which contains the bulk of the diagnostic, therapeutic, and physiological information. As the Desi tells us, commentaries on the *Instructional Treatise* were exceedingly rare.[207] The Dalai Lama gave the order to begin the project; it was commenced in 1678 and finished the next year. The blocks for printing were carved by 1681.[208] This work's completion is another example of preserving the old while simultaneously creating something new, in this case offering new readings on any number of medical issues.[209] *Ancestors' Advice* came to exert enormous influence, including on the Desi himself, whose own commentary on the *Four Treatises*, the *Blue Beryl*, written ten years later, copied many long passages from it, including from the parts written by Darmo and his team.

Finally, the importance of one more key medical achievement under the Great Fifth, the production of a definitive blockprint edition of the *Four Treatises*, cannot be gainsaid. Here the inextricability of old from new becomes more tangled yet. There had already been several blockprint editions of the root medical text, and before that several manuscript versions, but their accuracy seems to have been a recurring point of contention.[210] We can also detect the prestige that the Desi associated with producing the most correct version. The effort to republish the classic began while he was young and seems first to have been a matter of preservation, when at the urging of Jangopa, the Dalai Lama asked Döndrup Pelwa and others to reproduce the old Dratang edition. This appeared in 1662 under the direction of Jangopa, with a new colophon by the Great Fifth himself.[211] The Dratang had been the first blockprint of the work, edited and produced by Zurkharwa in the previous century.[212] Zurkharwa's own efforts to locate the original manuscripts (*phyag dpe dngos*) and commentarial golden notes (*gser mchan*) of Yutok and his successors as a "witness" (*dpang*) attest to the antiquarian mentality already in place a century before the Dalai Lama's initiatives.[213] Since then several other editions had appeared. But when the Desi came, in the course of his own medical education, to study the latest version of the Dratang that his elder colleagues had reproduced under the auspices of the Dalai Lama, he found interpolations and omissions. He wanted to return to the other editions and to do better.[214] Recounting the story, he reviews with subtle disdain the qualifications of Jangopa, calling into doubt his pronouncement that the Dratang *Four Treatises* was "very excellent."[215] The Desi characterizes his doubts in terms of the discrepancies he found regarding certain medicinal plants' habitats; he also complains that some of the outline structure of the text was out of order and that words had been dropped. But I don't think it is entirely accidental that the Dratang he critiqued so strongly was originally the handiwork of Zurkharwa.

Whether or not the Desi's push for another edition had to do with rivalry, it represents a rare case in which he suggests that even the Dalai Lama was mistaken on something and indeed that his lord's knowledge of medicine was wanting. As he says so diplomatically, "My lama the Great Fifth is omniscient with respect to the definitive meaning of all things and can see everything there is to be known in terms of how they are and what they are in reality, like the shape of a myrobalan in the palm of the hand. But as for the perspective of people with lower merit, it appears in his precious biography that he only memorized the *Root, Explanatory*, and *Final Treatises* and the way to listen roughly to the outline and the mnemonic diagrams. He shows himself as otherwise not knowing medicine in detail." In fact, the Desi avers, the Great Fifth was wrong in supporting the Dratang edition and writing a colophon to it. He whispered this assessment to the ruler himself, letting slip into his ear that the colophon was "a little over the top."[216]

The comfort with criticism and candidness in the Desi's day made this breach of protocol feasible. Even so, it remains astonishing that the Desi not only could make that comment orally but also wrote about it, delicate locutions notwithstanding. It is pertinent to note one other ingredient in this episode that marked the climate of academic medicine and its patronage in the period, at least at this uppermost echelon: the evident intimacy and trust between the Desi and his master. We can appreciate its significance in light of what Steven Shapin has shown to be at the heart of the growth of science in seventeenth-century England.[217] The close bonds marked by civility and shared elite social status in the Dalai Lama's circle might also explain how the Desi could be so forthright in his written assessments of his other colleagues that he shares in retrospect with his readers. As for the event in question, the underlying trust meant that the Great Fifth was comfortable enough—and committed enough to the quest for truth—to brush the insult off with grace. The Desi reports that the ruler responded "with some amusement, 'If we set aside what Jangopa and the others have said, and since what you are saying is true, if you could do a bit more research and comparison that would be good. So do it!'"[218]

And so the Desi got a chance to put his mark on the prized root medical text. Years after getting the Dalai Lama's approval, and after writing his own commentary to it, he finally completed a new edition of the *Four Treatises* itself, in 1690.[219] He writes in retrospect how he produced what he calls the Dzonga edition with a committee of colleagues, looking into all the versions known to him.[220] He also examined the "manuscript chapters written with an iron pen," and especially "the inner old *Treatise* box of Zurkhar and the hand-smudged manuscript" that had been discovered by Zurkharwa.[221] The Desi assures his readers of his own thorough and detailed research, in which he did not take the "easy way." As he adds,

> regarding the verses that were missing in the Dratang *Four Treatises*, I checked with my own reasoning on what would be appropriate from the old texts. For example, points that are in the summary are incomplete in the detailed section. Or the signs of an illness will be explained but the treatment is missing. And there are also confusions in the outline. There are some erroneous letters that I determined neither by approximation nor by just concocting something, but in accordance with authoritative sources and with reasoning. All the old words that are hard to understand

> were clarified, emendations and editing were well done, and then it was carved for printing.[222]

Thus does the Desi indicate that he also relied on his own judgment and reasoning to change and augment the great medical classic.

The projects initiated by the Dalai Lama reached toward a range of kinds of newness and suggest broad participation with other movements in the early modern period in the region.[223] The Desi and his colleagues at court were invested in the recuperation of old knowledge, but also in the importation of new knowledge. They made new compilations of data about the past; new codification and completion of older works; new corrections of mistakes in past work. There were assumptions about the virtues of the ancient but also concerns about consistency and proper organization. It is not entirely clear how much of the Desi's new edition of the *Four Treatises* was seen as restoring the original and how much as representing new improvements. There can be no doubt at least that to correct the location of medicinal plants, which is one of the things he says he did, would have constituted additions that, as we saw in the last chapter, were based on new empirical knowledge from the field.[224] Here as elsewhere, the invocation of the empirical seems to require no justification at all. Especially in the Desi's commentary to the detailed chapter on *materia medica* we find him repeatedly describing the occurrence of things in the natural world that have no standardized description, i.e., from the old systems of classification (*'khrungs dpe*). He can dismiss the lack of classical prototype for what he and his colleagues were newly recognizing, on the grounds that the specimens in question were easy to classify now.[225] Here the authority of what is seen in the world trumped other considerations—certainly it trumped the textual, a point we will see again.

Thus did the medical mentality not only foster an urge for realism, an urgent desire to get the facts right, and a willingness to criticize peers and even elders or the venerable writings of the past if they did not measure up. It also brought into high relief a tension between what was encountered empirically in the world and what the classical sources said. And yet the prospect of the *Four Treatises* as correctable remained a source of tension. The root text had by this time acquired an authority on par with scriptural works in the Buddhist canon. The fact that the Desi considered the *Four Treatises* to be Buddha Word alerts us to the delicacy of introducing any corrections to it. No doubt there was grand prestige in having produced the definitive edition, but there was simultaneously a need to be very careful.

RUMINATIONS ON COMMENTARY: MISTAKES AND INNOVATION

The creative tension between the new and the old emerges in spades in one further great literary accomplishment in medicine during this era: the Desi's magnum opus itself, his *Blue Beryl* commentary. He completed it early in his writing career, just a few years after the demise of the Great Fifth. It became the definitive gloss on the *Four Treatises* for academic Tibetan medicine.[226]

Writing a commentary on a work is different than editing it. But it still does not constitute something completely new. Rather, it occupies an ambiguous place in between. One is commenting on, and thus constrained by, an authoritative text, but surely one is saying at least *some*thing new—otherwise why write it at all?[227] To be sure, writing a commentary to the *Four Treatises* had long provided an arena to express dissent and differing opinions, not to mention new ideas. However, the very question of the grounds for and especially the implications of writing commentary had not been meditated upon before, at least not in such personal detail.[228] The Desi took up such questions overtly, apparently obsessed by them. His complex discussion at the end of the *Blue Beryl* is a good place to see how his nimble negotiation of newness, innovation, and criticism allowed him to advance medical knowledge and the place of authority therein.

The Desi is clearly conflicted about the status of the new. There seems to have been an earlier moment in his career—or perhaps he only portrays it as such in retrospect—when he was quite confident of the newness of a treatise on astrology that he had composed. Speaking of his *White Beryl,* written between 1683 and 1685, he can taunt potential critics to "scrutinize it carefully, clarifying whether or not it is an elegant teaching without precedent, coming out of the energy of intelligence. And if you have doubts, I have evidence to offer through authority and reasoning."[229] The *White Beryl* did produce controversy, prompting the Desi to write two further treatises with even more detailed specifications.[230] But the very idea of an "elegant teaching without precedent" boldly signals his bravado in characterizing his ideas on astrology and divination as issuing out of not only authoritative sources (though he is quick to add he can supply those too) but also his own genius.[231] And yet he is saying these things in the context of much more complicated feelings on the occasion of concluding his medical commentary; his satisfaction with the *White Beryl* serves to highlight a contrast with his quite uncertain sense about the *Blue Beryl.*

The Desi is both detailed and inconsistent. He can cite the obvious motivation—rife in the medical circles of his day—that to flesh out medical knowledge was to "save" medicine from decline.[232] But how does he think that this can be done, and *was* done, by himself? The problem of newness now takes center stage.

The Desi evinces an ethos that gives highest credence to tradition and authority. He also expects to be criticized if he is thought to be introducing something new. And yet he still manages to claim to be doing just that—while protecting his tail in the same stroke. In his extraordinary conclusion to the *Blue Beryl*—extraordinary, that is, for the way it waffles back and forth—he ends up having it both ways. He writes,

> As for this *Blue Beryl*, others may assess it to be a new exposition, but from my perspective it is entirely grounded in the old. It is others who, in their assessment of it, are distinguishing it as a new breed of knowledge. To be sure, it is not easy to achieve the standard of having the wherewithal to compose an exposition.[233]

Here the Desi is clear in his deference to tradition and in denying any transgression. But as he continues to discuss the sources for the *Blue Beryl*, he leaves open a space for his own originality. In the next sentence he admits that "about half of [my commentary] on the *Root* and *Explanatory Treatises* summarizes word for word those by Jang and Zur. The rest, my own words, are little more than half, so it is not merely my own composition."[234]

He is defending against a charge that he simply made up the information in his commentary, but deploys a half-full/half-empty strategy: the *Blue Beryl* is only half my own words; don't blame me since I did not make it all up, but do give me credit since I did make some of it up. Elsewhere he notes he actually had an associate, Chakpa Chömpel, write up the first draft of the *Blue Beryl* based on the Desi's notes, since he himself was so busy.[235] He also refers to the *Blue Beryl* as something that was "compiled" (*phyogs bsgrigs*) rather than using one of the usual words for composition.[236] But he does consider the commentary to be his own. To return to the statement we just looked at, what he is really concerned with is the issue of innovation as such, rather than whose it actually is. And on this we find him talking out of two sides of his mouth at once. When he moves in the next sentence to critique his predecessors, he is justifying the need for something new: "Moreover, I did not find a satisfactory way to introduce these topics from explanations by others that I could trust definitively, nor did I find greatly reliable and complete commentarial texts."[237] And yet in the next sentence he returns to defending against his imagined critics, displaying the conventional humility and showing that he is not conceited enough to trust his own ideas.

> As for it not being appropriate to manufacture [knowledge based on mere] conception (*rtog bzo*), Śāntideva has said, "That which has not appeared previously will not be here explained. Skill at composition is not something I have."[238] So it says. I have put this into my mind.

By the next line he has found a happy medium. He has done as much research as possible, then gone further based on what he knows, but he is always careful to check it. He is referring to his immediately preceding discussion of his research into various old botanical classification texts along with consultations and examination of actual plants, eventuating in his adjustment of and additions to the information in the plant recognition chapter of the *Four Treatises*.[239] And yet he still would like it to be known that he is not quite comfortable, not as confident as he is about his previous work on astrology.

> So I investigated the portions of reliable manuals mentioned above. On top of that, whatever I was certain about based on my own reasoning, I checked. Still, I did not have confidence like I did about the astrology upon which I wrote [in the *White Beryl*], and in my own mind I was not satisfied [with the *Blue Beryl*].[240]

Yes, he's doing something new:

> Previously there was some explanation from Jangopa and so on, but other than a few scattered teachings, there has been no tradition of making a manual like this that compiles the treatise and commentary as illustrations that allow one to introduce by pointing a finger.

And yet he isn't. In the next sentence his work is only based on what was already there and is not a new creation: "There is no one up to par who is working hard to learn what already is there from the past, so for me to do something more than that would seem only to make me tired."

He gets to have it both ways. As we go on through this closing statement, though, and also other statements, we find a further and important dimension to the problem he is facing. He is worried not only that he will be accused of adding something new but also about the possibility that he will make a mistake and say things that are not true. While the exact parameters of the Desi's standards of such truth have yet to be discerned—and remember, at least regarding knowledge of *materia medica*, he even thought that the *Four Treatises* itself had room for improvement—there is a clear sense of a standard out there in the world. This palpably undergirds his critiques of the unsatisfying commentaries of the past. Other than a few shining luminaries, "the other earlier ones diluted [the knowledge in the *Four Treatises*] like milk from the market, such that it morphed into all sorts of things."[241] The Desi does identify those "earlier ones" who actually could be trusted, namely the Jangpa masters Namgyel Drakzang and Mi Nyima Tongwa Dönden, and the founder of the Zur school, Nyamnyi Dorjé. But he was eminently comfortable critiquing other teachers of the past—far beyond his usual whipping boy, Zurkharwa Lodrö Gyelpo. This is why new commentary is needed: the existing commentaries are not reliable.

The Desi is critical in particular of the few existing commentaries on the *Instructional Treatise*, which provides the bulk of the practical instruction.[242] Indeed, issues of procedure opened up the biggest space for revision and innovation. A special genre of medical writing that addressed the diagnoses and therapies of the *Instructional Treatise*, but which was not a commentary as such, had already begun to appear several centuries before the Desi's time. The source of information was cast as the clinical experience of the physician. Such works were called, appropriately, "writing from experience" (*nyams yig*).[243]

A key ingredient in the attitude that experience and innovation are requisite for effective medicine is a concept of error—especially the ability, or license, to acknowledge it. The Desi avers that many of his shorter writings on medicine were efforts to correct the errors and confusion of others.[244] He maintains that it is inevitable that a later generation will find things to correct in writings from the past. "When an earlier work is examined by later people, they see faults—this is natural," he maintains.[245] He even allows this observation to account for some of the criticism that his nemesis Zurkharwa leveled at his own predecessors.[246] Apparently mistakes have to happen; they are part of the way things are. Several times in this discussion, he uses a classic term (*chos nyid;* Skt. *dharmatā*) that in Buddhist philosophy denotes the most basic and unchanging level of metaphysical reality. For the medical mentality evinced here, it is change in knowledge over time that is deemed to be "natural." That would mean that this law will also pertain to his own work: "So since I myself am not so confident, I hope that more mistakes will be pointed out."[247] He knows that he too will have gotten some of it wrong.

The assumption that mistakes will always occur is coupled with its social implication: the need to exhibit humility and openness to critique. The Desi had already showed his *White Beryl* to others for comments; now he is recalling how he did the same for the *Blue Beryl.* But it is telling how he characterizes what people told him. The Desi cannot resist a jab at his own reviewers, questioning whether they are really up to the task. He reports saying to Lhünding Namgyel Dorjé, Nyemo Tseden, and Namling Panchen, "If there are any doubts or faults I need you to send them my way." But then he adds, "And so they acted like they were assessing it, but they could not manage it."[248]

The Desi shows that he can actually receive criticism in his account of ten responses that apparently he received on what he might add or cut in the *Blue Beryl.* One point had to do the meaning of a single arcane syllable in a particular medical concoction that the Desi had glossed as meaning "smoke," but that was not supported by the root text. He also mentions a few further small revisions that he assented to, including corrections that were advised regarding the paintings that illustrated the *Blue Beryl.* He clearly feels the need to record these issues. And yet he also is motivated to point out that he rejected some of the other suggestions. One issue, already mentioned in the last chapter regarding the shape of a particular leaf, was decided in a way that was quite definitive: the Desi adjudicated it by sending someone to get the real thing (*ngo bo*) from the field. In that case the proof of what was correct and what was a mistake was out in the world.[249] We will return to this point repeatedly.

Overall, the Desi remains "unsatisfied" and "unconfident" about the *Blue Beryl.* This may be as much a display of etiquette as a fully representative report. But etiquette says a lot about the parameters of discourse within which the Desi had to operate. He projects a critical audience of powerful scholars whom he entreats for their tolerance and from whom he feels subject to criticism.[250] But as much as he expects criticism, he also feels entitled to give it. At one moment he evinces a real sense of being alone, of having no one out there to receive his work and judge it properly, such that producing his commentary is like "doing transference for a corpse without a head," as he quips (the esoteric "transference" rite involves expelling consciousness out of the top of the skull as a person dies: there would be no point in doing it for someone who lacked a head). Despite his lack of confidence, there is no one who can really assess its mistakes.[251] In another moment, now at a late point in life and writing about his other main medical work, his history of medicine, he reverts to the conventional diffidence. Here he imagines his judgmental audience as a bemused group of slightly critical sages who spy "out of the corner of their laughing eyes at the one who composed this festival of attachment-making, just like Brahmā with the girl who issued from his own imagination." But in the very next lines he is in yet a third, quite balanced and reasoned mood, praying for the good effects of the work and urging his readers to use their analytical powers to assess his history without bias.[252] He can say something like that about his *Blue Beryl* too: "I don't have a lot of capability, but I am not very ashamed of this either."[253]

What is common in all of these moments of self-reflection is the powerful ingrediency of a critical readership on his horizon. The question remains of how distinctive were the Desi's dispositions toward self-correction and revision, compared to the larger world of religious scholastic culture in Tibet. Other cases outside the domain of medicine where scholars sent provisional drafts of their writing to colleagues for comments and critique, as the Desi did on several occasions, are not unknown.[254] The same can be said of posting challenging statements in public places, although I am not aware of such statements posing technical intellectual questions, as those of Zurkharwa and Lhündingpa did. But even if it is not unique, we can recognize a signature expression of the Tibetan medical mentality with respect to disputation and authority in the foregoing passages, most of all in the last sentence of the Desi just cited. There, leaving behind both bravado and diffidence, he forges a middle position marked by cautious confidence . . . and a deep investment in being realistic. This reasonable—and human—attitude is discussed in the last chapter of this book, at the heart of a formative account of professional medical ethics. Once again a related disposition may be recognized in broader practices in Tibet around writing autobiography—a genre also fraught with concerns about etiquette and a tension between deference to tradition and the impulse to write realistically.[255] But the expressions of such concerns take on a particular inflection in medical discourse, especially by the Desi's day.

Most of all, we can understand the Desi's critically probative attitude as a function of his desire to get it right in medicine. A large part of what that would mean would be to adhere to publicly observable criteria. That is what the adversion to empirical evidence amounts to; it is behind the all-important question of how to handle the potential death of a patient that takes up much of the medical ethics, and it made both for a certain caution in one's own claims and a license to pounce vociferously upon the shortcomings of others.

But if allegiance to physically and publicly discernible verities sometimes meant transgressing the protocol of deference, it also had the potential to put medicine at odds with some of the most powerful arenas of authority in Tibetan society, the intellectual practices and ritual institutions of Buddhism. Much delicacy was required to negotiate such potential tension, and perhaps no one appreciated the scope of the problem more than the Desi himself. The last section of this chapter broaches one very key way that the tension became clear to him. It has to do with a literary and conceptual domain in which Tibetan medicine was distinguishing itself as a tradition in its own right. The Desi's critical acumen again comes to the fore, and yet the implications of his rhetoric begin to slip out of his control.

THE DESI AND MEDICAL HISTORIOGRAPHY: A SPACE APART

One of the surest signs of medicine's self-conception as a tradition unto itself is the growing practice of writing specialized histories of its origins. Often labeled with a special genre title—*khokbup* or *khokbuk*—many of these accounts stretch back to India, and several even cover some of the same ground as Tibetan histories of Indian Buddhism do in the course of giving particular attention to the Buddha's teaching on medical topics. They contain much detailed material unique to medicine, both in India and especially as it reached Tibet. They give a palpable sense of a long development with its own trajectory, but not directly mappable onto the history of Tibetan Buddhism.

What's more, by the time that the Desi is writing his own history of medicine, there is second-order reflection on what it means to write such a work. The elaboration of standards to which medical history writing should be held had already begun in preceding centuries. But an exceptional passage in the final section of the Desi's own *khokbuk* amounts, for the first time, to a veritable history of writing medical history. As might be expected from the Desi, this proceeds in exceedingly critical terms. And yet his efforts to delimit the telling of Tibetan medicine's history run up against a conflicting need to place it within a larger Buddhist universe.[256]

Many of the conventions of medical historiography are familiar from the more well-known histories of Buddhism in India and how it came to Tibet. The medical histories also draw on other Tibetan narrative genres, including accounts of ruling families, histories of other clans and families, and other kinds of Tibetan genealogical discourse, memorably characterized by R. A. Stein as "necessary for upholding the order of world and society."[257] Two of the earliest Tibetan histories of medicine were published in the *Eighteen Pieces from Yutok* collection and are close to contemporary with the composition of the *Four Treatises* itself. One of them, entitled *Soaring Garuda* and already labeled as a *khokbuk,* is taken up with justifying the status of medicine, and then the *Four Treatises* in particular, as a teaching of the Buddha.[258] As Frances Garrett has noted, *Soaring Garuda* addresses only in the briefest terms the main topics proper to the history of Tibetan medicine as found in other works from the same period.[259] But its strenuous effort to provide medicine's Buddhist pedigree suggests an early felt need to counter a competing assessment that medicine was *not* fundamentally a product of the Buddha's dispensation.[260]

Other early works do provide a conventional history of medicine in Tibet, recounting the various medical practitioners and scholars who came to the court during the imperial period to teach and translate medical treatises, the formation of early schools, and new systematic and practice-based works composed by Tibetan doctors. The other medical history in the *Eighteen Pieces* collection, *Crucial Lineage Biography*, is an important example.[261] It appears to have been written, at least in part, by the principal student of Yutok, Sumtön Yeshé Zung, and most of it focuses on the early Tibetan physicians who inherited the *Four Treatises* from Yutok. Another important early *khokbuk* history, *The Way That Medicine Arrived,* probably written in 1204 by Chejé Zhangtön Zhikpo, deals primarily with the figures associated with the Indian *Aṣṭāṅgahṛdaya.*[262] It also mentions teachers and teachings entering Tibet from all seven schools of medical learning in the world according to the author's knowledge, including Tangut/Xixia (Tib. Mi nyag), Khotan (Tib. Li), Trom, and China. This work displays a conception of the genealogy of Tibetan medicine that has little to do with the importation of Indian Buddhism. It was well known to the Desi, who cites it as one of his own sources.[263]

A key sign of the growing sense of an autonomous medical tradition was the very use of this distinctive genre term *khokbup,* or *khokbuk.* It appears in the titles of two of the three early works just mentioned, and the third glosses itself with the same phrase.[264] To be sure, the term is not always included in the titles of medical histories, and even when it is, it is not used consistently, in that the contents and structure of works so named can vary widely. The term is also occasionally used in other literary contexts quite apart from medicine, for example to name a liturgical manual. Still, *khokbup/k* has a special resonance for medical historiography, and it is part of the title of the two most influential such works, those of Zurkharwa and the Desi. Both look backward to a whole series of what they call *khokbup* or *khokbuk* texts as the models for their own histories.[265] I suggest that the adoption of this moniker served to distinguish medical history from other kinds of histories, especially those of religion, and suggested an alternate textual space.[266]

The label itself is curious. Among other things, the two main variants have different literal meanings: *khokbup* (i.e., *khog 'bubs*) translates as something like "pitching, or framing, an interior," while *khokbuk* (i.e., *khog 'bugs*) would mean "piercing the core."[267] Modern scholars have puzzled over the term, and the genre.[268] I would argue that both spellings/senses refer reflexively to the genre as such. To put it another way, both metaphors reference what it means to be a book. This is especially so for *khokbup*, which contemporary Tibetan lexicographers understand to be the main term, and which most commonly occurs in titles.[269] It is listed as one of five ways of presenting a book's content, specifically, that which lays out the general structure of a topic. While *khokbup* more colloquially refers to the pitching of a tent, in book terms it is that which sets out a structure of classifications, or more generally, covers a topic. The metaphor is felicitous for the purposes of forging medical tradition. The writing of *khokbup* would mark the pitching of a tent or raising of a roof over an interior space thereby created—in other words, a space apart. Its further sense of framing or laying out highlights the fact that classification is a key part of what it means to write medical history. Indeed, a key question was how, and if, medicine and its root texts are to be classified as the teachings of the Buddha.

We can also see this reflexivity in the variant spelling, *khokbuk*. *Soaring Garuda* defines it as "that which reaches all understandings of the meaning, once one knows all the ways to explain." This describes what the text is doing, both for the classification of medicine within the five sciences and for a "piercing of the core of this very *Four Treatises*, since it is the essence of all medical knowledge."[270] In that the goal of *Soaring Garuda* is to demonstrate that medicine in general, and then the *Four Treatises* in particular, is a genuine teaching of the Buddha, the work "pierces the core" of its fundamental concern—the heart of the matter, as it were—about the nature and authority of medical knowledge.

It is especially instructive that the *Soaring Garuda* and Zurkharwa's history of medicine use both versions of the genre term as a verbal noun, or even an indicative verb.[271] This suggests that the genre accomplishes something by virtue of its very writing—not unlike what a speech act does, quite apart from whatever semantic content a statement like "I do" denotes. Thus did the *khokbuk/p* accomplish a kind of "book act." Its very writing established a zone of medicine with its own history and its own core virtues.

The medical histories by Zurkharwa and the Desi were landmarks, authoritative sites for establishing medical tradition, and also arenas where key historical issues were contested. Both owe much to the earlier tradition of writing such works, especially a medical history by the fourteenth-century Drangti Pelden Tsoché that provides important evidence of the influence of Greco-Arabic medicine in Tibet in the royal period, along with detail on the early history of Tibetan medicine.[272] But Zurkharwa's rendition expanded the purview of medical historiography exponentially.[273] Zurkharwa couches the history of medicine in a detailed account of the entire world as he knew it. In many ways he phrases this in Buddhist scholastic terms, drawing on Abhidharma discussions of cosmology and materiality. Still, his account is inflected with a special interest in the origin of physical matter, and he treats questions like the relation of sentience to materiality at some length. He also shows far more interest in the "defiled world" than in the "purified" one, i.e., that which would intersect with Buddhist soteriological concerns. In fact, he characterizes his comments on the purified world as an "aside," and they take up but a single page.[274]

Zurkharwa continues with the development of medicine as such, including much of what Tibetans knew about the history of medicine in India. But he interjects into that a lengthy (over 70 pages in the modern Chinese edition) account of the Buddha and the history and categorizations of his teachings. He gives a history of medicine in the heavenly realms, then in the realm of humans, then in the Buddha's teachings, and finally, on page 246 of the 384-page modern edition, begins the story of Tibetan medicine in the royal period. This goes down to the life of Yutok; I will say more about his treatment there in the next chapter. Zurkharwa closes his history with a lengthy section on how to educate oneself in medicine and its practices.

The Desi's own *khokbuk* history, written about 130 years later and at the end of his career, seems to reverse this trend, restricting its topics to medicine but reaching new levels of detail as well as self-consciousness of what medical history should be doing. The Desi reviews the teaching of medicine from the gods' realms to Indian medicine, the medical teachings of the Buddha, and their transmission in Tibet. Far more than anyone before him, he goes on to provide a sustained historical account of the development of Tibetan medicine after the time of Yutok the Younger, with detailed biographies of the major figures of the Jang and Zur lineages, the history of institutions of medical learning, and the life of the Great Fifth. The Desi also treats himself as a topic of discussion, adding a 24-page section (in the modern edition) on his own contributions to medicine.

To provide biographies of key figures is one of the conventions of Tibetan Buddhist historiography, but the Desi also takes up problems and specificities particular to medicine. He gives second-order attention to the nature of the field as such, beginning with a short introduction on "what is special about medicine." He also includes a long and interesting section at the end of the book about how to study medicine, and the relation of that to kinds of training familiar in Buddhist traditions, such as the systems of Vinaya, bodhisattva, and tantric vows. A good historian, he finishes with a lengthy account of his sources that turns into a study of the writing of medical history in Tibet.

Here in the Desi's review of earlier medical histories are telling signs not only of a self-consciousness of medicine as a separate domain of learning and practice but also of the delicate issues surrounding such a proposition. It is surely a testimony to the nervousness about separating medicine into its own tradition—which proves that such a separation was already under way, or at least being contemplated—that so much attention was devoted to showing how medicine does fall under the umbrella of the Buddha's dispensation. The vehicle of the five sciences served this effort well. This means that a key piece of Tibetan medical historiography involved providing a typology of the teachings of the Buddha. In beginning his review of his own sources, the Desi singles out *Soaring Garuda* approvingly for "establishing medicine as Buddha Word," a phrase that came to have much mileage in Tibetan writing about medicine.

The Desi goes on to work through about twenty such histories, written from the eleventh century up to his day. But he is not merely listing his sources, as Tibetan authors often do. He is using the occasion polemically to assess the relative merits of these works and to place them in a tradition whose standards are important to articulate and to advance. Indicating that medical history is a domain unto itself, the Desi expresses disapproval of examples that spend too much time on irrelevant topics.

He is especially explicit in critiquing his nemesis on just this point. Zurkharwa's medical history is also the only one of his sources for which he gives a chapter-by-chapter summary. The Desi is overtly disdainful when he says that Zurkharwa's history of medicine "ends with a three-part practice outline that he set up, and a calculated explanation of the five sciences in brief, and an explanation of letters, [but given the previous chapters about] the world and its contents, and the deeds of the Buddha, and so on, it is a huge heap of words about everything that he himself saw, flaunted as if it belonged to the history of medicine."[275] The same criticism is echoed in the Desi's resumé of Zurkharwa's student Ruddha Ananta's even more expansive medical history. This work discusses the classic twelve deeds of the Buddha, the Buddhist canon, and the future appearance of Maitreya; it's another "huge heap of words."[276]

At this point it is not a complete surprise to find another case of criticizing something in a context where critique is highly unusual, i.e., in listing one's sources at the end of a work. But there is an interesting twist that goes quite beyond the simple protest we might lodge that actually Zurkharwa's medical history is significantly shorter than the Desi's, and hardly such a "huge heap of words." Rather, from the Desi's description of the other histories of medicine in this section, the ones he doesn't critique but just summarizes, they too apparently had lots of what surely would be extraneous material, on the Buddha's teaching and the Dharma in general and so on, at least if we are to believe what the Desi says about them.[277] So a good guess is that the real reason he singles out the Zurkharwa writers for their verboseness is his far more momentous irritation with them for maintaining, as he says here, that "the *Four Treatises* were taught by Yutok," i.e., a historical human author.[278] In contrast, the histories that he doesn't critique are the ones that make the *Four Treatises* the Word of the Buddha. In fact, the main information that the Desi provides about these more acceptable medical histories is *how* each of them makes the *Four Treatises* the Word of the Buddha—whether they argue that the Buddha Śākyamuni and the medical Buddha Bhaiṣajyaguru are the same; whether they say that Buddha taught as Vajradhara or not; and so on. Their stance on the status of the *Four Treatises*' ultimate expounder seems to be the main criterion for his approval.

And so when the Desi takes to task the Zurkharwa histories for trying to "attach a nonmatching lion's upper jowl to the lower jowl of a camel,"[279] he is saying that despite their flamboyant display of Buddhist learning, their denial nonetheless of the *Four Treatises*' status as Buddha Word is a move in the opposite direction, creating an unacceptable distance between the field of medicine and the Buddha's teachings. Hence the discrepancy between the two kinds of jowl. In other words, these Zurkharwa scholars can say all they want about Buddhist topics, but they hit a false note, and they're not fooling anyone. But note the ironic upshot: the Desi is singling out the Zurkharwa writers for talking too much about general Buddhist topics in their histories of medicine. Even as he insists that the *Four Treatises* is Buddha Word, he is arguing that medicine deserves a separate kind of historiography that just focuses on medicine. The Desi, it seems, was up to his chin in lion and camel jowls too.

AN IMPERFECT UNION

The question of Buddha Word and the *Four Treatises* had been a serious point of contention well before the Desi. We will follow the fortunes of this defining debate in the next chapter. The Desi's interventions came at the end of it, after the most interesting cultural and intellectual issues had been thoroughly thrashed out. His own actual position on the matter is examined in the Coda following chapter 5. Surprisingly, when he gets down to substance, he mostly agrees with Zurkharwa. Nonetheless, the animus in his blustering rhetoric to the contrary bears our attention. It has everything to do with the special burden that medicine's new political capital meant for the regent. Even though he himself had been involved in a slew of activities that pushed medicine toward autonomy—founding the first freestanding medical academy in the capital, directing a monumental set of medical paintings that relativized Buddhist images, arguing for the authority of the evidence of the physical world, arguing that the truths of old could be subject to improvement—he tried to hold the line in insisting on the Buddha's authorship of the *Four Treatises*. It is all about authority. The foundation of the new state rested on the absolute truth of intentional reincarnation and the accompanying yogic power and apotheosis of the Dalai Lama—all deeply dependent on the magisterial authority of the Buddha's teachings.

This is a double movement that we find repeatedly. The Desi moved medicine further into a critical and empiricist direction, but tried simultaneously to keep it domesticated under the sign of Buddhism. If it is sometimes hard to keep track of the Desi's position, the same is true for Zurkharwa, to whose own wrestling with the problems of science and religion I will turn in the next three chapters. I would actually concur with the Desi that Zurkharwa inappropriately padded medical history with extraneous information about Buddhist teachings, no doubt fueled by a similar pressure to present his ideas under the aegis of the Buddha's teachings even as he, perhaps even more than the Desi, tried to come to terms with the independent witness of direct observation and empirical verities. We will see in the next chapter how Zurkharwa continuously blurred his own positions. Just when he made his most radical moves that would truly create an epistemic space apart for medicine, he also obscured the effect, precisely, I would say, so as to effect an allegiance to a Buddhist universe on every other front.

No one—not the Desi, not Zurkharwa, not the many other medical scholars who also struggled over this issue—doubted the deep imbrication between Buddhism and medicine. None of them doubted that the Buddha was a master healer and that his teachings incorporated considerable medical knowledge. None was oblivious to the long-standing place, in Buddhist scholasticism, of medicine as one of the five sciences. All were the product of a Buddhist education with shared assumptions about root texts, the purposes of commentaries, and the exemplary role of biography, not to mention the very practices of bookmaking.[280] And then there were the particularities of the Tibetan state that cast both Buddhism and medicine as valuable foreign imports to advance civilization. This was true in imperial times, and it was repeated during the era of the Desi, for whom the fate of medicine was very much framed in the same terms as the fate of the Buddha's dispensation]: both were in danger of falling prey to the degenerate age, both were in need of constant renewal, purification, and preservation. We have seen the parallel between the modus operandus of medicine and the Treasure tradition, which provided the perfect schema to justify the *Four Treatises* as Buddha Word. Equally momentous was the ethical homology vis-à-vis the state: both medicine and the Buddha's teachings are good for the people. The ruler, be he a Dharma king or the Dalai Lama or a lay regent, has a responsibility to take care of the people, just as a bodhisattva or buddha does for sentient beings. Both Buddhist teachings and medical care figure as ways to do that.

And yet despite all these reasons for alignment, it did not entirely work. The lion's and camel's jaws had not quite matched for centuries. Tibetan medical scholars had long noticed discrepancies and incommensurabilities in very basic things like standards of truth and conceptions of what is real. A distinctively medical mentality had been percolating for generations before the Dalai Lama and the Desi tried to create the conditions for advances in medicine. This mentality found itself in tension with certain kinds of Buddhist scriptural authority, and sometimes even with verities from the most exalted Buddhist traditions of meditation. The tension is nowhere more overt than in the debate on the authorship of the *Four Treatises* itself. And while the fact that some resisted medicine's ultimate origins in the teachings of the Buddha irked the Desi until the end of his days, even his own backbends to make this particular problem right could not remove entirely the evident cracks in any totalizing vision of Tibetan Buddhist culture.

◎ PART II ◎
BONES OF CONTENTION

3

THE WORD OF THE BUDDHA

When Zurkharwa Lodrö Gyelpo quipped that "in this snowy land of Tibet, as soon as three or more get together . . . they discuss it," his levity belied the gravity of the matter at hand.[1] The discussion to which he refers is something he cared about greatly. Not only did he codify the most radical stance of all in the heated debate about the authorship of the *Four Treatises* (in which many more than three participated); he was also exceptionally bold in saying what no one else dared to venture.

The Desi's nemesis and medical predecessor by over a century, Zurkharwa picked up on a baldly empiricist reading of certain incommensurabilities in the root medical text that had been brewing in medical intellectual circles for centuries. But Zurkharwa's rendition worked out the implications with rhetorical finesse, ending up with a historicist account of epistemic authority that was unprecedented in scope or hermeneutical precision. More than the proposals of the Navya "innovative traditionalists" in the same centuries, of whom Yigal Bronner writes, the kinds of issues that were being raised in the medical Buddha Word debate, and which Zurkharwa brought to a head, had the potential to recast the entire edifice of canonical status and the grounds of truth claims widely assumed across Tibetan scholastic worlds.[2]

The work at the heart of the matter had a status on a par with canonical Buddhist scripture, and challenges to its authority were momentous. Clearly compiled in twelfth-century Tibet, the *Four Treatises* is cast as the

"Word of the Buddha" (Skt. *buddhavacana*), and borrows key genre features that point to that status. Such a move is not new in itself; many apocryphal works in Tibetan Buddhism were equipped with a Sanskrit title and couched as a sermon delivered by the Buddha.[3] Even Chinese astrological calculation was given a Buddhist pedigree at a certain point in Tibet.[4] But what draws our attention to this gesture in the case of the *Four Treaties* is its reception, the way that its Buddhist and Indic provenance was refuted by some of the Tibetan intelligentsia in overtly historicist terms, and at a level of specificity rarely found in Indic or Tibetan Buddhist writing.[5] This is notably different from the debates around the Tibetan Buddhist Treasure tradition, to which the *Four Treatises* apologists were in some respects indebted. The Treasures were defended in metaphysical terms that blurred all significance of conventional time and place, in order to protect the ultimate author function of the Buddha. In striking contrast, medical commentators of Zurkharwa's ilk were comfortable with dismantling the Buddha's authorship on empirical grounds. And even more unusual for the Tibetan cultural sphere, such criticism did not dent the prestige or value of the work.

The medical Buddha Word debate was introduced to Western scholarship by Samten Karmay more than twenty years ago.[6] Reviewing the documents that Karmay surveyed, along with certain key sources that have come to light more recently, shows how the argument came to entail a reckoning of scriptural authority with empirical evidence. In this way, akin to the history of the "radical" wing of the European enlightenment, the critical turn in *Four Treatises* scholarship made for a felicitous rapprochement, whereby key theological concerns were respected even though a rift between science and religion had clearly been suggested.[7] Even if such a rift never developed to an order of magnitude anywhere near that in the West, the fact that it was on the table at all marks an exceptional moment in Tibetan intellectual history.

Zurkharwa played the leading role in this history. In doing so, he has to have been one of the deftest debaters in all of Tibetan polemics. He both delivered the coup de grace by laying out the true origins of the *Four Treatises* and then quickly domesticated those origins in Buddhist ethical terms. In effect, he forged a way to have a text of great value that need not be dubbed sacred scripture (hard as that might be for a Tibetan audience, as Zurkharwa noted acerbically) or violate historical verities.

In many ways my own historiography follows Zurkharwa's lead, and often agrees with it. This is not only because of being able to relate to him as someone who takes history seriously and can mobilize a probative approach toward authorship claims and disparate evidence. Nor is it just in admiration of the ways he could draw upon Buddhist resources in innovative ways to serve his scientistic aims. I also appreciate his effort to articulate enduring moral values for the Buddhist world in which he lived. He has something to offer me as a scholar of religion. This and the following two chapters look at how Zurkharwa showed a way to honor the realities of religion, including its impact on bodily experience, without losing sight of another, scientific perspective that remains committed to empirical accountability and a critical assessment of how religious practice works.

The last chapter began to broach the episteme of Tibetan medicine through the lens of Desi Sangyé Gyatso's writings. This extended to its impact on the larger intellectual climate in the seventeenth-century capital, including the self-conception of the newly centralized government of the Fifth Dalai Lama. The fact that medicine created a separate history and literary canon for itself, not to mention a separate institutional base, suggested a certain distance between medicine and the epistemic grounds of Buddhist knowledge and authority in Tibet. In this and the following chapter, I look at a few debates in which such issues come to the fore. This requires going back into the history of medicine in Tibet and poring systematically through the early documents, in order to appreciate subtle shifts and their growing implications. This will culminate in the work of Zurkharwa, and his sometimes straw-dog, sometimes real interlocutors, quintessentially represented by Jangpa Tashi Pelzang.

Chapter 2 already studied the Desi's account of Zurkharwa's life, which, despite the scathing rhetoric, captures most of the available historical information. One of the Desi's main sources was the opening pages of Zurkharwa's own responses to the medical questions he had posted in Lhasa early in his career. In an essay he entitled *Old Man's Testament*, Zurkharwa begins with a review of his education. There is nothing in the way of biographical information beyond what is included in the Desi's sketch. But what is notable in *Old Man's Testament* is the way Zurkharwa looks back on his life, specifically his education and his colleagues, and his rhetorical style in representing it.

These pages readily demonstrate how much of the arresting irreverence in the Desi's writings was already alive and well a century and a half earlier. Zurkharwa may recount his own accomplishments with more sympathy than the Desi was willing to muster, but he is even more caustic than his haughty successor with respect to his own colleagues. He rails in detail about each of those who could not even understand his questions, let alone produce a decent response. Indeed, much of what he receives is the "vomit of conceit, and does little more than to make dogs happy."[8] He goes on and on like this. He directly addresses one of his responders, Nangso Dönyö, as "friend," telling him that his answers are good, better than the rest, and yet he has displayed no sign that he understood the questions. One of his best colleagues, the great scholar Kyempa Tsewang, gets through to Zurkharwa that the latter's earlier commentary on his own questions was itself hard to understand. Thus Zurkharwa is gradually convinced to write his own response.[9]

Much else of what we know of Zurkharwa's approach to medical learning anticipates that of the Desi too. Zurkharwa studied prodigiously in all of the medical traditions, gathering information eclectically for his own edification. He may have been the first of the medical scholars to post challenging questions on a public pillar in Lhasa, helping to create the agonistic attitude in learned medical circles upon which the Desi built his own career. He did his own antiquarian research, probing the same old medical histories that the Desi would later analyze. He produced the first blockprint of the *Four Treatises*, which the Desi would later endeavor to correct; although he had difficulty garnering patronage, that sponsorship illustrates the larger dependence of medical learning on power that would become so central by the Desi's time, in patterns closely similar to the patronage of Buddhist scholarship. Zurkharwa's life corresponds to the heyday of the Tsangpa strongmen Tseten Dorjé and his son Karma Tensung Wangpo, two generations before the final defeat of Karma Tenkyong Wangpo by the troops of the Fifth Dalai Lama. His projects were underwritten by a variety of powerful supporters, and he was closely aligned with the Karma Kagyü throughout his life. His connection with the waning Rinpung warlords hardly match the spectacular position that the Desi would achieve, but the overall picture exemplifies the conditions for many Jang and Zur luminaries of the period.

Zurkharwa's two seminal works on medicine, his *Ancestors' Advice* commentary on the *Four Treatises* and his history of Tibetan medicine, had enormous influence on Tibetan medicine's story of itself and its reading of the *Four Treatises*, not the least of which is to be seen in the Desi's wholesale appropriation of many passages in his own work. Their effects are still being felt in contemporary scholarship, inside Tibet today as well as further afield.[10] I will be drawing on them extensively in the next three chapters, following Zurkharwa in both his impertinent assessments of wrong ideas and his delicate negotiation of some thorny issues at the crossroads between medical and Buddhist traditions of knowledge.

Whereas the Desi singled out Zurkharwa for special invective, Zurkharwa's vociferous arguments often light on Jangpa Tashi Pelzang, who flourished in the second half of the fifteenth century, about eighty years prior to Zurkharwa's own career.[11] Justifiably or not, Zurkharwa used the Jangpa exegete's work to epitomize an approach to medical history and theory against which Zurkharwa defined himself in contrast. His frustration with Tashi Pelzang is most explicit in debates about pulse diagnostics, to be studied in the following chapters. But he critically addresses many of the points raised in Tashi Pelzang's own milestone essay on the Buddha Word issue too.

Tashi Pelzang's family line in the north of Dokam was traced by the Desi back to imperial times. He was the son of the outstanding scholar Mi Nyima Tongden, who was in turn a disciple of the Jangpa school's founder, Namgyel Drakzang. All three were closely related to the Jangpa nobility of Northern Latö, powerful rulers in the fifteenth century. Tashi Pelzang himself was involved with the Treasure-based lineage of Tibetan medicine.[12] Like Zurkharwa in the Zur lineage, Tashi Pelzang represented a brilliant younger generation that succeeded a learned and venerable founder of a school. The Desi speculates that Zurkharwa likely received his own Jangpa transmissions from a lineage stemming from Tashi Pelzang.[13] Regarding the debate about pulse and gender, Tashi Pelzang was far more subtle, innovative, and empirically grounded than Zurkharwa allows. Once again, Zurkharwa anticipates the moves of the Desi against himself, pillorying Tashi Pelzang on a trumped-up charge while appropriating his ideas. But elsewhere, and certainly on the issue examined in this chapter, Tashi Pelzang was extremely conservative. This is a side of Treasure tradition not often recognized.[14] Tashi Pelzang strained to maintain the Indic origins of the *Four Treatises* and what he took to be the literal meanings of the authoritative texts, sometimes at the expense of all common sense.

The debate on the Buddha Word issue that Zurkharwa represents between himself and such predecessors serves well to introduce the direction in which figures like him—and in the end, the Desi too—were leading Tibetan medicine as it came into its own. But the path along which the argument proceeds is not always direct. Authors go out on a limb and then immediately retreat and cover their tracks . . . and then venture back out again. Statements making almost the exact same point wind up, by virtue of very tiny shifts, at two diametrical ends of an argument. It is surprising how often these shifts occur, and how many times positions change.

THE FOUR TREATISES AS THE WORD OF THE BUDDHA

Whatever else one might want to say about the Buddha Word debate, it must first be noticed that the *Four Treatises* itself takes steps to insert itself into a Buddhist canonical context. In this it is to be distinguished from its Ayurvedic counterparts—and all other known full-service medical works in Asia—by framing itself as a teaching of the Medicine Buddha. While the Indic medical classics ultimately attribute Ayurvedic knowledge to deities like Brahmā and Indra, the texts themselves are clearly authored by humans.[15] No Indic or Chinese work dedicated to a comprehensive study of medicine presents itself as a teaching of the Buddha.[16]

There is at least one full-fledged medical text—covering a full range of anatomy, physiology, diagnostics, pharmacology, and therapeutics—that was available in Tibet at the time the *Four Treatises* was composed and is attributed to a Buddhist deity.[17] This is *Sman dpyad zla ba'i rgyal po* (a.k.a. *Somarāja*), which probably dates from the Yarlung period.[18] It presents itself as a teaching by Mañjuśrī delivered on Wutai Shan. But its scriptural genre apparatus is nowhere near as intricately developed as it is in the *Four Treatises.* The other early Tibetan medical works that are available, such as *Bi ji'i po ti kha ser*, offer prostrations and praise for a deity in their opening lines, but are presented as written by humans.[19] Even the *Four Treatises*' author's own earlier medical writings lack the genre-making frame of a Buddhist scripture.[20] Clearly the *Four Treatises* represents a new and deliberate effort to read as a teaching of the Buddha—on the part of its author or the disciples responsible for its initial propagation.[21] It bespeaks an intention to participate in a larger Tibetan literary milieu, that of the revered and successful Buddhist canonical scriptures. The *Four Treatises* jumped genres, as it were, in this joining other apocryphal works with high ambition, such as the Tibetan Treasure texts, which also framed themselves as the Word of the Buddha. They thereby gained legitimacy and authority. Zurkharwa himself will make this point about the *Four Treatises.*

The *Four Treatises*' appropriation of scriptural genre conventions starts directly after its initial lines of homage with the formulaic phrase, "Thus have I heard at one time." Although this line was tampered with in later versions, our earliest witnesses show the text opening its teachings in the classic and widely recognizable Buddhist revelatory fashion.[22] Then it moves directly into a point-for-point appropriation of what Tibetan scholars call the "five excellences." These are standard to the basic setting (*gleng gzhi*) of a work representing Buddha Word: place, teacher, time, audience, and teaching. Their excellence presages the excellence of the work itself.

The five excellences in the *Four Treatises*' basic setting begin with the place. The opening pages provide an unusually detailed description of where the teachings are occurring.[23] It is a palace in Tanaduk, a city of medicine. Tanaduk is adorned with wonderful medicinal plants and substances from the four mountains that are arrayed around its four cardinal directions. The Medicine Buddha, Bhaiṣajyaguruvaiḍūryaprabha, who is residing there is mentioned several pages later. He is sitting on a throne in the palace, surrounded by his retinue. Then he enters deep meditative absorption. A light emerges from his heart. After it purifies the illnesses of all beings, it emanates the manifested teacher Sage Intelligent Gnosis, who hovers in the sky and speaks to the other sages who are present, exhorting them to study medicine for the benefit of all. Then the Buddha emanates healing lights from his tongue, and a manifestation of his speech, Sage Mind-Born, appears. Mind-Born beseeches Intelligent Gnosis for those medical teachings. And so Intelligent Gnosis starts to teach. As a sign of how well this basic setting has been integrated into the work, all 156 chapters of the *Four Treatises* open and close by reminding us of this scene, and cast the entire teaching as a dialogue between the master, Sage Intelligent Gnosis, and his interlocutor, Sage Mind-Born.

These familiar signs of a Buddhist canonical text decidedly place the work in the category of Buddha Word. Although some early medical scholars were concerned about this, there is nothing unusual about its being taught by a buddha other than Śākyamuni, i.e., the Medicine Buddha. Nor is it odd for him to manifest or empower someone else to actually preach the teaching: the same pattern can be observed in many Buddhist canonical sūtras and tantras. And as for the rest of the five excellences, there is long precedent in the canon for Buddha Word to have been preached in many places, for many audiences, at many times, and with special contents tailored for those circumstances. Most of all, the very rehearsal of the excellences itself, setting the stage in the familiar way, along with other common sūtraic elements such as the Buddha entering meditation and emanating light rays, all would have given a clear signal to readers that the *Four Treatises* is indeed Buddha Word.

That said, discrepancies in the work's own portrayal of this primal scene are readily evident. Quite revealing is the *Four Treatises*' audience. It consists of four groups, including not only "insiders," i.e., Buddhists,[24] a common Tibetan rubric that here refers to the bodhisattvas Mañjuśrī, Avalokiteśvara, and Vajrapāṇi; Ānanda, the famous disciple who recited all of the Buddha's teachings after his death; and Jivaka, the Buddha's physician. The other three groups are the gods, sages, and *tīrthikas* (non-Buddhist Indian teachers). Some of the sages (Skt. *ṛṣi*) are central figures of mainstream Ayurvedic tradition, such as Ātreya and Agniveśa. The gods include Brahmā and others, also cited in Ayurvedic works. Although it is common for Buddhist canonical works to have diverse audiences, this one seems tailored specifically to reference the broad Indic medical tradition upon which the *Four Treatises* actually draws.

A second statement near the end of the *Four Treatises* goes even further. This time, a crack in the entire mythic edifice appears when reference is made to historical, geopolitical entities. It is an odd passage, coming at the end of the penultimate chapter and immediately following a survey of various therapies. Suddenly the frame story is invoked again, and Mind-Born asks the work's expounder whether there is anything in medical science that is not included in the *Four Treatises*. The answer is a list of medical teachings that manifestations of the Buddha have taught for the benefit of beings in varying contexts and with varying needs. In India they have taught the use of medicines; in China, moxibustion and purgatives; in Dölpo, bloodletting; in Tibet, pulse and urine diagnostics; for the gods, the text *Gso dpyad 'bum pa;* for the sages, the *Caraka* in eight sections; for the *tīrthikas*, the *Black Īśvara Tantra*; and for the insiders—i.e., Buddhists—teachings related to the three bodhisattva protectors.[25] All of these teachings are included in the *Four Treatises*, the text goes on to aver; there is no other medical knowledge that is not included here.[26]

Setting aside the documentary value of this overview of medicine—not to mention the hubris in making the Buddha the ultimate author of the *Caraka* and teacher of Chinese medicine—the text seems to be communicating something about the origins of its own contents. The statement would seem to represent an admission of its disparate sources. These extend not only into Buddhist scriptural passages but also the South Asian mythological world of gods, *tīrthikas*, and sages, as well as across specific areas in the rest of Asia, even mentioning the Himalayan region of Dölpo for some reason. The passage readily appropriates the long-standing Buddhist hermeneutic of the Buddha's multivalent skillful means in order to make the work ultimately the Buddha's teaching, but at the same time to account for the diversity of knowledge systems upon which its author drew.

In this book I am following Zurkharwa and like-minded Tibetan commentators in assuming that the *Four Treatises* was indeed compiled from a variety of medical traditions, and composed, originally in the Tibetan language, by Yutok Yönten Gönpo ("the Younger") and his circle in the twelfth century. As evidence that there was some hesitation in locating the *Four Treatises* entirely within the literary universe of Buddhist scripture, this late passage in the work indicates that someone felt compelled to own up to the role of Ayurvedic and other sources of medical knowledge outside of the Buddha's dispensation as well.

Despite these signs of ambivalence—and not to mention other telltale signs of the work's true provenance, which Zurkharwa's predecessors will dryly point out—there was widespread acceptance of the work's own claims. Tibetan Buddhists, at least in the Old School to which many of the early propagators and commentators on the *Four Treatises* belonged, were used to issues around apocryphal scriptures. The *Four Treatises* partook in many of the strategies of the Old School's Treasure tradition, which was just getting under way during this period and also produced native Tibetan works that were cast as Buddha Word and purported to be translations from Indic texts. And that is quite apart from the use of the Treasure mode to solve some of the work's transmission problems, which were noticed as soon as it appeared.[27]

EARLY RECEPTION

Despite the *Four Treatises*' own devices in casting itself as Buddha Word, the claim still needed bolstering. One sure sign of an ongoing defensiveness is found in the early historical account called *Soaring Garuda*, probably written by a disciple of Yutok Yönten Gönpo.[28] In the last chapter, we saw this early example of medical historiography work singled out by the Desi for its virtues. In fact the entire work is dedicated to establishing the *Four Treatises* as Buddha Word—which no doubt explains both the Desi's approval and Zurkharwa's withering critique of it.[29]

The polemical intent of *Soaring Garuda* is clear enough in its closing statement, where it refers to itself as a response to objections and as a means to reject criticisms (*rtsod zlog*), a genre term repeated in later apologies. *Soaring Garuda* provides evidence that the debate was already in progress close to the time of the *Four Treatises*' creation. Everything it discusses, from the nature of the five sciences to the Buddhist scriptures that touch on medical topics, sets the stage for placing the *Four Treatises* within the category of the Buddha's teachings. It also provides a typology of three kinds of Buddha Word: that which was "actually spoken" by the mouth of the Buddha; that which has the express "blessing" or inspiration of the Buddha; and that which was preached by someone else who had the "permission" of the Buddha to utter something that would serve as Buddha Word.[30] Drawing on a long tradition in Buddhist scriptural history whereby types of Buddha-inspired speech (Skt. *pratibhāna*) pronounced by figures other than the Buddha himself were still presented as Buddha Word, this tripartite scheme would be invoked repeatedly in the coming centuries.[31] *Soaring Garuda* identifies the *Four Treatises* as the second kind of Word, that which has the Buddha's blessing, on the logic that the Medicine Buddha entered meditation and then created Intelligent Gnosis and his interlocutor Mind-Born for the purposes of delivering the teaching.[32] Otherwise, *Soaring Garuda* avers, since Intelligent Gnosis is a manifestation, or emanation (*sprul pa;* Skt. *nirmāṇa*), of the Buddha, the *Four Treatises* could go into the category of that "actually spoken" by the Buddha. It also goes out of its way to deny that the *Four Treatises* is instead a *śāstra*, a far more general category that in Indo-Tibetan Buddhism denotes a composition by a teacher other than the Buddha.[33] Apparently someone else had already contended that this was what the *Four Treatises* really was. *Soaring Garuda* evinces anxiety about such a claim.

While this early attempt to reject criticism argues from the perspective of the original preaching scene, another work contemporaneous to *Soaring Garuda* takes up the *Four Treatises*' propagation in Tibet. This is the *Crucial Lineage Biography*, also produced by the generation immediately succeeding Yutok, most likely his chief disciple, Sumtön Yeshe Zung, and his student Zhönnu Yeshé.[34] It forms a pair with *Soaring Garuda,* in a manner similar to the legitimating apparatus of the Buddhist Treasure tradition. *Soaring Garuda* is like an "origin account" in that it tells the mythic circumstances of a work's original preaching. The *Crucial Lineage Biography,* in contrast, works like a "revelation account," relating the events by which the work came to be transmitted in the world of humans.[35] This is one more example of medical literature mirroring patterns being established in the Buddhist Treasure literature during the same period.

The *Crucial Lineage Biography* simply assumes that the *Four Treatises* is a teaching of the Buddha Bhaiṣajyaguru, although the details are interesting. It specifies that the medical city Tanaduk, in which the original *Four Treatises* teaching took place, is in Oḍḍiyāna, the famous land to the southwest of Tibet that is closely associated with Padmasambhava, the founder of Tibetan tantric Buddhism and the principal teacher of the Treasures. Again, the introduction of Oḍḍiyāna strongly suggests Old School lore. But most telling is the wholesale adaptation of Treasure tradition in the story that is supplied about how the *Four Treatises* was transmitted after its original preaching. The text was passed down to a variety of Indian figures, through two principal lineages: one from Ātreya, the other from Jīvaka (the physician to the Buddha in many Buddhist sources). The latter's line eventually extended to Candranandana, the Indian author of the principal commentary to the *Aṣṭāṅgahṛdayasaṃhitā*. Finally, we get the story of the *Four Treatises*' transmission in Tibet as a Treasure text; this would be repeated many times in medical historiography. The work was obtained by the eighth-century Tibetan master Vairocana (well known more generally to the Treasure tradition), who brought it to the Tibetan king Tri Songdetsen. The latter hid it in a pillar at the first Tibetan Buddhist monastery, Samyé. After three days for the gods of the desire realm (very long in human time),[36] the work was rediscovered as a Treasure by Drapa Ngönshé (a historical figure from the eleventh century) and soon passed down to Yutok Yönten Gönpo.

A key to what else the *Crucial Lineage Biography* is up to comes next. It states that Yutok judiciously passed the *Four Treatises* to a single disciple, the first-person narrator of most of the *Crucial Lineage Biography*, [Sumtön] Yeshé Zung.[37] There are repeated injunctions not to share the *Four Treatises* with others, and repeated reference to the fame that will be attained by those who possess it. The self-serving agenda is pretty thinly disguised. Yeshé Zung in turn passed it down to his disciple Zhönnu Yeshé, who also speaks in the first person and whose name appears again in the colophon to the work.[38] A premier concern of both authorial voices is to legitimate themselves as the bona fide inheritors of the *Four Treatises*, to the exclusion of others. Yeshé Zung speaks in detail of his own education and practice and then his own writing and note-taking. He even equates his teacher, Yutok, with the very expositor of the *Four Treatises*, Sage Intelligent Gnosis. And then he equates himself with the *Four Treatises*' interlocutor, Sage Mind-Born. But his rhetoric is interesting: he actually only says that Yutok Gönpo is *like* Intelligent Gnosis, and that he himself is Mind-Born, *he thinks*.[39] There is a bit of hesitancy about the apotheosis the author is pursuing for himself and his teacher.

Two things to note from this short but pithy work: first, it seems to be our earliest source that provides a Treasure narrative for the *Four Treatises*' transmission in Tibet. Importantly, Treasure transmission allows a historical hand (actually two—the discoverer, Drapa Ngönshé, and then the putative codifier, Yutok) in the work's formation while still making the original author the Medicine Buddha, and the work itself an Indic composition that had to be translated into Tibetan. If there is anything to my speculation that the formulators of the *Four Treatises* were anxious about the fiction they had created regarding its authorship, they might have assuaged their unease by acknowledging part of the fact of the matter while still preserving the prestige and authority of Buddhist and Indic origins. In addition, the Treasure narrative handily allows the *Four Treatises* to appear publicly in the Tibetan world relatively close to the time that it actually did, in the twelfth century. Historical common sense is not violated. Indeed, as Zurkharwa complains later, if the *Four Treatises* came to Tibet from India during the imperial period, why is there no evidence of its presence in Tibet before Yutok? A good start in response would be that the text was in hiding at Samyé. But more on this later.

The second noteworthy realization from reading the *Crucial Lineage Biography* is that to lift up the *Four Treatises* as originally the teaching of the Buddha would also lift up the *Crucial Lineage Biography*'s narrators, Yeshé Zung and his disciple, twelfth- to thirteenth-century medical practitioners who seem bent upon constituting an exclusive "lineage of one." This may have a lot to do with the original casting of the *Four Treatises* as Buddha Word in the first place. The *Crucial Lineage Biography* bears striking witness to the means by which the early possessors of this medical treatise managed to create a buzz around it (in contrast to other medical treatises in circulation in Tibet at the time, which had no such enlightened frame story) while still keeping the text to themselves. *The Crucial Lineage Biography* is a reminder that Tibetan medicine was a highly competitive field.

One more important early account also mentions the origin of the *Four Treatises*, but complicates the picture in a surprising way. This is the *Heart Sphere of Yutok Story.*[40] It is part of the *Heart Sphere of Yutok* cycle, the set of tantric teachings associated with Yutok's legacy mentioned in the previous chapter. The cycle itself is closely aligned with other Old School healing practices, and its rituals are not of immediate concern here.[41] However, the work that presents the "story" of the cycle is very pertinent.

The *Heart Sphere of Yutok Story* has some relation to the *Crucial Lineage Biography*. It is not really much about the *Heart Sphere of Yutok* as such, but rather focuses on Yutok Yönten Gönpo's teachings and life. It too is bent on bolstering its author, the same Sumtön Yeshé Zung, as the sole inheritor of the *Four Treatises*.[42] Sumtön speaks frequently in the first person, naming himself explicitly. Also like the *Crucial Lineage*, the *Heart Sphere of Yutok Story* incorporates some possibly genuine autobiographical material from Yutok's own hand.[43] But what is most important about this work, and would be duly noted by the sharpest of the later medical commentators, is the way that Yutok is glorified as an exalted conveyor of Buddhist teachings, in statements cast in both the first and third person. Critically, he is even on occasion referred to as a "manifestation."[44] The use of this ambiguous epithet here probably just reflects the growing tendency in Tibet to recognize great teachers as manifestations of the Buddha. Indeed, the text also makes it clear that despite his high accomplishments Yutok was not actually a buddha as such.[45] The Desi nonetheless would later invoke this early epithet to bolster his contention that the *Four Treatises* is really the Word of the Buddha, playing on the slippage that the preacher of the work, Intelligent Gnosis, was himself pointedly called a manifestation of the Buddha. But a century before the Desi, Zurkharwa already picked out this same statement in the *Heart Sphere of Yutok Story* precisely to disaggregate such an elevation of Yutok's virtues from any apotheosis as a buddha. In this Zurkharwa was a more faithful reader of the early document, given that its own narrative already seems largely to assent to Yutok's human authorship of the *Four Treatises*. Indeed—and this is very significant—in the midst of giving a long account of Yutok's career, a statement put in the mouth of Yutok himself explicitly uses the verb "composed" (*brtsams*) to describe what Yutok did with respect to the creation of the *Four Treatises*.

That telling verb is astounding. It is cited pointedly by Zurkharwa and the others who questioned the *Four Treatises*' Buddha Word status. And in the same breath, the *Heart Sphere of Yutok Story* also describes the *Four Treatises* as a medical *śāstra*.[46] It characterizes Yutok's authority to compose the work as based on the permission granted to him by the *yidam*, or "tutelary" deities. "Permission" names one of the three kinds of Buddha Word, but it does not seem that the *Heart Sphere of Yutok Story* is using the word here to signify that. Far more revealing is the fact that it refers several times to the *Four Treatises* as a *śāstra*, a very loaded term that precisely contrasts with Buddha Word.[47] Indeed, the very same key sentence that uses the verb "compose" and calls the work a *śāstra* glosses the *Four Treatises* further as "having not even the slightest difference from the blessings of the Secret Mantra unsurpassable tantras."[48] The fact that this needs to be said already indicates that the work is actually *not* one of those tantras, i.e., Buddha Word as such.

The *Heart Sphere of Yutok Story* is nonetheless overflowing in its praise for Yutok. Yutok attains the two kinds of powers; then, based on the prophecy and permission that he receives from the deities, he composes the *Four Treatises*, which is no different from the highest Buddhist tantras; as soon as he is finished, he receives offerings and praise from the bodhisattvas; then there are more portents of the work's future destiny and an exhortation to the reader to hold and read it in order to reach nirvana. The entire scene it paints is redolent of what is so familiar in Mahāyāna Buddha Word scriptures that proceed through the Buddha's blessings or permission, as when Śākyamuni praises Avalokiteśvara for his utterance of the *Heart Sūtra*. And yet the narrative also makes clear that the *Four Treatises* is not actually Word: it is *śāstra*.

It is puzzling that the same person, Sumtön Yeshé Zung, argues for the Buddha Word status of the *Four Treatises* in one work, while in another, otherwise overlapping account he lets slip that Yutok composed it, and that it is a *śāstra*. This may represent a shift on the part of Yeshé Zung. In the course of writing the *Heart Sphere of Yutok Story*, he goes over the top in his praise and respect for the *Four Treatises*; he also repeatedly makes the strong case that the "True Dharma" of the Buddha and medicine should be seen as thoroughly integrated and stridently attacks those who try to separate them.[49] Still, he stops short of making it Buddha Word. And yet in another moment of increased boldness, i.e., on the occasion of writing the *Crucial Lineage Biography*, he does confer that highest status to the *Four Treatises* and also places it into the Treasure tradition.[50] The discrepancy is confusing to Tibetan scholars too, and some posit that these two seminal accounts were written by different authors with the same name.[51]

In the end, Yeshé Zung's inconsistency is part and parcel of a tug of war already under way. In the narratives being offered for the *Four Treatises*' genesis, heavy machinery brought in from Buddhist theories of enlightenment and enlightened preaching, along with the transmission wizardry supplied by the Treasure narrative, would have obviated any historical discrepancies apparent to some. Once attention turns to the historical personage of Yutok Yönten Gönpo, however, there is some ambivalence, although he is in any event elevated to high spiritual standing in Buddhist terms. That can make him either a true emanation of the Buddha or merely a latter-day follower who composed the work, albeit with the most enlightened intentions. The difference is slender, but much significance attends which way the attribution tips.

COMMENTARIAL THINKING, HISTORICAL THINKING

Commentaries on the *Four Treatises*' medical teachings also appeared soon after the work's codification. At first the Buddha Word question seems to have been muted, and either no note was taken of it or the matter was dispensed with swiftly. One early commentary invokes the Word/*śāstra* distinction and labels the *Four Treatises* as Word by virtue of a blessing.[52] But it moves through these topics quickly and seems to be repeating an established position rather than arguing anything new. Several other early commentaries simply gloss the opening elements of the *Four Treatises*' basic setting, noting that the title provided at the beginning of the work indicates that it was translated from an Indian language.[53]

But in assessing commentarial polemics, we need to keep in mind how the very nature of commentary affects the way the root text's claims are received. The root text commences with the Medicine Buddha empowering Sage Intelligent Gnosis to preach; the commentaries must start with that basic setting too. It is hard to imagine a commentary—at least in the Indo-Tibetan literary world—that would call into question the very genre features that make a work worthy of a commentary in the first place. Johannes Bronkhorst has gone so far as to wonder if commentarial practice placed a severe restriction on scientific innovation and imagination, at least in South Asia.[54] Even the powerful Desi was circumspect in the commentarial arena. Certainly many medical commentators simply accepted their root text's claims, especially since the main concern for many would have been to get beyond the introductory sections to the medical knowledge in the work. Even when the larger intellectual and perhaps political implications of the *Four Treatises*' status came into focus, the boldest of the commentators—Zurkharwa himself—still had to respect the opening lines of the original text.

One of the earliest commentaries we have that provides substantive comment on the opening passage of the *Four Treatises* was written by the intriguing fourteenth-century medical historian Drangti Pelden Tsoché, mentioned in the last chapter.[55] Drangti also implicitly accepts the *Four Treatises*' status and basic setting, but he betrays awareness of arguments already in progress. He seems to have been the first to weigh different options regarding something that would later become a key point of debate, namely the medical city Tanaduk, site of the *Four Treatises*' preaching. The nature and location of this place seem to have been a concern even for the root text itself, which addressed the topic in unusual detail. Drangti takes up the question of where this city might be and goes through a number of opinions. One is Oḍḍiyāna, also the claim of the *Crucial Lineage Biography*; another is Mount Sumeru; another is a paradise mentioned in the *Avataṃsakasūtra*.[56] Drangti ends up placing Tanaduk in Akaniṣṭha, an important paradise for Tibetan Buddhism, and staunchly defends this conclusion.[57]

Tanaduk was probably a problem because it is not associated with medical teachings in any Indic work.[58] The name itself (Skt. Sudarśana) is known, however, as one of the golden mountain ranges around Sumeru in Abhidharma cosmology.[59] We will see how this curious issue about the location of Tanaduk *qua* site of the *Four Treatises*' preaching sets the stage for the real stripes of the medical mentality to show in Zurkharwa's treatment. For now we can note that Drangti's earlier solution, to locate it in a classical buddha field, is no surprise, since he is clearly a strong supporter of the Buddha Word position and the *Four Treatises*' Treasure transmission in his more famous work, his history of medicine.[60] Although Drangti gives a lot of valuable historical detail about medicine in Tibet, when it comes to the *Four Treatises* he seems uninterested in taking issue with the work's status. He contrasts the category of Buddha Word with *śāstra* but does not acknowledge that anyone has actually argued that the *Four Treatises* itself might be *śāstra*.[61] Such a move will be used by later authors too, and sometimes becomes a strategy that allows key points to be scored without admitting that anything is amiss.

Another *Four Treatises* commentator added further nuance to the Buddha Word position. This was Namgyel Drakzang (1395–1475), founder of the Jangpa school of Tibetan medicine.[62] He specifies that the *Four Treatises* is the kind of Buddha Word that is "spoken by a manifestation," rather than "actually" by the Buddha.[63] He is referring to Intelligent Gnosis, the preacher of the *Four Treatises*; there would not be a move to capitalize on Yutok's own identification as a "manifestation" for another two centuries. Namgyel Drakzang specifies that the *Four Treatises* is sūtra, and belongs specifically in the *gāthā* section of the Buddhist canon. He is clearly aware of a variety of debates on the whole issue, and summarizes them on careful and scholarly grounds. While some say the *Four Treatises* is not a scripture since it was not spoken directly by the Buddha, Namgyel Drakzang observes that such a standard would mean that many other scriptures would also not be Buddha Word, including the exalted *Hevajratantra*.[64] He also argues convincingly that the Buddha emanated countless manifestations to help him in his work for sentient beings, so the fact that the work was actually uttered by a manifestation hardly disqualifies the *Four Treatises* from being Buddha Word.

Namgyel Drakzang knows of a position claiming that the *Four Treatises* was actually created in Tibet, although there is no such statement in any of the early works currently available to us. He is not motivated to take it up and simply dismisses it with a vague gesture to reigning consensus: many learned people believe that Vairocana translated the text into Tibetan from an Indian language.[65] It would only be Namgyel Drakzang's successor, Tashi Pelzang, who would betray the extent to which a critical and empiricist argument was in full swing—which the conservative Jangpa exegetes were endeavoring to forestall.[66]

Tashi Pelzang is solidly in his predecessor's camp but is far more polemical. Something has shifted in the landscape. We do know that a momentous liberty had been taken at some moment prior to when Tashi Pelzang wrote a free-standing essay dedicated to proving the *Four Treatises*' status as Buddha Word.[67] The liberty involved but a single word, but it strongly suggests that some concerns had emerged in the medical community's view of the *Four Treatises.* To wit, the emblematic line "Thus have I heard at one time" had been shifted to read instead "Thus have I explained at one time," at least in some versions.[68] The new phrase is a rather glaring sign that someone wanted to communicate something special about the authorship of the *Four Treatises.* To say "explained" would be an acknowledgment that the work is not, as the standard formula would have it, a record of a teaching of the Buddha by a mnemically gifted scribe such as Ānanda, but rather is something that the purported scribe *himself* had originally "explained." Canonical Buddhist scriptures can either name an interlocutor/scribe or, more commonly, simply remain silent on who committed their content to writing, but the content itself is still cast as a firsthand recounting of a buddha's teaching in oral form. In contrast, the new opening phrase for the *Four Treatises* would appear to collapse scribe and teacher/author, and thus perhaps not be Buddha Word at all but rather a mere worldly composition by the work's scribe.[69] Although Zurkharwa would later argue that it doesn't make much of a difference which verb is used, Tashi Pelzang's vociferous defense of the *Four Treatises*' canonical status shows his cognizance of formidable opponents to his conservative views.

Samten Karmay calls Tashi Pelzang's arguments "narrow and dogmatic and most of the time very naïve." Harsh words from a modern scholar, but they are probably deserved.[70] One's position on the *Four Treatises*' status maps onto one's willingness to take into account empirical evidence, even if it contradicts received tradition.[71] And clearly, Tashi Pelzang was not uncomfortable ignoring empirical evidence.

Tashi Pelzang's complexly argued work is a fascinating artifact, and deserves more detailed study than is possible here. Among other things, he may have been the first to name as such the problem of whether there was actually ever an Indian text of the *Four Treatises*, which at least suggests a willingness to consider questions of historical evidence.[72] But Tashi Pelzang dodges the implications. Instead, he reviews the entire nature of the translation of Indic Buddhist scriptures into Tibetan and then lays out the Treasure version of the *Four Treatises*' later transmission.[73] He claims that it was due to jealousy of Sumtön Yeshé Zung's singular inheritance of the work that certain individuals first began to claim that the *Four Treatises* were really written by Yutok himself.[74] This proposal confirms the image of Sumtön as embattled, but it turns on its head the idea that Sumtön felt vulnerable about the Buddha Word claim. Tashi Pelzang submits instead that such doubts were the work of disgruntled colleagues.

Although we are presently missing most of the works that put forward the arguments against Buddha Word status to which Tashi Pelzang is responding, those who made this point, whose names were later supplied by Zurkharwa, were not specialists in medical theory or history but rather outstanding scholars of Buddhist practice and thought. They include the prolific scholar Bodong Panchen Choklé Namgyel (1375–1451);[75] the Sakyapa scholar-saint Taktsang Lotsawa Sherap Rinchen (1405–?), who did author, among many other things, a survey of medical practices;[76] the brilliant and critical Sakyapa exegete Shakya Chokden (1428–1507);[77] and one Dzagön Gyawo. Zurkharwa also mentions the great historian Butön Rinchen Drup (1290–1364) as someone who considered the matter but did not come to a decision.[78] The list in itself illustrates how widely the history of medicine was being discussed, and how widely (that is, at least among intellectuals) critical and historicist criteria were being brought to bear. A brief statement that survives in a letter by Shakya Chokden—written to Namgyel Drakzang, in fact—illustrates how far the thinking had come.[79] Shakya Chokden does not raise the specter of Tibetan authorship but merely focuses on the impossibility that the text was taught by the Buddha. He is well aware of Indic Ayurvedic tradition and the fact that its teachings are attributed ultimately to the god Brahmā.[80] He is savvy enough to ask why, if Brahmā and the Buddha are synonymous, and the *Four Treatises* and Ayurveda were originally preached by the Buddha, the great Ayurvedic commentary *Aṣṭāṅgahṛdayasaṃhitā* does not cite the *Four Treatises*. Nor does the *Aṣṭāṅga* mention the Buddha in its history of medical teachings. Shakya Chokden also points out that if the Buddha taught the four Vedas (including Ayurveda), such a scene would have preceded his turning of the wheel of the Buddhist Dharma, which he finds implausible. It would mean that the complex Ayurvedic *śāstras* such as the *Aṣṭāṅga* would have preceded the clear and easily understandable Buddha Word scriptures, reversing the ordinary sequence of things whereby the lucid teachings come first. He even notes that the attribution of non-Buddhist medical teachings to the Buddha destroys any distinction between what is Buddhist and what is not (using again the monikers "insider" and "outsider" to denote such a distinction). Such powerful arguments with clear investment in textual history and coherent sequentiality would have posed an estimable challenge to those who sought to defend the *Four Treatises* as Buddha Word.

Tashi Pelzang does not name his interlocutors in his defense of the *Four Treatises*' status, but does work through key arguments that various doubters have raised.[81] One takes up the unconventionality of the opening line of the *Four Treatises* that reads, at least in his version, "Thus have I explained at one time." Tashi Pelzang plausibly responds that this anomalous opening has to do with the ambiguity of identity between the teacher of the work and its compiler (both Intelligent Gnosis and Mind-Born are manifestations of the Buddha), and adds that a similar ambiguity obtains in other canonical works too.[82]

We begin to see Tashi Pelzang's approach to inconvenient empirical facts in his response to another critical point raised by some: that "there is no certainty that the city called Tanaduk exists here."[83] "Here" would seem to refer to the world in which we live. The main force of the question has to do with the problem that the city is nowhere to be seen. In his answer Tashi Pelzang adverts instead to the magical power of the Buddha, adding that many places mentioned in the sūtras are not visible, especially to those with bad karma, and especially in these degenerate times. Tashi Pelzang says that Tanaduk was a city that was magically produced by the Buddha at Varanasi in India, and that buddhas regularly produce things magically. He closes his response by insisting, "just because you don't see it does not mean that it does not exist." Thus would he take the steam out of any argument based on eyewitness.

What has shifted most, visible in many of the rest of the points against which Tashi Pelzang attempts to defend, is the move into Tibetan specificities alongside the strictly Buddhological issues. This signals a growing sense among Tashi Pelzang's interlocutors that there are actually two issues at stake: not only is the *Four Treatises* not Buddha Word, it wasn't even translated from an Indian language at all and is actually a Tibetan composition. For example, Tashi Pelzang's sharp-eyed opponents compare the medical system in the *Four Treatises* to that found in the *Aṣṭāṅga* and note that the main "eight sections" in the Indic texts and the *Four Treatises* are different.[84] For his response, Tashi Pelzang is shrewd enough again to go back to his Buddhological repertoire, maintaining for example that the Buddhist sūtra *Suvarṇaprabhāsottama* also has a different list of these sections, and asking sarcastically if that then means that the *Suvarṇa* was composed in Tibet too.[85]

But in several of the Tibet-focused arguments, Tashi Pelzang cannot always take recourse to the Buddha's infinite flexibility. Sometimes he must appeal instead to a strategy on the part of the imputed translators. For example, when a tellingly Tibetan feature of the *Four Treatises* is raised—namely, that the work mentions the Bön religion—his response is to suggest that the translators compassionately rendered certain non-Buddhist (again, "outsider") rites that appear in the Indian version as Bönpo rites, so that Tibetans could have access to such worldly rituals, including those connected with gods like Brahmā and Indra, when they needed them. What is more, such means are used in many of the Buddha's teachings, such as doctrines of the *tīrthikas* in the *Kālacakra*.[86] Or again, on the point that the *Four Treatises* betrays its Tibetan origins when it mentions tea, since it is known that there is no tea in India, Tashi Pelzang floats the idea that the translators rendered what was *paṇ* (i.e., *pān,* betel leaves) in the original Indic text as the Tibetan for tea (*ja*), judiciously deciding that it was the best analogue to *pān* in the Tibetan context.[87]

Sometimes he is even forced to admit a possibility that is at odds with his own convictions. When in another, very clever salvo the opponent points out that the system seen in the *Four Treatises*, wherein vitality resides in the head and the pervading wind resides in the heart, contradicts what is found in the *Kālacakra* and other tantras, and therefore proves that the *Four Treatises* was composed in Tibet, land of ignoramuses, Tashi Pelzang is reduced to averring that actually it is rare to find any Buddha scripture that has no errors.[88] In another case, his refusal to acknowledge the evidence is starting to wear thin. An opponent shrewdly points out that the system of five elements found in the *Four Treatises*' pulse system is aligned with Chinese "black divination," which would again make it unlikely to have been composed in

India.[89] Tashi Pelzang can only invoke the compassion and magical power of the Buddha again, wanly pointing out that the Buddha is omniscient and knows all systems, and so of course will use whatever will help all beings.[90]

Worse yet, in response to the astute observation that since the *Four Treatises*' diagnostic systems include images of a supine tortoise, the work is indebted to Tibetan conceptions not known in other Buddha Word, Tashi Pelzang asserts categorically, without providing any evidence, that the supine tortoise can be found in both Buddha Word and *śāstra*.[91] Or again, when he responds to a point that since the text refers to porcelain and porcelain is not known in India, it can't be an Indian work, he pronounces that indeed porcelain actually appeared first in India.[92] And in response to the observation that would be the delight of later critics, namely that the *Four Treatises*' reference to the quintessential Tibetan food *tsam pa* makes it pretty likely that the work was composed in Tibet, Tashi Pelzang simply ducks the point, challenging his opponent instead: Is it also the case that because the *Four Treatises* mentions tea, it was composed in China? Or since it advises the consumption of beer and meat on occasion, that it was written only for rich people?[93]

In the end, his defense comes down to either invoking the omniscience of the Buddha that anticipates the needs of all sentient beings, ignoring the evidence and claiming an expedient translation/adjustment to fit the Tibetan situation, or simply denying the issue outright. His approach helps identify a critical distinction. From the buddhalogical perspective, it does usually work to cite the Buddha's omniscience and compassion to smooth over all evident incommensurabilities.[94] In this way, Buddhist hermeneutics introduces a wild card: anything can happen. But Tashi Pelzang's opponents are not interested in such a solution. They are more interested in establishing the reality of their root text's origins on empirically verifiable grounds.

Note that in invoking empirical realities rather than the Buddha's infinite powers, these opponents don't have a wild card of their own. But what is telling for our purposes is the kind of resources they bring to bear instead. The arguments about tea and porcelain and calendars and *tsam pa* and the Bönpos all bespeak an astute sensibility about cultural and even climatic difference. They are about historical and regional specificity, not at all about a universalist dispensation, as in the common Buddhist notion of "all sentient beings" or the like. Most of all, they are about Tibetan particularities—which is most appropriate, since they are trying to demonstrate the Tibetan provenance of the work—and here I would note again the shift from the issue of Buddha Word to country of origin. We can begin to discern in Tashi Pelzang's putative opponents indications of a scientistic mentality that values observation of local realities over received tradition.

As we continue to track this growing mentality, we should also note how the field in which the medical Buddha Word debate played differed significantly from the one in which the Treasure scriptures were defended as Buddha Word, at least in one important respect. The critics of the Buddha Word status of the *Four Treatises* not only came from outside its own lineage and tradition, as they did in the Treasure case; some were also fellow medical theorists, fellow commentators on the text. Strikingly, those medical scholars did not for that reason reject the *Four Treatises* as wrong or not valuable. Not being Buddha Word was starting not to be an intolerable flaw.

EXPLICIT DISSENT

So far we have noticed some defensiveness or muted discomfort among those who positioned the *Four Treatises* as Buddha Word, and some probing criticisms of that position on empirical grounds, albeit reported mostly secondhand. A century after Tashi Pelzang, two outstanding *Four Treatises* commentators wrote long and considered firsthand statements arguing that the *Four Treatises* is *not* actually Buddha Word, and is instead a Tibetan composition. This allows a direct look at medical writers making a case for the importance of historical verity. They are far less blunt than what Tashi Pelzang represents. Indeed, they have to make their case very cautiously, carefully preserving their own Buddhist sensibilities and loyalties. Nonetheless, they begin to lay out a different set of criteria for credibility—a key but vague factor, as Steven Shapin has shown, in all scientific discourse—than what was assumed so widely in the intellectual history of Tibetan Buddhism, namely, that the Buddha's revelation has ultimate epistemic authority over all knowledge.[95]

Kyempa Tsewang and Zurkharwa Lodrö Gyelpo were contemporaries, and both were part of the Zur medical lineage. They knew each other and had some interchange; Kyempa was the elder.[96] They represent a moment of second-order awareness of the implications of the challenge to Buddhist authority in the *Four Treatises* Buddha Word debate, a growing sense that there might be more than one system of knowledge and all truth need not be grounded in the Word of the Buddha.

At least three separate discussions of the matter by Zurkharwa are extant, written over the course of many years, and very complex in their rhetoric. We might look at his elder Kyempa's statement first. While Kyempa is not quite as bold or detailed, he decisively sets the stage for his younger colleague's salvos.

Unlike many other medical writers, Kyempa wrote a commentary on the entire *Four Treatises*, and it has fortunately survived.[97] This fine work has only recently been recognized by modern scholarship. It was certainly known to Tibetan academic medicine in the centuries after it was written, but it seems never to have been published in blockprint form.

It is a measure of how far the *Four Treatises* expository tradition had come that Kyempa could venture into intellectual issues quite beyond the business of commenting on the words in the root text as such. This is especially evident in the first section of the work, "an explication on the general meaning of the text," which amounts to a full essay in three chapters on the relationship of medicine and the Buddha's teachings. Given the delicacy of the issue, Kyempa seems aware of all he needs to do to contextualize what he is going to say, and he draws freely upon advances made in other parts of the Tibetan intellectual world.

Kyempa backs way up to get the big picture. The first chapter reminds the reader of the place of medicine in the five sciences, the larger benefits of studying medicine, and an array of Buddhist perspectives on the importance of compassion and other virtues for both teachers and students. This discussion goes on for a full eleven pages in large-size Western-style book format, in a work that often otherwise moves quickly from point to point. Kyempa displays an impressive facility in the history of Buddhist thought. He makes a compelling case that medicine is clearly within the purview of Buddhist tradition, is of great value, and preserves much useful information about student-teacher relations and a range of important ethical questions. Practicing medicine will even lead eventually to enlightenment.

The second chapter goes into the weighty issue at hand. Kyempa clears the stage by launching into a survey of the entire distinction between "Word and *śāstra*," starting with the kinds of Buddha Word. He duly preserves the three subtypes already laid out in *Soaring Garuda* and supplies examples for each from canonical Buddhist works. Then he defines kinds of *śāstra*, also drawing from classical Buddhist sources. He displays further judiciousness in naming three kinds of lineages associated with the *Four Treatises* itself, based on authorization, explanation, or meaning.[98] And then he finally springs his own assertion: the *Four Treatises* is the third type of *śāstra,* a compilation of many textual traditions rolled into one by Jetsün Mahāguṇa (a hybrid Tibetan and Sanskrit moniker for Yutok Yönten Gönpo).[99]

Kyempa compellingly reiterates his claim for Yutok's authorship in the last chapter of this introductory section when he cites a long passage from the *Heart Sphere of Yutok Story.* This is the passage flagged above, the one that used the telling word "compose." It is also the same passage that explicitly calls the *Four Treatises* a *śāstra.* This statement was already notable in its original context, but now, in the midst of a concerted debate, it is all the more striking. After all, these sentences were uttered by Yutok himself, as Kyempa himself reminds us.[100]

Throughout his discussion Kyempa provides a variety of defenses. He is clearly expecting a rebuttal. He invokes an argument that had been made before, and would seem to be unwarranted: it would not be appropriate, Kyempa insists, to say the *Four Treatises* is the Word of the teacher Medicine Buddha and his manifestation Intelligent Gnosis, citing the recurring phrase "for one teaching two teachers don't come."[101] But he also has a slew of historicist points. We don't find the city Tanaduk as the site of preaching other works that are Buddha Word. Nor does its physical description match what we know either from authoritative sources or reason, he adds.[102] It is only Zurkharwa who would actually unpack why it is not reasonable. Indeed, later on in the commentary Kyempa himself cites Zurkharwa on the question of Tanaduk, indicating that what Zurkharwa wrote set the standard on this issue.[103]

Kyempa insists that the *Four Treatises* is heterogeneous in its origins and combines many medical traditions from both India and China.[104] He draws on the work's own statements about its fourfold audience—gods, sages, *tīrthikas,* and Buddhists—to raise a question about the temporal relationship between the *Four Treatises* and the medical works of other traditions. Like Shakya Chokden, Kyempa was well aware that Ayurveda was ultimately said to have been preached by Brahmā, and he thematizes the problem specifically with regard to the timing of the teachings of the Buddha. The *Four Treatises* itself states that it compiles all medical teachings, including the "eight sections of the *Caraka*" taught for the sages. Given that it presents these teachings in the form of a compilation put together later, i.e., after their initial creation, Kyempa maintains that the *Four Treatises* must be *śāstra.* If it had somehow been in existence before the *Caraka* and so on were compiled, i.e., had been the source of the *Caraka* and Ayurveda more generally, it would not fit with the temporality of Ānanda, the disciple of the Buddha and the imputed compiler of all Buddha Word.[105] In other words, the Buddha could not be the ultimate origin of these teachings since they already existed before he lived. Kyempa also points out here that if they were taught at the same time, it would be confusing to consider the sage tradition (i.e., Ayurveda) and its compilation (i.e., the *Four Treatises* itself) as different. In this he displays his conviction that classical Ayurveda and the *Four Treatises* are not identical.

Once again we might notice the disparity between such a concern and appeals to the all-encompassing and timeless magical power of the Buddha on the part of Tashi Pelzang. The fact that Kyempa, and Shakya Chokden before him, were paying attention to when the Buddha and Ānanda would have lived vis-à-vis Ayurvedic textual history shows a keen appreciation of historical sequentiality in adjudicating questions of authorship.[106] Kyempa is adamant in this argument, insisting that it is not possible to reply to him.[107]

And yet once Kyempa leaves this opening section and commences the commentary as such, he is more cautious. He skillfully manages to follow the text in all its claims, but slightly shifts its points in order to avoid accepting its status as Buddha Word. Such a strategy is already evident in his initial list of the lineage of teachers prior to Yutok—those who putatively would have transmitted it from India down to Tibet. He is not yet commenting on the words of the text itself; instead he is invoking received tradition in the medical world about the history of the *Four Treatises qua* Buddha Word, transmitted from India to Tibet and then concealed as Treasure. He characterizes this as the lineage of "authoritative sources, the basis of composition (*rtsom gzhi*)," a category he himself had set up as one of the kinds of transmission associated with the *Four Treatises*. It almost appears to be a story that he himself is telling of the origin of the *Four Treatises*, except that he closes the whole section with the succinct but possibly skeptical salvo, "so it is said." Representing tradition does not necessarily entail Kyempa himself accepting it.[108]

Once Kyempa gets to unpacking the full title of the *Four Treatises,* he tackles the task at hand without noting the possible implications of the fact that a Sanskrit title is provided along with its Tibetan translation.[109] Kyempa just discusses kinds of translation, using categories well known in Tibetan Buddhist scholasticism, but skirts the question of whether indeed *this* work is such a translation.

Apparently, unlike the version that Tashi Pelzang was working from, Kyempa's text of the *Four Treatises* did begin "Thus have I heard at one time."[110] This clarion call of Buddha Word would have been one more element of the work that he must duly analyze phrase by phrase, part of his duty as a commentator on the text at hand. While he does not accept the implications of that loaded introductory sentence, as a commentator he could hardly claim that the text is lying, so he has to use other interpretive strategies to bring out his critique.

In all of this we can see the force exerted by the genre of commentary itself. But it is also the case that this force can be subverted even as its conventions are heeded. One such occasion presents itself as Kyempa goes through the requisite "five excellences" of the basic setting—the Medicine Buddha in Tanaduk, the preaching by the manifested sage, and so on. Here he does signal his dissent, albeit elliptically:

> [The *Four Treatises*] is arranged with a basic setting of five excellences, in this according with the way Buddha Word is explained, so that disciples will come to hold it on their crowns. If smart people analyze the meaning of the words carefully, they will come to realize what is the special meaning to be understood.[111]

This is subtle, but there can be no mistake about what he is saying. The *Four Treatises* is set up to look like Buddha Word so that people will honor the text. That's why the five excellences introduce the work, and that's why Kyempa as commentator will duly unpack them. Having given a clear signal of how "smart people" should understand the artifice, Kyempa launches into explaining the details of that original scene of preaching, without another word to the contrary, going through the detailed description of Tanaduk and all its various plants with little hint of doubt—save a brief comment at the end referring the reader to Zurkharwa's analysis, and noting again how "smart people" will check into the sources.[112] But prior to that concluding caveat he seems to be having a good time with all the botany, evidently enjoying the occasion to display his knowledge of herbs.[113]

As would Zurkharwa, Kyempa wants to honor the pretexts of the root text without undue disturbance. Perhaps in this he is not unlike modern scholars: able to hold apart critical assessment of the historical claims of a story even while appreciating the narrative on its own terms, relishing its inner logic and the experiences and ideas it explores.

ZURKHARWA'S CAUTION

A distinction between the world inside the text and the world outside it is indicated even more keenly by Zurkharwa Lodrö Gyelpo.[114] Zurkharwa goes to great lengths, and quite self-consciously, to pay due respect to what is credible in a Buddhist world as this is represented in the basic setting of the *Four Treatises*. But he also makes sure not to violate the sensibilities of the medical intelligentsia, who could not ignore the external criteria that make the *Four Treatises* not attributable to the historical Buddha and not composed in India.

There is also a second tier to Zurkharwa's negotiation of the Buddha Word debate, a move to a higher level of analysis. Who the author is ultimately does not matter; what really matters is the value of the work. Anything that is of such fundamental benefit to people could be said to be on a par with the Buddha's dispensation. So even if the *Four Treatises* is not historically Buddha Word, it might have a similar status in this more important sense. Zurkharwa adverts to Buddhist hermeneutics in order to make this point, yet effects a dethronement of the Buddha's unique authority in the process.

This second-order reflection bespeaks a modern sensibility. Our appreciation of value and meaning need not be undermined by historical knowledge, in this case, who the author of the work really was. Rather, it is precisely this knowledge that can elicit the higher-order perspective. Long before, in the *Heart Sphere of Yutok Story*, care had already been taken to assert that the *Four Treatises* was no different from Buddha Word in the strength of its blessings. Zurkharwa draws on that move and runs with it. He eventually mounts a fully critical hermeneutic that understands the *Four Treatises*' opening setting in terms of psychological aspiration: the work's actual historical authors were fancying themselves as Buddha Word utterers. He also appreciates those historical actors' canny sense of what plays as credible in the Tibetan literary world, a reading so skillful that there is not the slightest hint of accusing Yutok and company of fabrication or dissimulation. But in all of this we can also detect the stakes in taking an anti-Buddha Word position, and how conservative and fierce was the protection of the all-encompassing author function of the Buddha. It is almost as if Zurkharwa could anticipate in advance the scathing critique that the Dalai Lama's powerful regent would level against him in the following century, when enlightened charisma and the Tibetan state truly became one. Thus did he become ingenious in protecting his back. This is not to suggest he was being disingenuous. It is eminently clear that he was both reconciling and making sense—in ways that were credible to *him*—of the growing epistemic disparities between academic medicine and the Buddhist establishment. His own enduring allegiance to the assumptions of Buddhist soteriology and praxis will especially come into view in the next chapter. But they are also evident in his careful approach to the loaded issue of the Buddha's authorship that we will follow now.

Zurkharwa's painstaking maneuvers can only be appreciated in light of his entire *oeuvre*, from which vantage the nuances of his line-by-line strategies—some of which we also must read line by line!—become clear. Zurkharwa discussed the Buddha Word status of the *Four Treatises* in at least three separate works.[115] In each case he handles the delicate matter differently. This in itself is interesting, for it shows the way that different literary contexts allow different levels of frankness. It also may mean that he was working out what he himself thought. Or then again, it might just be that he always knew what he thought, but took time to lay it out publicly. My own guess is that the latter is closer to the reality. We even have an early indication that Zurkharwa had grave doubts about the traditional attribution on ethical grounds as well: How can we think that the compassionate Buddha would have taught the painful procedures in the *Four Treatises*' therapeutics? he asks in one of the questions he posted on the pillar in Lhasa.[116] Such a point fits with Zurkharwa's general sense that medicine has ways and means that are distinct from those of the dispensation of the Buddha—the Great Medicine King's own concern to heal the world's ills notwithstanding.

Zurkharwa's *Ancestors' Advice,* his magnum opus commentary on the *Four Treatises*, appears to be his first substantial statement on the authorship of the root medical treatise.[117] This work is only beginning to get the recognition it is due, at least in Western scholarship. It offers brilliant and probing discussions of many key issues in the *Four Treatises*, often far surpassing anything that came before it. The previous chapter noted the Dalai Lama's efforts to complete and publish it, as well as the indebtedness of the Desi's *Blue Beryl* to it. Several other issues from *Ancestors' Advice* will be examined in the following chapters; as here, the history of their discussion reveals careful negotiation of the sometimes conflicting concerns for scientific accountability and religious sensibility.

Since *Ancestors' Advice* is Zurkharwa's main study of the *Four Treatises*, one might think it should be his definitive statement on the Buddha Word question, but in fact it is his most conservative. Most likely, the main reason for his circumspection is the expectations and demands of the commentarial genre. In fact, without the other two treatments it would be hard to discern what Zurkharwa is saying on the issue in *Ancestors' Advice*. With the exception of a few key lines, one could almost read it to be in the Buddha Word camp.

As in Kyempa's commentary, the Buddha Word question comes up right at the beginning. Zurkharwa quickly makes clear that he thinks the *Four Treatises* should be classed as *śāstra*. Granted, he does this largely by indirection. Discussing the very need for the *Four Treatises*, i.e., the basic reason it exists, he starts right up with a variety of classic sources on the standards for writing good and bad *śāstras.*[118] That in itself already says a lot, of course. Why start with an exposition on *śāstra* if the work he is about to comment on is Buddha Word? He never actually owns up to this, though, except to say rather unobtrusively that the classic quotes should be understood "here" as well.[119] Then again, that would actually say it all.

Zurkharwa goes on to address the topic of translation methods, and duly talks about the Sanskrit title, but like Kyempa, he adroitly avoids the question of whether *this* work was indeed translated. He seems to be accomplishing other aims in this section, such as showing off his knowledge of things Sanskritic.

When he gets to the all-important opening phrase, he betrays detailed knowledge of a long-running debate and the fact that different versions of the text vary in whether they read "heard" or "explained." Rather surprisingly, Zurkharwa argues that it *should* read "Thus have I heard at one time," i.e., the classic marker of Buddha Word.[120] His reason for preferring "heard" rests on the inner logic of the setting, and the importance of respecting the conventions of the five excellences.[121] And he is right. Given that the *Four Treatises* is set up in this way, with Intelligent Gnosis uttering the work and Mind-Born compiling it, the work should have been "heard," by the compiler; it does not make sense to say "explained." As I proffered above, "explained" seems like a guilty admission that something is amiss in the author function of the work. Zurkharwa seems to share this assumption, in his response to certain critics: "Those whose learning is inferior say, 'According to your way, all these terms of the basic setting like "Thus have I heard" would make the work Word. But that is not tenable, since it was put together by the compiler.'"[122]

Zurkharwa deems it an ignorant implication that for the opening line to read "heard" would inappropriately make the work Word. He seems to be arguing with others in his own camp: even if the work was put together by the compiler, he suggests, don't assume it should not begin "Thus have I heard." But the way he states his reason is tricky. He grants that the *Four Treatises* is not Word spoken by the mouth of or blessed by the Buddha, but seems to be supporting the idea that it is Word by permission:

> It is true that the words are not spoken by the mouth [of the Buddha], nor are they Word by virtue of a blessing. However, it is posited that it is Word by permission. From the *Karuṇāpuṇḍarīka*, "In future times, when you collect all of my Dharma, at that time first make use of the basic setting by virtue of [the phrase] 'Thus have I heard' and so on. In the middle put in all the arranged words. And at the end insert 'The Bhagavān fully praised this speech,' since he had given his permission."[123]

What does he mean by saying "it is posited" to be Word (*bka' yin par 'dod de*)? This could refer to his own viewpoint, but in light of the next lines, a more likely reading is that the *Four Treatises*' creator himself positioned it as Word by permission. In other words, Zurkharwa is still respecting the inner logic of the work. When he quotes the *[Mahā?]karuṇāpuṇḍarīkasūtra,* he would appear to be quoting an authoritative Buddhist canonical work, here giving instructions on how to compile Buddha Word in the future.[124] Every such work should begin by saying, "Thus have I heard," and end by representing the Buddha's praise of its contents, which will indicate that it has the permission of the Buddha. We cannot fail to notice the specification of "the future." The passage Zurkharwa cites reflects a movement that began in the early centuries C.E. to open the door to the existence of many buddhas in addition to Śākyamuni, living in many buddha fields. That in turn meant that new Word of the Buddha scriptures would continue to appear. In Zurkharwa's passage, the means to make such works conform to the appropriate genre conventions are clearly spelled out.

The critical issue of temporal discrepancy comes to the fore in the next lines. A different objector notes that if the compilation of the *Four Treatises* took place after the Buddha's nirvana, the compiler, who was manifested before the Buddha entered nirvana, would also have had to dissolve back into the Buddha beforehand.[125] Again, by "compiler" this objector is referring to Sage Mind-Born. Zurkharwa notes in annoyance that the *Four Treatises* says nothing of Mind-Born's reabsorption into the Medicine Buddha, as is specified for Intelligent Gnosis at the end of each of the *Four Treatises.* The exchange seems trivial, but later, in another work, Zurkharwa makes clear that he thinks that the real compiler and teacher of the *Four Treatises* were none other than Sumtön and Yutok, respectively—people who indeed lived long after the Buddha's nirvana. That subtext might inform the question he makes the objector raise here in *Ancestors' Advice*, but he does not say so explicitly. He only bats away this objector "who is lacking in analytic skills" by retorting that manifested beings like compilers who are bent on the benefit of all beings do not reabsorb. It is almost as if Zurkharwa was deliberately setting the stage for his position on the real scene of compilation that he would make years later.

Zurkharwa does not spell out his ultimate position on the Buddha Word question in *Ancestors' Advice*. But he does address one key piece of it in detail: the vexed matter of Tanaduk—its location, its status in the world. We already saw that Kyempa seems to have considered Zurkharwa's discussion to be the locus classicus for the right approach to the issue.

Zurkharwa's unusual and starkly empiricist reading rules out any possibility that Tanaduk is to be located in the real world at all. Strikingly, he rejects all such options that his medical predecessors proposed, such as Varanasi or Oḍḍiyāna.[126] As he says at one point in frustration,

> As for that, there is absolutely nothing that is correct. Why? It says nothing about Tanaduk being Varanasi, and it cannot be established by either authoritative statements or logic. Varanasi does not have four mountains around it. Moreover, if Varanasi were to have all those special features, you would have to accept that the people living there would have no time, no death.[127]

If a place called Tanaduk existed in the real world in the way that previous commentators believed the *Four Treatises* to be describing, it would not obey the normal rules about time and other natural processes. Much of this point has to do with incommensurabilities in how the mountains are described. One would appear to be always in the sun, one always illuminated by the moon. Zurkharwa also cites the "very long distance" between these mountains in reality, such that they could not surround a single town.[128] (Zurkharwa is referring to the actual geographical locations of the four mountains named here, which do indeed exist in the world, but he is distinguishing them from their invocation in the *Four Treatises*.) It is striking indeed to see him critiquing an exuberant and mythic picture of a place, so common in Buddhist scriptures, on the grounds that it is not plausible in terms of what we know about the real world and its geography.[129]

Zurkharwa's own take on Tanaduk is to regard the place as a kind of manifestation.[130] This was also the solution of Tashi Pelzang. But this is what I meant by saying at the outset that sometimes almost identical propositions have a vastly different sense in context. While Tashi Pelzang's line served to preserve the sanctity of Tanaduk by rejecting the direct evidence of what can be seen, Zurkharwa, I argue, moves Tanaduk out of the realm of the empirical in order to preserve the reality (if not the sanctity) of the everyday world. Making Tanaduk a manifestation—or even, as we will see later, a flight of fancy—serves to remove a vulnerability from the *Four Treatises*. It puts this and the other conceits of the "basic setting" into the realm of the magical, thus protecting the medical establishment from having to defend something that contradicts the everyday reality with which a medical scientist would be most concerned. As he insists, "If you say it had such qualities in the past but now in the present they no longer exist through the force of time, you can't say that either. . . . Therefore wherever you put that city in this present world you will have a problem."[131]

Zurkharwa is unrelenting in his resistance to any effort on the part of previous commentators to locate Tanaduk in the real world. Now he specifies that while the names that the *Four Treatises* gives for the four mountains surrounding Tanaduk are in fact the names of actual mountains in the real world, they are not what are referred to here. "All these are mere names. Their specific qualities with respect to the town don't make sense. Even the other words are nothing but false constructions. There is nothing real there that is actually being touched, and so [those words] are superficial."[132]

What he argues instead is that those names are just being used figuratively to indicate the kind of herbs that grow there, but the town and the mountains could not actually exist in the world in the way that the text describes. Note also the striking statement that there is nothing real there that is being touched. He's talking about the lack of empirical referent. He gets more specific, going through each of the four mountains, in each case granting the existence of a real place with that name, but then insisting that that's not what is being referred to here.

> First of all, the mountain to the south is called Vindhya. But that is just a designation; it is not the real (*dngos*) Vindhya Mountain. It is called that because on its side there is light and the power of the sun and all the plants that are made to grow there have salty and sour tastes. . . . In the same way, the mountain to the north whose side is in the shade and has the power of the moon is called Himavat but it is not the real Himavat.[133]

Zurkharwa argues further, and again surprisingly, that the *Four Treatises* is not really describing four mountains at all. That to which these various mountain names are figuratively assigned are actually just the four sides of one mountain, on top of which Tanaduk is perched. Zurkharwa notes that previous generations of scholars did not realize this.[134] Now, it is a bit perverse to read statements from the *Four Treatises* like "To the south of that town is a mountain with the power of the sun called Mount Vindhya" to be referring to a mountain slope running down from the southern side of the town, and so on.[135] It is furthermore not clear what is gained by this correction, other than to distinguish the originality of his reading and also perhaps to make the entire place slightly more plausible, with the sun reaching one side of a single mountain and another side always in the shade. But he also continues to maintain that Tanaduk can't be real and is not a buddha field either, adding that the latter proposals bespeak an inability to "find grounds to recognize it [elsewhere] and so they just run to that."[136]

What he does do, however, is offer a bone to the devout reader who wants some resolution, and this admittedly challenges our line of interpretation. At the start of his "own position" on the matter, he ferrets out a quote, which he attributes to a work he calls *Vibhāṣa*, regarding the various places in which the Buddha lived and taught. One line refers to the Buddha living in a medicine forest for four years. Zurkharwa takes this to be an authoritative reference to Tanaduk, although the quote does not name such a place.[137] Zurkharwa maintains that at the center of the medical forest was a mountain with four sides, and on its top was the manifested town Tanaduk.[138] No one else had been able to find any hint of a place devoted to medicine in canonical sources. Such a salvo would behoove someone who is arguing that the *Four Treatises* is indeed Buddha Word, and it seems not at all consistent with Zurkharwa's otherwise critical attitude to the verity of the town as described. But keeping in mind his main point that the town is actually a manifestation, the citation serves to locate it in the credible world of classical Buddhist literature even while confining it to that realm and resisting its existence otherwise, outside the texts and on the ground. In the process he is also positioning himself as an expert, even on a position he does not believe in.

We are not going to understand what Zurkharwa really means in all this until he is out of the commentarial context. But we have already seen him suggesting that the *Four Treatises* is *śāstra*, even while he contends that the self-presentation of the work makes it Buddha Word by permission. He alludes to the need to obey the genre conventions of Buddha Word and considers the time discrepancy between the lifespan of the manifested compiler and the Buddha from whom he would have manifested. Most of all, we just have to marvel at the treatment of Tanaduk. For one thing, there is the sheer amount of mileage that it gets: the *Four Treatises* itself already devotes a disproportionate number of lines to the place of its preaching. For his part, Zurkharwa spends a full twenty-two large Western-style pages on the "excellence of the place," in contrast with one paragraph on the teacher, six pages on the retinue (also a conflicted issue), one paragraph on the time, and then six pages on the Dharma being taught (mostly motivated by the confusion about the multifarious retinue). The attention to Tanaduk shows an exceptional interest on the part of the medical thinkers in place as such, seen especially in how much Zurkharwa and others are taken up with the medical botany and the appropriate geography and climate of particular plants. At the minimum, the debate about Tanaduk evinces a characteristically medical emphasis on material conditions—the biosphere, if you will.[139]

Zurkharwa's treatment helps us understand why the Tanaduk question was such a flashpoint for the Buddha Word debate. It is here that the nagging question of whether the *Four Treatises* is really the teaching of the Buddha or was made up by someone else comes home to roost. Discomfort is felt most keenly about place—if the place is not real, it can't be a genuine teaching of the Buddha, somehow.[140] What emerges in the process of investigating Tanaduk is a standard by which truth will be determined: whether or not the description of a place matches how things really are in the world. And this is where Zurkharwa has already taken a major step, even if he does not yet tip his hand on where he is going with it. He has at least argued definitively in *Ancestors' Advice* that Tanaduk as described is not a worldly place.

It is remarkable to see a vociferous insistence that the place where the most precious text of a tradition was preached must not be considered of this world. We might compare here the arguments put forward by certain Japanese intellectuals in the eighteenth and nineteenth centuries, in light of modern geographical knowledge, that the descriptions in the Buddha's teachings of Meru, the central mountain of the old Buddhist cosmos, should not be taken literally but rather were metaphorical, and merely represent the cultural knowledge of the Buddha's audience.[141] Zurkharwa's treatment of mythic place comes prior to Tibetan contact with European map-making, but already represents an impulse to relativize religious imagery so as to clear space for scientific knowledge.

THE HISTORY: ZURKHARWA PITCHES HIS TENT

Before looking at the free-standing essay in which Zurkharwa returns, now more boldly, to these same issues, his other major medical composition, entitled *The Pitched Interior of General Knowledge, What Doctors Are Not Allowed Not to Know,* merits our attention.[142] This is his *khokbuk* history of medicine that the Desi critiqued so stridently. Here Zurkharwa considers other key pieces of the Word debate, which complement his discussion in *Ancestors' Advice.* He is still being cautious, if on different grounds, but even so makes headway in dismantling the *Four Treatises'* status as Buddha Word.

Zurkharwa's *Pitched Interior* was a major advance in medical historiography. I summarized its overall structure in the previous chapter. Just like the Desi complained, it spends a lot of time on the Buddha and his canonical medical teachings, making a big point of accepting the proposition that the Buddha taught medicine. But when it finally turns to medicine in Tibet, it takes a more critical and historicist tone.

Ever the good historian, Zurkharwa begins his account of Tibetan medicine with a fair and lengthy survey of the views of others, starting with *Soaring Garuda.* This is followed by a long rehearsal of medical lineages and activities in Tibet. Then he returns to question many of the supposed medical teachings of the Buddha in the views he just discussed.[143] He rejects a lot as faulty or untrue, not to be found in the long accounts of the life of the Buddha, and asks his interlocutors where they found this information. He also criticizes claims that he says are associated with Bön and the Old School, and disparages *Soaring Garuda* and other works. He further mentions the difficulty of properly determining whether certain works are really Buddha Word spoken by mouth, or are blessed by the Buddha, or have the permission of the Buddha, or are *śāstra,* or what, quoting another work to say that apart from what a text says about itself, there is often no other evidence.[144] Again and again, he seems to delight in displaying his critical acumen in dismissing as groundless certain historical claims of others.

Then he gets to the stories that the *Four Treatises* was translated into Tibetan and then either hidden as Treasure and later taken out by Drapa Ngönshé, or passed down orally in a lineage of one to Mutik Tsenpo. And now Zurkharwa begins to show his stripes.

One big flaw in those stories would emerge if it could be shown that the *Four Treatises* was not in existence in Tibet before the time of Yutok Yönten

Gönpo (i.e., the Younger; Zurkharwa evinces no knowledge of an "Elder" Yutok in imperial times). The claim that the work is Buddha Word rests on the premise that it was brought to Tibet from India by Vairocana during the imperial period, centuries before Yutok lived. As already suggested, the entire reason the Treasure theory was brought into the *Four Treatises*' transmission history in the first place was to address the lack of evidence of the work's existence in Tibet during the period from the time of Vairocana down to Yutok. The answer that the Treasure narrative proffers is that no one knew about it because it was hidden in the pillar at Samyé and discovered by Drapa Ngönshé at some point close to when Yutok lived.[145] So to discredit this story would eventuate in the conclusion that it was written in that latter period, i.e., the eleventh to twelfth centuries.

This is what Zurkharwa is up to in this passage. He mounts a variety of arguments, including calculations of how long it had been from the time the Buddha died until the reign of Tri Songdetsen. One of his most effective arguments is to cite the biography of Drapa Ngönshé. That clearly mentions his study of medicine as a youth with his uncle and much other activity connected to medicine, Zurkharwa avers, but there is "not even a little bit" about him taking out the *Four Treatises* from Treasure; in fact, several of the elements in that account are mutually contradictory.[146]

In the same way, Zurkharwa goes on, we need some signs that the *Four Treatises* was known in Tibet prior to Yutok, but there is not even a mention of such a name. He also points out that one of the Treasure biographies of Padmasambhava says that the Chinese doctor Heshang Ma[hāyā]na and the Tibetan doctor Tsenpashilaha translated the *Bdud rtsi snying po gsang ba man ngag* (these terms are similar to part of the *Four Treatises*' title) in 156 chapters; if that were true, it would suggest that the *Four Treatises* is a translation from Chinese.[147] Zurkharwa is bent on illustrating that there are any number of reports about the origins of the *Four Treatises*, making the theory that it was discovered as Treasure by Drapa unlikely.

Zurkharwa goes on to mount this powerful *coup de grace* to destroy the entire idea that the *Four Treatises* was ever hidden as Treasure: "It does not make sense to have to bury it as Treasure on account of it being profound!"[148]

"Being profound" is one of the main reasons teachers, quintessentially Padmasambhava, are said to bury Treasures. Because such teachings were too esoteric or complex for the people living in that era, the typical story goes, the teacher concealed them until a time in the future when the appropriate disciples who could appreciate and practice the teaching would appear.

But, Zurkharwa cannily points out, how would this be an appropriate plan of action for a medical treatise?

> And it does not make sense to distinguish near from far, i.e., between the beings of that time and those of the future, who, from the perspective of compassion, will need it later. Hence anyone who is intelligent will find it is easy to realize that these stories are like the six lies of the owl.[149]

This is a convincing argument. Unlike in the standard Buddhist Treasure rationale, it would never make sense for some compassionate teacher to think that medical practice is not appropriate now and only will be appropriate in the future. Surely the king would have wanted doctors to start practicing its techniques immediately.

Strong words, calling the Treasure story about the *Four Treatises* a pack of lies. And let us not miss the arresting difference in all this between medicine and the esoteric spiritual exercises described in the Buddhist Treasure scriptures. Medicine is always needed, and for everyone. It would never be restricted to special circles, nor earmarked for particular epochs.

Having dispatched this broadside, Zurkharwa ratchets his tone down. After a lengthy rehearsal of the efforts on the part of the Tibetan kings to bring medical teachings to Tibet from a variety of places in the known world, he moves to the postimperial period.[150] Here he provides details on the translation of the *Aṣṭāṅgahṛdaya* into Tibetan and the many lines of medical teachers in the following century, including the ones reaching Yutok.[151]

Finally, he gets to his "authentic account" of the *Four Treatises* itself, which he opens with another clever quip.[152] The question of whether this work is Buddha Word or not or not is debated by the wise, the foolish, and everyone in between. Actually, the quip introduces exactly the approach he will take. Rather than leveling more caustic criticism, he will go back over the gamut of views on the topic, now simply representing them as a historian would: some say this, some say that.

That so many believe it is a teaching of the Medicine Buddha, originally translated by Vairocana and then transmitted as a Treasure or as an oral transmission (*bka' ma*), motivates a judicious comment from Zurkharwa. It will require effort to cast sufficient aspersions to get people to accept that the *Four Treatises* are not Buddha Word. Even then, they will find reason to make it Word. But there also are many who claim it is fully a Tibetan *śāstra*, Zurkharwa goes on to aver.[153] While allowing that some don't have any good reason for this and just like to hold on to their own position and denigrate others, he recaps the plausible arguments that have been suggested. Some have based their conclusion on their examination of the way the *Aṣṭāṅgahṛdaya* is structured and presented; others on incriminating evidence in the substance of the *Four Treatises*' content, basic setting, and overall intention. Then he pulls out the trump card that we lingered over and Kyempa also duly flagged: the powerful evidence of the *Heart Sphere of Yutok Story* passage, with its damning phrase "he composed the medical *śāstra,* the *Four Treatises*."[154]

Now Zurkharwa is really racking up points for the Tibetan *śāstra* side, displaying powerful evidence. But he keeps to a stance of impartiality and reasonableness. He reviews the ideas of scholars such as Nyamnyi Dorjé, who cited one Ja Mipam Chökyi; the Sakya scholar Taktsang Lotsawa; and others who ventured scenarios of how the *Four Treatises* was indeed authored by Yutok Yönten Gönpo, teamed variously with Drapa Ngönshé, Üpa Dardrak, and Tönchen Könkyab, or even by Vairocana himself. He can quote no less a luminary than Bodong Choklé Namgyel (1376–1451), who submitted that "The diagnosis of disease based on pulse and urine is not explained in the methods of the Indian scholars. [The *Four Treatises*] was made by a Tibetan doctor bodhisattva," providing an important precedent for the empiricist arguments to which Tashi Pelzang tried to respond.[155] But Zurkharwa himself does not totter from his objective perch. In the end, he takes refuge in the uncertainty of another great light of Tibetan historiography, the polymath Butön Rinchen Drup. "As for me," Zurkharwa says, "just as in the words of Omniscient Butön, I can't make a decision."[156] He is being judicious. He has presented damning evidence, especially against the Treasure theory, but he will remain open to the possibility that other considerations may prove him wrong after all.

Although he didn't nail the coffin on the Word theory, what he goes on to say in conclusion is extraordinary. Even if the *Four Treatises* is *not* the Word of the Buddha, that is not of much moment. Rather, it is the content of the *Four Treatises* and its contribution to the world that is most important. As the *Uttaratantra* says particularly well in a quote that Zurkharwa offers here, "If one explains something with an undistracted mind merely under the influence of the Conqueror's teachings, it will be in accordance with the path to attain liberation. Therefore I hold it to my head as if it were what the Sage [i.e., the Buddha] said."[157] Zurkharwa takes the statement to apply to the *Four Treatises*; "as if it were" would be of course the critical phrase.

Note that Zurkharwa is citing a Buddhist source to make this point. He is drawing on hermeneutical criteria long developed in Buddhist contexts to determine authentic Buddha Word based on the truth value of a text's content, but he uses the strategy to come to a different conclusion about the *Four Treatises'* author.[158] He hasn't quite admitted straight out that he thinks the *Four Treatises* is not Buddha Word. He sidesteps that, but assures his reader that he will hold it in the same esteem even if it isn't Word.

> Moreover, there is no second work more beneficial, useful, and competent that is precise with regard to the place, time, situation, and people of in this Snow Land than this [*Four Treatises*]. For that reason, it is not allowable to depend on mere disputes about terms, which destroy the fruit. Considering this in itself as good meaning, we need to hold it very dear."[159]

The confluence of climatic and cultural specificity that the *Four Treatises* achieves is unparalleled. Therefore Zurkharwa sagely recommends avoiding superficial disputes, which will obscure what is really important about the work: its medical knowledge. That is what needs to be protected, not a mere taxonomical question of whether the text is actually Buddha Word.

Zurkharwa's presentation is impeccable in his history of medicine: cautious, specific, critical, and yet generous. It is hard to see how the Desi could have been so enraged by it. Perhaps Zurkharwa was worried about the repercussions of coming out too clearly with his true opinion in his most visible compositions, his commentary and his history.[160]

SHINING A LIGHT ON THE DARKNESS

It is telling that Zurkharwa worked out his most frank assessment of the Buddha Word debate in a short free-standing essay devoted to the question. Here he revisits many of the same arguments he had already broached in his longer works, but in a far more candid and ultimately risky manner. *A Lamp to Dispel Darkness* was written close to the end of his life, when he was sixty-four, in 1572.[161] I have not been able to determine whether it ever was carved for blockprint reproduction. It is probably safe to say that it was distributed on a smaller and more selective scale than his substantial commentary and history. And yet it appears that, a century later, the Desi did read it.

Zurkharwa does little beating around the bush in this brief work. He jumps right into the fray after the quip that opened this chapter, about how the *Four Treatises* was the hot topic of the day:

> All some do is insist that it is really Word. Some say it is really *śāstra*. But if one doesn't connect with authoritative sources or reason, this is only insisting. Therefore, for the benefit of my own students, I will expound a discourse that is connected with authoritative sources and reason, such that if scholars see it, it will be easy to understand, and easy to explain to the common people—a discourse that aspires to be definitive.[162]

Zurkharwa is still taking pains to paint himself as judicious. Even though he is going to argue that the *Four Treatises* is a *śāstra,* he is not the kind who just insists without reasons.

That said, he immediately comes out with his most radical claim, one of the main statements that roused the Desi's ire:[163] "Here there are three ways to explain: explaining [the *Four Treatises*] from the outer perspective as Buddha Word; explaining it from the inner perspective as a pandit's *śāstra*; explaining it from the secret perspective as a Tibetan *śāstra*."[164]

This is very explicit. The outer/inner/secret taxonomy is common in the Tibetan literary world. Everyone would know that the secret is the truest account of the three. And so in truth, Zurkharwa says, the *Four Treatises* is a Tibetan composition.[165] But this is quite an exceptional way to say it. Zurkharwa has turned on its head the standard progression, which in religious writing inevitably moves from the "outer" exoteric level to the more internal and esoteric, spiritualized versions of something.[166] Here, he has the outer version posit a buddha as expositor of the *Four Treatises*, whereas the secret truth represents the prosaic view of an empirical, historical, and human composition. This stunning reversal is emblematic of the deep difference between the medical mentality that Zurkharwa is helping create and the assumptions at the center of Tibetan Buddhist doxography.

As Zurkharwa proceeds, he softens his statement a bit. For example, he makes the same point as in his history of medicine, about meaning and function being more important than who actually taught something.[167] He further recaps his argument in *Ancestors' Advice* on Tanaduk. But when he repeats his citation of the Abhidharma text that mentions a medicine forest where the Buddha stayed for four years, adding more detail on its location near the famous Deer Park in India, it is clearer what he is doing.[168] We were right to suggest above that he was simply working through the presumptions of the *Four Treatises*' narrative on its own terms, for here he only presents this point in the context of the outer view, which for him is the least accurate. Nonetheless, on that view, if the *Four Treatises* is to be Buddha Word, then the place where the text was preached ought to be identified according to the authoritative source he has located. He adds the further proposition that while the Tanaduk mountain still is there, the city and also the preacher sage and audience were all just manifestations and now have disappeared. Admittedly, this would grant some empirical location to Tanaduk in the past, i.e., during the life of the Buddha, and it is not consistent with what Zurkharwa says elsewhere.[169] But again, this is only the outer view—not the one where Zurkharwa will plant his flag.

Zurkharwa's treatment of the inner view is brief; he is not interested in the position that the work could be a pandit's *śāstra*. He only gives a brief review of kinds of *śāstra* and then a few lines on who might be the *Four Treatises* author on this view. The possibilities range from Padmasambhava and Candranandana to a variety of Tibetans. Zurkharwa dismisses them all, insisting that such positions only serve to advance their proponents' desires.[170]

Zurkharwa's use of categories is telling for the secret view that the *Four Treatises* is a Tibetan *śāstra*. He divides his discussion into citations of other scholars and explanations of their essence. Elsewhere Zurkharwa usually divides his discussion into the views of others that he will refute, and then his own view, which he will advance. But here the views of others are in accord with his own.[171] He still seems to want to shore up his position by reminding the reader of others who have made the same radical point that the *Four Treatises* is the composition of a Tibetan. Then he ventures just who this Tibetan person would be, and introduces the figure of Yutok. But even when he finally gets down to the actual secret position toward which he's been building, he opens with one more acknowledgment of the reasonable concerns of his opponents: "Might it not be the case that there is a problem [with your position], given the way that the *Four Treatises* sets up as its foundation that it is the Word of the Buddha? Well, this is what should be explained."[172]

Here is his clincher in response to his own rhetorical question: "If it were not made to be as if it were Buddha Word, then Tibet's scholars, dolts, and everyone in between would all have a hard time trusting it."[173] This is an astounding salvo, not seen previously.[174] It bespeaks a critical distance uncommon in his time and place. He characterizes the people of Tibet, learned and ignorant alike, as having a certain predilection that must be taken into account by any author who wants his writing to be taken seriously. Tibetans want everything authoritative to be the Word of the Buddha.

Zurkharwa's statement implies that the *Four Treatises*' status as Buddha Word is a fiction, deliberate and calculated: "For those who don't investigate the meaning and only worry about terms, if it were not made to be as if it were Word, it would not be acceptable." Perhaps this serves to justify Zurkharwa's own foregoing attempt to make it all work, *as* Word, for just such people who don't investigate the meaning.

Zurkharwa evinces knowledge of the various arguments rejected by Tashi Pelzang about the obvious Tibetanness of the *Four Treatises* and in contrast, finds them convincing: "Moreover, the references to tea and pottery and the tradition of black divination in the context of checking pulse and urine, and the sounding of the voice of the cuckoo, etc., make it very clear that this is a Tibetan *śāstra*." This leads him to the following magisterial interpretation about the grounds for Yutok's fabrication, as he finally ties up his argument. "Moreover, regarding its being set up to be like Word, you need to know this way of explaining."

When Zurkharwa says "like" Buddha Word, he means that the *Four Treatises* is *not in fact* Buddha Word. He continues,

> You need to explain from the perspective of the foundational idea, necessity, and what will disprove its reality.[175] So, the foundational idea for it to be made as if it is Word is that the strength of the sun and moon at Yutok Yönten Gönpo's birthplace has excellent efficacy for both hot and cold [plants]. [176]

Although stated briefly and elliptically, there is an unmistakable point here that is his lynchpin for understanding the *Four Treatises*' entire basic setting, including the location of the text's putative preaching over which scholars had been puzzling for so long. It also resolves everything Zurkharwa has been implying in his other writings by insisting that Tanaduk is not a real place in the world.

Although Zurkharwa does not say more than the lines just translated, a fuller statement of the very same point was made by a contemporary who might well have been a colleague, the great historian Pawo Tsuklak Trengwa (1504–66?).[177] The historian is in the midst of telling the life story of Yutok. He is actually quoting someone else, one Jé Trinlé Zhap.[178]

> To the east of where he lived were mountain meadows with many blue medicines growing that were similar to Mount Gandhamādana.[179] And to the south grew hot blue medicines, and to the north there were snow mountains and cold medicines growing and to the west there was a forest, etc. The parts are all in accordance [with the description of the four mountains around Tanaduk in the *Four Treatises*]. And so he styled his own place of residence the medical city Tanaduk.

Tanaduk is really an exuberant rendition that Yutok "styled" based on an actual mountainous area around his home.[180] While Trinlé Zhap represents the more conventional understanding of Tanaduk as a city surrounded by four mountains, he otherwise is making a very similar point to what Zurkharwa intimated. If we have the right identification of Trinlé Zhap, it seems already to have been suggested a century before Zurkharwa and Tsuklak Trengwa. And this domestication of the *Four Treatises*' basic setting would reach greater specificity yet, a century after the Desi, with another great critical medical mind, Lingmen Tashi (b. 1726), a close student of the polymath Situ Panchen Chökyi Jungné at the outlying medical center at Pelpung in eastern Tibet. The lengths to which he would go in the demystification of Tanaduk are exceptional. Lingmen states that Yutok's conception of Tanaduk had to do with his way of perceiving the market town of Gurmo in Tsang.[181] Here, rather than the old Buddhist idea of manifestation, we have a psychologized notion of a kind of euphemistic projection. Or perhaps in light of the reference to "styling" or "making," we might say a literary flourish, or a kind of magical realism.

If Zurkharwa and colleagues' reading of the true identity of Tanaduk were not enough to bring down the entire edifice of the *Four Treatises*' pretensions, a second point made by Zurkharwa is even more bold. It starts with a claim that goes with the first. Just as the place where the work was taught is really Yutok's hometown, the main actors are actually Yutok himself and his student Sumtön. Zurkharwa writes,

> As is set up in the *Rnam thar bka' rgya ma*, "The Lama is the real Intelligent Gnosis. I myself am Mind-Born, I think." Just as that says, [the *Four Treatises*] is set it up such that Yutok himself is Intelligent Gnosis and Sumtön Yeshé Zung is Mind-Born.[182]

We ourselves already noticed a similar statement to the lines Zurkharwa is quoting here in the *Crucial Lineage Biography*; it is another sign of our confluence of historical method.[183] But while in the early source such an exalted claim about both his master and Sumtön himself fit seamlessly into the rest of the self-enhancing tenor, Zurkharwa puts it to a different use. The very claim of enlightened authorship, so common in the Treasure tradition, becomes here the very proof that it is really just the fanciful imagination of the real-life author. When Sumtön claims that his master is Intelligent Gnosis and he himself is Mind-Born, he thinks, this shows Zurkharwa and his critically minded compatriots not the true spiritualized identity of the two Tibetan physicians but rather, and more simply, that they set up (*bkod pa*) the work in such a way that they would figure thusly in the *Four Treatises*.

Once again we can turn to Pawo Tsuklak Trengwa's quote from Trinlé Zhap, for it articulates this second point of Zurkharwa's more specifically. It also breaks apart the very process of visionary inspiration and authorship in a way rarely seen in Tibetan scholastic writing.

> And he styled the heart realm of reality as the Buddha Bhaiṣajyaguru, the variety of conceptions as the four kinds of retinue, the thought that desires to compose as Sage Mind-Born, and basic intelligence as Sage Intelligent Gnosis. And then he expounded the *Four Treatises*.[184]

Here Trinlé Zhap recaps all the main elements of the basic setting and associates each with one of the cognitive processes involved in writing. That the basic ground of such writing is equated with the exalted, enlightened state of the "heart realm of reality"[185] fits with the ongoing high respect for Yutok, even from the most critical commentator, and the repeated use of the label "manifestation" or "manifested body" to refer to him.[186] Accordingly, Yutok's basic inspiration for composition is, quite appropriately, figured as the Medicine Buddha. The issues to which the urge to write is responding are personified as the audience of the work—the gods, sages, Buddhists, and *tīrthikas*. The basic urge to write, i.e., the pretext, is the interlocutor. The intelligence that creates the teaching is Sage Intelligent Gnosis. It's all Yutok, though, working with his student: that's the point.

Tsuklak Trengwa continues with his own comment. There is nothing terribly transgressive in Yutok presenting both an ordinary appearance and then a pure appearance for disciples. Indeed, the trope of the great teacher who has various aspects depending on who is looking is nothing new in Tibet. Speaking for himself now, Tsuklak Trengwa avers:

> For that ordinary appearance to be a pure appearance for students in the way it is explained in the basic setting—I think this is not a transgression. Moreover, if you were to ask if crafting the basic setting as if it were Buddha Word is wrong, [I would reply that] it is a great marvel that [the *Four Treatises*], having been blessed by the power of the Buddha, shone forth spontaneously in the heart of this manifested body.

Tsuklak Trengwa is very comfortable with inspired writing. And note, by the way, that neither he nor the others arguing against the Word thesis are noticing the possible implications of calling Yutok a manifested body. That is not read as rendering Yutok a buddha or his writing Buddha Word. Perhaps this shows how generalized the *tulku* label (i.e., *sprul sku,* lit., manifestation body) for reincarnated masters had already become in Tibet: it was a sign of great esteem, but it need not imply that everything such a master wrote was Buddha Word in the strict sense of belonging in the canon of the Buddha's teachings. The same applies to Tsuklak Trengwa's invocation of the "blessings" of the Buddha. That too can have a general meaning without necessarily taking on the technical connotation associated with kinds of Buddha Word.

Tsuklak Trengwa goes on to summarize the various views of those who believe the *Four Treatises* to be Word, but then points out that even though many do believe this, the two diagnostic methods of pulse and urine found in the *Four Treatises* are distinctively suited for the time and place of Tibet. This brings him to the same move that Zurkharwa made, to finally blur the whole issue away:

> Examination of the definite feeling of symptomatic pulse and looking at urine is not in other works, and therefore there is no better medical work in the world than the *Four Treatises.* In particular, the kindness of its suitability to the place and time of Tibet has no comparison. Therefore I raise it to my crown as if it were Buddha Word.[187]

The "as if" phrase stands out again. Really it is hardly noticeable; the sentence could almost be read to say "I raise the Buddha Word *Four Treatises* to my crown." Medicine could almost be Buddhism. But when scholars like Tsuklak Trengwa and Zurkharwa assure their readers that the work is as good as Buddha Word, they are making an important move that has two inseparable but importantly distinguishable parts. They have subtly concluded that the

Four Treatises is *not* Buddha Word and is rather the composition of a Tibetan scholar, but that does not matter. That means there is a more important issue at stake: the value of the work, its practical virtues *qua* medical textbook. This issue emerges precisely by virtue of relativizing the Buddhist dispensation. Let us recall at this juncture the vantage point and cosmopolitanism in the Desi's medical paintings, whereby illustrating the medical classification of knowledge relativized religious practices and symbols. Already, a century earlier, the mentality percolating in Tibetan medicine was facilitating such moves to higher epistemic ground.

Zurkharwa concludes *A Lamp to Dispel Darkness* by returning to the other two categories that he had said were needed to explain how the *Four Treatises* were constructed as Buddha Word, beyond the "foundational idea" just reviewed. The second, "necessity," is quickly dispensed with, but it underscores Zurkharwa's position. The work had to be Buddha Word "so as to lead along those people who are hard to satisfy." And the third, "what will disprove its reality," is what he already showed in *Ancestors' Advice*. The mountains named in the *Four Treatises* are not the real ones. Rather, the names are based on the equivalent strength and quality of the mountains in the vicinity where the author was writing. This is what "doctors of today" need to know, Zurkharwa concludes. These current-day physicians have not trusted the scholars who have it right, and their mindset has been under the power of "old habits."[188] Such doctors don't understand the point of the work and can't distinguish what is from what is not.[189]

There is again a modern sentiment in this. In fact, Zurkharwa thematizes a difference between past and present knowledge on several occasions. While to be sure, he sometimes participates in the usual rhetoric bemoaning the fallen state of current medical scholarship (that is, other than his own), he also frequently berates a general category of "scholars from the past" and insists that the current generation can do better.[190] This seeming lapse of etiquette is one of the reasons the Desi purported to be so upset with him. I pointed out in chapter 2 that it is hardly the case that Tibetan Buddhist scholars never criticize the views of their predecessors, for they do; what's more, the Desi himself broke with tradition far more significantly by launching *ad hominem* attacks on his *living* teachers. Of course, for both men much of the bravado can be read simply as an effort to distinguish their own knowledge as superior. Still, alluding to a generic category of previous knowledge and old habits does take a step beyond the more conventional practice of critiquing a particular view of a particular scholar from the past. Zurkharwa sometimes is talking about an inferior pastness as such, and that stands in sharp tension with the prevailing Tibetan Buddhist rhetoric about how the present is a degenerate time, a fall from a prior golden age.[191] His characterization of medical knowledge's problems in temporal terms and accompanying urge to get medicine up to speed in the present is unusual and worth noting. It is one more aspect of the distinctive mentality that his self-positioning suggests.

In sum, there are several distinctive things to notice in the medical Buddha Word debate. There had certainly been debates about the provenance of a purported Buddhist scripture before, and whether it was indeed originally from India. But the critical wing of the medical debate pursued this question on empirical grounds regarding climate, culture, and time period, rarely invoked elsewhere. There are also questions about the reality of place in other Tibetan literary contexts, as well as early attempts to describe the world in geopolitical terms, but few in which historicist arguments grounded such discussions.[192] One outstanding exception, an early and seemingly realist appeal to the empirical qualities of place, is in the virulent critique leveled by Sakya Paṇḍita at attempts to locate Indian sacred geography in Tibet, studied by Toni Huber.[193] Motivated by a quite different concern, i.e., the threat to monastic control of tantric practice represented by pilgrimage cults, Sapan warns his readers to distinguish factual descriptions of places from poetic characterizations of their attributes, anticipating some of Zurkharwa's logic.[194] But perhaps because of the difference in motive, Sapan's argument ends up mostly avoiding the more basic question of the status of these places as such. As long as there is no argument about their location in Tibet, he seems to have been satisfied. In so doing he seems to have laid his argument open to the vociferous response over the centuries, a robust assertion of these places' location in Tibet after all, based in large part on a hermeneutical argument about kinds and levels of perception that essentially make any appeal to empirical evidence moot.

The turn to a hermeneutics of perception, which shifts attention from the reality of a place to the qualities of the perceiver, is seen elsewhere in the history of Tibetan religion too. John Newman reports on a transformation in the conception of the mythical land of Shambhala in *Kālacakra* tradition, and here the difference from the way that Zurkharwa handles the Tanaduk issue is telling.[195] An early travel guide to Shambhala from the thirteenth century represented the route as passing through commonly known towns and geographical areas, but there is a gradual spiritualization of the journey in subsequent descriptions. By the eighteenth century Newman finds in Panchen

Lozang Pelden Yeshé's guidebook, written two centuries after Zurkharwa lived, an awareness of a clash between this imagined place and modern knowledge of geography. Newman shows that Pelden Yeshé opted to render the journey in spiritual terms in order to "protect the sacred utopia with a veil of ritual magic."[196]

We might think that this is of a piece with Zurkharwa's insistence that Tanaduk should be moved out of the domain of everyday reality, but the idea is really very different. In Zurkharwa's case the urge is not to lift the *Four Treatises* to ever higher levels by making Tanaduk a manifested, magically created place. Rather, his approach serves to keep the mythical confined to its own domain, that of the "outer view," which for him represents the common expectations of average Tibetans who need such a fiction. This frees up medical historiography for the real business at hand, which is to determine the historical circumstances around the writing of the *Four Treatises*. Such a strategy to clear the space for something more scientific will be even more apparent in his handling of the channels in the body, to be studied in the next chapter.

Questions about the authorship of a purported Buddhist scripture are also well known in the Buddhist Treasure tradition, but again the differences from the medical debate should be noted. Whereas in the Treasure tradition the critique came from the outside, i.e., from those bent on delegitimizing works that are not really Buddha Word, in the medical case the skepticism comes equally from within. Critiques like Zurkharwa's were not at all about trying to bring the *Four Treatises* down. Rather, they are about trying to get the story right. And when it turns out that what is right is that the work has human and historically locatable authorship, it does not mean that the work is any less authoritative. No, the author's knowledge was based on real study and eclectic research, and if Yutok's basic inspiration was lifted up rhetorically to make him indeed inspired by the Buddha, that is not a problem. The conceit is well motivated, understood as a fortuitous idiom to fit reigning conceptions of authoritative knowledge.

Such recognition of—and, once it has been duly noted, comfort with—mythologization and artifice is ironically part of the scientific mentality these scholars were helping to foster. It did not undermine their strict attention to empirical matters in other domains, i.e., in the medically relevant material itself. But it might indeed have been part of the double strategy in which scholars like Zurkharwa were invested, to protect against the protests of more conservative colleagues who were endeavoring to maintain medicine's Buddhist identity, while preserving a space apart for the development of medical knowledge as such. Even the exceptional subversion of the usual outer, inner, and secret heuristic in Zurkharwa's final essay, whereby the secret or highest truth is that which is most down to earth, can be read in this pragmatic sense. Beyond making clear that the Buddha's authorship is only the outer version of the origin story, that which the common man would hear, there is probably something quite strategic in making the real truth "secret"— not in the esoteric sense but with the very straightforward purpose of keeping such a controversial view under the radar screen.

The next chapter will follow another debate in which Zurkharwa completes the circle. Once again we will appreciate the vulnerability of the empirically based critique of religious truths and see even more caution and deftness from Zurkharwa. Once again too we will see him cordoning off a spiritualized, more salvific dimension of the matter at hand, acknowledging its importance, but making room thereby for another set of standards that are subject to ordinary sensory inspection. The turf will shift a bit, however. Zurkharwa still wants to keep alive a religious vision of the body's channels for salvific purposes, making for a much more ambivalent and delicate negotiation. The stakes for medicine are higher too. Now the issue of accuracy concerns the very anatomical map upon which the physician must depend in order to practice his craft.

4

THE EVIDENCE OF THE BODY

MEDICAL CHANNELS, TANTRIC KNOWING

Another medical star from the heyday of the Fifth Dalai Lama's rule is surely Darmo Menrampa Lozang Chödrak (1638–1710). An elder colleague of the Desi, he was at the helm of several of the Great Fifth's medical initiatives in the mid-seventeenth century. But right at the start of his long collaboration with the Dalai Lama, Darmo performed an experiment, apparently on his own initiative. It would have constituted one of the boldest challenges to textual authority—Buddha Word or not—in all of Tibetan medical history.

The incident is especially striking in light of the larger history of science and the public anatomy lesson.[1] For at some point around 1670, apparently perplexed by the differences and vagaries in various medical works around the number of bones in the human body, Darmo decided to gather his students together and let the body speak for itself. Setting themselves up in Luguk Lingka, a park in Lhasa, he and his acolytes proceeded to dissect a sampling of human corpses—one old man, one old woman, one young man, and one young woman—and count their bones. Thus did they reach a precise count, he avers, by virtue of "vivid demonstration."[2] Acknowledging that the standard accepted number was 360, Darmo reports that he and his disciples counted instead 365, adding that the discrepancy had to do with the way they divided the fissures in the skull.[3]

Darmo's career marks a watershed in Tibetan medical history. He wrote prolifically in a range of instructional genres, frequently in quite idiomatic style, on anatomy, plant recognition, compound preparation, and therapeutics, often drawing on his own clinical experience. He also wrote a history of medicine, along with any number of commentaries on the *Four Treatises.*[4] The Desi speaks approvingly of his "medical merit," and of his close relations with the Dalai Lama.[5] But it also appears that the Desi sought to detract from his fine reputation, recounting how Darmo abandoned a certain Mongolian lama as a patient (whom the Desi himself then duly retrieved and healed)[6] or was too busy to pay close attention to the words of the texts.[7] The Desi seems to have deliberately played down Darmo's role in what were actually his greatest accomplishments.[8] Was he being a little competitive? A modern-day historian of medicine is driven to complain that Darmo's career has been unfairly eclipsed by the Desi's own achievements.[9] Clearly Darmo had the Dalai Lama's confidence. The Great Fifth not only entrusted him with the completion of Zurkharwa's *Ancestors' Advice,* the editing of the *Eighteen Pieces from Yutok*, and the writing of the hagiography of Yutok, but also put him in charge of one of the medical schools he established. It was Darmo whom he set up to study with the foreign physicians brought to the court. And he was the one whom the Dalai Lama allowed to operate on his own cataracts.

Darmo's bone-counting endeavor encapsulates well the progressive side of medicine in the seventeenth-century capital. It intimates that textual knowledge could be bettered or even contravened by empirical evidence. It also suggests that present findings could supersede the authorities of the past. There had long been differences of opinion about the count of bones in the medical texts of old. Much of South Asian medicine seems to have agreed that the number of bones in the human body is 360, but the great Ayurvedic classic *Suśruta* avers instead that there are 300.[10] In Darmo's time the authoritative texts in play would have included the *Four Treatises* itself, *Aṣṭāṅgahṛdaya,* and Indic Buddhist sources such as *Kālacakra* and the Tibetan tantric classic *Profound Inner Meaning,* all of which repeat the classical number.[11] But when he writes about the matter Darmo reports that he has not been able to study Indian and Chinese works. He says he is concerned largely with writings by Tibetans, whose account of this part of the anatomy left much to be desired, in his view, tending either to leave matters implicit, to be very imprecise, or to dodge the issue altogether.[12] Darmo reviews several sources on how to count the bones, but in the end feels that it is not sufficient "to leave the matter by figuring out a mere count." Instead, he has "resolved the matter on the basis of being able to recognize by pointing a finger."[13] This is the same metaphor that the Desi used to indicate the superiority of visual images over words.

Darmo's desire to point directly to what he is trying to know favors material display over discursively articulated system. It also separates two disparate kinds of epistemic authority—the textual and the empirical—which are accessed in different ways. In particular, the empirical for Darmo is accessed by "vivid demonstration" (*dmar khrid,* lit., "red instruction"), i.e., something which facilitates fine detail; the crimson metaphor is often taken to imply "naked" or "raw," or perhaps even—as in this case—the blood that must accompany such a direct look.[14] We should also not miss the implications of specifying that his findings depended upon the way the counting was done. As Darmo goes on to say, if instead of counting, as he and his students did, only the bones that measure between "the span between fingertip and elbow, and the width of a finger," one were also to reckon "the numerous small bones that are merely the size of a roasted bean," a higher number yet would be reached.[15] His point bespeaks the inchoate intractability of the material world, and its multifarious nature. Its specification depends upon the angle from which it is viewed. This approach is very different from the predictable and rational categories of system usually represented in the textbooks.

Darmo acknowledges the open-endedness of the body's actual disposition. He notes that others have pointed out that there might be more or fewer bones in the body than the standard account would have it; even the *Four Treatises* leaves open the possibility of extra bones or teeth and so on. Then he adds that if someone comes along in the future who can provide a vivid demonstration (again, he uses this key term) of a different number of bones, we shouldn't try to point out his faults with an ill-disposed conception.[16] For Darmo, greater precision in medical knowledge is still possible, and he will be open to it if it is presented clearly and convincingly. His concern to account for unanticipated variation in individual specimens is echoed too in his effort to study four samples from the human population—as if he thought that the standard might not apply to all people. His suspicion is especially notable for considering the possibility of gender difference, an issue in medicine to which we will return later.

Darmo's exercise represents a radical edge of the Tibetan medical mentality, and a prime example of the impulse toward direct observation. It is especially significant that it was performed in public: outside in a park, witnessed at least by the doctor's students, if not others.

We have yet to discover a contemporaneous biography of this important figure that might add further details on the event, its motivation, and its reception at the time. However, it does appear that Darmo's efforts to examine the empirical body on this and perhaps other occasions informed his writings on human anatomy, in which he corrected and expanded information in the *Four Treatises* and its earlier commentaries.[17] But he was hardly a thoroughgoing empiricist in every dimension of his medical career. He played a key role in the construction of the myth of the Elder Yutok. He also was less than faithful to the evidence of direct observation with respect to parts of human anatomy that were not as immediately evident as ligaments, tendons, and bones. On at least one occasion when he takes up another dimension of the body directly implicated in religious practice, he seems much more ready to accommodate doctrinal system.[18] Thus does this key figure exemplify a growing quest for empirical evidence in seventeenth-century Tibetan medicine, yet at the same time an equally growing impulse to reconcile science and religion in the Fifth Dalai Lama's court.

This chapter takes up the controversy around that religiously significant part of the anatomy just mentioned, for which we have a substantial historical record. It revolves around a salient discrepancy between the empirical body and a different kind of anatomy represented in certain authoritative Buddhist scriptures and engaged by virtuosi in their meditation practice. This is the long-running inquiry into the channels of the body.

In brief, whether or not the *Four Treatises* itself is really the Word of the Buddha, a few of its passages show influence from other scriptures widely considered in Tibet to be the Buddha's enlightened Word. One of the main examples is its description of the channels that run through the human body. Its quite original system of four kinds of channels has little analogue in Ayurvedic sources or any of the other medical traditions upon which it otherwise drew.[19] Rather, it provides an eclectic mix of straight-ahead description of the body's veins, arteries, and nerves, and a set of notions about a more subtle system of channels and the energy substances moving within them. The latter are indebted both to tantric Buddhist ideas and to old Tibetan conceptions about life and energy.

The controversy around the channels did not originally center upon the *Four Treatises*' system per se. It had to do with the discrepancy between the picture of the channels portrayed by Buddhist tantric scriptures and used as the basis of yogic practice and what is found in the physical body, that is, the human corpse. This discrepancy was noted early on in Tibet. But when medical theorists began to take it up, they did so in the context of writing *Four Treatises* commentary. This meant that the *Four Treatises*' categories and conceptions colored the conversation.

The issue of the channels thus put at least *three* kinds of authority into play: the knowledge represented by the tantric scriptures, what can be seen and touched in the body itself, and the *Four Treatises'* system. The *Four Treatises* really served to mediate the discussion, not only because it was medical theorists who became exercised about the physical location of the tantric channels but also because the medical work itself shows some debt to Buddhist tantric systems, however vaguely. The *Four Treatises* itself does not acknowledge its tantric debts, nor does it use overtly tantric language. But those few elements of its system that seemed amenable to tantric conception were not lost on the later medical commentators.

The history of the channel question reveals, among other things, the way that the *Four Treatises* functioned as an authority—almost like Buddha Word, if you will, but a bit more susceptible to an occasional factual doubt. We can discern a general expectation that the *Four Treatises'* anatomy must describe something observable in the human body. A large part of the work's weight comes from the virtues of its author, the idea that its words issued out of the direct experience of a great physician—be that Yutok Yönten Gönpo or indeed the Buddha himself. And yet the *Four Treatises'* own heterogeneity, particularly in the channel section where it mixes yogic tradition with medical knowledge, can make its descriptions hard to parse. In the end the *Four Treatises'* system did not settle the problem for these writers. As the attempt to find the tantric channels in the empirical body gathered steam, it became clear that a reconciliation of tantric conception with the experience of medical practice would not be easy to achieve.

The history of the channel discussion goes to the heart of this book's questions about the relations, and disjunctions, between medical ways of knowing and ways of knowing fostered in Buddhist regimes of personal cultivation. It also speaks directly to the medical concern to accord with empirical observation and to question idealized system—whether medical or religious. It seems to be decided in advance that the evidence provided by the directly observed body has an unquestionable authority of its own. In this the medical mentality I am tracking is not quite the same as what the modern-day Fourteenth Dalai Lama would seem to have in mind in his own reckoning of the authority of empiricism. There, true empiricism has to do with the deep knowledge revealed by the Buddha, based on the Buddha's own enlightened realization, and is only directly knowable by him and similarly advanced yogis. For the Dalai Lama and indeed much of Buddhist epistemology, trust in these enlightened realizations trumps any possibility of proving them wrong; indeed, scientific testing will only prove them correct. While the Dalai Lama's concern in the current religion and science exchange has more to do with the ethical and spiritual implications of scientific research than with the nature of the physical world as such—on which he is says he is willing to acknowledge errors in Buddhist tradition—the existence in Buddhist doctrine of soteriologically based constraints on what constitutes correct direct perception is important to note.[20] In contrast, this chapter documents a moment in Tibetan history where medical theorists sought evidence for the truth of the Buddha's Word in the physical evidence known to medical practice. In this atmosphere a discrepancy between such evidence and Buddha Word was not to be explained away merely as a function of the ordinary deluded mind of the physician.

As far as I have determined, the earliest scholar to report on the discrepancy between tantric tradition and the observable body was a tantric theorist in the thirteenth century, Yangönpa Gyeltsen Pel. His solution was to separate the two conflicting witnesses, in effect creating two kinds of bodies. But the medical writers took a quite different approach, trying to reconcile the contradiction by locating tantric channels in some part of the everyday body after all. Once again, it was Zurkharwa Lodrö Gyelpo who provided the definitive stroke, in effect stemming this trend and proposing yet a third kind of resolution. The latter, complex as it was, would hold sway for the rest of the life of traditional Tibetan medicine.

Zurkharwa ended up driving a wedge, epistemically, between tantric and medical vision, albeit on different grounds than what the tantric theorist proposed. In achieving this solution he is as difficult and sometimes contradictory as he is canny and deft. In addition to protecting a space for medicine on its own terms, he was also clearly invested in maintaining some legitimacy for the tantric anatomy. This was not merely for instrumental aims—to avoid criticism from the Buddhist establishment, or to please his patrons, or even to prove his own scholastic prowess in all things tantric, although all three were surely at work. It also appears that Zurkharwa wanted to keep the door open for tantric meditation to have material efficacy in the body. But this was difficult to realize in empirically accountable terms, and it left him palpably conflicted.

Most challenging of all—for both Zurkharwa and our own analysis—is the fact that medicine could not entirely disavow the tantric anatomy in any event. Tantric meditation involved detailed visualizations and manipulations of psychophysical energies in the body. Its early adepts had produced maps of the body based on these exercises. Although Tibetan medicine largely eschewed those schemes, some of it was of interest, especially regarding those loci in the body, such as the heart and the sexual organs, that require attention to the interaction between body and psyche to be fully understood. So even while Zurkharwa worked to contain tantric knowledge, some of it had already left an indelible mark on medical anatomy in Tibet, not only in the *Four Treatises* but also in the larger repertoire of knowledge that physicians were bringing to their craft. In fact, the entire question of whether the perdurance of the yogic imagination in Zurkharwa's vision of the body represents an uncritical repetition of old tradition, or a rethinking and recontextualizing of such ideas for medical purposes, is not easy to answer definitively.

Zurkharwa ended up with an ingenious account of the body's channels that well served the needs of medical science . . . in a traditionally Buddhist world. It is a tribute to his nuanced writing and studied ambiguity that the breaks his strategy represents were not entirely appreciated by his successors. The Desi virtually repeats Zurkharwa's account of the channels verbatim, and so subtle is the resolution that it was read in some quarters as merely affirming the empirical reality of the tantric anatomy. The upshot can be discerned in the tone of a twentieth-century commentator's warning to those who would question the reality of the channels as described in the tantras: "If one does not accept the disposition of the body's channels as it is explained by masters who have realized the profound tantras, and rather explains it in another way, such an attack on authoritative scripture and reasoning will not stand. So it is better to abandon arrogant originality and follow the masters."[21] This modern commentator is citing as his authority the very reading of the channel system by Zurkharwa studied in what follows. He fails to note that not only was Zurkharwa indeed arrogant enough to call into question the masters who preceded him, he also relegated the tantric scriptural accounts to a corner of the *Four Treatises* system that has little to do with the everyday "disposition" (*gnas lugs*) of the body that it attempts to treat. We can also note the evident anxiety in the contemporary recap, no doubt in response to a new materialist critique emerging in Lhasa today, especially among medical theorists eager to accord with modern biomedicine.[22]

In what follows, I will endeavor to define, as well as possible, the terms and distinctions I have already begun to invoke—including "the empirical," "the medical," and "the everyday." However, these and other distinctions will be hard to maintain neatly. There is a large difference between the rhetorical invocation of empirical verity and actually working through the implications of something presented to the senses. Moreover, none of the arguments was made exclusively on a principled empirical basis—in any sense of what that might mean. None of the commentators looks exclusively to the empirical, or indeed the rational, or even the authoritative as their sole grounds for argument. Even Zurkharwa—or perhaps especially Zurkharwa, given all the fish he is trying to fry at once—freely twists or ignores, on occasion, what he knows to be true about the body, or what the *Four Treatises* says about it, just so he can make a point that he feels he must for other reasons.

But none of this should detract from the master stroke whereby Zurkharwa was able, in a very contested terrain, to carve out a space for doctors to engage the body's prosaic channels on their own terms. His premises and intricate strategies merit close attention, starting with the rich history of medical speculation that set the stage.

THE CHANNELS OF THE BODY

The *Four Treatises* describes the body's channels in the fourth chapter of its second book, the *Explanatory Treatise.* This comes right after the chapters on embryology and a brief excursus on similes for body parts. The chapter introducing the channels is entitled "The Disposition of the Body," and it amounts to an overview of anatomy. This includes bones, orifices, and vulnerable points, but the chapter spends most of its time on the channels that transmit substances and energies through the body. Here is the entire passage on the channels:

> The teaching on the disposition of the connecting channels is as follows.
>
> There are four: growth channels, channels of being, connecting channels,
> life channels.
>
> The growth channels ramify into three from the navel.
> One channel goes upward and grows the brain.
> Stupidity, in dependence on the brain, abides there;
> it gives rise to phlegm, which abides in the upper body.
> One channel reaches into the middle and grows the vital channel.

Hatred, in dependence on the vital channel and blood, abides there;
it gives rise to bile, which abides in the middle of the body.
One channel reaches downward and grows the secret [genital region].
Desire abides in the secret [region] of males and females;
it gives rise to wind, which abides in the lower body.

The large channels of being are four.
[One] channel makes the objects of the senses arise;
it is in the brain, where it is surrounded by 500 minor channels.
[One] channel makes the objects of memory appear;
it is at the heart, where it is surrounded by 500 minor channels.
[One] channel makes the body grow;
it is in the navel, where it is surrounded by 500 minor channels.
[One] channel makes for offspring, the propagation of the family line;
it is in the genitals, and is surrounded by 500 minor channels.
Above, below, and straight ahead, these take care of the whole body.

The connecting channels are two, white and black.
The [black][23] vital channel is the trunk of the channels;
like branches, they ramify upward from that.
Twenty-four large channels produce flesh and blood.
Eight covert large channels connect inwardly with the vital and
hollow organs.
Sixteen visible channels connect outwardly with the limbs;
from those ramify the 77 bloodletting channels.
112 key channels are vulnerable.
189 channels combine.
From those, there are 120 channels in the outside, inside, and
intermediate areas.
The minor channels ramify into 360,
from which ramify 700 minor channels,
from which more minor channels connect around the body as a net.

From the great ocean of channels in the brain,
[the white vital channel] reaches down like a root,
[from which ramify] nineteen water channels that make for movement.
Also, connecting inside the vital and hollow organs
are thirteen covert channels [like] silk tassels

and six visible channels connecting out to the limbs,
from which ramify sixteen minor water channels.

There are three channels for human life.
One abides pervading the entire head and body.
One moves in association with the breath.
One is like the *bla*-soul and wanders.

Since these [four kinds of channels] connect to all the openings
through which move wind and blood, both within and without,
and abide producing the body
as well as being the root of vitality,
they are called channels.[24]

This complex picture has no known precedent. Unlike other parts of its physiology, the *Four Treatises*' channel system does not originate in Ayurvedic tradition.[25] It is clearly trying to account for a wide variety of functions that such channels would serve. It incorporates knowledge from Buddhist tantric tradition, from early Tibetan conceptions of vitality and life, and from clinical experience of various channels in the human body based on practices of bloodletting and moxibustion. The four main channels that the text lists are basically four *kinds* of channels, although some overlap is evident.

1. The first group consists in three initial "growth channels" (*chags pa'i rtsa*). These first appear in the fetus in the sixth week of gestation. They make for the primary formations that define the basic structure of the body.[26] The embryonic growth channels emerge out of the navel and ramify into three. One ascends and forms the brain. One operates in the thoracic cavity and gives rise to the "vital channel" (*srog rtsa*), a key but very ambiguous term. A third descends and is responsible for the development of the reproductive system.

We can already recognize here strong influence from Buddhist conceptions. The three classic Buddhist cognitive/moral afflictions—ignorance, greed, and hatred—are part and parcel of the *Four Treatises*' understanding of the initial growth of the body. The brain is identified as the site of ignorance, or stupidity. It is also what gives rise to phlegm. Thus do the *Four Treatises* relate the affliction of ignorance to the medical humor of phlegm. In similar fashion, the vital channel that arises as the second growth channel and is associated primarily with blood is also the site of the second affliction, hatred, and gives rise to the medical humor of bile. The third growth channel is the site of the affliction of desire, abiding in the secret parts of males and females. It gives rise to the third medical humor, wind. While the afflictions and humors had already been connected in certain Indic Buddhist works, the *Four Treatises* makes this anatomically specific.[27]

2. The *Four Treatises* calls the second group of channels the "channels of being" (*srid pa'i rtsa*). These amount to matrices of channels at the brain, heart, navel, and genitals. They are responsible, respectively, for perception, memory, growth, and reproduction. This is the segment of the *Four Treatises* channel system that most clearly echoes tantric physiologies, in that the brain, heart, navel, and genitals more or less correspond to the sites of four of the main yogic *cakras*, which are also matrices of channels. Though the *Four Treatises* does not use the term *cakra*, the basic idea of networks of channels clustered at key points at the center of the body along a vertical axis certainly recalls the classic tantric picture. The terminology of the channels of being is virtually absent from the rest of the *Four Treatises*, and has little discernible function in medical practice.

3. The third variety, called "connecting channels" (*'brel ba'i rtsa*), focuses on two main systems of "vital channels," one white and one black.[28] Their label recalls the initial embryonic vital channel, but these systems are distinct from that and develop later in gestation.[29] The white and black vital channels roughly correspond to the biomedical nervous and cardiovascular systems, respectively. Their smaller channels branch out and connect to organs, muscles, and bones all over the body.

Despite certain inconsistencies—all fodder for the commentators—the anatomy of the black and white channel systems is spelled out later in the *Four Treatises*, in various chapters of the *Instructional Treatise*.[30] The black channels divide into veins and arteries.[31] The white channels, descending from the brain, are what are responsible for the body's ability to move.[32] The term "water channel," though causing some confusion in the commentarial tradition, denotes the similarity of the white channels in appearance to ligaments and tendons.[33] Most contemporary Tibetan doctors consider the main trunk of the black vital channel to be the aorta and the main "root" or trunk of the white vital channel to be the spinal cord.[34]

In practice, the connecting channels have to do with the treatment of wounds and other maladies, along with the very central practices of bloodletting (mostly but not exclusively from the veins), pulse taking (on arteries in the wrist), and moxibustion (on certain white channels and other points in the body).

4. The fourth group is the "life channels" (*tshe yi rtsa*). As a further sign of the multivalence of the label "vital channel," these are also glossed at the end of the passage as "roots of vitality" (*srog gi rtsa ba*). The group is divided again into three, one that pervades the entire head and body, one that is associated with the breath, and one that wanders in the body like the *bla*-soul. The first would seem to be akin to the old Indic conception that essential energies circulate throughout the body on a monthly basis in accordance with phases of the moon.[35] The third one seems to describe something similar, but also betrays indebtedness to the old Tibetan notion of *bla*, which has to do with vitality or soul.[36] The second kind of life channel has something to do with breath, and some of the medical commentators take that as a reference to tantric yogic practice involving breath and the vital winds. Overall, the commentators have trouble understanding what the life channel category refers to and how the three subtypes relate to one another. If there are tantric resonances here, they are not explicit. Nor is it clear in what ways these three channels, ever connect to medical practice, or indeed ever come up again in the *Four Treatises*.

The many ambiguities of this system are to be attributed to the heterogeneity of sources on which the *Four Treatises* drew, among other factors. In any event, the channels received much commentarial attention. In large part these comments are concerned with locating the third group, the connecting channels, for the purposes of medical practice. But we can also track the commentators' gradual use of the *Four Treatises* categories to try to solve a more global problem regarding the human body.

THE PROBLEM OF TANTRIC ANATOMY

The tantric system of channels is pervasive not only in Tibetan Buddhism but also throughout South Asian yogic practice. It is described in a wide array of Buddha Word tantric scriptures, commentaries, and yogic practice manuals, along with many other works in Hindu and even Jain tradition.[37] These channels provide the somatic basis for adepts to achieve religious enlightenment.

Although details varied in Indian Buddhist tantra and in the early centuries of its reception in Tibet, the Buddhist tantric anatomy can for our purposes be summarized by this basic picture. A central channel runs from the brain to either the solar plexus or a central spot in the genitals.[38] It is flanked by two other channels, one on either side, named *kyangma* and *roma* (i.e., *rkyang ma* and *ro ma*) in Tibetan (Skt. *lalanā* and *rasanā*). These encircle the central channel at key points, forming wheel-like *cakras*, which either constrict or facilitate yogic practice. They also facilitate more basic functions in the body, including the flow of ordinary substances and the elimination of waste.

It is the central channel that is the main focus of tantric meditation. Yogis attempt to mobilize and move special kinds of spiritualized substances and vital winds into this primary pathway, in order to engender special meditative states linked to salvific realizations. Note that average people don't access the central channel or those refined materials and vital winds at all. The materiality of the central channel is considered very subtle in any case.[39]

The earliest sign that I have found of a perceived problem with the tantric channel system in Tibet is in tantric writings. Apparently it was noticed that the classic picture from the scriptures—a central axial channel with two side channels entwined around the flowerlike circles of knots along its length—is not to be seen when one looks inside the thoracic cavity of a dead body.[40] We may be reminded of a momentous scene recounted far more recently, when the Hindu reformer Dayānanda Sarasvatī (1824–83) tore up his copy of the *Haṭhayogapradīpikā* after pulling a corpse from a river, dissecting it, and failing to find the *cakras*.[41] The potential for the demise of religious truths would have been similarly apparent in the Tibetan instance, if perhaps not so dramatically. While evidence of an epiphany at any particular point in time is lacking, one can assume that Tibetan society preserved good knowledge of what the inside of the human body looks like. The practice of dismembering corpses in order to feed them to vultures is an old and widespread way to dispose of dead bodies in Tibet. Techniques to examine the human corpse for anatomical information were also known to early Indian medicine.[42]

The early tantric theorist who reports on the issue and defends the tantric vision is the Drukpa Kagyü yogi-scholar Yangönpa Gyeltsen Pel (1213–58), writing soon after the time of the composition of the *Four Treatises*. His treatise *Secret Explanation of the Vajra Body* considers the relationship between the physiology of the enlightened Buddha as it is portrayed in Mahāyāna and tantric sources and the everyday body of the average human being.[43] His notion of that average human's embryology already betrays heavy tantric influence, far more than anything we can recognize in the *Four Treatises*.[44] But his discussion of the tantric "central channel" will directly inform the problem that later came to the fore in medicine.[45]

> Some say this central channel is an imputation. If that were so, then all the qualities that the central channel brings together would become imputations. One would not be able to traverse the Secret Mantra path.
>
> Some posit it to be the vital channel. [But] if wind and mind come together in the vital channel, one goes insane. Since the central channel only gives rise to good qualities, even if you bribe it, it will never make a problem. So that cannot be right.
>
> Some say it is the spinal cord, but that does not have all four features of the central channel.[46] Since it is missing the key points of the channel, that cannot be.
>
> Thus those who recognize the central channel are few. If you do not know it but would do technical path meditation, that is laughable.
>
> So this is how the central channel abides: it is called "uncommon secret." From the *Mahāyogatantra*, "It stays inside of the soul stick.[47] It has thirty-two knots. Finest of the fine, it is perfectly precious. It has neither outside nor inside."[48]

Yangönpa is concerned to refute opinions that he considers wrong. He even knows of some who would deny the material existence of the central channel altogether and make it a mental imputation instead. It is not clear if such a group represented a radical branch of medicine or skeptics from another corner, or even if the position is just an imaginary one against which Yangönpa is defining his own. Nor should it be assumed that the various opinions he portrays were written up or argued in any cogent detail. They may have simply been ideas that were floating around in conversation. In any case, note Yangönpa's assessment of the upshot of the most radical position among them. If the central channel were only an imputation, totally lacking material existence, that would disqualify the viability of the Buddha's tantric teachings—i.e., the "Secret Mantra" and the special yogic practices of the "technical path" (*thabs lam*)—*tout court*. In other words, Yangönpa insists that the central channel does have some sort of material dimension.

Yangönpa is satisfied to account for the central channel's physical existence in terms of the old tantric idea of an intermediate and very fine or subtle kind of materiality that does not exhibit the normal features of material things. In contrast, the ensuing medical discussions take just the opposite tack, working to establish the tantric channels' conventional material existence. This reflects the medical theorists' sense that material substantiality is the high bar for demonstrating the truth of something. But it also shows their continuing interest in the tantric vision of the body and a need to account for it.

In fact, none of the medical theorists whose writings are known to me propose that the tantric channels are a mere figment of the imagination. All the other arguments that Yangönpa proceeds to consider, however, mirror moves seen in later medical writing. Some do, as Yangönpa notes, try to equate the tantric central channel with the so-called "vital channel," a multivalent category that can refer either to the *Four Treatises*' embryonic growth channel or to one or both of the two main trunks of the cardiovascular and nervous systems. At least one later commentator tries to equate the tantric channel with the spinal cord. But Yangönpa rejects all options that would find the central channel in the ordinary body, on the grounds that they would violate the way this channel functions according to its scriptural specifications. He also adds that if one attempted to draw the powerful yogic winds and mind substances into the vital channel (it is not evident what he means by that, but he is clearly referring to something in the ordinary body), as is done in tantric yoga with the central channel, it would cause insanity.[49] As we will see, Zurkharwa makes a similar point in refuting these medicalist moves and even quotes Yangönpa on occasion, though his larger project is still trained on medical science, in contrast to the tantric apology in which Yangönpa is engaged.

Yangönpa is invested in dismissing any location of the central channel such that it could be ascertained by an ordinary observer. His tantric Buddhist perspective is rather concerned with a central channel that defies everyday physical laws, even if it has some material basis. He characterizes its special mode of materiality as uncommon, secret, fine, and locatable neither outside nor inside. His concern is not to reconcile such a way of being with ordinary material reality, but rather to relegate the latter to a lower level than the domain of enlightened bodily existence, or what he calls the enlightened body's "substantial nature" (*dngos po'i gnas lugs*).[50]

Yangönpa thinks there are two levels of somatic existence—one accessed by virtue of tantric practice and the other observable to the ordinary sense organs. This is different from the old supramundane/worldly distinction in Buddhist thought and even certain tantric attempts to account for ordinary bodily functioning. The idea that the central channel's mode of existence is of a different order than the everyday body's objectively observable materiality would provide a way to contain any materialist critique of tantric anatomy on the grounds that the latter is not visible. But the solution is a double-edged sword. It could also suggest a vulnerability—or at least a lack of comprehensive purchase—for the tantric picture of the body. Such an upshot might have been coming into view particularly in medical circles, where the exigencies of the everyday body held sway. In any event, instead of accepting the incommensurability that Yangönpa suggests, a slew of medical commentators would attempt to reconcile the two visions. But as Yangönpa anticipated, this is hard to achieve.

SOMATIC MODALITIES

Moves that would bring the tantric channels into sync with the *Four Treatises*' understanding of the body's channels can be recognized from the beginning of the commentarial tradition. The commentators gradually slip in tantric nomenclature even if at first they don't name the problem or what they are trying to do.

A few terminological distinctions are in order. It is evident that disparate kinds of bodies are referenced in these passages. Although they are not always so labeled, we will do well to try to distinguish them, although there is overlap. There is the tantric body (what Yangönpa and Buddhist tantric tradition often call the *vajra* body): that which is described in the tantric scriptures and mobilized by yogic virtuosi.[51] That is to be contrasted with what both Yangönpa and Zurkharwa dub the "everyday body" (or "ordinary body," *tha mal kyi lus*), which would name the body of an average person who has not practiced tantric yoga.[52] The latter is not necessarily the same as what I will refer to as the "empirical body," although the everyday body would indeed be the subject of direct perception as well as medical scrutiny. The everyday body can in turn be distinguished from the "medical body" (or "body to be healed," *gso ba'i lus*), a term sometimes singled out in medical writings, but which I will use broadly to refer to the body on which medical procedures as described in the *Four Treatises* are performed, and to which its theories of physiology, illness, and health refer.[53] There is again some overlap with the empirical body: medical practice must assume that the vein to be bled, for example, is indeed a channel that can be seen and touched by an instrument. But I will try to use the term "empirical" only when the commentators themselves make explicit reference to the possibility of direct observation. I am more interested in the history of how a notion of the empirical came to be distinguished and touted than in empirical reality per se.

We can further discern in these passages an idea of the "embryonic body" (*chags pa'i lus*), another term sometimes distinguished in Tibetan but which I will use more loosely as a heuristic to analyze Zurkharwa's discussion.[54] Such a category would be a moving target since the body in the uterus changes dramatically over the course of gestation. In fact, on my reading the ambiguity of whether the early embryonic growth channels are still there once the organs and connecting channels are fully formed in the fetus is a critical piece of Zurkharwa's strategy in the channel debate. In any event, the appropriate terminology and assumptions will be specified as much as possible in what follows. I will also invoke a category of the "mature" or sometimes "adult" body (similar to the terms *grub pa'i lus* or *dar ma'i lus*), by which I mean the body's structures in their developed postfetal form, which would begin at the moment the fetus leaves the mother and continue to be so structured throughout life. That is different from but related to the dead body, which also seems to have been the subject of some direct scrutiny. Overall, the difficulty in making ironclad distinctions across this diverse spectrum of somatic conceptions gives a good indication of how challenging it can be to parse various theorists' attempts to both connect and differentiate them.

MEDICAL SOLUTIONS: TRYING TO LOCATE THE TANTRIC BODY

At first, the appearance of expressly tantric channels in *Four Treatises* commentary may not have been connected to the problem that Yangönpa highlighted. For two key early medical works, it appears rather that tantric anatomy stood as an opportunity to provide more specificity and precision to medicine from another angle, outside the purview of medical sources proper. To pin down just what the *Four Treatises*' kinds of channels are—where they are, what they do, what they look like—the commentators began to turn to what they were familiar with from other contexts: the anatomy conceived in tantric tradition.

One of the two early medical commentaries to do so is known as *Black Myriad*. The other is often called *Small Myriad* and is part of the *Eighteen Pieces* collection.[55] Both are attributed to Sumtön Yeshé Zung, the disciple of Yutok Yönten Gönpo.[56] The two works are closely related studies of the *Explanatory Treatise*. When they get to the description of the channel systems in the fourth chapter, both seem drawn to notice the similarity of the schema of the *Four Treatises*' "channels of being" to the tantric channels. In particular, they add detail to the root text's description of the channel matrix at the heart, which is connected to the faculty of memory. The matrix at the heart abides in the middle of a seminal sphere (called *tiklé*, already a tantric, not medical term) that collects five kinds of highly refined distilled fluids.[57] It is itself a distilled channel, and it is called "crystal bamboo tube." White in color, and as slender as a hair from a horsetail, it abides on the basis of foundational consciousness.[58]

These statements betray an obvious debt to the special physiology of the tantric visionary practice of the Tibetan Great Perfection tradition.[59] That literature describes a crystal bamboo tubelike structure that connects the heart to the brain.[60] The medical commentaries that draw on that tradition also trade on a notion of basic consciousness underlying all of the body's manifestations; the *Four Treatises* never does this, but it is standard in Buddhist philosophy and the tantras.[61] The motivation for providing a Great Perfection gloss to the *Four Treatises*' heart matrix is not entirely clear, but there is no evidence of controversy or anxiety about it.

The next pertinent commentary is the one on the *Explanatory Treatise* by Jangpa Namgyel Drakzang, examined in the last chapter. Namgyel Drakzang duly has a section on the channels, where he is very dependent on *Black Myriad*, citing it for the matrix of channels at the heart and invoking the same white crystal bamboo channel.[62]

But Namgyel Drakzang's real interest is not in the channels of being, but rather in the connecting white and black vital channels. He spends far more time on them—twelve pages in the modern edition of the text—than his one page on the growth channels, two pages on the channels of being, and one page on the life channels. His principal concern is the location of the connecting channels. Namgyel Drakzang associates the white vital channel with the spine.[63] He also disagrees with earlier commentaries on other channels.[64] It is clear that the precise anatomy of these channels was subject to dispute in his day.

Beyond what he takes from *Black Myriad*, Namgyel Drakzang also adds a tantric layer in his brief section on the life channels, the fourth kind of channel in the *Four Treatises*. Rather starkly, he claims that the life channel that is the breath's pathway is associated with the breath that passes through *ro* and *kyang* and goes out from the nostrils.[65] These are the tantric *roma* and *kyangma*, the well-known pair of pathways to the sides of the tantric central channel. But the stipulation is really in passing, and again it is hard to say why he makes this leap. His statement gives the impression that *ro* and *kyang* are more familiar locators than is this curious breath-related life channel. Once again, medicine is resorting to Buddhist knowledge systems to translate something rather arcane, i.e., the *Four Treatises*' category of life channels, into something more familiar, the tantric anatomical system.

As with the Buddha Word dispute, it was a while before someone in the medical lineages shows awareness of a debate in progress. The Jangpa commentator Sönam Yeshé Gyeltsen,[66] son of Tashi Pelzang, the sometime nemesis of Zurkharwa, is the first medical writer I have encountered who does so.[67] It is not known if he is original in this, since we don't have all the sources to track the history.[68] But we clearly see not only a further stage in the creeping insertion of tantric language into the *Four Treatises*' channel picture but also a hint of self-consciousness that there are epistemic and hermeneutical issues at stake.

Sönam Yeshé Gyeltsen sprinkles tantric references liberally onto the *Four Treatises*' channel system. For one thing, he makes the tantric *roma, kyangma,* and central channels the foundations for the *Four Treatises*' three embryonic growth channels.[69] This is where, several generations later, Zurkharwa will take his stand with much ado. Although Sönam Yeshé Gyeltsen gives no further comment, it would appear that he is trying to insert tantric language just to make the point that there is such a bridge, since the gloss he provides does not suggest any discernible advantage for either medical knowledge or tantric functioning. It just lines up two systems. He also upholds the *Black Myriad* tradition that understands the channels of being at the heart in terms of Great Perfection terminology.[70]

But when he gets to the connecting channels, Sönam Yeshé Gyeltsen has a more advanced kind of move to make. It comes up in his very first line for this section: "The channel with the nature of the fire element and that conducts blood is known as the outer *roma*, the black vital channel. The channel with the nature of earth and water and that primarily conducts wind is the outer *kyangma*."[71]

This equation of the two side tantric channels with the black and white vital channels is again just slipped in. He is not arguing, but simply asserting the equation.[72] His concern, especially over the next ten pages devoted just to the black vital channel, again seems to be about location and counting the channels. But it is momentous that categories like "outer *roma*" and "outer *kyangma*" have been introduced into the discussion. The move draws on a larger, very standard hermeneutical strategy in Tibetan Buddhist scholasticism to hierarchize things as outer and inner, or outer, inner, and secret.

In the last chapter we saw an example of this hierarchy in the Buddha Word debate, along with Zurkharwa's inversion of it. Here for Sönam Yeshé Gyeltsen the heuristic is not ironic but quite serious. And the implications are considerable. To talk of an outer *roma* implies that there will also be an inner *roma,* a more real or pure *roma*, more authentic. The notion of an outer *roma* thus becomes a mediating device. It participates simultaneously in the world of ordinary bodies—this would be the outer part—and in the world governed by tantric soteriology, i.e., the real *roma* as such. It also implies that *roma* can have both a realized aspect in the body of an enlightened person and an everyday samsaric side. In this Sönam Yeshé Gyeltsen is drawing on an old portion of Indic tantric physiology that assigned ordinary functions such as the movement of the humors and the elimination of urine, feces, and semen to some parts of the tantric channels.[73] But in that context the tantric physiology is really focused on the processes of yoga and enlightenment; the everyday bodily functions seem added so as to account for the rest of human existence, but with little anatomical specification or precision. In Sönam Yeshé Gyeltsen's case, it is just the opposite: it is the tantric nomenclature that is added as a kind of afterthought. But in the process Sönam Yeshé Gyeltsen provides an answer to the problem that the tantric channels are not to be seen in the thoracic cavity. Actually they could be seen, if one only knew how to recognize them. The black vital channel (which, if we can assume refers here to its main trunk, would be equivalent to the aorta) *is* the (outer) *roma*.[74]

Sönam Yeshé Gyeltsen also takes care of the central channel, *dbu ma*. He locates again an "outer central channel," within the category of the *Four Treatises*' life channels rather than the connecting channels. Apparently he is still not willing to put this most exalted piece of tantric physiology, even in outer guise, into the most overtly medical segment of the *Four Treatises*' channel system. But he leans on the description of the first of the three life channels, the one that is said to pervade the body, to make a very significant leap forward:

> There is such a one that sits in the body from the crown of the head to the heels of the feet. Its width is a mere one hundredth part of a horsetail.

> Clear and luminous, it sits straight. It is called the outer central channel or *avadhūtī* and it is not equivalent to the other common channels, for it lacks the two extremes. Moreover: "The life channel is said to have three types. One sits in the body from the crown of the head to the tip of the feet. That is the outer central channel. It is not the same as the other channels." Thus does it say.[75]

For the first time among the available commentaries, the tantric central channel, explicitly so named, has been brought into the *Four Treatises*' categories. Not only has Sönam Yeshé Gyeltsen inserted it into the embryonic body, as seen above; now, more significantly, he also finds it in the life channels of an adult. He even provides a relatively clear anatomical description. He is quoting some unidentified earlier commentary, but the fact that the picture has already shifted substantially from that of his predecessor Namgyel Drakzang, who only connected the two tantric side channels, *ro* and *kyang*, to one of the life channels, indicates that the issue was a live one. The move that Sönam Yeshé Gyeltsen represents is major because the central channel is the main site for tantric practice. Unlike *ro* and *kyang*, which are also associated with the functioning of the everyday body, the central channel is only accessed in high-level yogic practice. Again, Sönam Yeshé Gyeltsen and his source employ the hermeneutic device of the "outer" rubric, but the description sounds very much like the full-fledged tantric central channel. Note too that its physical location as here specified is quite at odds with the *Four Treatises*' own statement that this channel pervades the body everywhere. And while it has been placed in the elusive life channel category, whose actual applicability in medical practice is minimal at best, there can be no doubt that, as in the move to identify an outer *roma* and *kyangma*, the central channel has been located in the everyday adult body. It would be potentially available to direct inspection, even though it is specified as being exceptionally fine.

KYEMPA TSEWANG, CONSUMMATE EXEGETE

By the time that Zurkharwa's elder colleague Kyempa Tsewang was writing, a much more self-conscious juxtaposition of tantric and medical knowledge was under way. One sign of how far down that path medicine had ventured is the frequency, in Kyempa's own trudging through the *Four Treatises*' channel system, with which he cites the influential Tibetan survey of Buddhist tantric practice, the *Profound Inner Meaning* by Rangjung Dorjé (1284–1339).[76] At the same time, Kyempa is expressly aware of the significance of juxtaposing overtly Buddhist materials with the medical. He realizes that this is already happening in the *Four Treatises,* as when it brings together the three classic Buddhist defilements with the three humors in the discussion of the growth channels. So, for example, in his comments on these channels, Kyempa distinguishes the medical phlegm that he is careful to specify as "*equivalent* to the dharma of heaviness-stupidity and so on," instead of just lumping phlegm and stupidity together without notice.[77]

More significantly, Kyempa creates his own new juxtapositions. He goes on in the next lines not only to quote the *Profound Inner Meaning*'s very tantric account of the central channel, stretched from the crown to the genital region, but also to own up to the fact that he is speculating: "That [*Four Treatises* account] is in accord with what [the *Profound Inner Meaning*] says, I think."[78] Later on, after suggesting something about the location of tantric channels in the body, he admits that "All that juxtaposition of Secret Mantra is an extraordinary move, but it makes sense. Nonetheless, for now my comments are in accord with medicine (*gso rig*)."[79] Thus does he name the "juxtaposition" he is proposing—*sbyar bshad*—and indicate that he knows it is a special measure.

In his comments on the growth channels, Kyempa does not come out and actually identify the initial embryonic vital channel as the tantric central channel mentioned in the *Profound Inner Meaning.* But the citation of a passage from that work in close proximity to his comments means that the implication is there. As he moves on to the channels of being, he builds on the overlap between tantric and medical descriptions of these four matrices hinted at by earlier commentators, now not just citing *Black Myriad*'s account of the crystal tube channel at the second matrix but also providing an enhanced picture of the first matrix at the brain. He explicitly glosses it as the tantric central channel and specifies that there is a letter *haṃ* at the brain, clearly borrowing the *haṃ* often visualized at the top of the central channel in tantric yogic schemes.[80] He also brings the central channel into his discussion of the matrix at the reproductive organs, where again he provides further specificity by recourse to tantric terminology relating to the central channel's bottommost point.[81]

What stands out the most is Kyempa's awareness of what he is doing. To close his section on the channels of being, he avers,

> As for the nature of the channels of being, if one considers the *Four Treatises* as the principal resource, one would only explain the overt channels. But from the perspective of their function, you also need to posit implicitly what is said in the *Profound Inner Meaning*, i.e., "channels, wind, and *tiklé* together with mind."[82] Here, I have explained the general situation in accordance with medicine, but the detailed situation in accordance with Secret Mantra.[83]

It is striking both that Kyempa considers the two systems, medicine and Secret Mantra, to be separate, and that he simultaneously recognizes how the latter might be capable of giving more specific information on what is but a general picture in the former. He is also aware of the limitation imposed by the fact that he is writing a commentary on the *Four Treatises*; hence his distinction between what is explicit and what he deems to be implicit. Unlike previous commentators who had a similar urge to juxtapose, Kyempa feels comfortable owning up to this move overtly. The upshot of his statement is that the channels of being are included in the *Four Treatises* so as to set the bodily stage for tantric practice. But the medical treatise is not going to give details on that practice; one needs to go to a different kind of source for those.

Like his predecessor Sönam Yeshé Gyeltsen, Kyempa also finds tantric channels in the connecting channels. But, as evidence of the patent sense of freedom to innovate, Kyempa does it differently. He may be the first medical commentator to really wrestle with how to locate the tantric central channel in the everyday body of an adult human. And he does so through meticulous scrutiny of the medical categories of the *Four Treatises*.

Actually, like all of his commentarial predecessors, Kyempa is most interested in locating the black and white vital connecting channels, and spends many pages on pinning down the anatomy of the various parts of these two channel networks. He begins with the black vital channel, but prefaces this with a consideration of the larger category of vital channel as such, the confusion surrounding which we have already noted:

> "Vitality" is the basis for the survival of the body. The channel in which wind and mind abide together, and which has been dubbed the vital channel, that very one is the foundation for the ramification of channels, like a trunk. It is spoken of as two, the black vital channel and the white one.[84]

Like others, Kyempa understands the rubric of vitality (*srog*) to refer to the vital role of blood and wind for the survival of the body. But he also seizes an initial opportunity here to adumbrate the tantric connection. When he characterizes that "which has been dubbed the vital channel" as the "channel in which wind and mind abide together," he would be referring to the tantric central channel, not the medical one, for the *Four Treatises* never speaks of the mind flowing in either the black or white vital channel. Rather, "mind" (*sems*) would be a synonym for the tantric *tiklé* substances, often glossed in tantric literature as "mind of enlightenment" (Skt. *bodhicitta*).[85] In this passage he has just asserted the very identification of the adult vital channel with the tantric central channel that Yangönpa had refused.

But the foregoing is just an initial and rather subtle suggestion. Kyempa's real concern in this passage is not with tantric matters but with confronting the ambiguity of the vital channel category in the *Four Treatises* itself, where it is used for both the embryonic channels and the connecting channels. Here he shows his hermeneutical acuity. The problem for *Four Treatises* commentators is that the text makes the vital channel singular in the embryonic stage and double in the mature connecting channel stage. Moreover, it also makes several further statements later in the work about the main "trunk" of the black and white channels, which taken together are themselves confusing, and it is not clear if the work's author sees the trunk/s as one or two. Nor is the trunk/s' precise location in the body made entirely clear.

Kyempa says explicitly that there are two connecting vital channels, and indeed, that is what the *Four Treatises*' initial discussion suggests:

> The connecting channels are two, white and black.
> The [black] vital channel is the trunk of the channels;
> like branches they ramify upward from that.
> Twenty-four large channels produce flesh and blood. . . .
>
> From the great ocean of channels in the brain,
> [the white vital channel] reaches down like a root,
> [from which ramify] nineteen water channels that make for movement.

In his comment on these lines Kyempa takes the opportunity to cite a further statement on the black and white channel trunks from later in the root text, in the *Instructional Treatise,* which adds some additional information. But this second statement seems to suggest that the two trunks run together, and the number of the noun is not clear. Kyempa writes,

It says in the *Instructional Treatise*,

> The trunk[s] of the inner channels,
> the white and black vital channels,
> stand[s] up inside the spine like a pine tree.
> From that, all the branches, the minor channels, ramify.[86]

On that, *Black Myriad* says, "The white vital channel travels upward from inside the spine, to which it is attached. The upper part of the channel ripens in the heart as a fruit. The lower part of the channel ramifies and gets thinner, and goes down in the spine[87] to disappear in a hole in the tailbone. The black vital channel stays without being attached anywhere, going upward inside the [thoracic] cavity." Thus is it explained.[88]

Here *Black Myriad* seems to contradict the *Four Treatises*' original statement in the *Explanatory Treatise*, where the white channel is clearly said to descend from the brain, and surely it would be the black channel that ripens in the heart. In considering *Black Myriad*'s statement, Kyempa shows his awareness of the confusion about where the main trunks of the black and white channel actually are, and whether they really are one or two. He also addresses the same statement about the pine tree in his commentary on the *Instructional Treatise* itself, where he repeats his position clearly that it refers to two distinct channels, but specifies for the first time that the two are attached to each other but do not intermix, and also that the black one runs in the front and the white one in the back.[89] At another point he also addresses yet one more relevant statement in the *Instructional Treatise* on the black and white channels, which reads,

> There are two kinds of channels that are vital channels.
> The black one comes from the vital channel and ramifies upward.
> The white comes from the brain and reaches downward.[90]

This brings Kyempa to specify that the vital channel runs in the spine, standing up like a pine tree, from which the black blood channel ramifies upward, whereas the white wind channel runs downward out of the brain and is like a ligament.[91] Kyempa thus makes clear that he understands the *Explanatory Treatise*'s initial description to mean that both connecting vital channels run in the spine, although it does not say that. It also only calls the black channel a trunk.[92]

But the imagery of a trunk, or a pine tree, or the straightness of the spine itself is important, for they all suggest a likely candidate for equation with the tantric central channel. Kyempa puts together the *Four Treatises*' elliptical statements on the vital channels to make some specifications of his own in precisely this direction. In his comment on that last *Instructional Treatise* quote, he goes on to claim that the vital channel from the *Four Treatises*' embryonic channel discussion, the one that develops in the sixth week of gestation, is the same as the combined trunk of the black and white connecting channels.[93] No one had made that statement before, at least in what we now know of *Four Treatises* commentarial history, and it would seem to contradict the *Four Treatises*' own specification that the connecting channels develop in the seventeenth week of gestation, i.e., eleven weeks later.

But perhaps floating that equation supports his boldest claim of all, whereby, back in his commentary on the *Explanatory Treatise,* he goes on to identify the white vital channel with the tantric central channel. He has already suggested that the embryonic vital channel is the tantric central channel. Here is his companion assertion about the tantric identity of the mature connecting vital channels:

> The white vital channel is in accord with how the central channel is discussed in Secret Mantra. It is the basic foundation upon which depend all the channels—growth, being, and life channels. It appears that the black vital channel, that in which blood flows, is in accord with *roma.*[94]

Here Kyempa makes his tantric specifications clear. The white vital channel is in accord with the tantric central channel and the black vital channel is in accord with *roma.* Kyempa has left a small escape hatch by adding the qualification "is in accord with." The same can be said of the further qualification "it appears," regarding the concordance of the black channel with *roma.* But when he exploits one more quote that he also finds in the *Instructional Treatise,* which contains specifications about moxibustion at the bottom of the white vital channel, it is evident that he really considers the white vital channel itself to *be* the central channel:

> In the *Instructional Treatise* it says,
>
> Treatment for the king of channels, the vital channel, includes
> burning at four fingers' length below the navel,

at the red eye channel, at the *aso* channel,
at the cavity of the collarbone, and at the sixth [vertebral] digit.
Bleed the *long rtsa, rtsa chen, rtse'u chung* and the *snod ka.*[95]

Saying to burn at four fingers' length below the navel of the white vital channel has the same meaning as the bottom point of the central channel. To burn the others, since the arteries block the path of wind, [the text] explains that one should bleed the veins such as *rtse'u chung* and *long rtsa* et cetera, which are parts of the black vital channel.[96]

He is pointing to the fact that the *Four Treatises* suggests that the main trunks of the white and black vital channels run together at some point, as he reiterates later in his comments on the *Instructional Treatise.* Again, his main concern is to provide a detailed description of where these prosaic medical channels are. But he also notes in passing a concordance with the tantric central channel. The logic seems to be that the specifications he has found in the *Instructional Treatise* for where to burn the bottom of the white vital channel match descriptions of the bottom of the tantric central channel, which is also often said to be four finger-lengths below the navel.

Kyempa does not locate the central channel further in the *Four Treatises'* statements. He just proceeds for another six long pages in the modern edition to specify the location of all of the smaller white and black connecting channels. Like his predecessors, he mentions bridges to the tantric world only incidentally. But his position remains clear: the tantric central channel and *roma* are in close conformity with, if not equal to, the main trunks of the white and the black vital channels, i.e., the nervous and cardiovascular systems, in the mature human body. He looked hard for whatever he could cull from the *Four Treatises* to locate these tantric channels in a spot recognizable to medicine (curiously, though, he does not seem to be concerned about finding the *kyangma*). In this effort, he appears to be the first to discover the tantric central channel in the *Four Treatises'* connecting channels. He has located it far more centrally within the orbit of the actual practice of medicine than among the much vaguer life channels, where Sönam Yeshé Gyeltsen tried to find it.[97] Now the tantric central channel has been equated with what is clearly part of the empirical body, the spinal cord, although Kyempa does not invoke observability as such in the course of this argument. But his solution would mean that the aporia created by any potential examination of the thoracic cavity vis-à-vis tantric knowledge systems would be erased. You would be able to see the tantric central channel: it would be the spinal cord.

In sum, everything we have seen so far does not show the medical commentators to be challenging the material existence of the tantric anatomy in the adult body, on the order of what was suggested by Yangönpa's unnamed opponent. Quite the contrary, there was a growing effort to find that material existence by lining up the medical and tantric systems. Kyempa brings in tantric information the most freely, but at the same time he is aware of the leap in doing so. He is not just making natural juxtapositions, as seems to have been assumed previously.

With Zurkharwa's intervention, this awareness intensified. Now there was a sense that one had to defend bringing the two systems together. And although Zurkharwa does provide such a defense, he does not seem so confident that the two systems are entirely compatible. We start to detect a turnaround, whereby the tantric is something that needs to be accounted for, but not in terms of the medical and empirically based description of the body—at least as much as that is possible to avoid. There is also for the first time an explicit acknowledgment of the most direct and radical form of the challenge, namely, why the tantric channels cannot be found in the human corpse.

ZURKHARWA LODRÖ GYELPO, AND A SPACE APART FOR MEDICINE

Zurkharwa is far less controversial on the channels than he is in the Buddha Word debate. But his position is even more complex, if that is possible. This no doubt has to do with the fact that he himself is split on what he wants to say. His extended comments betray a desire to provide a plausible account of the tantric channels, which he thought had real influence upon the body and demonstrable efficacy in the salvific domain of yogic practice. But he is extremely cautious and controlled. His commitment to empirical accountability is strong, and he is loath to posit things that cannot be seen. Much as he wants to account for it, he is also very bold in limiting tantric purchase on the medical body. What we see then is a man on a tightrope, or serving two masters at once. Zurkharwa's discussion provides a prime example of how science and religion collide, and yet how, with the greatest of care, they can be made to work together . . . or almost.

One thing is sure: Zurkharwa exceeds his medical predecessors by far in the amount of time he devotes to the tantric channels. That in itself is a sign of the importance he ascribes to getting it right. He spends a particularly long time refuting the many wrong views he knows of, including several of the moves just chronicled.

Strikingly, Zurkharwa bucks the trend in medicine to find the tantric channels in the physically observable body. If he sometimes falls short, it is because he can't get rid of certain arcane specifications from tantric tradition about the channels around the heart, or because it is all too tempting to borrow tantric specifications about the female sexual organs, on which medical knowledge was wanting.

But far more than the few instances where tantric channels briefly show up in his account of the functioning mature body, Zurkharwa's main strategy is to locate them in the early weeks of the life of the fetus. It is there in the embryonic phase that he identifies the tantric channels in detail, and in just the terms that the tantras describe them. Surprisingly, Zurkharwa demurs from locating the tantric channels in either the channels of being or the life channels, as some of his predecessors did. Perhaps that is because they are low-hanging fruit. Or perhaps it is because both hardly ever, if at all, figure in medical practice or theory apart from the *Four Treatises*' initial description.

I think that Zurkharwa favored the embryonic moment to locate the tantric channels in the body for several reasons. Though he was likely informed by a few Indic and Tibetan tantric works that had already described the embryological origins of the tantric channels, that is not Zurkharwa's concern here, and he does not cite those passages. Rather, he depends, like Yangönpa did, on tantric statements about the mature anatomy.[98] It is the mature tantric channels that were at stake in the issue we have been tracking, and it appears to be these that Zurkharwa is locating in the *Four Treatises*' gestation stage, the unlikely logic of such an idea notwithstanding. But doing so, I would submit, serves to render them unavailable to empirical scrutiny. After all, the fetus at six weeks, which is when the tantric channels would appear, is ensconced in the mother's womb and is also rather tiny, little more than a blob.[99] Thus this fetal stage becomes a safe space, so to speak, to air the tantric system and ostensibly to account for it. Yet the tantric channels' appearance at this originary moment does allow them to have an impact on the body as it develops, especially on the channels that really do have medical functions and are observable in the adult, namely, the *Four Treatises*' connecting channels. The relationship between the tantric channels and the medically accessible connecting channels thus becomes one of causality—and of constituting an ultimate foundation to which Zurkharwa will be able to point for a variety of purposes. And if we are never told just how the tantric channels themselves function after their initial appearance, and whether they are still standing in the mature body as described in the tantras, that is because Zurkharwa elides such questions and, apart from the vaguest gesture, never pins these matters down. His skill in ambiguity is nowhere more evident than here.

Zurkharwa makes it entirely clear, however, and on expressly empirical grounds, that the tantric channels are not actually to be equated with the connecting channels, that is, the black and white system of veins, arteries, and nerves seen and accessed by the physician. So even if these bear a debt to the tantric channels of the fetus, Zurkharwa has managed to keep the two systems separate. He thereby preserves a space for the connecting channels to be considered on their own terms, which has to do with everyday human physiology and the practice of medicine. In short, he has managed to simultaneously accommodate and divide two separable but related systems of anatomical knowledge.

Note the difference from Yangönpa's strategy, which also separated two domains. Not only does Zurkarwa pay intricate attention to the everyday medical body that Yangönpa largely dismissed; he also demurs about Yangönpa's principal claim that the tantric channels are invisible because they defy normal rules of observable matter. Zurkharwa does once take recourse to the subtlety of the tantric channels, but again, this is in the embryonic phase, and is likely influenced by the many tantric scriptural passages that he cites. There is even one sentence where Zurkharwa sounds like Tashi Pelzang in maintaining that just because you can't see something does not mean that it is not there. But he largely refuses this potentially powerful argument. So it may well have been that he believed that the tantric channels continue to abide in the adult body in subtle and imperceptible form but was very reluctant to make such a case, save in the embryology and in the voice of the tantric scriptures. I think he was circumspect because the tantric picture was at odds with the medical mentality. Subtle or not, either something is there or it isn't, and if it is, it should be locatable. But that is something that Zurkharwa would barely venture for the tantric channels, and even then only in a vague, noncommittal way, quite in contrast to Yangönpa's successors, who actually pinpoint the location of the tantric channels in the everyday body.[100] Zurkharwa's strategy seems rather to make the tantric channels merely implicit in the adult body. This would be an implicitness that looks to both their imprint from the past (in the fetus) and their power for the future (when yogic meditation is performed). He never works out how meditation mobilizes that power, but he implies it. That, apparently, is the most he will do.

In the end, it is yet another effort to move in two directions at once: a classic case of trying to have cake and eat it too. Yet I submit that Zurkharwa still tips the scales in the direction of the empirical accountability of medical science. Despite his detailed attention to the tantric channels in the embryonic phase of human life, he barely returns to them in his medical masterwork again. When they briefly resurface, they do not figure in practices of diagnosis or therapy but serve rather as an explanatory device, to account for the gendered style and orientations of the body's dispositions.[101] In placing the tantric channels' physical force in the embryo, Zurkharwa gave them influence on the patterns of human embodiment. But he got them out of the way of the practice of medicine, which very much depended on the other kind of channels, the ones that needed to be located in ordinary physical terms.

SELF-CONSCIOUS OF HIS METHOD

Zurkharwa's self-consciousness of the liberties he is taking on the channel issue may be his most significant advance of all. He broaches the question several times, but most prominently in the following defense of using tantric knowledge to unpack the *Four Treatises*' embryonic phase. He is using the same verbal root that Kyempa deployed to name the act of juxtaposition (variants of *sbyar/sbyor*). It is a sign of his scholarly sophistication that he can readily acknowledge that different streams have fed medical knowledge but can also show why they can still mix together.

> Someone with an analytical mind might say, "It is not right to be juxtaposing all this information with the Secret Mantra, since this text of yours explains the Causal Vehicle."
>
> If such comments were to arise, this would be the rejoinder: I may be unpacking a great Causal Vehicle textual tradition, but even so it is not improper to juxtapose thereby all its meanings with the Secret Mantra. For example, just as it is all right to combine kinds of medicine coming from Vedic texts, Vinaya references, *Suvarṇaprabhāsottama*, and *Kālacakra*, and so on, here also it is simply a valid juxtaposition, all in one place, of

> authoritative sources that represent a consolidated understanding of all the scriptural resources. In particular, this text [i.e., the *Four Treatises*' verses on the growth channels] appears to be in rough accordance with the channel, wind, and *tiklé* system of the Secret Mantra tantras. But if it were to have explained that overtly, using the terminology that is well known to Secret Mantra practitioners, there would be some confusion that [the *Four Treatises* is synonymous with] the Resultant Vehicle, and so it discusses [that material] in a covert way.[102]

Zurkharwa here suggests a conception of the *Four Treatises* that accords well with our own sense of its historical background. As in the Buddha Word debate, he references the variety of sources to which the medical knowledge in the *Four Treatises* is indebted, including Vedic tradition (down to and including Ayurveda), the Buddhist monastic code, and both exoteric and tantric Buddhist scriptures. He makes this point rather obliquely, referring to some imputed practice of mixing together medical preparations described in those disparate sources (this itself is odd; at least, there are no medical recipes in the *Suvarṇa*). But what is significant is his proposition that the *Four Treatises* itself was self-conscious in its presentation of the body's channels. Zurkharwa, like Kyempa did briefly, suggests that the work's author was deliberate in disguising some of its sources that were incongruous with what readers might expect. The *Four Treatises* represented tantric knowledge only implicitly, in order not to confuse medical knowledge, which Zurkharwa classes as part of the exoteric Causal Vehicle, with the Resultant Vehicle, i.e., tantric teachings. That does not mean, however, that the perceptive exegete can't bring the latter out.

But even while paying homage to the special knowledge preserved in tantric tradition and its possible relevance for medicine, Zurkharwa has separated the two taxonomically. Tantra and medicine *can* be mixed, when needed, and have in any case long been compatriots in the complex history of the understanding of the body. They are different streams nonetheless.

THE TANTRIC CHANNELS OF THE HUMAN FETUS

That the issue of the tantric channels is a major concern for Zurkharwa may be seen from the fact that he addresses the problem at the very first moment that it can come up in a *Four Treatises* commentary: not the chapter that lays out the types of channels, but the chapter on embryology. Some of Zurkharwa's predecessors—Sönam Yeshé Gyeltsen, and then Kyempa—already gestured in this direction for a place to find the tantric channels, although in very minimal form. Nor did they take great care, as Zurkharwa does, to avoid conflating the embryonic vital channel with the mature "connecting" vital channels.

Zurkharwa is determined to work out the matter of the tantric channels as much as possible in the context of the embryo. He brings it up at the point when the embryo emerges from a bloblike set of stages that the *Four Treatises* dubs with onomatopoeic terms like *gor gor* and *mer mer.* This is in the fetus' fifth week. Zurkharwa starts, as Kyempa did, with the ambiguity of the vital channel rubric.

> At the fifth week the [fetus] passes beyond its previous forms such as *gor gor* or *mer mer*, et cetera, or whatever, and becomes solid and hard, for which reason it is called the day of becoming hard. At that point, the navel, which is like a root that gives rise to all the channels, winds, and *tiklé* along with the vital channel, gradually comes to grow.
>
> With regard to this, we see that this thing known as the vital channel is discussed in a variety of ways. So we might examine it a bit.[103]

Zurkharwa has already tipped his hand in suggesting that the classic tantric trio of channels, winds, and *tiklé*—terminology that never appears in the *Four Treatises* itself—are born from the embryonic navel. But first he has to tackle the category of vital channel. He is aware of its complex history, but the ensuing discussion considers not the medical vital channel at all (as did Kyempa's parallel remarks), but rather the tantric central channel. It is as if Zurkharwa has already assumed that the vital channel and the tantric central channel are synonymous in this context. But he is also clear that there is a question about the real existence and availability of the tantric central channel to empirical observation, in what he specifies as the everyday (or ordinary; *tha mal*) body.

> First, [the refutation of wrong views]:
>
> Some Secret Mantra practitioners say that the discussion from the tantras laying out *ro, kyang*, and the central channel is only pertinent to meditation but these are not there in the everyday body.

> This is not valid. When one abides in the everyday body, even if you meditate constantly on channels that have no substantiality, you will not be able to bring them into existence.[104] For example, if you meditate that there is a wild yak's horn on the head of a horse, there will never be a time that it will actualize.[105]

Zurkharwa represents here the most radical skeptics on the material existence of the tantric channels, who would allow that the system of tantric channels is relevant to meditation but has no ongoing substantial existence (*gtan med*) in the everyday body. This is further glossed by Zurkharwa's own example of the horn of the wild yak on a horse. The putative objector is suggesting that the tantric channels are incommensurate with the material nature of the body. The point parallels Yangönpa's representation of an argument that the tantric channels are just an imputation, but Zurkharwa provides a fuller rendition. Such imputation would consist in meditative visualization.

We know that Zurkharwa was using Yangönpa's old essay—to my knowledge, Zurkharwa is the first of the medical commentators to cite him—and in many ways their arguments are parallel.[106] Here Zurkharwa elaborates what is largely the first point that Yangönpa made. Interestingly, on Zurkharwa's reading it was *tāntrikas* themselves who were struck by the invisibility of the tantric channels and tried to retreat to a position that would consign that anatomy to the projections created by meditative practice.

It is significant that Zurkharwa does not take that bait, as we might expect an empiricist to do. Instead he, like Yangönpa, refuses this solution to the tantric anatomy problem, and on similar grounds, although he makes the point in more physicalistic terms. Zurkharwa insists that the tantric channels must have some substantial existence in their own right: consigning them to the imagination is not acceptable, for they would remain confined there. Reading this together with Yangönpa's conclusion that such a move would prevent one from traversing the tantric path suggests that for both Zurkharwa and Yangönpa, the tantric channels have to have some physical purchase on the body, since tantric meditation is indeed efficacious. This betrays an empiricist presumption in itself, i.e., that tantric meditation works in discernible ways.

Zurkharwa is indeed concerned with dealing with the problem in empirical terms. He has the imagined critic spell it out very clearly:

> But he responds: "The channels as they are described should be able to appear in actuality in the dead body. But they do not so appear, and therefore [I voiced this opinion]."
>
> To that, some say that since the channels, winds, and *tiklé* are mentally based, they stay for as long as you are alive, but when life's time is finished, they disappear, along with the mind, like an autumn rainbow. Therefore they do not appear [in the corpse].[107]

The objector's statement that the tantric channels should be able to "appear in actuality" can also be translated as "appear to direct perception," if we take *mngon sum* in the way it is used in Buddhist epistemology. The objector goes on to distinguish what is actual in the body and what is merely imagined in the mind, offering the possibility that imagined tantric elements could be effective and have a kind of reality of their own, but they would disappear with the death of the mind.

In Zurkharwa's negative response to this ingenious solution, he really does sound like a *tāntrika*:

> However, my position is not like that.
>
> From the *Commentary of Avalokiteśvara*, "The *bodhicitta* sprout becomes the support for the vital and other winds as the ten kinds of channels in the environs of the heart. These [channels] are extremely subtle, the mere size of a hair." And, in the *Sampuṭatilaka*, "The channel that abides inside of that has the nature to manifest as not-manifest. It has no parts and is exceptionally fine. It has a light of great splendor, and abides always in the center of the heart." And in the *Mahāyogatantra*, "It has thirty-two knots. Finest of the fine, it is perfectly precious."[108]
>
> Just as that says, since it abides as fine, for those who look toward it, it is not possible for it to appear, but other than that, it is not the case that it is not there, even in the dead body.

In adverting to Indic tantric sources, Zurkharwa relies on just the argument—and at least one of the same tantric sources—that Yangönpa invoked, which marks the entire tantric position. This maintains that the tantric channels stem fundamentally from *bodhicitta,* i.e., the most exalted seminal substance that is synonymous with the mind of enlightenment. The channels are very fine, or subtle, described in terms that belie conventional physical existence. This intermediate category of subtlety of course decides in advance that the evidence from direct perception will not be relevant. Zurkharwa goes on to assert that even though the tantric channels are not available to empirical observation that would look "toward" things—i.e., as objects in front of one—that does not prove they are not actually there, thus holding on to some claim for their physical existence after all.[109] The passage is surprising for a medical theorist. At least in this moment, he is deploying the kind of imperious argument that Tashi Pelzang used to defend the Indic origins of the *Four Treatises*.

Zurkharwa is complicated. But he is not done yet. In the very next line, he rejects any location of the tantric channels in the organs or channels of the mature body, as his medical predecessors had been trying to establish. "As for those who posit that those [tantric] channels always remain in the body, their positions are not in accord, as follows: . . ."[110] This statement is key to Zurkharwa's strategy. Rebutting the position that the tantric channels "always" remain in the body—in its fully developed form, throughout life—still leaves open the possibility that they are there in the embryo. But Zurkharwa is not ready to make this point explicitly. First he will disqualify the various solutions that his medical predecessors offered for finding the tantric channels in the mature human body. These don't "accord" with the way the tantric channel is defined in authoritative scripture.

> Some posit that [the tantric central channel] is the spinal cord. Some posit that it is inside the enclosure of the vital channel. Some posit that in the distinction between the white and black vital channel, it is the white one. So there is a variety [of positions].
>
> As for those, if it were as the first one posits, then it would contradict the fact that the central channel is explained to be hollow and to touch the center of the heart. If it were as the second one posits, it would contradict the fact that the aspect of it that moves as vital wind (*srog rlung*) in the form of an *A* is explained as being Rāhula. If it were as the third one posits, it would contradict the fact that the white vital channel conducts the white element and is thus explained as a water channel. So all of these are not seen to be valid.[111]

Zurkharwa eliminates all positions known to him that would equate the tantric channels with features of the medical body, that is, the mature postnatal body that is available to medical practice and inspection. The first is that the central channel is the spinal cord, as Kyempa proposed. But Zurkharwa disqualified this because the spinal cord is not hollow and does not do what the central channel is supposed to do, according to tantric description, which is to touch the center of the heart.

The second idea, more arcane than the others, is that the tantric central channel runs inside the "enclosure" of the vital channel—the sense of which is unclear here.[112] Evidently he is referring to some part of human anatomy—perhaps the spinal column—and the impossibility of engaging it for tantric practice, in which vital wind moves in the tantric central channel together with mind (i.e., in the form of the letter *A*, frequently glossed as the Indic astrological planet Rāhula).[113] Zurkharwa is thus refusing the idea that this enclosure is equivalent to the tantric central channel.

The third idea, which was Kyempa's position, is to identify the tantric central channel with the white vital channel. This too will not work because the latter, according to Zurkharwa, conducts water, rather than the wind and mind that are supposed to travel in the tantric central channel. This is a surprising claim, and Zurkharwa will make it again. But it is patently false. It is indeed the case that the *Four Treatises* calls the white vital channel a "water channel."[114] But that term is really a reference to the white channels' appearance like a ligament (*chus pa*), which is indeed white, and the word for which is homophonous with the word for water (*chu*). As the *Four Treatises* itself makes clear on several occasions, wind runs in the white vital channel; this is the basis for the entire idea that the white channel system is responsible for bodily movement.[115] The fact that Zurkharwa twists the facts in this way is a sign of how much he is invested in keeping the white vital channel separate from the embryonic vital channel. This is the first of several places where he maintains forcefully that neither of the connecting vital channels in the mature body is to be equated with tantric channels.

Zurkharwa now goes to his own position on the embryonic growth channels: to locate tantric elements at the formative moment of the first appearance of the navel and the embryonic vital channel. This allows him to say anything about their existence in that originary phase that he wants to, since those channels would not be subject to inspection. What Zurkharwa is suggesting—and what the *Four Treatises* itself indicates, despite Kyempa's claim to the contrary—is that the embryonic growth channel is a temporary structure that gives rise to the mature connecting channels and other elements of the body as the fetus grows and eventually leaves the mother's womb.

Zurkharwa opens his account of his own position with a brief caveat on the covert place of tantric physiology in the *Four Treatises*, anticipating his fuller methodological defense later that tantric knowledge is implicit in the *Four Treatises*' channel discussion.[116] Indeed, implicitness is what Zurkharwa's reading is going to be all about. Far more than leaning on an idea of subtle physicality, as *tāntrikas* like Yangönpa do, Zurkharwa makes his primary solution about basic nature, or a generative association.

Zurkharwa betrays this in the very next lines. The mature white and black vital channels are on his reading somehow the result of the embryonic tantric channels—and yet not those channels themselves. The statement is a masterwork of ambiguity. The two side tantric channels, *roma* and *kyangma*, are "the foundation" (*gzhi*, a word that is already open to a range of interpretations) to which are "related" (*'brel ba*) the white and black vital channels. Here and later, the key term "related" could be read instead as meaning "connected," in the sense of physically connected, which in turn could mean that the tantric channels do remain in the adult body in some overt form. It also plays on the moniker "connecting channels" (*'brel rtsa*) that names the black and white vital channels. But Zurkharwa never makes either point expressly or with any specification. Rather, as his treatment unfolds, the ambiguity seems quite deliberate and strategic.

> Moreover, below [in my commentary on] chapter 4 [of the *Explanatory Treatise*] there will be an explanation of the two, *ro* and *kyang*, which are the foundation to which are related the white and black vital channels, and an explanation of the channels of being, which are discussed in terms of the meaning of the *cakras*.

In saying this Zurkharwa is aware of the importance of context. He is alerting the reader that he will also reference the tantric channel system later in his discussion, in his commentary on the connecting white and black vital channels, and will further analogize the matrices of the channels of being to the tantric *cakras*. Both moments will reference the mature body—but neither the white and black nor the channels of being will amount to the tantric channels. In any event, for now, Zurkharwa is only going to talk about "this context," i.e., the embryonic phase:

> In this context, that which is called "vital channel" is really the central channel itself. This is because it is that which is the basis for the vital wind to have the shape of *A*, and is the vital axis (*srog shing*) of the four or six *cakras*. And also since it is the channel that develops alone, the first of all the channels.

Zurkharwa says in no uncertain terms that the *Four Treatises*' embryonic vital channel is to be equated with the tantric central channel. In qualifying his assertion with the terms "really" and "itself," he stresses the incontrovertibility of the identification.[117] There is no implicitness or other kinds of vagary or avoidance. He also adduces two brief reasons in support of this identification. One is not really an argument, but merely equates the embryonic vital channel with the main elements of the central channel in tantric sources: it is the basis for the vital wind, often visualized as the letter *A*, and it has a number of *cakras* along its length. The second reason associates the fact that the embryonic vital channel is the very first channel in the embryo's body with the fact that the central channel is the most important and basic tantric channel. He seems to be trading on an assumption that first and most basic are the same.

Zurkharwa then launches into another set of tantric quotes, providing more standard fare on the central channel.[118] One of them, from the Tibetan work *Profound Inner Meaning*, also mentions that the central channel is the first to develop in the body, but curiously, Zurkharwa does not cite that work's embryology chapter, which does support his claim about the central channel in the fetus. Instead he cites its chapter on the channels in the fully developed body, the same quote provided by Kyempa, in fact, where he suggested, albeit only elliptically, the identification of the embryonic channel with the tantric one.[119] Among other things, the passages supply a measurement: the channel is twelve and one-half finger-widths long between each of the *cakras*. The work's self-commentary, which Zurkharwa brings in here too, also says that the central channel is the basis for foundational consciousness as it "abides in the present time." Both quotes would be talking about the central channel in the adult body, and suggest what in any case most tantric sources presume: the tantric channels and *cakras* remain in the body throughout life. Zurkharwa never says that in his own voice, however. And in any event, he is ostensibly talking about the embryonic phase.

It is possible that Zurkharwa uses these quotes to support other elements of his own account. They confirm that it is wind, or vital wind, that travels in the tantric central channel, a crucial point that disqualifies for him any identification of the central channel with the medical white vital channel such as previous theorists had asserted.[120] They also attest to the occurrence in tantric works of the ambiguous term "vital channel" as a synonym for the central channel.

In the following section, Zurkharwa turns in fact to a tricky issue around his equation of the vital channel with the tantric one. He is responding to a critique akin to what Yangönpa had claimed: if the central channel were the vital channel, then what yogis try to do—bring the vital winds into it—would be dangerous and would cause insanity and disease.[121] This is a potential sticking point that could derail Zurkharwa's larger strategy. He *is* saying that the central channel is the vital channel, but he means the vital channel in the embryo, in accordance with the *Four Treatises*' account. Later, when the vital channel rubric pertains to the mature body, it refers to two different channel systems, the black and the white, which Zurkharwa makes very clear are *not* the tantric channels.[122] But apparently for Yangönpa and perhaps others, "vital channel" refers to a single channel running in the adult body, into which a yogi must not draw the vital winds.[123] Zurkharwa appears now to be addressing this alternate sense of the term. He goes to some length to distinguish two kinds of vital winds, those that are disturbed and bring illness and those that are not. He adverts to a disparate point drawn from medical knowledge that there are in any case various kinds of winds in the heart—including vital winds. These do not make people ill, he avers, as long as they are not disturbed.[124] Zurkharwa maintains that the yogic practice to bring winds into the central channel only deals with winds that have already been pacified through that very yogic practice, and thus would have no negative impact on the body.

In making this case, Zurkharwa is indeed talking about a central channel that is available for yogic practice in the adult body (since it is generally adults who do tantric yoga, not fetuses). But that is not Yangönpa's concern; rather, he is worried about doing yogic practice with his own notion of the vital channel, which for him is a part of the ordinary body, and not the same as the tantric channel. So apparently Zurkharwa also felt the need to consider the implications of this alternative conception and address the worry that an ordinary part of the body would be vulnerable to confusion if it contacted the vital winds. Perhaps Zurkharwa is just trying to cover all his bases, given the ongoing ambiguity of the vital channel rubric. Even if the tantric central channel were equivalent to a part of the ordinary adult body, there would not be a problem if the vital winds entered it.

In Zurkharwa's final section on "the meaning of the words," he recapitulates his position, and once again he seems to be splitting the difference, or addressing several perspectives at once. In the sixth week of gestation the "central/vital channel" (*dbu ma 'am srog rtsa*), the receptacle for vital wind, ascends upward. It goes up twelve and one-half finger-widths and forms the *cakra* at the heart. Zurkharwa goes on to specify that although the *Four Treatises* does not say this, in the seventh and eighth weeks the vital channel continues upward another twelve and one half units and forms the *cakra* at the throat, and then again up to the crown, where it forms a *cakra* and the basis of the eye organ, and it also goes down to form the secret *cakra*. All the *Four Treatises* says in this context is that the vital channel arises in the sixth week and the faculty of the eyes arises in the seventh week.[125] Zurkharwa is clearly drawing in information from the *Profound Inner Meaning,* but again, it comes from that work's chapter on the mature body, not the embryology. The latter in fact measures the entire central channel of the fetus as twelve and one-half finger-lengths.[126] What's more, if the finger-length unit refers to the fingers on the body it is measuring, as it usually does, it would make little sense here since the fetus does not have any fingers at all until week sixteen, at least according to the *Four Treatises.*[127] Or if he is referring to the adult measurements, why give them here in the embryology section? It really seems as if Zurkharwa is just loading the entire tantric anatomy into the embryology, and rather rotely, just to account for—and perhaps confine—it there.

In any case, by the ninth week, Zurkharwa continues, the overall shape of the body has been formed with the exception of the four limbs. All the winds have been formed as well. He closes the entire section by saying that "at this time, through the power of the vital/central channel, the body is formed into a single stem, and comes to be like a fish, and so it is called 'the period of the fish.'" He is referencing the old system related to the ten avatars of Viṣṇu as mapped onto the embryo, commonly invoked in tantric works, as in Yangön-pa's essay.[128] The function of the vital channel cum tantric central channel in Zurkharwa's system is to form the basic contours of the fetus' body.[129]

EMBRYONIC CHANNELS REDUX

Further testimony to how fraught the matter is for Zurkharwa is the fact that he adds another seven long pages on the relation between the embryonic and tantric channels when he gets to the *Four Treatises*' main discussion of the channels of the body.[130] Now, perhaps because the *Explanatory Treatise*'s fourth chapter has to do with all four kinds of channels and mostly with the mature body, Zurkharwa is more precise and also more overtly concerned with what is empirically evident and what is not.

Zurkharwa first rejects, as he is ever wont to do, others' wrong views on the growth channels of the fetus.[131] He notes that generations of old did not explore the matter in detail, but in later times some scholars offered some ideas. There is no space to go through each of these technical discussions here. But one interesting point is his refusal to connect the *Four Treatises*' equation of the embryonic brain with stupidity to the tantric central channel, on the grounds that the wrong things, i.e., things other than wind, flow in the portion of the embryonic vital channel that gives rise to the brain. This leads him to refine the position he presented in the embryology section: really, the embryonic vital channel may only be said to be the central tantric channel in its lower portion, where indeed wind does flow. Apparently he is balking at having to associate the tantric central channel with stupidity (even though he allows it to be associated with sexual desire).[132]

When Zurkharwa goes on to provide another extended defense against equating the mature connecting vital channels with either the embryonic channel or the tantric central channel, it becomes one of several occasions for him to reject the possibility that a tantric channel, in this case *roma,* could be empirically observable (*mngon sum du mthong rigs pa*). That is a move he will not allow, and that in any case contradicts the way *roma* is described in the tantric sources. These do say things about *roma*'s location in the adult body, viz., on the right side of the body, not attached to anything at all, and very subtle. In referring to this description, however, Zurkharwa is quick to add the qualifier "so it is explained," displacing its authority to tantric tradition.[133] What Zurkwarwa himself wants is to disqualify any equation between the mature connecting channels and the tantric channels as they are defined in tantric sources. Most of all, his discussion shows that he regards the black and white vital channels as observable, a trait not shared with the tantric channels. As he states,

> If for you, that thing called vital channel is an outer *roma* in which blood flows, that is to propose that the black vital channel explained in the context of the connecting channels is itself *roma*. That would also mean that the central channel and *kyangma* are in size like *roma*, and empirically observable, but that is not the case, and so it is not valid.[134]

This is the point that Zurkharwa wants to drive home. The tantric channels—even as mitigated by the "outer" rubric such as Sönam Yeshé Gyeltsen suggested—can't be located in any sort of observable form.

Now Zurkharwa begins to spell out his own position, although it has changed from his prior discussion:

> My own basic idea is posited as follows: Because the first of all the channels develops alone, it is said [in the *Four Treatises*], "The growth channel ramifies into three from the navel."
>
> Now the central channel above the navel has already been explained in detail in the section on the development of the body.[135] As for this quote, it shows, in order, *kyangma*, *roma*, and the central channel below the navel.
>
> The channel that gives rise to phlegm, et cetera, is *kyangma*, which is a water-conducting channel. The channels that give rise to bile, et cetera, are *roma*, which is a fire-element-conducting channel.[136] The channel that gives rise to wind, et cetera, is the wind-possessing channel, which is the part of the central channel that is below the navel. The bottom tips of *ro* and *kyang* are taught implicitly.[137]

He is now accounting for the other two tantric channels, *roma* and *kyangma*, in his picture of the initial growth channels. The central channel is actually only the lower portion of the embryonic vital channel.

In making room for the other two tantric channels, Zurkharwa forges a crucial bridge between the embryonic phase and the prosaic functions of the mature body to come. As already noted, *roma* and *kyangma* are associated with the everyday mature body in Indic tantra, which attributes the elimination of feces and urine to the work of these two side channels. Zurkharwa now cites the *Kālacakra* and its commentaries to lay out the mature tantric anatomy of the three channels on just these points.[138] And yet the fact that he is still doing so in the context of the six-week-old fetus—and only in the voice of scriptural citations—remains curious. Certainly when he considers the channels that conduct waste out of the body later in *Ancestors' Advice*, now outside of the embryonic context, he never uses these tantric terms to name them.[139]

What's more, when he turns to fending off critiques of his own position again, he resists the temptation to provide actual anatomical specifications for the tantric channels. Representing one of his imputed interlocutors who objects to his insistent separation of *roma* from the black vital channel, Zurkharwa acknowledges that the *roma* indeed "reaches into the middle of the body," i.e., in that formative moment of the embryo's sixth week, but he distinguishes that from the black vital channel that conducts digested food substances into the liver, where they collect and make for the growth of flesh and blood. He cites here the *Instructional Treatise* for its specification of the anatomy of the black channel's connection to the liver.[140] In fact the *Four Treatises*' embryology has the liver and other solid organs developing their basic shape for the first time in the twelfth week of gestation, and they are complete and functioning only in the twenty-fourth week, a full eighteen weeks after the growth phase in which Zurkharwa places the *roma* and the other tantric channels. The black and white connecting channels themselves are only said to develop in the seventeenth week.[141] All this suggests that when Zurkharwa goes on to associate—but not equate—the black and white vital channels of the mature body with the embryonic vital channel *qua* tantric channels, the association is a potential that will only be fulfilled later. Things like the black vital channel are not yet in existence at the moment that the embryonic vital channels *qua roma* and so on are in play.

He goes on to sum up his position this way:

> Putting it all together, if you summarize all the kinds of channels, they are subsumed into three: wind channels, blood channels, and water channels. All wind channels ramify from the central channel, all blood channels from *roma,* and all water channels from *kyangma.*[142]

Zurkharwa is now consolidating the seminal role of the tantric channels in the embryo. They constitute basic forces that give rise to all of the channels and movement of substances in the body. Still, the verb "ramify" provides another fulcrum of ambiguity. It would seem to echo the *Four Treatises*' own language in describing how the embryonic channel first ramifies into three parts.[143] The statement could also be read to say that the three tantric channels remain in the mature body, with the other channels branching out from them. But such a reading is undermined by the fact that Zurkharwa never provides the specifications of such an anatomy: where the black blood channels connect to the *roma* or where the nervous system of the white vital channels branch out of *kyangma.* In contrast, he provides the anatomy of the prosaic black and white channels of blood and wind and water in great detail.

Yet another specification offered as an aside does locate one very particular intersection of the tantric channels in the mature body. Immediately following his pointed summary of the three main kinds of channels, Zurkharwa feels obliged to briefly mention a special subtype, the arteries, which are channels in which blood and wind mix together. These ramify out of a confluence of the three tantric channels at the heart, Zurkharwa maintains. There, under the influence of an opening for the karmic and deludedly impassioned mind to enter the heart, the heart is made to beat.[144] This is a critical point that accounts for the very life of a person. In fact, this "mind-opening" of the heart does come up in the *Four Treatises*, albeit in another context to which we will return in the following chapter. The root text itself never attributes this opening to the confluence of the tantric channels. But having brought up the topic of the arteries, Zurkharwa takes the further liberty of inserting a vestige of an old tantric notion, absent from the *Four Treatises* but deeply ingrained in Tibetan conception, and deemed a valuable addition to the medical anatomy of the heart.[145]

There is one more site where the medical anatomy cedes pride of place to tantric conception as Zurkharwa completes his account of the embryonic growth channels. Turning to "the meaning of the words," Zurkharwa reads the original *Four Treatises* passage one more time.[146] He starts by reviewing the embryonic channel's threefold development, now without the tantric language. He describes how the growth channels ramify and give rise to matrices of channels, organs, emotional obscurations, and, in due course, the black and white vital channels. The discussion seems to confirm that for Zurkharwa the embryonic channels are only at work at the outset of the body's life. He portrays a succession of cause, condition, and effect, starting from the primary embryonic developments all the way to the full-fledged result in the adult, whereby, for example, "we see that when someone suddenly becomes angry the middle of part of the body gets disturbed." This condition would hark back to those primitive moments when blood, hatred, bile, and the black vital channel came to be ensconced in the center of the body. Zurkharwa makes the same point about the brain, phlegm, and the fogginess that seem to be ensconced in the head. And finally male-female desire, based on wind and the seminal *tiklé* substances, are seen to abide in the lower part of the body. Then he adds that this lower passageway can carry either wind or the "primal-awareness substance," i.e., the specially refined substances deployed in tantric yoga.[147] He supports this claim by invoking tantric sources again, including Yangönpa's work, bringing out in particular the everyday functions of the tantric channels in the lower parts of the trunk.[148] The latter include the elimination of feces and urine, as well as the male and female sexual seminal substances.[149] Even here, where Zurkharwa has clearly shifted to the adult body rather than the embryonic phase that he is ostensibly talking about, robust anatomical specifications linked to spinal digit or organ are few and sketchy. But there is one exception, when he moves momentarily into a description of the female sexual anatomy.

It is another case of leaning on old tantric anatomy. Zurkharwa cites Yangönpa on the four kinds of female genital shapes—the conch, the deer, the elephant, and the lotus—based on an arcane connection between the lower tantric central channel and notions of the female vulva from *Kāmasūtra*-influenced tantric traditions. Zurkharwa goes on in his own words to add the *samse'u*, a Tibetan gonadlike organ in both males and females that is not found in Indic sources but was influenced instead by East Asian medicine.[150] He specifies that the male white refined substance flows through *kyangma* and enters the *samse'u*, where it collects, and that the female red refined substance flows through *roma* and enters the *samse'u*, where it collects. And on occasion, when male and female get together, both refined substances go down through the tantric central channel and make for the bliss of sex.[151] There is no question that Zurkharwa is now talking about segments of the tantric channels functioning in the everyday adult body. It all follows plausibly from the fact that the *Four Treatises* passage on which he is commenting ends with the lower part of the embryonic vital channel and its function to make for sexual desire in people. It is already clear that Zurkharwa is massively importing the tantric channels into this embryonic moment. Perhaps the fact that he is already in the discursive space of Yangönpa's and other tantric accounts elicits further anatomical specification in those same terms. In particular, it seems to me that the tantric information on the female genitalia represented welcome detail that was in scant supply in the medical treatise itself.[152] But whether this intervention indicates that Zurkharwa saw the full-blown tantric image of the channels at work in the everyday mature body is far from clear. Certainly when he goes on later to take up these topics in their appropriate place in the medical treatise—as where he comments in detail on the *samse'u* in the context of explaining menstruation and sexual reproduction—the tantric language or anatomy never recurs.[153]

At the end of the day, virtually everything that Zurkharwa will ever say on the tantric channels is laid out in the two sections of his *Four Treatises* commentary that have to do with the embryonic growth channels, in the sixth week of the life of the fetus. Perhaps he feels that the medical need for empirical accountability won't allow him to develop it elsewhere. Or perhaps he is displaying his knowledge of the tantric channels here in order to give a semblance of accounting for them, while really working to confine them to a space where they are irrelevant to anything medicine really touches.

A third possibility (and these are not mutually exclusive) is that Zurkharwa is protecting and preserving some physical reality for the tantric channels but can only remain vague on their specific existence in the mature human body. However, his solution to locate them in the early formative moment of the fetus does allow him to deeply associate—and yet not identify—the tantric channels with the mature black and white vital channels of the medically observable body. The upshot can be best appreciated when he finally provides his position on the black and white vital channels themselves. But let us first briefly review Zurkharwa's handling of the other two kinds of channels in the *Four Treatises* system, the channels of being and the life channels. He finds tantric elements in both of these—just not the tantric channels as such.

TWO OTHER POSSIBILITIES

In his two-page examination of the channels of being Zurkharwa is only addressing the flowerlike formations of the matrices, toward which he gestured in his discussion of the embryonic phase.[154] Now in the portion of his commentary where the channels of being actually come up in the *Four Treatises,* he unpacks their appearance in some detail. As he did in the section on the embryo, he uses tantric language and scriptural citations, but notes again that such tantric content is only covertly present in the medical text. In the course of his discussion he mentions that the various "petals" or "folds" of the matrices ramify from where the central channel, *roma*, and *kyangma* conjoin at the matrices' center.[155] He leans again on the verb "ramify" but leaves unspecified when this ramification occurs. It would certainly seem from his statement that the tantric channels are there in the adult body, somehow holding up these very *cakra*-like tangles described by the *Four Treatises.* But other than what he says about the matrices, he provides no anatomical description or location for the tantric channels themselves.

It seems that the channels of being's system of warrens at the brain, heart, navel, and genitals was so uncontroversial that Zurkharwa didn't feel the need to refute wrong views. In stark contrast to his discussion of the other three kinds of channels, where he supplies minute critiques, here he offers only a general discussion and then the "meaning of the words." Clearly he is trying to provide a medical interpretation of the channels of being—what he introduces as the "substance upon which 'being' depends and resides."[156] He even invokes a verse from the *Aṣṭāṅgahṛdaya* on the Ayurvedic notion of the arteries (*dhamani*), whose rendering in Tibetan as "great channel" (*rtsa bo che*) perhaps misleadingly suggests a large single trunk such as the Tibetan commentators were seeking as an analogue for the tantric central channel. The *Aṣṭāṅga* verse does vaguely suggest at least one part of the Tibetan channels of being system, i.e., the *cakras,* when it likens the cluster of arteries around the navel to a wheel.[157] But the Ayurvedic picture of twenty-four arteries otherwise has little in common with the picture provided in the *Four Treatises.* In any event, Zurkharwa does not attempt to identify this "great channel" with the tantric central channel. His concern is about the matrices themselves, whose intricate intersection of smaller channels take care of experiential and developmental functions of the body, from "the beginning of one's body through its middle and to the end."[158]

As his discussion ensues, there is a brief example of other Buddhist imagery serving medicine to make sense of the body. The channels of being's matrices are like a king, Zurkharwa suggests, functioning like the classic Buddhist *indriyas,* or sensory and sexual capacities, to rule and stand at the center of the organization of a country.[159] But while the *Four Treatises* characterizes these four matrices as controlling human sense perception, memory, consciousness, and desire, it offers little information on how that works. In the end the channels of being have virtually no purchase on the rest of the anatomy or therapeutics of the *Four Treatises,* and there is little anatomical specification save the ideal notion that each matrix has twenty-four main channels and five hundred minor ones. Zurkharwa himself adds information on the matrix at the heart, mentioning four channels that touch it from four sides. This amounts to another elaboration of the same locus around the heart that, as noted earlier, received much attention in both Tibetan medical and tantric theory. In addition to supplying a tantric source, the *Sampuṭatilaka,* Zurkharwa also finds one quotation from the *Four Treatises' Instructional Treatise* with a detailed description of three channels from the black vital channel system at the third digit of the spine, the center one of which is at the heart. There are three folds in that channel, which contain five wind channels, in the center of which abide four channels containing a mixture of wind and blood.[160] Zurkharwa cites this passage because it looks like the picture of the heart matrix in the channels of being system. He seems bent on referencing something from the cardiovascular system in order to give the channels of being—at least for the matrix at the heart—some medical credentials. Otherwise there is little reason to think they are meant to be used by physicians. In fact, Zurkharwa states expressly that the image of these matrices come from the tantras.[161]

The breadth of the speculative space is especially clear when Zurkharwa refers on several occasions to the "laying out of the four, or six, *cakras*" in those tantric sources.[162] The details presumably don't matter. Some systems have four *cakras,* some have six. Zurkharwa even goes on to add two other tantric *cakras* at the crown and the throat to fill out the *Four Treatises*' system of four matrices. He also provides a way of counting the tantric *cakras* as five.[163] This casual alternation provides a telling comparison with Darmo Menrampa's bone experiment. There, in recognizing that the count of bones would vary depending upon how one separates fissures and whether very small bones are counted or not, Darmo was aware of the limits of any strictly materialist description. He was coming to grips with the impossibility of demarcating the physical world without any conceptual intermediary at all. It is a realization reached at the end point of grappling with the bloody body itself—a postempirical turn, if you will. In contrast, Zurkharwa's demonstration that the tantric accounts line up with the *Four Treatises*' matrices is driven only by concerns of system, and from the outset. He is simply marking the fact that one system lines up with another. His "whatever" attitude to how many *cakras* there actually are betrays a lack of curiosity about the exact physical disposition of these matrices or *cakras* in the body. Indeed, as he says in the course of this discussion, he is merely providing a rough sketch. If one is interested in the details, they are to be found in the *Kālacakra* and other tantras.[164] The approach anticipates by a century the "whatever" view found in the Desi's painting set. Zurkharwa saves his energy for precise physical detail about the connecting channels. He ends his discussion of the channels of being with a note that it is important to take care of them so as to prevent untimely medical problems. But no instructions how to do that are forthcoming. As was perhaps already the case in the *Four Treatises* itself, for Zurkharwa the channels of being are mostly a place where tantric imagery helps us envision the somatic bases for experience and growth. Once that imagery has been given its due, though, medical theory sets these channels aside.

The last of the four kinds of channels in the *Four Treatises*, the "life channels," would also have been an easy place for Zurkharwa to locate the tantric channels. They were the first choice for Namgyel Drakzang, who briefly related the second kind of life channel's association with breath to *roma* and *kyangma*. Later, Sönam Yeshé Gyeltsen attempted to locate an "outer" tantric central channel in the first kind.

But Zurkharwa's take on the life channels is surprising. He seems to want to swat the category away from the channel issue entirely. He is well aware of attempts on the part of previous medical commentators to understand the first kind, the life channel that pervades the entire body, in terms of other medical categories, such as the black or white vital channels[165] or the "points of vulnerability" (*gnyan pa'i gnad*).[166] But he refutes these views stridently, sometimes on quite unconvincing grounds.[167] He further disqualifies previous attempts to connect the all-pervading life channel with aspects of the tantric channels, including the ambiguous category of the "soul place" (*bla gnas*);[168] or, again, an "outer" central channel, or a notion of an "inner Rāhula," which he berates as the talk of someone who has never heard tantric teachings.[169] The range of options he covers suggests there was widespread perplexity on just what the *Four Treatises*' author had in mind.

In his exposition of his own views, Zurkharwa is comfortable associating the first life channel with tantric substances that move about—and thereby pervade—the body on a monthly basis.[170] He even asserts in a general way that some tantric systems can map onto the medical system (*sman dpyad lugs*), but does not spell that out.[171] In the end, he just makes this first category about the tantric substances, rather than any channel as such.

The same tactic is even more explicit for the second life channel. He ends up claiming that the supposed channel that moves in association with the breath is only a case of using the term "life channel" to discuss the movement of breath in the body.[172] As for the third kind of channel, he tries to reduce it to the sense of another so-called "soul channel" (*bla rtsa*) that is mentioned several times in the *Four Treatises*' *Final Treatise* as one of the "water-bearing" channels running along the side of the arm.[173] Then he sums up his take on the entire category of the life channels in response to an imagined frustrated interlocutor. None of the three is really a channel at all. Rather, the text uses the term "channel" loosely to refer to what runs within them.[174]

It is a funny position, and he reiterates it at the end of his discussion. The life channels are about the life or energy that moves around parts of the body, to which the name channel has been bestowed, but the three have to do with the life part of that picture, not the channel part. This contradicts his making the third kind of life channel the prosaic "soul channel" on the side of the arm. But otherwise he studiously avoids the language of channels throughout this section. Even though he references the practices and presumptions of the tantras, he resists the chance to locate the tantric channels here as such.

And maybe that is just the point. Perhaps a channel is something that must be located. By stressing that this *Four Treatises* category is really about life or energy rather than channels, Zurkharwa renders it vague and unavailable for empirical confirmation. But perhaps also he is a bit cranky now and tired of the issue. The life channels come last, after a very long discussion of the other three kinds. When he gets to this fourth category, long a source of confusion for the medical tradition in any event, he makes the counterintuitive move of saying these are not channels at all, thereby batting it out of the range of possibility altogether. He has already given his magisterial statement about the embryonic phase, where he pretty much solved the larger issue at stake: where the tantric channels are if we don't see them. This means, most important of all, that he has cleared space to explore what is actually of concern to medicine regarding the channels of the body, the *Four Treatises*' third category, the connecting channels.[175]

THE CHANNELS OF MEDICAL PRACTICE

Of the four kinds of channels in the *Four Treatises*, the connecting channels are by far the main focus for medical practice and most in need of verifiability. For that reason, their identification, or not, with the tantric channels is most germane to the problem that we have been tracking and Zurkharwa is endeavoring to address.

Zurkharwa associated the black and white vital channels with the tantric channels in the course of his comments on the *Four Treatises*' embryonic growth channels. But this was vague, never pinned down. In his long discussion that deals with these two key channels directly, there is a stark difference between the tenor of this detailed anatomical investigation and everything he has said in fact about all the other kinds of channels.[176] The focus is now on location and function. He is finally in the domain of medical practice proper, and the information that will be critical for the central procedures of moxibustion, bleeding, and healing wounds.[177]

Zurkharwa's section on refuting wrong views is all about where in the body the black and white vital channels actually are. Their location is pinpointed according to what organ or digit of the spine they connect to. The discussion often hinges on what is known empirically. For example, citing again the critical line from the *Four Treatises,* Zurkharwa refutes a wrong view by insisting,

> Also, the black vital channel's way of staying that [you posited] is not correct. From the *Instructional Treatise,* "It stands up inside the spine like a pine tree."
>
> Just as that says, it is plainly visible that it joins to the very inside of the back.[178]

Refuting wrong views about the connecting channels' location also entails fending off mistaken attempts to bring in the tantric channels. Zurkharwa refuses once more categories like outer *roma* and outer *kyangma* proposed by his predecessors, invoking a variety of authorities, including the recurring trio of "scriptural quotes, reasoning, and what is directly evident."[179] He rides especially hard on the textual evidence—no such category has ever been articulated, he claims, nor are there any references to an outer central channel, which also would need to be identified if the "outer" rubric were to make sense. Another point that he critiques advances an idea that does not reflect "the way things actually are"; yet another is seen not to be well examined and must be checked. In short, the white and black vitality channels don't match the tantric channels in terms of what runs inside them, where they are located in the body, or the kind of channel they are.[180]

And yet, by virtue of what he accomplished for the embryonic phase, Zurkharwa has won the ability to gesture to the now-settled place of the tantric channels in the larger scheme of things. It still entails a certain sleight of hand. Even though he is presenting his own position on the questions of where the channels are, what flows in them, and how they are used in bloodletting and moxibustion, he makes repeated—if brief and nonchalant—lip service to the connecting channels' putative dependence on the agency of the embryonic tantric channels. In his use of glosses like "the *roma*-related black vital channel" (*ro ma dang 'brel ba'i srog rtsa nag po*) or "the *kyangma*-related white vital channel" (*rkyang ma dang 'brel ba'i srog rtsa dkar po*), he is playing the card he already briefly introduced in the embryology section.[181] This has to do with the term *'brel ba.* Rather than the more general idea of "related," as I am interpreting it here, once again *'brel ba* could actually mean "connected," in the sense it has when *'brel rtsa* serves as the rubric for the third kind of channel overall, the black and white vital "connecting channels" that link the body's organs and functions.[182] So when Zurkharwa mentions the *roma*-related (or -connected) black vital channel, it is not entirely clear what he means. Is the black vital channel physically connected to a *roma* presently locatable in the body? Or is it related in the sense that *roma* originally gave rise to the black vital channel in the embryonic phase, when it stamped its basic nature upon the prosaic functions of the body?

In a key statement in the next sentence, the term "later" suggests that he is indeed talking about a temporal progression, rather than positing that the tantric channels stand now in the mature body. I nonetheless provide both possible senses for this passage:

> As for the channels that are related/connected to *ro* and *kyang*: the two, the white and the black [come into being/exist] by virtue of the power of the white and red elements that run inside the main one. Among those two, [with regard to the first one:] later, the black vital channel is like the basic trunk from which are produced all the branch blood channels.[183]

He does not specify exactly what he means by "inside the main one," but it would appear to refer to *roma* and *kyangma*. The statement says that the white and red seminal substances that run within these tantric channels produce the black and white connecting channels, as he already established in his account of the embryology. Then "later," the black vital channel functions to produce the rest of the cardiovascular system. This implies that the black vital channel is not still "connected" to a putative *roma* in the mature body, but rather the two are "related" due to their original cause-and-effect relationship in the mother's womb. A similar implication can be found in Zurkharwa's gloss of the "*kyangma*-related [or connected]" white channel: "The brain that is created by the upper tip of *kyangma* is the foundation for creating all white channels."[184] This is a clear reference to *kyangma*'s role in the earliest moment of the production of the brain, in the sixth week of gestation, but then the brain takes over to create the white vital channels.

Zurkharwa's gloss of the black vital channel as created by *roma* supports this interpretation further. When he specifies that "that which is known as the black vital channel is the channel that, emerging out of the liver, is created by *roma* and becomes the proximate cause for the growth of flesh and blood," it appears that *roma* is in the background as the original cause but has ceded its function in the mature body to the black vital channel.[185] This reading thus suggests that Zurkharwa's picture of the relation/connection between the tantric channels and the mature nervous and cardiovascular systems is that the latter were originally brought into existence by the former. There is little indication that he is still locating those tantric channels in the mature body, physically connected to the white and black vital channels.

And indeed Zurkharwa scarcely mentions the tantric channels again in *Ancestors' Advice.* But the few stray ways they do appear are telling, and underline both his complexity and the delicacy of what he is trying to accomplish.

One of these occasions, just a single sentence, does stump the reader. It comes up later in Zurkharwa's comments on the same fourth chapter of the *Explanatory Treatise,* now regarding the various openings of the body. The passage lists the inner and outer openings and counts the inner ones as three: those that conduct vitality (*srog*), those that conduct the refined substances produced through the course of digestion, and those that carry waste.[186] While the outer openings that the text goes on to list are apertures, such as the mouth, ears, nose and so on, the inner openings would seem to work as channels.[187] Zurkharwa pays cursory attention to this portion of the text and only provides quick glosses. But astoundingly, he glosses the vitality opening, with no apology or further explanation, as the tantric central channel: "The opening through which runs the wind of great vitality is the central channel along with its branches."[188] In importing the tantric channel when it would seem to have been farthest from the original passage's intent, Zurkharwa is not anticipated by the much more tantric commentator Kyempa.[189] Zurkharwa's comment is hard to understand on several counts. For one, the tantric central channel is not usually said to have branches. It would seem that Zurkharwa should instead be referencing here the two vital channels, the black and the white, in whose channels vital things do flow and whose branches spread over the body. He glossed the term *srog* that way earlier himself.[190] But he has chosen instead to gloss vitality as the wind of the great vitality (*srog chen po'i rlung*), and that can only flow in the tantric central channel. And yet he does not also take the opportunity in this passage to name the tantric *roma* and *kyangma* as that which conduct refined substances into the *samse'u* and genital region, but just uses the normal medical terminology, as indeed he should. That underlines the anomaly of calling the vital channel the tantric central channel here. Is he just being ornery? Is he actually referring to the embryonic channel? Is he using the term "central channel" in an entirely unusual way to refer to one of the vital connecting channels? Or does he really imagine the tantric central channel sitting in the adult body? Given the absence of any discussion or acknowledgment of the liberty he has taken, this passage would be hard to put together with his earlier analysis, where he established that such a channel would have to be invisible and made clear that the *Four Treatises* never mentions tantric anatomy or practice.

However we choose to understand this odd comment, another brief reference to the tantric channels much later in *Ancestors' Advice* suggests at least that Zurkharwa envisions the tantric channels waiting in the wings in some sense. This comes up at the very end of the commentary, covering the single chapter of the *Final Treatise* upon which Zurkharwa was able to comment before abandoning the rest of the project due to illness. This is the chapter on pulse reading. In one passage Zurkharwa is rejecting—as he does repeatedly in this chapter—the wrong views of Jangpa Tashi Pelzang. In this instance it has to do with what time of day is best for checking the pulse of a patient.[191] In the course of his refutation, Zurkharwa announces that his opponent's idea that winds running in the three tantric channels are read during a pulse examination is nowhere to be found in either medicine (*sman dpyad*) or Secret Mantra.[192] Saying this serves firstly to support my contention that for Zurkharwa the three tantric channels are not engaged by medical practice; certainly not in pulse diagnostics.[193] And when he goes on to add that neither does wind definitely move in conjunction with breath in the three tantric channels during the day, night, or morning, as Tashi Pelzang purportedly proposed—and that other than the time that wind and primal awareness mix, the primal awareness wind is separated to the side and does not run in these channels at all—Zurkharwa is implying that the tantric channels are only activated when a person performs yogic practice, when primal awareness wind is compelled through special breathing practices to enter them.[194] "Separated to the side" is but another vague allusion, but perhaps it says something about Zurkharwa's idea of the tantric body. The winds mobilized in tantric practice and the channels they would enter are somehow available in potential, but do not constitute the body that functions on an everyday basis and is available to medical examination.

Zurkharwa's principal strategy on the tantric channels, which is to locate them overtly only in the early life of the fetus, may have also thus served, at least by suggestion, to render them available for religious practice later. In those early moments of human life, on Zurkharwa's reading, these channels exert a profound influence over everything that follows. They are powerful, and they are implicated in every function that the medical channels of the mature body go on to perform. And while they retreat from view in the everyday adult body, they remain in some sense accessible. Perhaps they can thus lie in potential by virtue of their subtle materiality, but that is a strategy to which Zurkharwa only takes brief recourse once, in the course of his comment on the embryology. The bulk of his discussion points rather to a kind of implicitness, which although vague, would seem best to characterize the way that the forces, tendencies, and potentials of the body's organs and functions remain forever stamped by the embryonic channels.

Zurkharwa never directly answers the problem long troubling tantric and medical commentators alike, why the tantric channel system is not visible in the human corpse. The closest he gets is with a double negative: "it is not the case that it is not there, even in the dead body." And yet, whether he believes that these channels actually stand in the body and in fact "connect" (*'brel*), in some unspecified way, to the observable black and white vital channels of the adult body, or would rather leave them "related" (*'brel*) to those observable channels simply as their ultimate foundational cause, Zurkharwa's careful treatment of the issue achieves something for medicine itself.

Most significant of all, Zurkharwa has separated two domains of practice and epistemic authority. The relation/connection slippage notwithstanding, he has forever invalidated the positions of his predecessors that fail the standard of observability. There is no question that he gestures to some impact of tantric truths on the physical body nonetheless. But unlike Yangönpa's separation of two epistemic realms so as to privilege tantric truths, the main thrust of Zurkharwa's project was to contain the tantric system—despite its flickering appearances later—in the embryonic stage, so as to free up the medical channels to be considered on their own terms. It was Zurkharwa's medical predecessors who threatened to muddy the waters of medical practice with a tantric overlay, and Zurkharwa stemmed that tide. In fact, those prior medical theorists also respected the incontrovertible evidence of what is observable directly, and seem merely to have wanted to lift tantric truths to that same existential certitude. We might actually say, then, that Zurkharwa represents a kind of postempiricist turn. He both advanced the exactitude of the standard of observability and went on to recognize the power of other types of knowledge in order to broaden our understanding of what embodiment entails.

The coda following the next chapter will look at the reception of Zurkharwa's intervention on the channel question in the succeeding centuries. His account was read variously in different quarters. At least one voice from the radical end of the medical spectrum in the eighteenth century, clearly standing on Zurkharwa's shoulders, could vociferously deny any purchase of the tantric channels on the material body at all. But quite beyond Zurkharwa's hermeneutical feats to preserve an inchoate imprint of the tantric channels on the body and simultaneously to create a space apart for the practice of everyday physical medicine, there is a third way that his treatment of the channels served in the ongoing negotiation between medicine and religion. Zurkharwa's work also shows how the tantric mentality, if you will, can help everyday medicine to better conceptualize the body than it might do on its own.

The *Four Treatises*' own vision of the channels of being, otherwise useless to medical practice, already provided a potent image akin to the tantric *cakras*, helping doctors envision the confluence of arteries, veins, and nerves at the brain, heart, navel, and genital regions that make for human perception, memory, growth, and reproduction. For his own part, Zurkharwa also capitalized on tantric ideas about the female sexual organ to augment medical knowledge of that part of human anatomy. But far more momentously, in bucking the ill-advised trend in medicine to look for the tantric channels in the observable body, Zurkharwa preserved their explanatory power for other purposes. As described in the last chapter, he was ever vigilant to prevent the religious imagination from being confused with what is observed in the light of plain day. Yet having achieved such a division, he went on to let that same imagination—once identified and put in its proper place—help medicine understand what it might not detect in a superficial look. The following chapters find a few more cases where Zurkharwa invokes the generative power of the tantric channels, this time to explain certain gendered patterns in the body. In the debate tracked in the next chapter, he uses the tantric channels to save the *Four Treatises* from its own empirically untenable claim. In the same stroke, he demonstrates the special resources that tantric anatomy offers to account for gender difference—something that he finds in any case to be moot and relative at best.

5

TANGLED UP IN SYSTEM

THE HEART, IN THE TEXT AND IN THE HAND

One further anatomical perplexity lucidly demonstrates how the physician's experience could play against—and with—received textual system. This time the discrepancy involves not Buddhist scripture (notwithstanding some bids to make it so) but the *Four Treatises* itself. A claim the work makes as part of its theory of pulse examination seems blatantly at odds with what is known empirically. The commentators are uncomfortable with it, and they subtly modify a key detail of the authoritative text. Their skillful means in doing so illustrates how Zurkharwa's preservation of the tantric channel system in the formative first weeks of life could help medical theory account for hard-to-perceive dimensions of human existence. More important, the history of this contested matter speaks to a signal development in the nature of medical knowledge as such. In a key set of passages that will emerge over the next few chapters, Zurkharwa pointedly resists the idealized polarities that *either* certain tantric or medical binaries might suggest—be these with respect to pulse theory, gender difference, or basic dispositions of the body, all at stake in the case at hand. This demonstrates well how medical theory could be critical of received tradition and yet still use its categories to rethink a problem that its systems entail.

The *Four Treatises*' puzzling assertion seems to say that the heart tips in opposite directions in the thoracic cavity for males and females. That affects how the pulse of people of each gender is read. In this the *Four Treatises* suggests an idealized system of gender difference, and Zurkharwa's nemesis Jangpa Tashi Pelzang tries to extend its reach further. But Zurkharwa resists the elaboration of system and mounts a considered critique of Tashi Pelzang's proliferating essentialisms. In the process he distinguishes the heuristic use of metaphors from a literalist reading that oversimplifies the pathologies to which those metaphors would refer. Once again his strategy is reminiscent of his deconstruction of the Tanaduk myth. But in this case even Zurkharwa is caught in a bind that the authoritative medical text created for its commentators.

The solution involves an ingenious substitution of another anatomical specification that saves the *Four Treatises* on medical grounds. As we will see in the following Coda, it is not at all clear that this resourceful move was first made by Zurkharwa, although he makes it seem that way. But let us first follow the argument as Zurkharwa presents it in *Ancestors' Advice*. We will not be able to work through every detail of the exceedingly complex pulse theory, but the thorny challenge faced by the medical intelligentsia should be clear enough.

YOUR TIPPING HEART

In the last book of the *Four Treatises*, the *Final Treatise*, the first chapter takes up the body's channels again, but now in terms of diagnostics. Pulse examination is the principal diagnostic tool in Tibetan medicine, along with urine inspection.[1] A comprehensive set of instructions is given in this first chapter of the *Final Treatise* on how to discern the quality of pulsation in the channels at various spots around a patient's forearm that correspond to major organs in the body. The system has close connections to Chinese pulse diagnostics, at least in its broad outlines.[2]

One very reasonable question that the commentators raise about this chapter (and one of many points on which the Tibetan pulse practice differs from Chinese tradition) concerns why it states that doctors should reverse the protocol for which forefinger they use on male and female patients, and also which arm of the patient they check, when doing the first part of the exam.

This is the *Four Treatises*' statement:

> Check the male's channel on his left [arm] and the female's channel
> on her right [arm].
>
> First check the [male] patient's left arm with the physician's right [hand]:
> Under the index finger, check the heart and small intestine channels
> Under the middle finger, the spleen and stomach channels
> Under the ring finger, the left kidney and *samse['u].*
> Examine the patient's right arm with the physician's left [hand]:
> Under the index finger check the lungs and large intestines
> Under the middle finger, the liver and the gall bladder channels
> Under the ring finger, the right kidney and the bladder.
>
> For females reverse the index fingers and the two channels [that they
> examine] from right to left.
> Why? While the lungs and heart are not on the right or left sides,
> The tip of the heart faces [differently in males and females] in that way
> [one to the right, one to the left].
> The rest of the foregoing [instructions] are the same, since the locations
> are [similarly] disposed [in males and females].[3]

The passage introduces two very clear distinctions between males and females. One regards which arm is used in pulse examination, and the other has to do with the position of the heart in the chest. It speaks well of the critical sensibilities of the medical commentators that they question such a stark and seemingly arbitrary gendered opposition. Contrasting gender characteristics, often drawn from Buddhist tantric ideology, are common and widely applied in any number of Tibetan cultural formations. One could well imagine Tibetan doctors simply accepting that pulse examination and its corresponding anatomical references would follow suit. In fact, one other theorist for whom a *Final Treatise* commentary survives, Kyempa Tsewang, does accept it without question. But in this case Kyempa is out of the loop of a pointed debate to which Tashi Pelzang and Zurkharwa give witness.[4]

Before going further, we need to unpack the *Four Treatises*' picture of pulse examination a bit more fully. The physician holds one of the patient's arms and then the other. He uses, in turn, his forefinger, middle finger, and ring finger to feel the pulse on predetermined spots in the channels as they are available on the arm near the wrist. Then he switches the arm of the patient and the hand he is using, and feels three more spots on the other arm. At each spot the physician checks the pulse of a channel associated with a pair of organs. The commentaries specify further that he should use one side of his finger to detect the functioning of the first organ and the other side to feel the second.

The *Final Treatise* writes a gender difference into this practice. The physician should begin with the male patient's left arm and the female's right arm. He should also reverse whether he is using his own right or left index finger for the first pair of organs that he examines on each arm, which are the heart and the small intestines, and the lungs and large intestines. The doctor checks these with his right index finger for males and with his left for females. The text stipulates that the rest of the examination is exactly the same for the two genders. So why the switch for the heart exam?

The *Final Treatise* answers this question itself, with a second gender difference: the tip of the heart (conceptualized as the bottom point of the heart) faces (*bstan*) in opposite directions in males and females.[5] Apparently this necessitates a corresponding switch in which of the patient's arms will offer information about the condition of the heart. The *Four Treatises*' account was understood to entail that the heart actually leans, or tips, in different directions. Indeed, for the tip to point or face in different directions would affect the entire angle at which the heart sits in the chest. The commentators work strenuously to avoid this implication.[6]

ONE SYSTEM BEGETS ANOTHER

Zurkharwa considers the pulse exam's gender difference for several pages. He is especially exercised by the comments on the same passage by Jangpa Tashi Pelzang. In representing Tashi Pelzang's ideas, Zurkharwa quotes several versified lines.[7] It turns out that Tashi Pelzang himself cites these lines in his own commentary on the *Final Treatises*. So it is unclear if the verses are his own or cited from yet another source. In any event, Zurkharwa attributes them to Tashi Pelzang.

The lines in question attempt to explain the gender discrepancy by launching into an arcane explanation of the overall theory behind pulse diagnosis.

> Each pair of solid and hollow organs are connected to light and shade,
> since the hollow organs follow on the solid ones.
> The way these abide is evident and then not evident
> like the inside and outside of the curve of a rainbow.
> The reversing of method and primal awareness with respect to the
>   heart tip
> is due to the power of *roma.*[8]

The statement addresses the gender difference in pulse examination—here dubbed "method and primal awareness," a standard tantric euphemism for male and female qualities—by turning to other categories of pairs. It refers to the fact that each set of two organs listed for each finger in the pulse protocol includes one "solid" and one "hollow" organ (*don snod*). This is a pervasive method in the *Four Treatises* for classifying kinds of organs, also found in Chinese medicine. In these pulse instructions, the heart, spleen, right and left kidneys, lungs, and liver are solid organs; the small intestine, stomach, *samse'u*, large intestine, gall bladder, and bladder are hollow organs. (The *samse'u* would correspond to the testicles in the male and ovaries in the female.)[9]

The statement also references the fact that each finger used by the physician has two sides, an upper side and a lower side. Each side feels one of the two organs that are paired at each pulse examination spot. The organs of each pair, and by extension the upper and lower sides of the fingers that feel them, are described as "light and shade" (*gdags sribs*)—a taxonomical device in the *Four Treatises*' pulse theory that is analogous to the Chinese notions of *yang* and *yin.*[10] Tashi Pelzang's lines then follow light and shade with yet another heuristic pair, evident and not evident (*gsal mi gsal*), which he maps onto the solid-hollow and light-shade pairs. The issue of evident or not is indeed a concern in pulse diagnosis, having to do with how the physician should hold the patient's arm so that the pulse rhythm can clearly be felt, but it is not part of the *Four Treatises* passage under discussion, so Tashi Pelzang or his source has adduced it as relevant to the matter at hand.[11] Then the statement provides yet another pair of putative opposites: the inner and outer sides of a rainbow. Finally it mentions the method and primal awareness pair, or male and female, the differences between which the *Four Treatises* claims necessitate a corresponding difference in pulse-reading practice, in turn reflecting the opposite directions in which the heart tip tilts. The concluding line of Tashi Pelzang's verses takes the extra step of attributing the gendered reversal in the heart tip to the power, or force (*dbang*), of *roma*—one of the three contested tantric channels.

In these few lines, Tashi Pelzang's verses reference a range of subtle concepts from pulse tradition and beyond. The root text is laconic on these concepts, and his verses are laconic here. The entire pulse procedure as it is articulated in the *Four Treatises* is extremely suggestive theoretically and would well reward a close study. Short of that, it is worthwhile just to outline the problem for which Zurkharwa takes Tashi Pelzang to task. It will reveal a key signature of Zurkharwa's effort to inch Tibetan medicine away from ideal system and into an appreciation of asymmetry and irregularity.

In brief, Zurkharwa's response attacks both Tashi Pelzang's uncritical conflation of disparate sets of polarities and his attempt to use them to reinforce the main polarity under discussion, that between male and female. Zurkharwa is already worked up from refuting Tashi Pelzang's predecessor Jangpa Namgyel Trakzang's misreading of the original *Four Treatises* statement. The latter apparently understood the root text to mean that only the side of the index finger that reads the heart should be reversed between male and females. But that would introduce major incommensurabilities in the entire pulse practice. "Absolutely nothing has been understood, only wrecked!" Zurkharwa exclaims. "Those big people would control even the root text and wreck it."[12]

His irritation growing, Zurkharwa goes on to ridicule the younger Jangpa Tashi Pelzang's mindless piling on of parallel pairs that are not really parallel. He is especially disturbed by the notion of evident/not evident, maintaining that while such a distinction might be discussed in general, it is not relevant to the way that the pulse throbs for solid and hollow organs: it is not the case that the solid organ pulses are evident and the hollow organ pulses are not. Rather, both beat together like a wound multicolored thread. Nor is the example of evident or nonevident colors of the rainbow appropriate for pulse, since pulse is not something that is an object for the eye, as a rainbow is. (It is an object to be felt by the fingers, as the general phrase for the entire chapter [*reg pa rtsa*] makes clear.) As for the idea that one can see what is inside and outside of a rainbow, that is "something marvelous," Zurkharwa says sarcastically.[13]

Zurkharwa's real problem with the evident/not evident binary is that the two are polar opposites. As such, they are not appropriate to the connection between solid and hollow organs, nor the nature of light and shade as conceived here. Part of the issue has to do with Zurkharwa's sense that pure opposites can't be contiguous, since if they were they would cancel each other out (as he puts it, if they are like hot and cold, then they can't abide next to each other).[14] And while it is true, as his own comments show a few pages later, that the medical notions of hot and cold are also mapped onto the solid and hollow organs, that means something very specific, and the relation in each of these pairs is far more complex than one of simple opposites.[15] This, it seems, is his main point. Each of the pairs of organ channels in the pulse exam are connected in the sense that one "fills up" the other, such that the solid organ is felt on the exterior of the pulsing channel and the hollow organ on the interior. What's more, the hollow organs do "follow on" the solid ones, but that has to do with the fact that their elements are in harmony.[16] Tashi Pelzang, in contrast, seems to understand the way that the hollow organs follow on the solid ones as meaning that one channel spot follows another, or is next to it, on the arm—which again would result in a situation of two opposite and separable things being contiguous, rather than the way that the organs' pulses really are, according to Zurkharwa: intertwined and closely dependent upon each other.[17]

As Zurkharwa says, the indications of illness in solid and hollow organs are not in opposition to each other.[18] Furthermore, a proper understanding of hot and cold in medical parlance does not entail that the two cancel each other out. Zurkharwa notes that there are people who have a heat condition in their upper bodies and a cold condition in their lower bodies, contiguous areas to be sure. This already tells us that such pairs are not exactly opposites in actuality—rather, monikers like hot and cold here are indications or rough metaphors that really describe dynamic complementarities.

But as Zurkharwa goes on to add—and this is really important, I think—it is not possible for someone to have a cold condition in their upper bodies and a heat condition in their lower bodies. In short, the hot/cold binary—much like upper body/lower body, and indeed, solid/hollow, light/shade, and, as we will see, male/female itself—is asymmetrical. It and the others do not represent pairs of reversible mirror opposites.

Most significant of all is Zurkharwa's conclusion to the passage, which drives home a very critical point: "The explanation of the way to examine—be that in general, in specific, roughly, in fine detail, or individually—should not be mixed up. Even the finest points need to be understood from the perspective of details and making distinctions."[19]

What is really at stake is specificity; it's all in the details. General categories or ideal system, especially if essentialized as simple binaries such as

Zurkharwa finds Tashi Pelzang suggesting, don't really suffice for precise medical knowledge.

HOLE IN THE HEART

This larger epistemic point underlies Zurkharwa's overall treatment of gender difference, which is what sparked this issue in the first place. But it will take the combination of his response to Tashi Pelzang here and on another dispute about the pulse, which I will address in chapter 6, to reveal its full contours.

Tashi Pelzang's set of analogies were to explain the relationship between the two genders, and why the heart tip faces in opposite directions.[20] When Zurkharwa finally takes aim at the final point in the cited lines—that the gender difference is all a question of the position of *roma*—Zurkharwa responds that this would mean that *roma* is on the right in males and on the left in females.[21] But that is not in accord with either medicine or Secret Mantra, he fumes. While the *Four Treatises* does say that the heart tip faces differently in males and females, the tantras, according to Zurkharwa, say nothing about *roma* being on opposite sides in males and females.[22] In backing up this contention, he only provides a quote about *roma* from Secret Mantra again. As in the channels debate, Zurkharwa is still passing on responsibility for the actual somatic location of the tantric channels to the tantras. Nonetheless, he will now allow an important ingrediency of the tantric channels in the disposition of the heart.

This emerges when Zurkharwa begins his own reading of the root text. He goes systematically through the procedure by which the doctor takes the arm of the patient and uses the soft pads of his warm three middle fingers to feel the status of the various organs. On the female, the physician checks her right arm first to check her heart and small intestine, and her left arm for her lungs and large intestine. The reason is not that the lungs are closer to the left on the female and the heart is closer to the right. It is only the disposition of the heart tip, as the *Four Treatises* discussed, Zurkharwa avers.

And what is that disposition? Zurkharwa repeats that there is no difference in the situation (*gnas tshul*) of the *roma* and *kyangma* for males and females. However, he continues,

> at the tip of the heart there is the mind-entrance opening, and that is directed toward *kyangma* in the male since he is of the nature of method,

> and in the female it looks to *roma* since she is of the nature of primal awareness. This appears among the issues discussed in Secret Mantra.[23]

Zurkharwa, like his predecessor Tashi Pelzang already did, is invoking the tantric channels to explain the *Four Treatises*' anomalous gender disparity. And it does seem that in rejecting Tashi Pelzang's idea that *roma* and *kyangma* are on opposite sides of the body in males and females, and thereby implying they are on the same side of the body for both genders, Zurkharwa is still physically locating them, at least in broad strokes. Once again he is not clear whether the heart faces one or the other channel due to their impact on the formation of the body in the early weeks of the fetus, or whether they are exerting this influence throughout life. But perhaps this distinction is immaterial here.

The larger issue that Tashi Pelzang and Zurkharwa are addressing is most definitely about anatomy, the right and left sides of the fully formed body and a putative anatomical difference in the position of the heart between males and females. One has to provide something by way of explanation for this difference. The tantric universe, with its very central thematics of gender difference, is the obvious place to go for Tibetan scholastics, and neither Tashi Pelzang nor Zurkwarwa resists doing so. Zurkharwa's acceptance of tantric tropes is also reiterated in his association of the male with "method" and the female with "primal awareness," again, language already used by Tashi Pelzang. What is interesting is that these very fundamental gendered tendencies have implications for the ordinary anatomy of the heart.

But note at least one crucial difference, in line with what I argued above. While Zurkharwa uses the tantric channels to explain gender difference, he still resists a model of polar opposition. Tashi Pelzang has the tantric channels on opposite sides of the body in males and females. For Zurkharwa the channels would be on the same side of the body in both genders, with the only difference having to do with the way that the two genders relate to them. Drawing on associations long adumbrated in Indian Buddhist tantras, Zurkharwa suggests that part of what it means to be female is to be disposed to *roma* by virtue of her primal awareness, and part of what it means to be male is to be disposed to *kyangma* because of his affinity with method.[24] This does not make the two genders mirror opposites, even if they are complementary; they are just different, reflecting qualities and substances that tantric tradition also associated with the two channels. While the vision remains arcane, at least we can say that resisting simple opposition is a promising beginning for gender theory.

But let us not lose sight of the original problematic, namely the untenable entailment in the root text's picture that the heart leans in opposite directions in the body for males and females. In the foregoing lines Zurkharwa also tampers with what the *Four Treatises* said on that point. And he does so in a way that is even more momentous than buying into tantric anatomy when convenient.

Zurkharwa suggests that it is not the tip of the heart per se that is meant when the *Four Treatises* uses the locution "tip of the heart" to talk about it facing in two different directions. Rather, it is an opening (*bu ga*), located somewhere around the tip of the heart (later the Desi will make this even more explicit when he says "near" the tip).[25] That is what faces in two different directions for males and females—not the tip as such.

This is a rather momentous move in that it recasts what the root medical text says. But in doing so it fully neutralizes the dubious implication about a differently leaning heart in males and females. If it is only an aperture at some point around the tip that faces in different directions, it eliminates the picture that the heart overall is tipping. It would only be an aperture—and a very small and subtle one at that—located differently on the heart. Note too that in presenting this idea Zurkharwa has changed the verb that indicates the directionality of the opening. In providing the alternate terms "directed toward" (*gtod*) and "looks to" (*blta*), he now seems to be underlining his reading that the root text's verb "facing" (*bstan*) really just means facing, and not a leaning or tipping as such. The revised scenario still serves to justify the anomalous pulse practice of switching between males and females. It just takes away the implication of a grossly inaccurate anatomical picture as its explanation.

The revision effected by introducing the heart's "mind-entrance opening" into the discussion trades on a specification made elsewhere in the *Four Treatises*. In one passage regarding dreams and another on insanity, the *Four Treatises* mentions that there is an opening in the heart to which a special channel is connected, and through which the mind travels.[26] The basic idea can be traced to Ayurvedic tradition.[27] But analyzing it in other contexts, medical commentators had also located tantric channels around this opening, drawing on information from tantric tradition about why the heart beats.[28] (As noted in the last chapter, tantric images helped the commentators understand the channels of being around the heart.)[29] It appears that Zurkharwa or some predecessor ferreted out the *Four Treatises*' notion of the heart opening and applied it to the problem of the tipping heart too, adding also the tantric dimensions of the opening's anatomy from other sources. Indeed, the brief surfacing of the tantric channels in this part of the anatomy served Zurkharwa well, providing a critical additional feature that needed to be ascribed to the heart opening for the present purpose: a basis for the gendered difference in its position. Such a point had not been made about the heart opening elsewhere.[30] But in the current context, introducing the extra specification of the gender-coded tantric channels becomes the key piece that allows the mind-entrance aperture to replace the heart tip as that which faces in specifically gendered directions.

HEART IN THE HAND

The heart's mind opening is imported into the present discussion for good reason. I want to jump ahead to the stridently empiricist comments on the issue by Lingmen Tashi, whose incisive comments on the authorship of the *Four Treatises* were noted in chapter 3. His interventions on the tipping heart matter drive home my point about the import of direct observation. Surpassing Zurkharwa in his bold willingness to set aside tantric anatomy altogether, Lingmen maintains that the idea that the heart leans in different directions is unreliable.[31] He also labels as "a commentary of fool's talk" or "a guess by fools and old people" the proposition that because males are of the nature of method, the tip of their heart leans right, and because females are of the nature of primal awareness, their heart leans left. He continues with this stark assertion: "As for me, I have seen the dissected corpses of many males and females. I myself have held the hearts [of people killed by a] knife and have seen that the heart tips of all males and females lean a little toward the left wall of the chest."[32]

Lingmen effectively ends any possibility that the heart leans differently in males and females by merely stating what he knows from his observation of dead bodies. Indeed, the heart does lean to the left (as modern biomedicine confirms).[33] But there are no gender differences. What allows him to say this is his own direct experience as a physician. That gives him an epistemic certitude that trumps textual authority. And while it postdates Zurkharwa by about two centuries, his confident assertion and impatient tone allow us to see what motivated Zurkharwa to insist that the *Four Treatises'* heart tip is actually a hole somewhere near the true tip. He thereby protected the *Four Treatises* from implying something that directly contradicts clinical experience.

Lingmen is a master wordsmith in all this himself. Throughout his discussion he uses the verb *bsten*, a form of *sten,* "to rely on" or "to lean,"[34] instead of its homophone and similarly spelled *bstan*, or "face," that the *Four Treatises* itself uses.[35] In fact, in referring to the *Four Treatises*' original statement, he seems to misquote it, shifting it to say "lean" instead. Perhaps he is caricaturing what he considers to be a wrong reading of the original passage.[36] Or perhaps he is using the same verb differently in different contexts, illustrating the possible slippage between facing and leaning. In any event and most importantly, Lingmen's clinical experience does not have to contradict the root text itself: once the addition that what it is discussing is the heart hole, not the tip, has been put in place, the root text's statement does not have to imply leaning after all. Lingmen affirms that there is such a thing as an opening in the heart, and even that it displays a gendered difference. He also notes that there is a notion in the tantras that the direction in which the heart leans, or faces, is connected to the position of that opening. Lingmen himself does not accept this idea from the tantras, but he does develop the notion of the heart opening itself, pulling together a variety of statements from *Four Treatises* commentarial history.

In so doing, Lingmen uses only information from medical sources. He provides a fulsome explanation of what the heart opening is, referring in general to the *Explanatory Treatise*, and working through the entire anatomy of the tangle of channels around the heart. One of these channels is called the mind-entrance opening. Lingmen makes the further clarification that the opening is actually a channel,[37] drawing on anatomical specifications about the mind-entrance that had already been made by *Four Treatises* commentaries in other contexts.[38] Lingmen drops these commentaries' allusions to the tantric central channel and adds the gender specification, which is not to be found in the other contexts. But unlike Tashi Pelzang and Zurkharwa, Lingmen offers no explanation for the heart's gender difference. He merely asserts that the mind-entrance channel connects to the heart on the left, but in females it connects on the right.[39] Apparently it is simply an idiosyncratic fact.

In short, like his predecessors, Lingmen supports the *Four Treatises*' gender difference in pulse-reading practice but manages to recast its empirically unacceptable suggestion about the position of the heart. He was comfortable doing so without resorting to tantric imagery. For Zurkharwa, writing two hundred years earlier, it was necessary to invoke some version of tantric system, having to do with the predominance of method or primal awareness and the impact of such a signature on the body. While motivated to question received system and to resist simple stereotypes, Zurkharwa was not able to throw all of it out.

Or perhaps he didn't want to. Standing at a different moment in medical history, and no doubt a different cultural and political context than Lingmen Tashi, Zurkharwa took on the complex task of protecting both the empirical accountability of medicine and its deep imbrication in a tantric conceptual universe. It seems in the case at hand, the tantric channels and their associated gender codes worked well for him. They helped to characterize a larger point about the human body that he thought relevant to pulse diagnosis: gender difference is accompanied by differing orientations to experience, emotions, and ways of conceptualizing, including directionalities within the body. This is just what Buddhist tantric scriptures lay out in detail: male-coded substances in the head; female-coded substances in the groin; male-coded tendencies to think and act effectively ("method"); female-coded tendencies to take account of an all-encompassing emptiness at the heart of all phenomena ("primal awareness"). In another debate with Tashi Pelzang that we will follow later, Zurkharwa makes such a point himself—that these differences are coded, and that the association of substances and channels and tendencies with a gendered primal awareness and method is really a convention, a taxonomical heuristic.

I will turn to this larger set of issues about gender and medicine in chapter 6, and even find aspects of medical theory that undercut a simple binary division between male and female *tout court*. For now, on Zurkharwa's reading of the gendered heart opening, the tantric channel explanation means that the minds of people with female tendencies tend to enter the heart from the right, due to a female affiliation with the *roma* channel, and the minds of people with male tendencies tend to enter the heart from the left, due to a male association with the *kyangma* channel. To be sure, this still begs the question of what those channels really are and when and where they operate. But at least they were a ready-to-hand scheme for Zurkharwa and his colleagues to account for the spectrum of styles and orientations across humankind.

Let us not fail to recall that the *Four Treatises* itself fell prey to stereotypical gender difference in its specifications on pulse examination. Zurkharwa's reading of the passage not only explains this difference by taking recourse to tantric tropes. He also manages to minimize the obviously unacceptable anatomy entailed by the root text's own explanation of its gender-differentiated practice. Zurkharwa displaced the entire problem to a well-nigh imperceptible opening in the heart near the tip, one that was already described elsewhere in medicine. In adding tantric channels to this picture, his predecessors had provided grounds—albeit also seemingly imperceptible ones—for this opening to have varying positions depending on the disposition of the person in which it is located. Drawing on these disparate points of knowledge about the heart, Zurkharwa thus found a way to make meaning and give coherence to the authoritative medical work, with much less obvious sacrifice of scientific accountability—although there was still some!—than its original explanation of pulse practice suggested.

In the previous chapter we saw the medical account of human embryology become a conveniently inscrutable place to locate the empirically questionable tantric channels. In this chapter, the force of these tantric channels becomes a shelter for an empirically questionable idea in medical knowledge. It also becomes a way to help medicine conceptualize gender. These are among the benefits of positioning the tantric channels in the seminal formation of the body.

CODA

INFLUENCE, RHETORIC, AND RIDING TWO HORSES AT ONCE

A small, further detail on the heart tip question provides a fitting way to close our study of the last three chapters' debates. It points to a critical dimension of the processes by which knowledge changes, exemplified not only in the tenor of Zurkharwa's interventions but also in the culture of medical learning by the Desi's day.

The detail is that while Zurkharwa's solution to the *Four Treatises*' puzzling assertion about the heart was ingenious, it wasn't really his. This small epiphany draws our attention back to the many skillful maneuvers in which Zurkharwa engaged. A key part of that maneuvering had to do with getting credit. And though it may not matter much for the purposes of this book exactly who introduced a new idea, since we are tracking the nature of innovation over the *longue durée*, it is important to note how important getting credit was to the actors involved.

This concern also calls attention to the central role of rhetoric in negotiating the fine line between scientific accountability and other cultural sensibilities, including religious ones. We saw the prominent role of rhetoric in the Desi's own career in chapter 2. Certainly the Desi, as much as Zurkharwa before him, would readily sacrifice conventional forms of deference, as well as full disclosure and even consistency, to make certain that his public face gave the appearance of mastery and credibility. Across the Tibetan intellectual scene, but most certainly in medicine, establishing mastery sometimes seems to have entailed a wholesale rejection, or even distortion, of earlier ideas, even if one was massively in their debt. In medicine this sometimes meant representing one's own position so ambiguously that it could be read in (at least) two ways, depending on how the author—or reader—chose to "ride" it (a metaphor that will become clearer in due course).

I would like to use this coda to return back to the Desi and his reception of the various issues negotiated by Zurkharwa. That will take us full circle to where we started in this book. The Desi's moment in the history of medicine marks a high point, but also is close to the limit for the conceptual advances in Sowa Rikpa brought to a head by Zurkharwa. Despite the Desi's innovative investment in natural history and commitment to critical assessment of received knowledge, he and his successors could only go so far down the path to empirical accountability and the autonomy of medical learning. Not, that is, unless they were to subvert the very premises of the Dalai Lama's apotheosis that underlay the Tibetan state.[1]

This short coda will be followed by two final chapters that fill out the picture further on what came before the Desi, disclosing two more pieces of the medical mentality that funded what the Dalai Lama and the Desi could achieve in the seventeenth century. Both also have everything to do with competition, rhetoric, and much else of what it took to get things right in the field of Tibetan medicine.

TASHI PELZANG ON THE TIP OF THE HEART

The revelation about the heart tip debate emerged when I endeavored to check Tashi Pelzang's own writing for the passage that Zurkharwa attributed to him. Fortunately, a *Final Treatise* commentary by Tashi Pelzang has survived and is held in Lhasa. Unfortunately, what I have is a dark, incomplete, and frequently illegible third-generation photocopy of a manuscript that is itself rife with spelling mistakes and smudges. Yet looking through a glass darkly still rewarded me—if perplexingly.

Zurkharwa represents Tashi Pelzang's interest in analogous binary pairs well. Tashi Pelzang does build an elaborate system of correspondences, lining up the solid and hollow organs, the light/dark binary, the male/female difference, and the question of whether pulse indications are evident or not.[2] Part of his system requires that the element of the heart is space, and he is aware that that contradicts the *Final Treatise*'s doctrine that the element of the heart is fire. He responds to potential criticism by arguing that such correspondences depend on what perspective (*sgo*) one takes and what one is trying to achieve. He notes that while the *Final Treatise* was concerned with the elements, he is concerned with the relation between solid and hollow organs.[3] This point—in service, to be sure, of cleaning up an inconvenient entailment of his argument—nonetheless evinces a keen awareness of the relativity of knowledge, dependent upon the perspective from which one is looking.

And yet while Tashi Pelzang betrays some ideological flexibility, he is still building a tower of ideal types to explain gender difference. In fact, he is far more overt than the passage Zurkharwa pulled out to quote. Tashi Pelzang says explicitly that the light/shade binary that corresponds to the solid/hollow organ relation and the evident/not evident relation is itself gendered: male and female are related in the same way as light and shade.[4] Thus he creates a far-reaching structure of mirroring *yin-yang* (or *yang-yin*) tropes, all in service of supporting the *Four Treatises*' implausible dictum on the position of the heart. It is quite a stretch, and we can sympathize with Zurkharwa's impatience.

But Tashi Pelzang's own comments on the lines that Zurkharwa cites already include a heart aperture theory. He is also concerned to clarify that the heart sits straight (*drang po*) inside the body, in all people. The gendered difference is only that in males there is an opening at the heart tip facing diagonally upward (*bsegs stod la stan*) to the right, and in females it faces to the left. This reversal in mode of facing (*ston lugs*) creates the reversal of the location of the heart pulse in males and females.[5] I do not find Tashi Pelzang glossing that opening as the place where mind enters, but that may simply be a function of the incompleteness of my manuscript copy, or perhaps of this early moment in the effort to deal with the *Four Treatises*' troubling assertion. Either way, it certainly appears that, save the gratuitous system of analogies, Tashi Pelzang's reading of the *Four Treatises*' gaffe is very similar to Zurkharwa's. The only real exception is that Tashi Pelzang does attribute the gendered difference to the opposite positions of *roma* and *kyangma* in the body in males and females—just as Zurkharwa claims.[6]

Thus must we correct the record on who introduced the heart-opening solution. Tashi Pelzang also explicitly refuses any obvious gendered difference in the disposition of the heart in the chest. Even if his picture of the heart as sitting straight is at odds with what is observed by modern medicine, he is as wedded as any of his colleagues to correcting what he understands to be an empirically unviable idea about the body. This certainly recuperates him somewhat from the dismal picture of his intellectual honesty on the Buddha Word debate. Now we see that even if the *Four Treatises is* Buddha Word, Tashi Pelzang is still motivated to adjust its statements to accord with what he takes to be clinical experience.

That leaves us with Zurkharwa's dissimulation in characterizing his differences from his predecessor. We can safely assume that Zurkharwa knew Tashi Pelzang's work well and likely read the very commentary that I consulted.[7] But Zurkharwa cites no source for the mind aperture solution to the heart tip problem and makes the reader think that it is his original interpretation. What's more, he exaggerates the illusions of his forebearer and makes it seem as though he had no concern for medical truths. Zurkharwa is competitive, and invested in getting credit for an effective solution to a delicate problem. His own massaging of the truth here only underlines how much he wants to paint himself as a reasoned knower of bodily realities. Whether he really deserves such a label in our own estimation is quite beside the point. What is important to note is the fact that there were great rewards in seeing oneself this way and presenting oneself this way to others.

RHETORIC ALL AROUND: ZURKHARWA AND THE DESI

Unrelenting in his criticisms of wrong views, willing on occasion to contradict himself, all so as to display his own prowess, Zurkharwa Lodrö Gyelpo still made momentous contributions to medicine. In his brilliant treatment of one of the greatest challenges of all, the potentially devastating break between the directly observable anatomy of the body's channels and that ordained by tantric tradition, Zurkharwa was so cognizant of the risk of alienating either his scientific audience or his religious audience—or both—that his overall solution can be read either to confirm the tantric authorities, to make them irrelevant, or to keep both medicine and tantra in their own domains but still mutually enriching. Zurkharwa can be said to have mastered the old trick of riding several horses at once until one sees which will serve best for the task at hand.[8] It appears that the maneuver worked, and Zurkharwa's imprint on this and many other difficult points of medical theory was secured. While the *Blue Beryl* of the Desi had unquestionably become the best-known medical treatise by the end of the seventeenth century, those in the know are aware that much of that work is lifted directly out of Zurkharwa's *Ancestors' Advice*.

The Desi's own character and medical legacy have a lot in common with that of Zurkharwa. Imperious, arrogant, anxious to display his acuity, he also made enormous contributions to the advancement of medicine in Tibet. Some of these stood on Zurkharwa's shoulders to be sure; others were a product of his own brilliance and verve. A brief look at how he received Zurkharwa's interventions in the Buddha Word, tantric channels, and heart tip debates in his own writings shows how this strident critic of Zurkharwa has the mark of his nemesis all over his own thinking.

The Desi's biographical pillorying of Zurkharwa seems to have had most to do with the Buddha Word question, and indeed that matter remained the least resolved. But consider first the simpler case, the heart tip debate. This is very straightforward. Short of a few inconsequential clarifications and his deletion of the sections critiquing other views, the Desi provides the very same reading of the passage as Zurkharwa, and in virtually identical language.[9]

Elsewhere in the *Blue Beryl* the Desi purports to know the true disposition of the heart, anticipating Lingmen in stating clearly that there is no difference between males and females, no doubt referencing the heart tip question.[10] As Lingmen would do, the Desi here affirms that the heart in both males and females faces downward and to the left, here using the different term (*bstad*) with the unambiguous sense of being turned to face. This comment comes in the course of making a fine distinction, for the purposes of visually depicting human anatomy, between how the body is described in the texts, how a diseased corpse might look, and how the heart looks in a

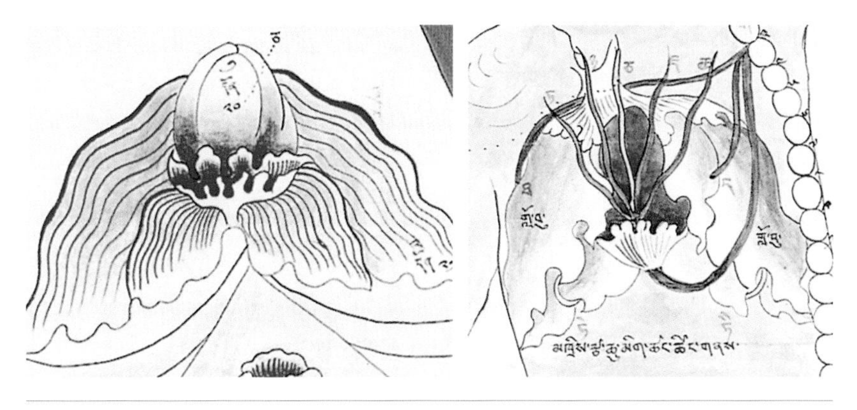

Coda.1 The heart and surrounding organs, as pictured in the Desi's medical paintings. *Plates 14 and 47, details*

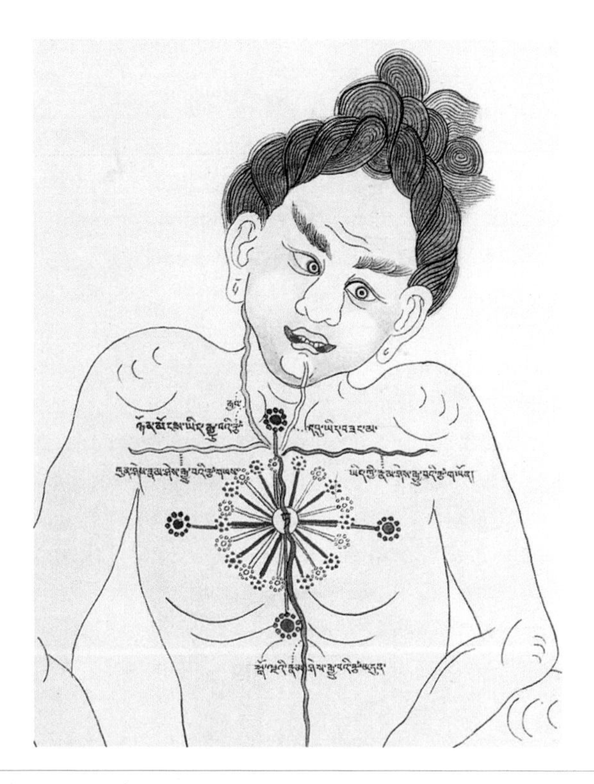

Coda.2 The system of channels at the heart that conduct kinds of consciousness. The mind-entrance opening is not pictured. *Plate 10 detail*

healthy, live person.[11] This does not necessarily mean that he had full knowledge of the body on the inside, Tibetan surgical and postmortem dissection practices notwithstanding. Nor did he portray the heart with what would be considered accuracy in the biomedicine of today. It is significant enough that he *claims* to know the heart empirically and will distinguish the grounds on which he knows it from what he takes from the texts.

THE FATE OF THE TANTRIC CHANNELS

The story is mostly the same for the Desi's reception of Zurkharwa's tour de force on the tantric channels. The Desi simply reproduces verbatim large swaths of Zurkharwa's commentary.[12] So it is quite striking that the Desi

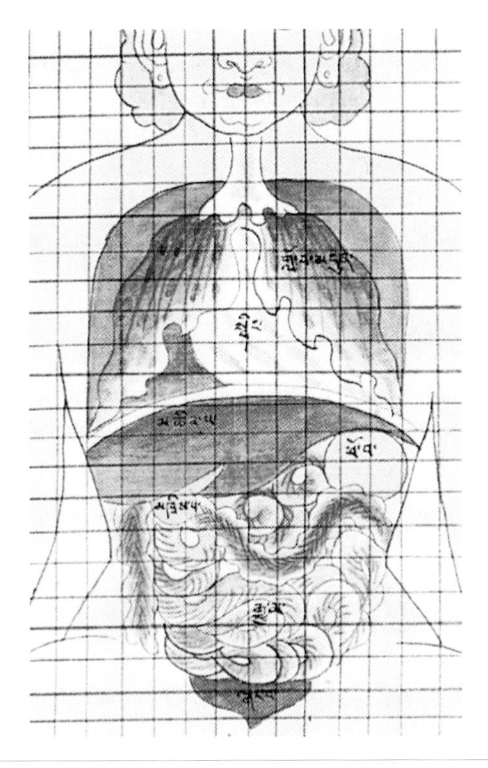

Coda.3 The heart and other organs, "based on Lhodrak Tendzin Norbu's observation." *Plate 49 detail*

makes a special point of pretending that this discussion is his own. After introducing the problem of the tantric channels, right at the beginning of his comments on the growth channels, he has the nerve to say, "Both Jangpa and Zurkharwa merely indicated and explained [this matter of the tantric channels] roughly. And so, to explain, including the fine points. . . ."[13] And then he launches into a near replica of Zurkharwa's "own position." Compared to Zurkharwa's use of unacknowledged sources, here the intellectual piracy is egregious. Did the Desi think that no one was reading *Ancestors' Advice,* when only a few years earlier the Dalai Lama had gone to such efforts to have it completed and published?

There is no question that the Desi fully accepts Zurkharwa's overall strategy to focus attention on what is most important for medical practice, the black and white vital channels.[14] In fact, the one place where the Desi does add something substantial to Zurkharwa's discussion is in his specifications about the minor channels, such as the channels that can be bled. Now the Desi, confidently resting on his own and especially his colleagues' clinical experiences, takes issue with many of his predecessors, including both Jangpa and Zurkharwa.[15]

And yet, given that some of Zurkharwa's boldest assertions come in the course of rejecting previous efforts to find the tantric channels in the functioning everyday body, it is possible that the Desi is resisting the more radical implications of Zurkharwa's work when he dismisses such refutations with a mere "be all those faults as they may" and simply deletes them from what he copies.[16] In this elision the Desi may have been tipping Zurkharwa's carefully crafted ambiguity back toward allowing the tantric channels overt existence in the body. But even so, there are very few references to the tantric channels in the Desi's own *Instructional Treatise* commentary. One regards their role around the heart, already noted as a holdover from old tantric tradition.[17] Another instance may even be a slip, because the Desi avers right beforehand that what he is talking about could be explicated by juxtaposing the uncommon explanations of the Mantra tradition, but since there are so many misgivings with respect to that, he will just proceed in accordance with medicine. And yet a few lines further, when he gets to certain issues about—again—the heart, he briefly glosses one of the wind channels as *kyangma*.[18]

However, outside the *Four Treatises*' commentarial context, we see significant divergence from Zurkharwa's cautious account of the tantric channels, at least by some of the Desi's colleagues. One lucid example is the free-standing medical treatise of Darmo Menrampa, cited in chapter 4, that reported on his bone-counting experiment. When Darmo gets to the contested channels of the body, he launches into a robust reconciliation of the tantric system and the four kinds of channels in the *Four Treatises,* taking Zurkharwa's suggestions about the embryonic stage and so on, but unpacking them in such a way that the tantric and the medical bodies are explicitly juxtaposed, with little hint of the studied ambiguity of his predecessor.[19]

A more ambivalent example of the fate of the tantric channels in seventeenth-century medicine in the Tibetan capital is the Desi's illustration project. Certainly the painting set would be the ideal place to depict tantric anatomy for all to see, and in vivid color, if one were so disposed.[20] The tantric channels are indeed portrayed in all their glory in the adult body—whether they are in theory invisible, or merely implicit, or not. And yet the way the paintings handle them indicates some hesitation.

Actually there is considerable vagueness in the medical paintings with respect to several of the large channels that are shown traversing the center

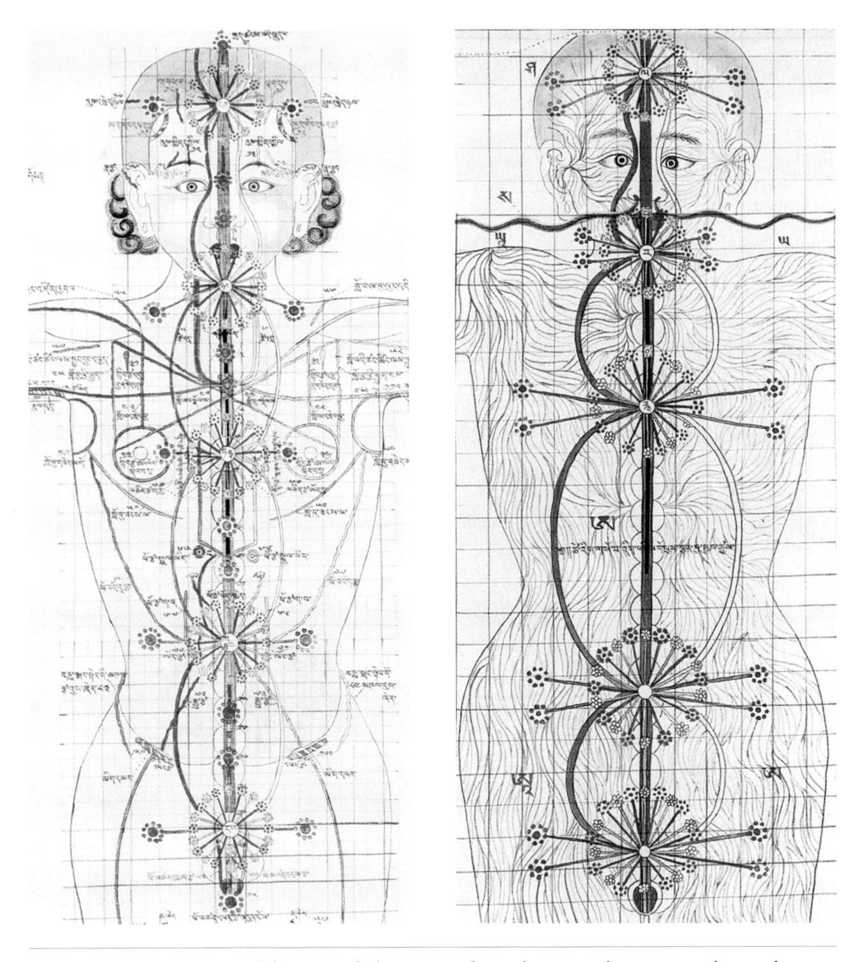

Coda.4 The central figures of plates 9 and 12, showing the tantric channels.

of the torso. Certain figures, such as on plates 10, 11, and 47, show a single central channel running from the top of the skull down into the tip of the penis, but these are not labeled and seem to represent a general channel associated with the spinal column. Whether they are meant to portray the long-ambivalent vital channel or something else is not made clear. In any event, out of the twenty-one plates devoted to human anatomy in the painting set, two, 9 and 12, do explicitly display the three tantric channels.[21] An indication of the tentativeness in so doing is the fact that neither is directly labeled with a caption, as are all other significant parts of the anatomical plates. But what these two instances demur about verbally is quite clear imagistically. Both overtly show the tantric central channel and the two side ones, meeting at intervals at the *cakras*.

The captions for the main figure of plate 9 only point to the bloodletting and vulnerable vessels. But the figure is also meant to illustrate the growth channels and the channels of being, although these are only mentioned in the colophon at the bottom of the painting, keyed roughly to alphabetic letters on the figure.[22] This colophon shows its indebtedness once again to Zurkharwa: "the three growth channels are explained in the tantras to be produced by the three tantric channels, the central one, *roma*, and *kyangma*." The colophon also adds that the petals of the matrices of the channels of being "ramify" from the three tantric channels—also familiar language from *Ancestors' Advice.* Neither of these points appears to be illustrated, however. Nor is a further claim in the colophon that the figure shows the way the black vital channel mixes with the tantric central channel from the front and the white one from the back.[23] But the overall image certainly shows what look to be the three tantric channels in the adult body, with a straight channel in the center and the two side channels converging at each of the *cakras*, the standard picture in tantric literature. In fact, toward the end of the colophon that continues on plate 10, it is mentioned, seemingly as an afterthought, that *ro* and *kyang* are pictured here, running on the right and left, with their "intrinsic marking colors of red and white."[24] This seems to say that the tantric channels indicate themselves. In any event, no caption labels them, as if there was some hesitation about being too overt in pointing to them so baldly.

The second instance in which the tantric channels are pictured in the mature body is on plate 12. Here the main image has ostensibly to do with the three kinds of life channels, i.e., the last of the four categories of channels, which Zurkharwa denied were channels at all. The plate includes captions that refer to all three life channels, but it is not at all clear what they are pointing to or how these life channels are illustrated. The figure also seems to be presenting again the standard tantric picture of the central channel, *roma* and *kyangma.* A caption toward the top of the plate acknowledges this, averring that *ro, kyang*, and the central channel are merely pointed out here, without detailed analysis. Noncommittally, this caption does not point to anything. The colophon acknowledges the figure's visual reference to the tantric channels once more, again noncommittally saying that the three channels are illustrated only roughly here.[25]

In short, we can see that the Desi and his artists endeavored to portray the tantric channels, but were evasive in doing so. They do not picture them on the plates devoted to the connecting channels or the organs or bones of the body, as Zurkharwa's discussion would certainly have cautioned them against.[26] But they did render them on the two plates that take up the embryonic growth channels, the channels of being, and the life channels—just the places that had long seemed apt for tantric juxtaposition.[27] Even here they were cautious, taking pains to note in the colophon to plate 9 that the tantric channels really come from another knowledge system, and reminding the viewer of their role in the embryonic phase from which the anatomy being illustrated originally sprung. In both cases they demur before providing captions that would "point a finger," suggesting some discomfort in straying into the supposedly subtle tantric anatomy and what might be more appropriately left to other representational venues. But given the temptations of such a beautiful new medium for anatomical depiction, how could they resist?

The significance of Zurkharwa's intervention on the channels is that he brought into focus the problem that tantric anatomy poses to any would-be empirical medicine. It also demonstrates the resistance of that problem to a clear-cut solution in the cultural climate of early modern Tibet. Most of all, it tells us what medical theorists felt compelled to address and what they could manage to say at a particular moment in the history of science.

The very virtue of what I take to be Zurkharwa's intentional ambiguity also means that it was subject to multiple readings in the years to come. At one end of the spectrum is a clear pronouncement in the eighteenth century that would separate medicine and religion with respect to the tantric channels. Perhaps because the status of Buddhist truths was most charged in the capital, the seat of the government of "religion and politics integrated," this comes from a medical center far outside the reach of Lhasa. Lingmen Tashi, trained at Pelpung Monastery in eastern Tibet, boldly claimed to have held human hearts in his hand and thus decisively capped the tipping heart debate. He also weighed in on the question of the tantric channels, going far out on a radical limb to fully disavow the relevance of tantra to medicine. Lingmen was well aware that the vital channel is sometimes called an "outer" central channel and so on; such a distinction aligns with his own general approach of understanding the ordinary human body differently from that of a buddha.[28] But regarding that ordinary body, Lingmen is interested in what is directly observable. He can declare categorically that the tantric channels are out of bounds in an anatomy of the "material" (*gdos bcas*) body; the tantric system is meant rather as a map for meditation.[29]

At the other end of the spectrum of Zurkharwa's legacy is the great nineteenth-century exegete and polymath Kongtrül Lodrö Tayé, who wrote a fully tantric commentary to the fourteenth-century Tibetan classic *Profound Inner Meaning*, yet recapitulated and even expanded the entire problem of empirical verification. Reviewing the wide range of views on this thorny problem—that the central channel is the spinal column, that it disappears at death, that it has no referent at all, or that it is a product of meditation—Kongtrül follows the view that there are three kinds of tantric central channels, a basis version, a path version, and a fruit central channel. The first is the ever-multivalent vital channel in the "impure" body. Through the devices of the "path" practices involving the tantric version of the central channel, one eventually produces the "fruit" central channel when one achieves virtuosity in meditation.[30] This scheme identifies a straight progression from the everyday body to the enlightened one, such as the medical commentators would not venture. Kongtrül goes on to give the kind of robust anatomy of the tantric channels that was never forthcoming in Zurkharwa's own voice. And yet, coming to grips with the centuries-old problem that Zurkharwa brought to stark awareness frames writing about the tantric body in the nineteenth century, that is, at least in the hands of a virtuosic scholar like Kongtrül.

From what I can tell, Kongtrül's confirmation of the tantric channels in the ordinary body represents the proclivity in most contemporary thinking, not only in tantric circles but also in medicine proper. We have already seen Tsültrim Gyeltsen's twentieth-century reading of the channels, which took Zurkharwa's treatment to support unambiguously the traditional tantric anatomy. One more permutation in the twentieth-century cultural politics of Tibetan medicine can be recognized in the medical scholar Tupten Püntsok's study, also directly dependent upon Zurkharwa's *Ancestors' Advice*. And yet in providing a contemporary biomedical illustration of the body that identifies the vena cava and the main aorta as two main trunks of the *Four Treatises*' black vital channel, Tupten Püntsok glosses these as *roma* and the tantric central channel—despite everything that Zurkharwa said to discredit such an equation! The old medical urge to find the tantric anatomy in the empirical body has come to full fruition, now mediated not only by the vocabulary of the *Four Treatises* but also by the anatomical idioms of modern biomedicine.[31]

The multiple readings, if not inconsistencies in these recent examples of medical thought suggest that the empirical visibility of the tantric channels is still in the air, when issues of science and religion have taken on new significance on the Tibetan plateau.[32] The attention to Zurkharwa in the recent literature, even if cast in ways that he would not have anticipated, only tells us how productive his treatment of the matter really was. It all helps us understand the Desi's own close appropriation of Zurkharwa's studied nuance, in a time when issues of science and religion also seem to have been afoot, if in different ways. When we look behind the rhetoric of one more issue, we will see that even there the Desi was closely informed by Zurkharwa's epistemic orientation—despite his vociferous refusal of it in other venues.

GOING BACK ON WORDS

The case of the Buddha Word debate is different. The Desi by no means replicates Zurkharwa verbatim. There are still many borrowed passages on the larger history of medicine, but the Desi has to make considerable adjustments of his own. Zurkharwa's commitment to scientific accuracy posed an estimable challenge to the Desi's own commitment to the *Four Treatises* most definitely being Buddha Word.[33] The issue thus differs from others in being less medical, with far more at stake in terms of cultural capital and symbolic implication. The Buddha Word dispute raised the issue of authority front and center, but it was not easy to adjudicate. While the Desi was a determined defender of the ultimate authority of the Buddhist dispensation, the evidence of the *Four Treatises*' Tibetan authorship was massive and impossible to ignore. And this is not to mention the Desi's investment in the Dalai Lama's own deep commitment to the advancement—and credibility—of medical learning.

Luckily the Desi had a watertight Buddhist hermeneutic to help him negotiate the perils. But when we look at his own words on the matter and see how close in the end he actually comes to Zurkharwa's most radical reading of the issue, we realize more than ever that the biographical invective was about public relations, making a show of beating up on the infidel. The Desi himself had apparently been largely convinced by all the evidence and turned to empirical verities, even on this signature issue.

Although he is very passionate about the Buddha Word question, the Desi is even more circumspect than Zurkharwa in the opening sections of the *Blue Beryl*. He just ducks the issue and does not take up the larger questions of

Buddha Word or *śāstra* there, as other commentators did. He simply assumes the *Four Treatises* is Buddha Word.[34]

The Desi only takes up the question overtly in his medical history.[35] Of a piece with the history of medicine in India and Tibet that he had already composed as an appendix to the *Blue Beryl*, this vastly expanded narrative is grounded in a careful rehearsal of the Buddha's medical teachings overall, an account that the Desi is far from the first to create but which he develops more than ever before.[36] And though for Zurkharwa, this larger Buddhalogical history demonstrated that he did not doubt that the Buddha taught medicine, for the Desi, as for many of his more conservative predecessors, this narrative served as a way to plausibly sweep the *Four Treatises* into the same category as the other Buddhist medical teachings that modern and traditional scholars alike would readily recognize.

Once again the Desi is closely dependent upon significant portions of Zurkharwa's work, in this case the latter's own history of medicine,[37] but also draws on other old accounts to engage in the hefty invention of history discussed in chapter 2, which is missing in Zurkharwa. This narrative endeavors to demonstrate that the *Four Treatises* originated in India and was translated into Tibetan during the imperial period. The Desi elaborates the story further, providing a fulsome biography of Vairocana, the imputed translator of the *Four Treatises*,[38] as well as of Drapa Ngönshé, who would have rediscovered the *Four Treatises* some centuries later.[39] Both accounts adduce information not associated with these well-known figures in other contexts and also develop significantly their bare mention in earlier medical works.[40] And, right in line with the Fifth Dalai Lama's investment in the creation of a full biography of Yutok the Elder, the Desi provides his own lengthy account of this mythologized figure.[41]

But the Desi's accountability to the historicist arguments aired by some of his forebearers begins to show in his treatment of Yutok the Younger.[42] Once again he tries to have things both ways. His ace in the hole is the claim that Yutok the Younger, like the Elder, is himself a manifestation of the Buddha—or even, as he also says, that Yutok the Younger is no different from the "actual" (*dngos*) Medicine Buddha.[43] That of course would mean that everything he writes is Word. But it also marks the entrée of the Desi as master hermeneut, for with a careful massaging of words, this becomes the occasion for the Desi to tilt to Zurkharwa's side after all.

Unlike Kyempa and Zurkharwa, the Desi uses the lines from the old *Crucial Lineage Biography* that suggest that Yutok is Sage Intelligent Gnosis,[44] and the statement from the *Heart Sphere of Yutok Story* that Yutok was prophesied by the *yidam* deities,[45] to argue that the *Four Treatises* really *is* Buddha Word. In doing so, he addresses the text-critical questions raised in Tashi Pelzang's work about elements of the *Four Treatises* that are clearly not Indian, such as its references to tea and porcelain. The Desi maintains that the fact that Yutok is a prophesied manifestation is exactly what explains why he deliberately made emendations in the text that would "accord with the place of Tibet."[46] In other words, the *Four Treatises*' Tibet-specific elements all amount to the compassion and skillful means of a buddha. Despite Yutok's "additions [of things that] do not exist in India, there is no difference [between the *Four Treatises*] and actual Word."[47]

The Desi also points out that it is nothing new to target an audience that is at odds with the place where the teaching is given. Just as the Buddha taught the tantras in the Heaven of the Thirty-three for the benefit of human beings, the *Four Treatises*, even though it was taught in Oḍḍiyāna, was actually meant to accord with the place and time of Tibet.[48] In saying this, the Desi is acknowledging the oddness of refusing to grant that the work was composed in Tibet even though it was clearly written for Tibetan circumstances. In short, the apotheosis of Yutok means that the fact that he composed the *śāstra*—now quoting those very loaded terms from the *Heart Sphere of Yutok Story*—should make no dent in our appreciation of the *Four Treatises* as having "not the slightest difference from the blessings of a tantra of the highest Secret Mantra."[49] It has certainly not affected his own reading: even when the very source he is citing calls the *Four Treatises* a *śāstra*, he declares that those who understand the root medical text in that way are mistaken.[50] Rather, "all the added words that [refer to things] not in India do not need a second thought."[51]

In these statements history itself, and anything empirically distinctive to local Tibetan conditions, have been subsumed under the larger umbrella of the Buddha's skillful means. This basic move allows the Desi to document with impunity all the parts of the *Four Treatises* to which Yutok supposedly made "additions": regarding the number of chapters in the *Root Treatise*;[52] the *materia medica* chapters of the *Explanatory Treatise,* where he also made many small adjustments concerning tea, medicines, and food; and many additions and corrections to the *Instructional* and *Final Treatises*, including material from the text *Somarāja* on pulse and urine diagnosis, the astrological/diagnostic systems having to do with "mother and son, enemy and friend," and again information about tea and porcelain, etc.[53] If we were to check these allusions in detail we would probably find that the Desi pointed to further Tibet-specific elements of the *Four Treatises* not noticed previously.

Note what the Desi has managed. In one and the same stroke, the Desi can exercise his historical scholarship and still keep it under the sign of the infinite compassion and skillful means of the Buddha. The medical text's place-based specificity is a function of prophecy, already intoned long ago in the *Yutok Nyingtik Story*: after the *Four Treatises* arrived in Tibet and was translated by Vairocana, the prophecy says, Yönten Gönpo would clarify its meaning.[54] Thus Yutok's efforts to change, add new sections to, or even "compose" the *Four Treatises* only prove that the work is an especially skillful Word that carries forward the compassionate dispensation of Buddha.

And yet, this is also the very move that flips. So ironclad is the narrative the Desi has invoked that perhaps he lets down his guard. His words get away from him. Words like "make" and "compose" really do suggest authorship. We might pass over the odd locution that "[Yutok] *made* a new translation [of the *Four Treatises*].[55] But when the Desi says, "[Yutok] *made* the entire [*Four Treatises*] *as if* it were Word," or mentions the necessity to present the work "in the manner of Word" due to the tendencies of beings in the degenerate time to disregard careful study, it is hard to miss how closely he is echoing both the wording and argument of Zurkharwa.[56] In short, shielded by the manifestation doctrine, the Desi made the space for himself to take account of historical specificity and historical actions, and to represent the *Four Treatises*' innovations accurately, perhaps even more fully than could his predecessors.[57]

One more sign of Zurkharwa's influence may be glimpsed in the Desi's detailed consideration of Tanaduk. Viewed from a distance, the Desi is pushing a far more conventional view than did Zurkharwa. But careful consideration of language and context reveals how close the Desi came to some of the most radical things that Zurkharwa himself said, although the Desi would probably be loath to admit it.

Addressing the Tanaduk question in both the *Blue Beryl* and his medical history, the Desi as ever puts his scholarly prowess on display.[58] He never finds a reference to Tanaduk in an Indic medical work, and we can trust that he or his aides really looked. Nonetheless, he asserts baldly that the Buddha taught the *Four Treatises* in Tanaduk.[59] He goes on to consider Zurkharwa's proposal that the place is referenced in the *Vibhāṣa* of Vasubandhu, and proposes instead a Buddhist *dhāraṇī* work as providing the more plausible scenario.[60] But this is just a friendly amendment. Recall, Zurkharwa's due diligence in finding a classical reference for the Buddha's teaching of medicine could be read—at least in that limited context—to corroborate the Buddha's ultimate creation of the *Four Treatises,* an implication that the Desi is glad to accept.

In fact, the Desi admits explicitly that he is in accord with *Ancestors' Advice* in critiquing all the wrong views about Tanaduk.[61] What's more, he has clearly appropriated Zurkharwa's idiosyncratic revision of the *Four Treatises'* suggestion that Tanaduk is surrounded by four mountains to the more climatically plausible view that it is perched on a single mountain with four sides.[62] He illustrates the medical pure land in just that way, with great fanfare on the very first plate of the medical paintings (see figure 1.9).

Notable as such influences are, and this time freely credited, a further, unacknowledged intellectual debt to Zurkharwa in the Tanaduk discussion points once more to the ineluctable impact of historical questions on the Desi's thinking. This regards the way that the Desi invokes the outer, inner, and secret heuristic to talk about Tanaduk. That was the move that Zurkharwa applied so effectively, and controversially, to nail the Word debate once and for all, and that sparked the Desi's furious invective in his biography of Zurkharwa.[63] So we might think that perhaps the Desi is trying to neutralize Zurkharwa's use of the heuristic in his own appropriation of its explanatory powers. Yet nothing really indicates that. It seems rather that the Desi is simply interested in taking up the heuristic himself. But in the wake of Zurkharwa's intervention, its implications may exceed the Desi's best intentions to present a consistent case that the *Four Treatises* was originally taught by the Buddha.

Actually, the Desi can cite several earlier scholars who applied versions of the outer, inner, and secret device to the Tanaduk issue. He draws on his colleagues' biography of Yutok the Elder, in which the Medicine Buddha addresses Yutok and tells him that the outer Tanaduk is any place where you can meet the Medicine Buddha.[64] That turns out to include all the places that scholars have proposed as the site where the *Four Treatises* were originally preached: Oḍḍiyāna, India, Akaniṣṭha, Sumeru. The inner Tanaduk is the town where you live, wherein you yourself become the Medicine Buddha. The secret Tanaduk is your own body, the parts of which are identified with the various mountains and places of the Tanaduk legend.

This statement is part of the larger invention of tradition spurred by the Dalai Lama and Darmo Menrampa. Tanaduk is thematized throughout the biography as a kind of Holy Grail. It is a place to which admission is reserved for the spiritually advanced; even Yutok the Elder has scarce access. The threefold interpretation marks a successive spiritualization of place, not unlike what was developing for Shambhala. Even at the outer level, it already does not matter where Tanaduk is/was. What matters is where prayer and a resulting vision of the Medicine Buddha occur. And at the secret level, all conventional sense of place is left behind by virtue of a familiar isomorphism between body and sacred place. That makes the debate about location more moot yet.

But the inner version of Tanaduk is interesting. This suggests that Tanaduk was just a meditative/imaginative rendition of where Yutok lived.[65] Now, even if originally addressed ostensibly to Yutok the Elder, how is such a suggestion different from what Zurkharwa and Tsuklak Trengwa proposed about Tanaduk, which rendered it as Yutok the Younger's hometown, and the entire myth as a literary trope coming out of his imagination? Actually the Desi's secret level suggests that Tanaduk is all a function of meditation too, since it implies that textual authorship is in the end just a matter of effective yoga. All three versions allow the *Four Treatises* to be the product of the visionary experience, if not agency, of Yutok—whether the Younger or the putative Elder, a distinction that in any event frequently disappears from view.

It is not at all clear that Darmo intended the heuristic in this way in the biography. Moreover, unlike Zurkharwa's bold measure to make the highest secret level of the *Four Treatises*' authorship its most mundane and historical reality, both the Desi and Darmo still make this secret dimension the most spiritualized, in line with Tibetan convention.[66] But in one more use of the taxonomy later on, when the Desi wrote his medical history, the context tips us back into wondering if it might really be about making space for everyday reality. Here the threefold taxonomy gets only a very abbreviated reference. Yet its placement, now in the middle of Yutok the Younger's biography, is telling.[67] We can already see that for the Desi, the real battleground on the Buddha Word debate centers on the younger Yutok, as he inserts a very lengthy discussion of notions of Buddha Word and *śāstra* right in the midst of this life story.[68] His brief allusion to the various levels of Tanaduk comes just at the end of that, and brings the Desi back to the biography and his defense of Yutok the Younger's reworking—if not fully composing—the *Four Treatises* in light of the specific conditions of Tibet already considered above. This aspect of Yutok the Younger's apotheosis most concerns the Desi, and the tripartite levels of place and personhood only facilitate a ready interchange. For the Desi the apotheosis is still ostensibly what makes the *Four Treatises* "really" Buddha Word. But the genius of Zurkharwa's own reading of Tanaduk, which tied visionary excess to a particular Tibetan time and place, makes the difference—between something really being the expression of a timeless enlightened realization, and something really being the expression of a historical person's enlightened realization—veritably negligible. In short, even though the Desi's rendering is couched in Buddhalogical language, and certainly does not go as far as Zurkharwa in suggesting that the *Four Treatises*' Buddha Word status is but pious fiction, the author and location of the medical root text's composition are there to be parsed as the historical personage Yutok the Younger in his hometown.

A timeless realization, when framed in terms of the actual person who had it, can be understood very specifically in terms of its historical place on planet Earth. Perhaps this ambiguity makes it all the more urgent for the Desi to demonstrate in other contexts that the latter part of the equation is *not* the way he most basically understands the origins of the *Four Treatises*. Perhaps it also makes it all the more clear that rhetoric, rather than doctrinal purity, is needed when the hairs being split are already very fine.

Or perhaps the Desi's bitter invective against Zurkharwa reflects personal stakes. Zurkharwa was a rival who anticipated many of the same challenges to medical science in a Buddhist world that the Desi himself sought to work out—and get credit for working out. Tibetan scholars of medicine today also often suggest that the Desi's animosity and lack of generosity toward his predecessor was occasioned by the political environment and the Dalai Lama's recent wars with the Karmapa's faction, with whom Zurkharwa had been closely associated.[69]

Most of all, we have to understand the competition as a sign of a certain atmosphere: a prevailing professional urge in medicine to show oneself as ready and able to muster the critical acumen to interrogate received tradition closely and to take stock of realities on the ground. As will emerge in the final chapters of this book, the reality on the ground for medicine has to do with both the patient's physical condition and with everyday social conventions and pressures. The next chapter shows more rhetoric and contradiction yet in early Tibetan writings on medical knowledge about women—along with some surprisingly forward-looking attitudes on gender and a decided commitment to getting things right. And in the last chapter, an equally early account of the personal qualities of the physician provides a startlingly candid set of instructions on how to flourish in the competitive scene of professional medicine. Both show the historical depth of the self-positioning strategies of the most gifted medical intellectuals, like Zurkharwa and the Desi. We will also discover what seems to come along with medicine's clear-eyed focus upon the vicissitudes of ordinary human life, namely, a deeply personal and embodied engagement with the art of healing others.

◎ PART III ◎
ROOTS OF THE PROFESSION

6

WOMEN AND GENDER

Gender and the status of women are complicated matters to take up in any context. Inevitably they entail contested issues in which not only the objects of our study but also we, the researcher or student, have personal stakes. Stakes can cloud our perceptions. Add to this the nature of gender itself—by which I mean the style and significance associated with sexual identity.[1] Such associations are often vague or inconsistent, if articulated at all.

The present chapter comes firstly out of a curiosity on my part. Should we expect the physician's immersion in medical realities to make a dent in the recalcitrant prejudices about women otherwise at work in a given cultural milieu? Does the putative focus on physical facts lessen the likelihood that medical theorists will participate in gender stereotyping? Anyone who has ever engaged in gender studies will readily anticipate the answer: mixed at best. For its part, Tibetan medical writing produces starkly misogynist passages on occasion. But there are also a few surprising moments when certain theorists soared above the usual consignment of the female to inferiority and even made liberative suggestions about gender. In addition to exploring how and when that happened in medicine, I also wonder how it compares to instances in Buddhist history where gender prejudice was addressed. I will only be able to gesture in such a comparative direction in the conclusion to this chapter. But I hope at least to lay the groundwork, from the medical side, for future exploration of the question.

Gender conception in any event parallels larger themes of this book. Gender as a category has everything to do with the relation between material existence and representation—the connection and yet also distinction between the physical body and its socially legible markers. As for other issues we have examined, powerful assumptions coming from Buddhist epistemology, ritual, ethics, and soteriology are well nigh naturalized for Tibetan medical writers on gender too. And yet considerations distinctive to the clinic sometimes conspired to trouble such naturalizations, at least by the early modern period. By the end of this chapter we will find Zurkharwa Lodrö Gyelpo once again trying to separate conceptual categories from the reality that they would represent, aware of the illusions into which such categories seduce us at the expense of accurate knowledge of individual particularities. He even develops a term to name the fact that real human beings do not necessarily display the qualities that their sexual identity would suggest. As he makes very clear, gender is but a designation, and is distinct from anatomy.

Issues around sex, gender, and the status of women add further dimensions to the already overdetermined character of Tibetan medicine. In previous chapters we have encountered both overlap and disjuncture between scientific aims and religious values, and seen the heterogeneous provenance of medicine in Tibet as it drew upon Indian, Western Asian, Tibetan, Central Asian, and Chinese medicine; upon concepts and values coming out of Buddhist scholastic and meditative contexts; and upon any number of other heritages, from Indian and Chinese cultural history to Bön. This chapter adds the forces of patriarchy, patriliny, androcentrism, and misogyny to the list. We have ample reason to think that the principal texts, institutions of learning, and centers of practice of Tibetan medicine were virtually always controlled by men, whose lives were funded by old traditions of regulating, demeaning, or simply ignoring women.

Beyond sex and gender per se, this chapter also attends more generally to social issues, rhetoric, and contestation in the formation of medical knowledge, also seen several times in the foregoing. Focusing now upon the *Four Treatises* and a few early commentaries, this chapter shows how long, and how fundamentally, instrumental agendas have affected medical writing in Tibet.

Finally, the material in this chapter provides a chance to track how Tibetan medicine received and responded to other Asian medical traditions with some specificity. Unlike the topics of the previous chapters, which were unique to debates initiated in Tibet and thus suggest only the broadest grounds for comparison (on the order, for example, of asking why Indian medicine didn't produce medical texts preached by deities, or invoke clinical experience to question yogic anatomies), the *Four Treatises*' handling of women's medicine shows close debts to Ayurveda, reproducing certain passages almost verbatim. But this allows us to notice what is nonetheless different in the *Four Treatises* and to ask why. Some of those differences have to do with Tibetan kinship patterns. There are also places where disparate medical streams have had some influence, such as in the introduction to an otherwise Ayurvedic passage of the gonadlike *samse'u* that is probably indebted to East Asian medicine; or the appearance in the *Four Treatises* of a uniquely gendered version of the Chinese diagnostic pulse tradition; or the occasional listing of blood as a fourth humor, which probably has a western Asian pedigree, and so on.

In looking for what is distinctive and historically specific, however, we should avoid any easy assumption that the parts of women's medicine in the *Four Treatises* that are influenced by medical streams from afar are for that reason not "Tibetan." In adopting and adapting Indic and other imported medical concepts throughout the *Four Treatises*—and Yutok selected and restated such material judiciously—the Tibetan doctors were making it their own.[2] The knowledge so deployed worked for them, and it accorded with what they already knew or suspected about the human body. Those occasions where Tibetan medical writers found the need to challenge what they were inheriting were rarely about foreignness per se (the Buddha Word debate does indeed raise issues about Tibetanness, but in service of a very different question).[3] Although the Fifth Dalai Lama and his court were aware of the value of searching abroad for new information and therapies, Tibetan medical tradition does not systematically distinguish foreign medicine from Tibetan medicine in just those terms. In short, our concern to discover innovation has to do with the conditions under which change occurs, not the cultural or national identity of those shifts as such.

With these several aims in view, the early materials we will look at, along with a few later interventions by Tashi Pelzang, Kyempa, and Zurkharwa, provide cause for both marvel and dismay. The following must nonetheless be limited in what it considers. Notions of gender and issues about the anomalous nature of women's bodies abound in Tibetan medical theory and practice. I will only take up a few interesting high points and some rather challenging low points—from a feminist perspective, that is—so as to puzzle over how to account for both.

I will not in any case be able to recover much of women's voices in all of this. Reports on experience by women surely fed Tibetan academic medical writing, but this is recognizable only through a close feminist reading, and only in the barest echo. What we see instead are (male) medical writers grappling with issues of sex and gender in ways that sometimes show their cognizance of women's experience. And that is not to mention a quite surprising insistence that it is important to account for women's bodies on their own terms. I will start with this encouraging, if rather puzzling, development first.

CREATING WOMEN'S MEDICINE

The *Four Treatises* diverges significantly from Ayurveda in at least one fundamental respect. It counts the illnesses specific to females as one of its eight principal "branches,"[4] respecting a long-standing Ayurvedic tradition that there are eight main branches, or sections, of medical knowledge. But it offers a different conception of what those eight sections are. In Ayurveda, obstetrics and embryology were part of the pediatric section of medical knowledge. Female medicine as such did not constitute its own branch. *Aṣṭāṅgahṛdaya* lists the eight branches as body (i.e., general internal medicine); pediatrics; demon possession; upper body; surgery; poison; geriatrics; and virility and fertility.[5]

The author(s) of the *Four Treatises*, in contrast, found it necessary to conceive of female pathology as its own branch.[6] They substituted it for Ayurveda's "upper body," which focuses on eye, ear, nose, and throat. In the *Four Treatises* system, that branch is incorporated into the more general "body branch." Thus the *Four Treatises* lists its eight sections as body; pediatrics; female pathology; demon possession; wounds and surgery; poison; geriatrics; and virility/fertility.[7]

Why the switch? And what did it accomplish?

The answers are not entirely clear. Some aspects of the rationale appear to be missing from the record. Still, we can tell a lot from what we do have. At least two early works provide sustained discussion of the assumptions behind the branches of medical science.[8] Ostensibly concerned with the more general question of why there are only eight branches (not more, not less), these pages show a recurring concern about the rationale for a female pathology branch.

It is especially striking to see the defiant response to the criticism that the shift deviates from Ayurvedic tradition and the *Aṣṭāṅgahṛdaya*:

> While in the *Aṣṭāṅgahṛdaya* an upper body branch is expounded, here in the *Four Treatises* a female pathology branch is expounded. If one were to argue that this contradicts the norm in medical science,[9] then the answer is: This so-called contradiction of the norm, is it that this *Four Treatises* is in contradiction to the norm, or is the contradiction of the norm in the normativity of there being eight branches, or is it that since eight branches are the norm it contradicts [other] normative treatises? A definitive argument has not been established. Since this *Four Treatises* is spoken by the Buddha Bhaiṣajyaguru, the four medical traditions are incidental. Thus a definitive statement on what are the normative eight branches has not been established. [Your wrong conclusion is based on assuming] that the sages' explanation is the most important one.[10]

In this clever and yet evasive rejoinder, the defense simultaneously contests that there is a single norm in medicine at all, and pulls out a trump card—just in case there is such a norm. It challenges the idea that the "sages' explanation," i.e., Ayurvedic tradition, is the most authoritative source of medical knowledge. Instead it not only invokes the heterogeneous genealogy for medical knowledge, referring to the four medical traditions, i.e., as represented in the *Four Treatises*' fourfold audience of gods, sages, *tīrthikas*, and Buddhists, but also goes on to trade on the *Four Treatises*' status as Buddha Word in order to relativize all of them. Indeed, here that status served medicine well, providing the rhetorical tools to proceed with what was thought to be a better way to organize knowledge. Thus do these defenders of the *Four Treatises* escape any pressure to conform to what in other contexts would constitute authoritative medical tradition, by virtue of the conversation-stopper that the *Four Treatises*—and its counting of female illness as one of the eight branches—is the Word of the Buddha. End of discussion.

Our two early sources nonetheless do go through a number of arguments, beyond this display of bravado. Still, they don't offer much reason for the shift other than a simple conviction—but perhaps this is a lot—that female illness is unique and needs to be treated on its own terms.

The discussion is notable for its overt acknowledgment of the androcentrism of received medical tradition. On several occasions the lynchpin is that the general "body branch"—which in the *Four Treatises* is said to consist of seventy chapters, far more than any of the other branches, and covering most workings of the human body—renders the body of the adult man as "primary."[11] Although the discussion grants that many features of the body are common to all people (such as general aspects of illness, vital fluids/bodily constituents, and imbalances[12]), the androcentrism of the body branch nonetheless becomes the main reason another special branch is *not* needed for the adult man, while it *is* needed for the female, the elderly, and children.[13]

Even though the (adult man's) body branch is understood to be basic to all the other branches, a key metaphor illustrates that the medical specialties those represent are also necessary. One passage analogizes the various branches of medical science to a situation wherein a father appoints his son to control eight communities of people. When this is seen to be too difficult, that son (i.e., the body branch of medicine) in turn sires seven sons of his own (the other seven branches) to complete the task. Such specialization is required in medicine since what is taught about the body in general does not suffice for all medical contexts, especially those illnesses particular to children, women, the aged, and with respect to sexual performance.[14] A second analogy regards "the eight major activities of human life," interesting in its own right for what it says about Tibetan society around the time and place of the *Four Treatises*' composition. These are agriculture; animal husbandry; explaining the Dharma; medicine; calculation; performing rites of aid; building and crafts; and making temples.[15] Here too the point seems to be that such activities are not reducible to each other and therefore require specialized expert knowledge.

Another pertinent discussion that turns on how much males and females have in common or not concerns whether the virility branch of medicine bears upon women. Here again is an admission that the concerns in this branch are androcentric. Yet since in the end, what are important are reproduction and especially the reproductive fluids, it is maintained that the virility branch pertains to both male and female. Both male and female sexual fluids ultimately have their source in the gonadlike *samse'u* organ in the body, these commentaries assert, and both males and females care about having children (or sons).[16]

Although this point might be read as a reason a special female branch is in fact *not* needed, it is used instead to make the opposite argument, i.e., that while most of the *Four Treatises* refers to humans in general, some issues, such as specifically female pathologies, still need specifically focused discussion. For example, one passage asks whether treatment for pathologies of the uterus should be considered part of the general body branch.[17] The critic argues that if indeed the body branch ignores female pathologies, then it is really but "half a branch." While this recognition that female bodies and their diseases should be as important as male diseases—and females make up half of the human population—is laudable, in rejecting its logic, the defenders of the *Four Treatises* are making a more basic assumption that, again, there needs to be a separate specialization in female pathology.

> Since the text of the *Aṣṭāṅgahṛdaya* does not include a branch to heal female pathologies, there is the upshot that its branches are incomplete. If you say it does have the healing of female illness, then is that contained in [its own] branch or is it contained by being subsumed into the body branch? If it is contained by being subsumed into the body branch then that goes against the explanation of healing twenty kinds of uterus disease in the *Final Treatise.* If it is not subsumed in the body healing branch, then it violates the premise that it is in the body. And yet [if it is in the body branch] there would be the problem that not all [uterus] diseases would be female illnesses.[18]

The passage is recondite in its language. It ponders how to best remedy a problem in the *Aṣṭāṅgahṛdaya*'s assumptions. Should female maladies be subsumed into the general body branch? If they are, then what about the specialized information provided in the *Final Treatise* on particular female diseases?[19] Further, if uterus problems were not put in a special section for women's illness, it would imply that they were not necessarily a female issue. In other words, all female illnesses should be in a special section, since female illnesses are not common to all bodies. And yet there is another category problem in tension with this point. If the female illness branch is *not* put in the body branch, then the very premises of the body category are violated, because this could suggest that female illnesses are not a human condition—a double bind also recognized by modern feminists.

Such a point is brought home further in another passage on whether it would be possible to reduce the number of branches to seven by leaving out female pathology as a branch altogether. Since male and female bodies are taught similarly, why the need for a branch exclusively on the female body? Moreover, would it not be but a half-branch (i.e., since all other branches relate to both sexes)? But the response is instructive. If a branch is organized around medical specialty with regard to the object to be treated, then such a specialized branch cannot be said to be incomplete or missing the other half of humanity, since it was already defined in terms of that specialization. If you had some sons and assigned two estates to each, but to one of them you gave only one, you would not say that was half an estate.[20] The example reiterates that it is legitimate to have a branch that is sex-specific. But note that this right seems to be granted only for females. Perhaps the largesse makes up somehow for the androcentrism of the main body branch.

Several other passages also insist that while the general body section is gender inclusive, specialized sections on certain topics, of which female pathology is most exemplary, are needed and justified. In contrast, there is no justification for a branch devoted to the upper body alone, such as the *Aṣṭāṅgahṛdaya* suggests.[21]

CATEGORY QUESTIONS AND ANDROCENTRISM

The arguments just summarized are not clearly conceived. It is not really specified why the female body should be singled out for its own branch. Nor is the reason the male body is the default norm for the body branch ever questioned or defended.

In any event, whatever principle is being articulated is not fully achieved in practice. In the actual organization of the *Four Treatises*, the female pathology branch is identified as chapters 74 through 76 in the *Instructional Treatise*. But information on female medicine is not confined to this section. There is also female-specific information in what is ostensibly the body branch: on conception, pregnancy, and menstruation in the embryology chapter; on conditions of the uterus, vagina, and menstrual irregularities in the virility and fertility chapters. Next to a chapter on the male genitals, there is also a separate female genitals chapter (which mostly repeats the content of the female branch chapters in briefer form).

The overlap and disorganization may indicate that the female branch chapters were added later as an afterthought, gathering together female-specific material from throughout the *Four Treatises* without adjusting the rest of the work accordingly.[22] Or it may just be one more sign of the general disorganization of the *Four Treatises*, whereby anatomy, physiology, and other topics are to be found in several different sections. And it is certainly not the case that all of the material in the body branch chapters is applicable to both sexes. Not only is there the female-specific material, as just noted, there is also the chapter on the male genitals that touches on penis malfunctions, as well as chapter 68 of the *Instructional Treatise*, which concerns the testicles.

Medical information on female conditions in particular was also disorganized in Ayurvedic texts. Martha Ann Selby writes that she has "gained no real sense of a coherent idea of gynecology in the Greco-Roman sense, or in the contemporary Western sense, but this has much more to do with the ways in which disorders and their treatments are taxonomically understood and arranged, rather than with a lack of information. . . . But for whatever reason, it remains true that the bodies of women seem somehow fragmented and dispersed across the wholes that are these two texts."[23] In fact, at least in the case of *Aṣṭāṅgahṛdaya*, the difference between its organization and that of the *Four Treatises*' presentation of women's medicine is slight. In the former, most of the information covered in what the *Four Treatises* calls its female pathology section is gathered in the first two chapters of the *Śārīrasthāna*. This suggests that the innovation that the early *Four Treatises* exegetes were touting may have been more about self-presentation and nomenclature than anything else.

Writing several hundred years later, Zurkharwa Lodrö Gyelpo comments pertinently on the wisdom of creating a female pathology branch of medicine and helps guide our own understanding. He cites a variety of criticisms of the *Four Treatises*' branch system, including one disputing the very idea of special branches, saying they should all just belong to the main body branch.[24] Zurkharwa incisively points to the confusion around the word "body" in the term "body branch." Does it refer to that which is basic to all medical issues, or to a branch that should be distinguished from the other branches? Does the latter option make the other branches thereby not part of the body?

Once again Zurkharwa wants to talk about "context."[25] The word "body" is used in different ways in different contexts. "Bodily illness," which names the subject of all the various illnesses for people "with bodies," is "just a designation; it is not the body as such." Adding that there is no need to articulate a category such as "illness of the adult" when what is discussed there is common to all illnesses of adults, Zurkharwa evinces again a probative attitude to language and the difference between heuristic categories and the reality that they would denote. He notes how the eight branches are construed differently by various sources, such as *Suvarṇaprabhāsottama*, as well as in varying contexts in the *Four Treatises* itself. These points serve the very important function of carving out freedom for medical scholarship: to reconsider categories, to organize them in ways that make sense, and to shift that organization when needed. This is not to say Zurkharwa does not betray his own unexamined assumptions about categories. When he argues that there is no male pathology that is parallel to the kinds of illnesses under the heading of female pathology, but makes no similar point the other way around, he seems to think that there is something intrinsically female about female pathology that has no analog in male illness. In this he seems to imply that male medicine can represent issues common to both males and females, while female medicine cannot.

But rather than getting sidetracked with Zurkharwa's own androcentrism, my point is simply to note how determined the medical scholars were to have a section of medical knowledge specific to females, even at the cost of reason. When Zurkharwa concludes his discussion with an insistence that this whole debate is meaningless since it is based on confusion about what the term "bodily illness" denotes, it reinforces how arbitrary the entire argument is.

And that only begs the question of why the Tibetan doctors insisted on this shift from classical Ayurveda in the first place. We might be tempted to read some proto-feminist motivation in it, to see it as a clear-eyed recognition that making the male normative for medical knowledge can be a costly error, as has been recognized in recent feminist history of science.[26] And yet the implications of having a special section on the female and not one for the male are multiple and potentially at odds. On the one hand it is understandable to want to isolate obstetrics as its own specialty, given the special role that the mother has in producing offspring, in turn so central to human flourishing. And yet on the other hand—and this again has been massively recognized in feminist thought—claims to uniqueness are a double-edged sword. They can correct a previous imbalance in the opposite direction, such as the androcentric bias seen in the *Four Treatises*. But they can also serve to limit gender flexibility, to inscribe gender essentialisms, and to underscore male privilege. Joan Scott puts the problem well: "women are either the same as men (but better represented by men) or they are too different to be taken into account. These contradictory assumptions have both resulted in the exclusion of women—from clinical trials and from a range of other things."[27]

It may well be that the resistance to isolating male pathology in its own section had to do with a worry that this would remove the male from his pride of place as the default normative body. Indeed, despite the notable self-awareness of medicine's androcentrism that we have just seen, the very visibly sexed renderings of medical knowledge in the Desi's seventeenth-century painting set still make that androcentrism, and especially the answer to the question of which sex can represent gender-neutral conditions (answer: the male), patently clear.[28]

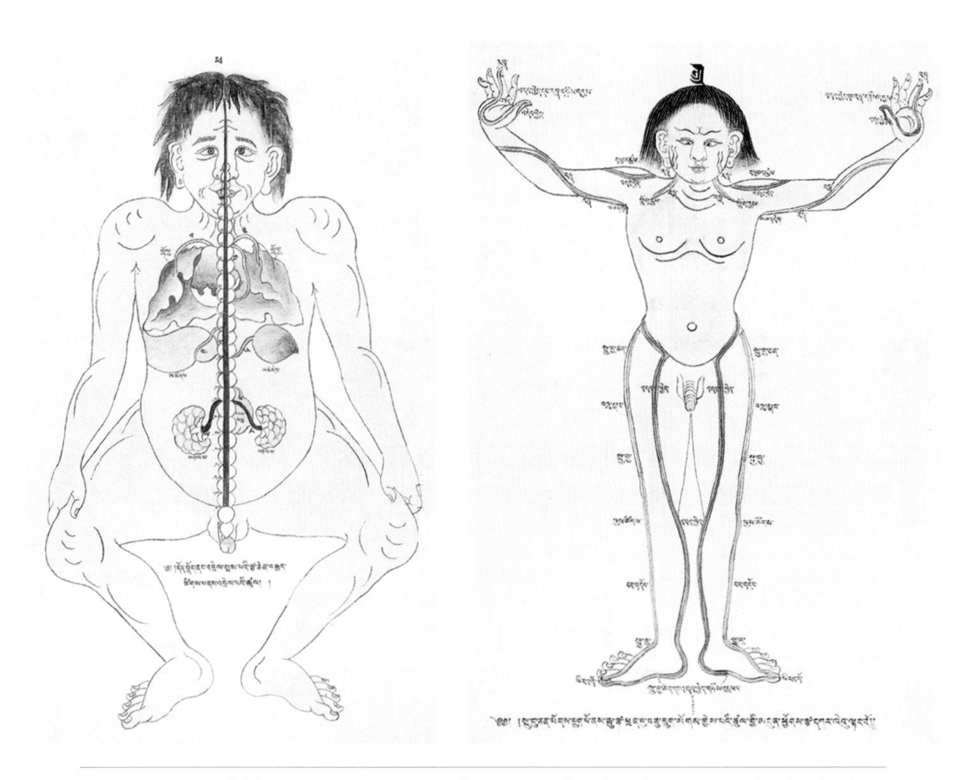

6.1 Anatomical illustrations gratuitously gendered male. The parts of the body they illustrate have nothing to do with the sexual organs. No anatomical illustration in the painting set is gratuitously gendered female. *Plates 10 and 13, details*

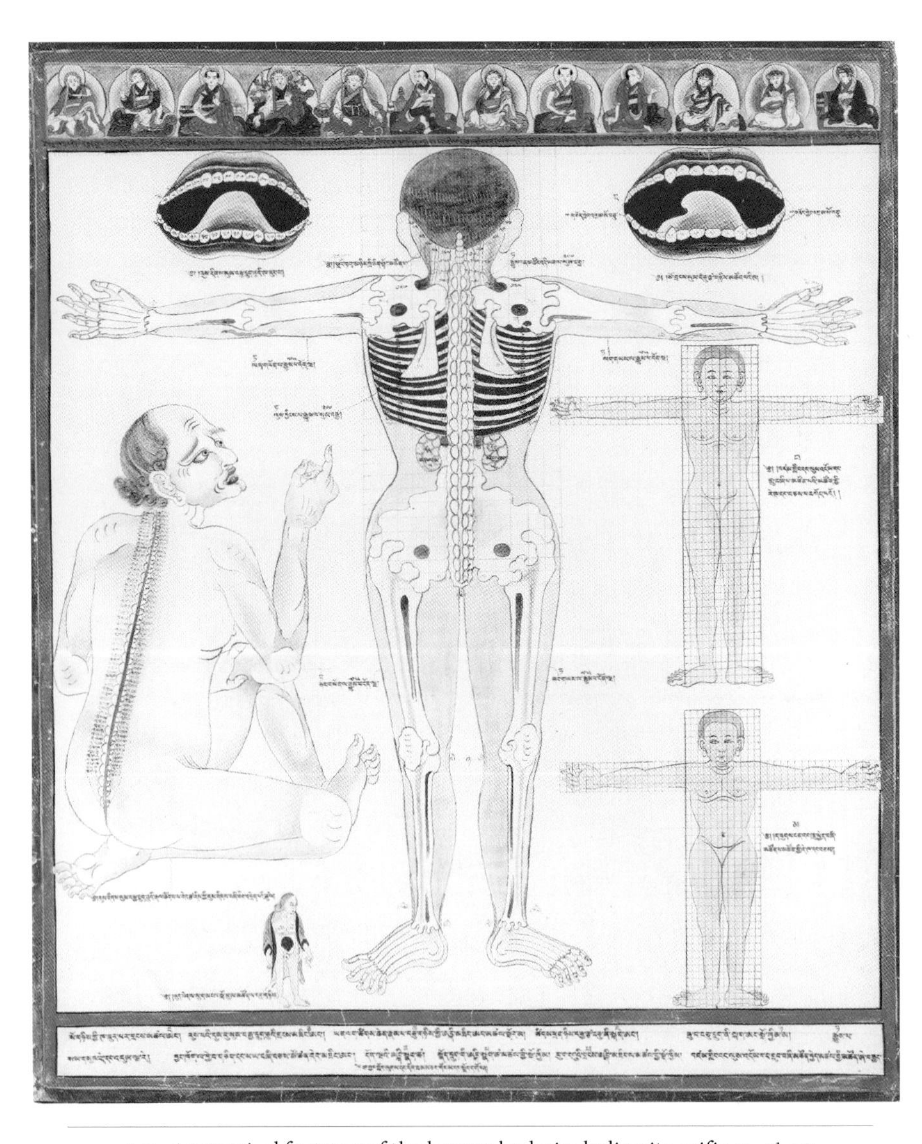

6.2 Anatomical features of the human body, including its orifices. *Plate 8*

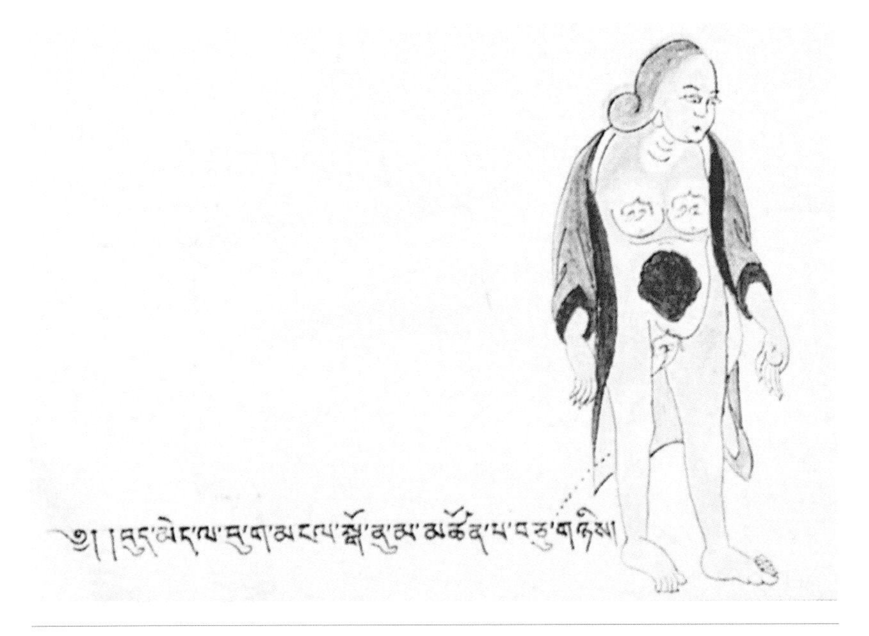

6.3 The "extra" orifices of the female. This small image is the most specific rendering of the female reproductive system in the entire painting set, in contrast to several detailed renderings of the male organ. *Plate 8, detail*

WHAT THE ONE HAND GIVES: MISOGYNY WRIT RATHER LARGE

Neither the lack of clarity about why this category shift was needed nor the inconsistency with which it was implemented undermines its significance. Quite the contrary, the shift's ultimately arbitrary nature makes it all the more telling. What remains when consistency and function fall away is rhetoric. Something—or perhaps several things—was gained by creating a branch of medicine devoted to females. It is possible that one of those things was to display a concern for scientific accuracy. But the shift might also have served to balance out the implications of another rhetorical agenda in the *Four Treatises*, which is nowhere more evident, ironically, than in the opening lines of the chapters that actually make up the touted female branch. As far as I know, the passage is unique to the *Four Treatises.* It seems to reflect a particular set of concerns of its author and his milieu.

The introduction to the first of the three female pathology branch chapters is stridently polemical. This is already evident in the way that the statement stands back to offer a general assessment of the female condition, in broad strokes and in terms that quite exceed the medical. Whatever good news there might have been in the recognition of a separate women's medicine is taken away in these opening lines. The statement trades heavily on the reigning Buddhist misogyny: to be born a female is the result of bad previous karma, excess desire, and a low store of merit.

> The body, made from the three poisons and the four elements,
> appears as male or female
> through the force of previous karma and desire.
> Through low merit one attains the deficient female[29] body.
> Breasts, uterus, and the monthly cycle are her special extras.
> The ultimate bodily constituents are the two fluids,[30]
> the white and the red.
> Red menses drip after she has reached thirteen.[31]
> She holds the semen inside the uterus,
> and the fleshly body [of the fetus] develops.
> The white spreads in the breasts
> and comes to nourish [the baby].
>
> In that, as a result of the conditions
> of previous karma, eating, behavior, and ghosts,
> there are five uterus illnesses, sixteen channel illnesses,
> nine tumor illnesses, and the two types of worm illnesses.
> The principal female illnesses are thirty-two.
> Along with the eight common illnesses, there are forty altogether.
> Since she is a low birth
> the woman's body has extra [illnesses].[32]

The statement makes various claims about female physiology. But seemingly gratuitous to the medical overview, the passage's misogyny could not be more overt. The female state that the ensuing chapters will address is the result of a low store of merit. It really looks like a deliberate chastening. Yes, we are going to take a special look at female medicine, but let us remind you at the outset that if there is anything special about being female, it is that it is a special misfortune.

Note that low merit, a classical Buddhist notion, has the medical upshot that the body is marked by the three "special extras" of the female: breasts, uterus, and the monthly cycle. This language bespeaks of course a fundamental androcentrism already evident in early Ayurvedic accounts of human anatomy, albeit without the misogynistic valence added by the *Four Treatises.* But in all such cases, the features that distinguish a person as female are something extra on top of the norm.[33] Elsewhere in the *Four Treatises* is further indication of the female's aberrational status: her extra fluids in the body, her extra flesh, her extra orifices (see figures 6.2 and 6.3).[34] Never is it thought to count the male's penis and scrotum as extras—as indeed they might be, if the perspective of Tibetan medicine were instead gynocentric.

It is not clear if we should recognize as a further medical misogyny the count of specifically female illnesses as forty. In contrast, the specifically male diseases are usually counted as seventeen.[35] Perhaps the discrepancy represents a genuine recognition of a greater number of ailments in women.[36] But the passage turns her elevated vulnerability to sickness into an affront: "Since she is a low birth the woman's body has extra [illnesses]."

Actually Kyempa provides some corrective when later on he notes a count of special illnesses for both men and women.[37] The early modern commentators also seem to show some discomfort with the *Four Treatises*' outright misogyny. In commenting on the introductory passage to the female pathology section, Kyempa, followed closely by Darmo and the Desi in the following century, provides two alternate views of the female gender. One concerns certain bodhisattvas who manifest as females and spread medicine. Here Kyempa is reminding his readers of a female who is essential to medicine's very conditions of possibility. But this is tempered by the further specification that Kyempa is at pains to make: that such bodhisattvas were actually males who had merely assumed the form of females. The second view has to do with "bliss-void method and primal awareness"—a euphemism for the male and female couple in tantric sexual yoga.[38] He does not specify whether he is referring to actual women or a visualized consort—tantric practice knows both—but the statement is fully applicable to real women with whom men engage in sexual yoga and who facilitate bliss.[39]

Such reparations are kicked up a further critical notch when Kyempa adds a final point: "One needs to understand in terms of the context."[40] This would remind the reader that women are precious in some contexts even if they may be inferior in others. It makes an important intervention by removing the option for an absolute female nature to be fixed for all time.

The statement fits into a growing pattern in medical thought already seen in previous chapters of this book. Standing back and recognizing that things have different meanings depending on the way they are being used opens up space for medical knowledge on its own terms. In this case, all three contexts under discussion—the medical vulnerability of women due to their inferior karma, the salvific female form assumed by certain bodhisattvas, and the key role of female consorts in sexual yoga—trade in Buddhist conceptions of the female. Recourse to Buddhist imagery and lore facilitates an epistemic position that knows there are various systems of thought and practice that can confer meaning; the picture of women represented by the *Four Treatises*' female pathology chapter is only one item on the menu.

MORE RHETORIC IN THE SEXUALITY

The interventions of the commentators are encouraging, but they don't overturn the harsh opening to the female pathology section. Nor do they explain why the female pathology was introduced in such a reprimanding way in the first place.

There is another opening gambit in the *Four Treatises*, at the beginning of the virility/fertility section, that also takes the occasion to make overtly androcentric and patriarchal, if not outright misogynist points. This time we can get a clearer picture of what was at stake, even if the rhetoric is more complex and harder to parse. It is worth going through in some detail, for it brings home the inseparability of medicine and certain instrumental agendas, at least with respect to women and gender.

Rotsa (i.e., *ro tsa*), an odd term in Tibetan (Skt. *vṛṣa,* or *vājīkaraṇa*), names one of the eight branches of medical knowledge in the *Four Treatises* and in Ayurveda.[41] The category is defined as addressing both sexual pleasure and fertility. In *Aṣṭāṅga* this set of topics is covered by one chapter, but the *Four Treatises* expands the *rotsa* section to two, one devoted to the man and one to the woman.[42] But despite the addition of a special chapter on the female, sexual pleasure is treated only in the male chapter. The female chapter is only about fertility. Moreover, the *Four Treatises*' entire view of fertility is presented very much from the perspective of male privilege and male progenitors. Unlike the female pathology chapters, where the misogyny is mostly gratuitous, in the *rotsa* section patriarchy and androcentrism are foundational to the organization of knowledge. Thus it is more understandable why the section's opening statement needed to address that explicitly.

In fact, this introductory passage suggests that there was some question afoot about why the androcentrism of *rotsa* medicine is so definitive. The opportunity seems to have been taken at the outset to foreground the very nature of virility and fertility, and the role of males and females therein. Once again the passage appears to be entirely the *Four Treatises*' original invention. The parallel chapter in *Aṣṭāṅgahṛdaya*'s virility/fertility section contains a lot of the same information as in the rest of the *Four Treatises*' male virility chapter, as well as overt androcentrism, if not a triumphalist male voice, but the Tibetan work's self-consciousness of its androcentrism and patriarchy is entirely missing in *Aṣṭāṅga*. Nothing in the Indic work is like the case the *Four Treatises* makes that the male must be the main focus of *rotsa*—or the distinction it makes between principal and auxiliary topics. This original passage in the *Four Treatises* thus reflects the social constructions of the particular milieu in which it was written.

O Sage Mind-Born!
Rotsa concerns the ability to perform one's desire,
and the propagation of offspring in one's family lineage.[43]

In that there are two aspects: principal and auxiliary.[44]
As for that:
through karma and emotional obscuration,
one is born as male or female.
The male man is the principal concern of *rotsa*.
If the man is not able to perform his desire,
then even if he is surrounded by a hundred women
the goal will not be accomplished.

When unflawed semen has multiplied in the man,
it is permissible to search for a woman in order to multiply offspring.[45]
For that reason, the principal concern of *rotsa* is the man.
The auxiliary concern is deficient woman.[46]
She holds the seed and is the basis for its development.
But since the daughter does not hold the generational lineage
of the father,[47]
she is not the principal.

[He] can do it with all [women]
but if [her] karma, power, and merit are low[48]
a son [or child] will not come.
In that case, the auxiliary, the means to search for [a cure for the condition of] no son [or, the means to search for a woman][49]
is precious.

For metaphors:
a field without seeds,
and a seed planted in a bad field.
It is the same as that.

[Problems of *rotsa* have to do with the condition of the] solid and hollow organs in general,
but are explained [in particular] as an illness of the *samse['u],*
which is the basis for the refined distillate, the white and red seeds.
For that reason, the illness is common to the two:
male and female.

And so,
rotsa for the principal, the man, is as follows:[50]

The last line actually introduces the rest of the chapter, now focused on the male. I include it for the irony of its insistence that the man is the principal topic of *rotsa*, despite the immediately preceding lines that have just spoken of the commonality between male and female reproductive issues. Indeed, the entire passage is an exercise in ambiguation, sleight of hand, and doublespeak. I have broken the passage into sections according to my reading of the individual arguments that it brings to bear.[51] Following its circuitous route through patriarchal privilege, sexual prerogative, various contradictions about conception and responsibility for the sex of offspring, and a host of aspersions cast at the female is an object lesson in how rhetoric can be twisted while consistency be damned. And yet it also affords a very clear view of what can fund medical rhetoric. The doggedness of the statement only underscores the power of the social realities to which it is so evidently speaking.

The passage identifies the two aims of *rotsa* to be the fulfillment of desire and reproduction. This is natural enough. But here both are put to work to defend patriarchy and androcentrism, as the passage moves into why the two sexes are unequal, one the principal and the other an auxiliary, in favor of the male.

It is a rather odd way to establish this male protagonist—making an egalitarian statement about karma and emotional obscuration as the common cause behind both male and female births. This is a different ploy from the opening of the female pathology section. The latter also attributes birth and sex to previous karma, but then goes out of its way to add that to be born female is a result of having particularly *bad* karma—i.e., a poor store of merit. We might suspect that the *rotsa* chapter's statement implies the old tropes as well, i.e., that good karma and fewer emotional stains make for a male body and bad karma and a lot of emotional stains make for a female. But it does not say that. On the surface it seems more like a rueful reflection on the samsaric condition of being born as a sexed being of any kind. What it seems to really say is that *while* the two are equal in coming out of the common human condition of karma and emotional obscuration, *nonetheless* the male stands out as the center of the *rotsa* teachings. This makes for an interesting comparison with the female pathology introduction, where the female is brought to the fore, yet has to be cut down to size first. In contrast, here in the *rotsa* chapter where the text is going to focus on his needs, it would appear that a bone is being granted to her in apology, admitting that at least from one perspective, the two sexes really are the same.

Anyway, the most telling feature of the passage is that it keeps waffling. First it implies there is no difference between the male and female condition, then it argues there is a kind of difference, then it keeps implying there is no difference, but then again difference is asserted. How to understand this? It is like the Desi ruminating over whether his *Blue Beryl* commentary is innovative, or is not, or is . . . or Zurkharwa saying that the tantric channels are not in evidence, but then again they almost are. Or that the *Four Treatises* are not Buddha Word, but then again they are equivalent to it. All of these passages betray exceptional caution, born out of an appreciation of complex issues at stake, requiring negotiation between the facts and conflicting social realities and bearing no easy resolution. Perhaps these writers are thereby creating a smoke screen that allows them to proceed with a shift in received tradition—or in this case, to assert androcentrism and patriarchy at the expense of medical reason.

After defining *rotsa* and then asserting its main thesis that the male is its principal concern, the statement's first main point is to defend that thesis in terms of performance. In order for sex, and reproduction, to happen, first of all he has to be able to perform. It's a *sine qua non*. Even if he is surrounded by a harem, it is all up to him. Therefore, *rotsa*'s primary task is to render him virile.

Performance is fair grounds for focusing on the male. But then the text adds a quite different reason he is central to *rotsa*. If indeed his semen is unflawed and plentiful, he can search for a woman to fulfill the desideratum to bear children. The commentators make this more specific. If for whatever reason, his partner cannot bear children, then it is permissible for him to find *another* consort.[52] And that too is why he is central. Here we can note the failure to consider that perhaps she, if fertile, might choose to find another partner herself. Ignoring the possibility that a fertile and ready female would abandon an impotent male and search for another one who could impregnate her, the text only considers his exclusive and asymmetrical social prerogative.

Yet on this very point, which would seem to be about patriarchy, the *Four Treatises* betrays several further ambiguities. For one, the very key word "son" or *bu* can mean either "child" or "boy" (the female *bu mo*, "daughter" is more specific, but its usage does not render all instances of *bu* necessarily male). Sometimes the sense in the passage is obvious, sometimes it is truly uncertain (and sometimes the commentators will endeavor to specify which is meant). In my translation I have tried to indicate when the alternate reading is also possible.

It may well be that in some cases the ambiguity is strategic. The same can be suggested for the phrase "the means to search for a woman."[53] In the passage translated above, that phrase names the auxiliary topic in *rotsa* medicine, the part that concerns the woman. But a major clue that something is amiss is the ambiguity regarding the title of the chapter that actually discusses that auxiliary topic. While the *Four Treatises*' own list of its contents duly refers to that chapter as "finding a woman,"[54] in the three earliest xylographic editions of the work, the chapter itself is labeled instead "how to search for [remedies for the condition of] no child [or son]."[55] The latter would be a more appropriate title and would match the actual content. This second, auxiliary *rotsa* chapter is all about the medical reasons women might not be able to bear children (or sons) and how to address those problems, in effect summarizing the content of the female pathology chapters, and also paralleling in large part the first *rotsa* chapter, which provides diagnoses and remedies for the afflictions affecting sexual reproduction in the male. But by the time that Kyempa was writing, the title of the chapter had changed to read "how to find a woman,"[56] despite the fact that its content has virtually nothing to do with that and the colophon to the chapter continues to call it "how to look for [remedies for the condition of] no child" in all of our available versions.[57]

The new rendering of the chapter title, in addition to reflecting the *Four Treatises*' initial list of its contents, is of a piece with the discussion in the first *rotsa* chapter, which introduced the auxiliary topic in just that way. In short, the *Four Treatises*—in all of its available editions—indicates not only a pervasive ambiguity in the usage of the verb *btsal,* which occurs five times in the two *rotsa* chapters and switches between connoting to look for and to attend to a problem. Far more slippery is the key term *bud med*, "woman." Pronounced *bumé, bud med* is homophonous with, and one consonant away from, *bu med,* "sonless" (or "childless"). By virtue of a certain sleight of hand, the auxiliary topic of *rotsa* is alternately labeled as a way to address problems in the female reproductive system and a way to look for (another) woman. It is hard not to see this as an intentional ruse, whereby the reasonable topic of how to address female medical issues is named by an overarching rubric that can, with the addition of a single consonant, be read instead to say how to find a new woman. But whether the pun is really deliberate or merely fortuitous, the point is clear. If the current consort's ills are not curable, go out and find another—indeed, as some commentators specify further, a very attractive one.[58] The pun thus became an occasion to reassert patriarchy through a sly confidence: we are supposed to be providing remedies for female reproductive problems, but we know what we can do if they don't work.

In any event, both readings of the auxiliary topic of *rotsa* render the female secondary. Medical science would address any problems with her body only secondarily to his ability to have sex. The next lines of the passage continue in the same vein, offering a few comments on the nature of women in general. Her role is to hold his seed and make it grow. This line takes away from the woman the function of having seeds for reproduction herself, despite the fact that the *Four Treatises*' own embryology chapter makes clear that she does have reproductive seed, and despite what the very same passage says just a few lines down.[59] This is perplexing. But the fact that she merely holds seeds (the modifier "merely" is added by all three commentators) is adduced as one of the reasons, again, why the principal of *rotsa* is the man.[60] The Desi makes it even more clear when he says that "any" woman can perform the function of holding the seed.[61] When the rest of the statement turns to a seemingly quite disparate point—that girls don't hold the generational lineage of the father—we start to see the logic (or shall we say again sleight of hand?). Even though the argument has switched from the mother to the daughter who can't hold the father's line, something fundamental about all females is being discussed, something that would reflect on the mother as well.

The point, and probably the entire passage, trades on a social reality: women don't hold the "generational lineage of the father."[62] The passage already invoked the importance of reproduction for the "family lineage," a term that also has patrilineal connotations.[63] Tibetan society, at least in the area around Yutok's homeland in the twelfth and thirteenth centuries, would seem to have been patrilineal. It is also explicit that the professional lines of inheritance in medicine were patrilineal.[64] So if females cannot hold the family line, they are not the main concern of fertility and virility treatments. Thus is *rotsa* about men making boys. The sleight of hand is that an already shaky claim about who is responsible for reproduction is bolstered by an irrelevant point about lineal descent—that is, irrelevant if you were only interested in the medical issue of reproduction, quite apart from inheritance and family issues. But of course reproduction and family are not separable. Not in the real life of societies, and not in medical knowledge. In this instance, patrilineal prerogative trumps biological facts.

The argument moves on to another point in the next line, and here the contrast is stark between the two sexes. When properly primed, he is a veritable sex machine. Meanwhile, on her side, another vulnerability actually renders her responsible for the failure to reproduce. Having given him the compliment of being able to "do it" (*spyad*) with them all—suggesting again that many woman are options, while he is at the center of the harem as agent and subject—suddenly the passage moves into the realm of moral quality. The reasons for her weakness are her poor karma, power, and merit. The term power (*dbang*) is interesting and would seem to refer to her lower social status, but the commentators understand it rather as an auxiliary to karma.[65] A discrepancy between the karma of the father and mother will have an effect on the sex of the child: even if the father's karma is right, if the mother's does not match it, no child (which likely means no son) will appear. Either way, the point is unflattering to her. Perhaps that makes up for the fact that the line nonetheless seems to presume that she has responsibility for the sex of the child, if *bu* here really means boy: if her karma is good it will be a boy, if it is bad it won't be a boy. But even if *bu* simply means child, and the line has to be read that no child at all will appear if she has poor karma and merit, would that not imply that she should be a main topic of *rotsa*? Shouldn't the chapter suggest ways to improve boy- (or child-) conceiving karma? No, the social reality kicks back in again in the next line. The only thing that her moral inadequacy means for *rotsa* medicine is that he is entitled to search for a better partner, as already stated.[66]

The two metaphors in the next line continue to take away from the female the fact that she too has seed that makes an infant. Here probably drawing on another old Indic conception, found for example in *Manusmṛti*, whereby the female is a field and the male the seed planted therein,[67] the metaphors as adapted here are at odds with the *Four Treatises*' rhetorical agenda to make the woman secondary, since the wording and rhythm of the line actually puts the parents in an equivalent position. The two things that can go wrong are that he has no seed and that her field is bad. That would seem to mean that even though the female does not produce seed, she still has responsibility for the success of the reproduction project. But this still does not raise her to parity in *rotsa*. Only the early modern commentators make the parity explicit, noting that the set of metaphors "looks back and forth" between both male and female. The commentators also make explicit that the first image refers to a flaw in the male, the second to a flaw in the female.[68]

By the last line of this opening passage it would seem that even the author of the root text has realized the leap of logic in insisting, against all evidence (including his own), that the male is the principal agent of reproduction. Or perhaps it is a return of the repressed. He is now bent on making a series of very explicit points about male-female parity with respect to *rotsa*. They both have gonads, they both have refined distillate, and they both have seeds, directly contradicting the metaphors just supplied. One's seed is white, one's seed is red. The ailments of *rotsa* are explicitly said to be common to the pair, male and female.[69]

And yet despite granting this point, *rotsa* is principally about the male. Period.

IT'S THE SEX

Social reality having won the argument over biological fact, the rest of the principal *rotsa* chapter is devoted to telling men how to fix their sexual maladies. Now the chapter draws on ideas and images from *Aṣṭāṅgahṛdaya*, but presents them in its own words and order.

> Someone, a man, relies frequently on his object,
> [but] the woman is like a tree with no branches.
> No signs whatever, be they felt, or visible, and so on,
> of there being something like a son [or child].
> Therefore that [man], wanting that [child], makes effort.[70]

In its endeavor to position the male's perspective as preeminent, the *Four Treatises* sometimes reconstrues the material it is taking from the *Aṣṭāṅga.* Whereas the latter makes the male the tree without branches, here it is the female, with the passage also suggesting that she is at fault. The scene painted by the *Four Treatises* suggests an anxious checking of the woman's belly for the first signs that she is pregnant, and desolation when it can only be concluded that she is not. Since "to rely" implies to have sex with, the problem here is not his performance, yet what ensues in the chapter is a set of instructions on how to have good sex.

There are conditions that will put him in the mood and get him ready. Once again, the imagery restates the *Aṣṭāṅga*'s conception of the conditions for sexual arousal. [71]

> Around his residence is a pond,
> or a forest of lotuses, a shady place, with sweet sounds.
> Moist and cool, his mind becomes happy.
> Then he will be capable of *rotsa*.
>
> The companion is grown up, beautiful and charming, and
> wearing ornaments.
> She has a sweet voice and soft words.
> She is appealing and her behavior is becoming.
>
> In preparation, cleanse with an oil massage and a purgative.
> Administer the *niruha* and *'jam rtsi* enemas.[72]

The image of the place of ardor suggests a balmy climate that can only be a fantasy for most Tibetans, while the next verse provides a well-nigh universal image of a beautiful woman, not a product of salacious imagination, but appreciative of female charms.[73] The purgatives added in the final verse are common in Tibetan medicine. I don't think these verses' Indic provenance takes away from their capacity to express basic conceptions of sex and gender on the part of the Tibetan author of the *Four Treatises.*

The passage continues with the problems the man might be having with either dried-up or overflowing semen. The remedies run from food and medicine to actions and special procedures; many of them have parallels in *Aṣṭāṅga,* but there are also differences.[74] It is touching to see, among the remedies for dried-up semen, aphrodisiac actions that in the modern West we usually think will cajole and arouse women:

> As for actions:
> exchange glances, kiss, embrace.
> Say nice words, and smiling, make friends.[75]

All of these efforts, including the administration of salutary foods and pharmacological substances, are geared for making male virility. If he follows the instructions,

> . . . every night he will be able to do it with a hundred women.[76]

> . . . even at the age of eighty he will be able to do it with everyone.[77]

> . . . even though he is an old man, he will do it like a youth.[78]

These fantasies of male potency—again echoing the words of the Indic *Aṣṭāṅgahṛdaya* but hardly anathema to Tibetan sensibilities—are promises that punctuate the various recipes for virility, and take over the horizon as the discussion proceeds. By the final summary of why *rotsa* techniques are needed—once more closely based on lines from the *Aṣṭāṅga*—the celebration of male sexuality is paramount.

> He who has no ailments and possesses the full ripeness of youth,
> will have unceasing sex in every season.
> His mind will be happy in the immediate moment,
> and a lineage of descendants will be born.
> Powerful one goes with all women without obstruction.
> Among objects of pleasure, lust is famed to be the best.[79]

6.4 Attractive women who will excite the male's virility. The last figure on the right is the man. *Plate 53, detail*

While the goal of reproduction has not fallen out of the picture, the emphasis is on carnal pleasure. In the end it would be difficult to say whether it is desire for sexual prowess or the maintenance of patrilineal and patriarchal power that drives the Tibetan medical discussion of fertility and virility. Suffice it to comment for now that with such goods hanging in the balance, the appropriation of pride of place by the male authors of the *Four Treatises* can hardly be a surprise.

MEDICAL KNOWLEDGE OF WOMEN

Whether the antifemale language that opens the female pathology chapters proceeds out of the same impulses as the fertility/virility introduction will have to be left to speculation. What is evident, however, is the very different tone of the rest of the female pathology section. Once it gets through its diatribe and down to actual medical description, all misogyny and androcentrism seem to be gone, and the discussion is observant and methodical about female experiences, symptoms, and therapies.

Many of the *Four Treatises*' issues regarding the female reproductive system and delivery problems are shared in broad strokes with *Aṣṭāṅgahṛdaya.*[80] But the presentation is different, in both organization and the *materia medica.* Probably much of the information in these chapters, unlike the embryology, which is closely influenced throughout by Ayurveda, represents Tibetan practices on the ground. I can only provide here a bare overview of what Tibetan medicine knew about the female reproductive system. Actually I am most interested in a few examples when the medical theorists are aware of what they don't know, and how they struggle to redress that. These are found not so much in the *Four Treatises* itself but rather in a few telling statements from the early modern commentators. In particular I will attend to those moments when they realize they need to resist ideal system and instead attempt to account for individual difference.

Most of the first two chapters of the female pathology, 74 and 75, are taken up with conditions of the uterus, or womb.[81] The chapters also address the effect of humor imbalances on the female reproductive system. Now we find a far more balanced etiology. No longer is the embodied female plight exclusively the result of her bad karma or moral disposition; it is equally the product of the food she has eaten and the way she has treated her body, not to mention the influence of demonic outside forces. These conditions are organized around excesses in the humors, the standard etiology of all kinds of disease throughout the *Four Treatises.* But note the exceptional fact that in the female pathology branch, blood is added as a fourth humor to the usual three. This rarely happens elsewhere in the *Four Treatises.*[82] I am not certain where the shift is coming from: it might have to do with the key role of menstruation in the female reproductive system, although what exactly is meant by blood as a humor and how that relates to both the menses and the blood that flows in the veins is a complicated question.[83] In any event, excesses of blood (*khrag tshabs*) *qua* fourth humor are traced, predictably, to menstrual irregularities, which will cause blood to collect abnormally in various organs. Symptoms include pain in the back and waist, sensations in the uterus and vulva, skin conditions, various excretions from the vagina, swellings of the face, general pains in the bones, dizziness and feelings of numbness and chill, and genital itches. Mental conditions are mentioned too: a sense of unhappiness, forgetfulness, and feelings of faintness.[84] Selby, speaking again of

Ayurveda, has speculated that such subjective descriptions of experience, as opposed to the external signs that the physician himself can describe, suggest women's contribution to medical knowledge.[85]

Treatments proceed largely in the form of douches, mostly administered to the vagina. The ingredients are specific to Tibetan food products and reflect detailed knowledge of animal husbandry.[86] The seventy-fifth chapter also discusses related ailments that afflict the body's channels, along with various kinds of tumors (*skran*), and finally kinds of parasites (*srin bu*, lit. "worm") that inhabit the uterus. The parasites in turn are divided into those that are "aroused" and those that have become "angry," or irritated. The way the symptoms of worms are described, including itching, unstable mind, a desire for sex with men, and a bad genital smell, echo familiar misogynistic imaginations of exaggerated female sexuality. While the same general class of parasites is also identified in the fiftieth chapter of the *Instructional Treatise* as the cause of other illnesses for both males and females, such as infestations in the stomach, intestines, genitals, and blood, and as inducing leprosy, their characterization in the female pathology chapters seems gender-specific. This is evident in the case of the irritated parasite. If such an infected woman does not have a chance to "meet" a man, she will use a finger or small piece of wood inside her vagina, making the parasite even angrier. The treatment is to have sex with men, as well as to apply certain compresses that contain semen.[87]

The third chapter on female pathology, seventy-six in the *Instructional Treatise,* has almost entirely to do with kinds of miscarriage, induced labor, abortion, and problems of delivery. It gives detailed recipes and instructions for the physician.

I already noted that the *Four Treatises* also addresses female conditions elsewhere than the three chapters labeled female pathology branch. The forty-third chapter in the *Instructional Treatise* on the female genitals is closely related to two chapters on the male and female genital area in *Aṣṭāṅgahṛdaya.*[88] It covers the female reproductive system as a whole, including conditions that impede pregnancy caused by too much sex and other improper behavior.[89]

Information on female pathology is also provided in the second *rotsa* chapter, already discussed. This reviews obstacles to conception caused by demon possession and by the application of birth control substances.[90] It also addresses barrenness and its particular symptoms in menstrual patterns. And finally, the embryology chapter too reviews a range of humor imbalances that render the reproductive fluids infertile, and also describes the appearance of normal female and male sexual fluids.

There is a plausible description of menstruation in the embryology chapter, suggesting some unblinking observation of this monthly condition of women.

> Between when a woman reaches twelve until she is fifty,
> the blood that is produced out of the refined substances collects monthly.
> It is dark and has no smell.
> By virtue of wind, it issues out of the two great channels
> into the door to the uterus
> and drips for three days.
> The sign that [she is menstruating] is that her energy is low,
> her face is bad, and her breasts, waist, neck, eyes, and belly swell.
> That she is menstruating is a sign that she desires a man.[91]

This knowledge too is closely based on Ayurvedic tradition; it basically puts together information from several verses in *Aṣṭāṅga.*[92] Much in this account accords with modern knowledge about conception, including a notion of something like the fallopian tubes, an idea that has a long history in Ayurveda.[93] For the Tibetan physicians the location of these two channels was the object of some confusion. The early commentary *Black Myriad* already tries to fill out the picture by maintaining that "two channels come out from the left and right of the *samse'u* and connect with the mouth of the uterus. [The menstrual fluids] pass through there, and then collect for a month in the uterus."[94]

Black Myriad expands the *Four Treatises*' statement by specifying that the two channels issuing into the uterus are connected to the *samse'u,* an organ not known to Ayurveda. The *Four Treatises* itself mentions this organ several lines later when the two channels come up again.[95] The *samse'u* was long a source of consternation for Tibetan medical knowledge, but in both the debate over the need for a separate female pathology section and the convoluted introduction to the *rotsa* section it is very clear that the *samse'u* is thought to be common to males and females.[96] In the present context, however, talk of channels from its left and right issuing into the uterus is specific to the female. By virtue of adding the *samse'u* the picture becomes even closer to the modern biomedical conception of the fallopian tubes, although the Tibetan ovary appears to be conceived in the singular.

The connection of this female gonad to the two channels and the uterus, and the role of all of them in menstruation and the conception of babies, remained sites of medical knowledge in need of further clarification. Kyempa spends time locating the channels leading out of the female gonad, but cites among other works an autocommentary to *Aṣṭāṅgahṛdaya* that does not refer to a gonad and instead places the source of the channels inside the uterus, claiming that these issue into the vagina (*mo mtshan*).[97] The picture painted by the *Four Treatises* itself is in any case not entirely in accord with the modern conception of the fallopian tubes. When the two channels come up again later in the passage, they are said to connect the uterus to the navel of the infant, thereby feeding it with the refined fluids issuing from the *samse'u*.[98] This leads Kyempa to specify that the two channels are themselves part of the *samse'u* and serve to connect this organ on the left and right to the inside of the uterus, where the fetus can access the gonad's nutrients.[99]

Zurkharwa for his part contests *Black Myriad*'s specification that would have the fallopian tube-like channels issuing into the "mouth of the uterus" (*bu snod kyi kha*). Instead he maintains that the mouth of the uterus faces downward, i.e., at the bottom of the uterus, which would be the cervix.[100] The *Four Treatises* statement itself seems to say that the blood arriving into the uterus through the two channels enters through the "door" of the uterus, not the mouth. Zurkharwa also contests the idea that the menstrual blood collects in the uterus over the course of a month. Rather, he says that such material collects in the *samse'u* itself, located above the uterus,[101] over the course of sixteen days. At that point it enters the "enclosure" of the uterus through the two large channels that run from the *samse'u* to the right and left of the uterus at its door. As soon as the blood enters, he specifies, the mouth of the uterus opens and the blood drips down, i.e., from the uterus into the vagina.[102] He also agrees that the same two channels that bring the menstrual blood into the uterus serve to feed the fetus when a woman is carrying it. That is the reason a woman does not menstruate when she is pregnant.[103]

Zurkharwa and his predecessors were refining knowledge already conceived in the *Four Treatises* and Ayurvedic texts, even if not with the exactitude of Gabriele Falloppio's (1523–1562) specification of the fallopian tubes, in turn suggested long before in Greek medicine by Herophilus and others. But at least Zurkharwa and Kyempa were seeking scientific precision. In other respects, however, Zurkharwa betrays a residual allegiance to ideal system, as when he ties the menstrual cycle to the phases of the moon.[104] When he maintains that the "red element" collects from the sixteenth day of the month until the new moon, when the menstruation process commences, he implies that all women menstruate on the same day, something that surely would have been contravened if he had cared to check women's empirical experience. Zurkharwa underlines his conviction when he also states that the connection to the moon's phases is why the female cycle is called menstruation, i.e., *zla mtshan,* lit., "sign of the moon."[105] He adds that the white element, which increases in the woman like the moon in the first half of the month, is emitted in the middle of the cycle, which shows that he was aware of the discharge that many women have at the time of ovulation, the peri-ovulatory mucus produced from the cervix.[106] The fact that this discharge is not mentioned in Kyempa's commentary or the others of which I am aware, including Ayurvedic works, might mean that it is an original contribution, based on Zurkharwa's or colleagues' clinical experience. At the least, we can say that knowledge of women's bodies continued to be refined in his day.

We have already seen a second description in the *Four Treatises*, at the beginning of the first female pathology chapter. Removed from its misogynist context, it represents another medical account of menstruation.

> The ultimate bodily constituents are the two fluids,
> the white and the red.
> Red menses drip after she has reached thirteen.
> She holds the semen inside the uterus,
> and the fleshly body [of the fetus] develops.
> The white spreads in the breasts
> and comes to nourish [the baby].

This description of menstruation differs from the one in the embryology chapter. It places the onset of a woman's period at the age of thirteen, at odds with the Ayurvedic idea that it begins at twelve. There was apparently some dissent in Tibetan medical circles on this question, and several editions of the *Four Treatises* have the female pathology description of menstruation reading twelve as well.[107]

Kyempa seems intent on getting it right. He brings up the statement from the female pathology in the context of commenting on the embryology, aware that there is information on female medicine in different parts of the *Four Treatises*. Kyempa corrects the age provided by the female pathology, which in his edition did read thirteen, and comes up with the new specification that menstruation actually begins at twelve and three months. His logic is that the claim that it begins at thirteen is based on counting age from conception, whereas he would count from birth.[108] Perhaps he is trying to reconcile the two statements in the *Four Treatises*, since twelve years and twelve years plus three months are almost the same. In order to do so, he has had to set aside the Tibetan convention of counting age from conception. In fact he has not really reconciled the two and still differs from the embryology chapter's count, if not by much, even though he displays a certain precision and originality in the process. In any event, he is still participating in ideal system. Like the root text, he is saying that all women get their periods at the same age (whatever that is), hardly a claim one would expect of an empirically based medicine.

And yet we also find a rather different epistemic position immediately contiguous to this discussion. Kyempa closes his section on menstruation with the following caveat: "However, the particulars regarding the amount [of menstrual discharge] that is produced, and the increases and decreases having to do with one's age, and so on, should be understood to be numerous, since they are a function of one's karma."[109] Here Kyempa has gone on to broach issues not in the root text at all, which never considers lightness and heaviness of periods and their shifts over the course of a woman's life. In so doing he points out a number of kinds of individual difference in menstrual experience. Zurkharwa too briefly considers individual difference, and flirts even further with questioning the root text when he adds the caveat that it is "*most* women" who menstruate from the age of twelve to fifty.[110] A small modification, but it does signal, once again, Zurkharwa's suspicion of generalized abstraction at the expense of individual irregularity.

Note too Kyempa's use of the category of karma. Unlike its invocation to naturalize gender hierarchy or to explain a particular pathological state, here a Buddhist notion is used to elucidate a scientific observation that the earlier medical system did not offer. In this instance, karma serves to account for a second-order principle of unpredictability and human variation; what is observed in the clinic is irreducible to ideal system.[111]

CONCEIVING A CHILD

The *Four Treatises* borrows wholesale from *Aṣṭāṅgahṛdaya* the idea that the day in a woman's cycle on which she becomes pregnant determines the sex of her child. A woman will conceive a boy on even days in her fertile period and a girl on odd days.[112] There is little evidence that the veracity of this dictum was questioned. We do have signs, however, of a probative attitude toward a few pieces of the system.

The *Four Treatises*' main statement on conceiving a child in the embryology follows directly on its statement about menstruation. Again, it is expressed in its own terms, though it is closely informed by Ayurvedic concepts.[113] Actually its wording is somewhat confusing, and Kyempa and Zurkharwa spill much ink making sense of it.

> After the uterus door opens, there is a period of twelve days:
> on the first three days and the eleventh she will not conceive a child.
> [A child conceived] on the first, third, fifth, seventh, and ninth
> will be a boy;
> on the second, fourth, sixth, and eighth it will be a girl.
>
> Like a lotus that closes when the sun sets,
> semen does not stay in the uterus after twelve [days] have passed.
>
> When there is a preponderance of semen, a boy will be born;
> when there is a preponderance of menses, a girl will be born.
> If they are in equal proportion it will be a *ma ning*,
> and if it divides, then twins will be born.[114]

Kyempa endeavors to show that the numbers these lines give for the days on which a boy or girl will be conceived are counted from the fourth day of the woman's cycle. That means that the statement that she can conceive a boy on the first, third, fifth, seventh, and ninth days actually mean the fourth, sixth, eighth, tenth, and twelfth days of her cycle, i.e., if counted from day one. The same adjustment applies to the days when a girl will be conceived.[115] Most of all, Kyempa's solution resolves the problem that the *Four Treatises* statement contradicts the *Aṣṭāṅgahṛdaya*, which claims that boys are conceived on even days and girls on odd days. The pertinent lines there read,

> The suitable period is twelve nights.
> The first three days are inauspicious, and also the eleventh.
> On even days [sexual union] will lead to a boy,
> and on the other days, a girl.[116]

In fact, that is what the *Four Treatises*' statement amounts to, if Kyempa's reading on when to start the count is correct. Zurkharwa agrees, and elaborates further, citing a variety of *Aṣṭāṅgahṛdaya* and commentarial statements.[117]

The upshot then is that both *Aṣṭāṅga* and the *Four Treatises* have boys conceived on the even days and girls on the odd days of a woman's fertile period. Other than the *Four Treatises*' disparity in wording, the Tibetan commentators don't question the main assumptions here from Ayurvedic tradition, such as the idea that a woman's fertile period begins on the fourth day after menstruation begins, quite at odds with the modern biomedical account that usually puts the earliest moment that a woman can become pregnant at around the seventh day. Only in the twentieth century did a prominent traditional Tibetan medical scholar think to correct the Ayurvedic and *Four Treatises*' large contravention of empirical evidence.[118]

I will continue discussion of the sex of the child, which also includes the third sex, the *ma ning*, in the following section. For now, note what we can extrapolate regarding the Ayurvedic/Tibetan picture of menstruation and conception in the context of the embryology. There is some knowledge of discharge mid-menstrual cycle, but neither Ayurveda nor Tibetan medicine appears to have an idea of ovulation. They explicitly consider conception to result from the union of the father's and mother's seed, and both discuss

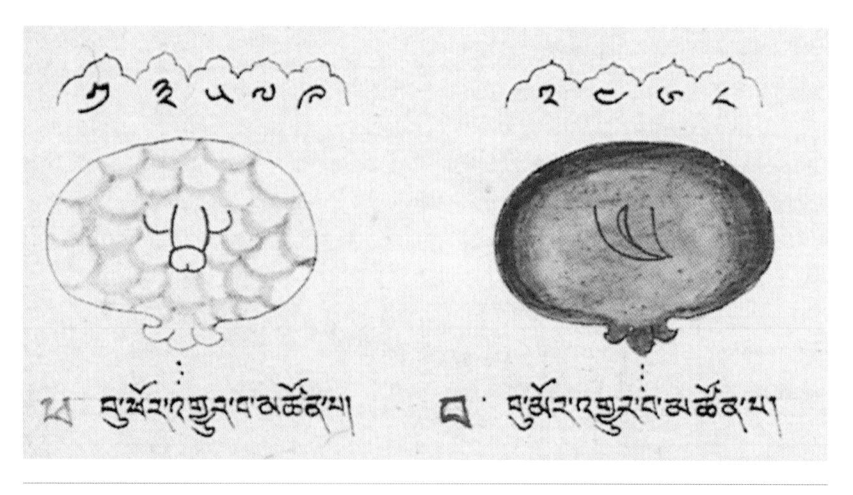

6.5 Potential fertilized eggs, one destined to be a boy, the other a girl. The numbers at the top indicate the days of the menstrual cycle on which boys and girls will be conceived. *Plate 5, detail*

6.6 Cutting the umbilical cord at birth. *Plate 5, detail*

how possible defects in either the male or female reproductive substances can impede pregnancy.[119] What these female reproductive substances are actually supposed to be, and what the relationship is between "blood," "menses," and the reproductive substance as such remain vague. But there is no question that both the *Four Treatises* and the Ayurvedic tradition consider the female substances, whatever they are, to contribute reproductive material, just as semen does.[120] Zurkharwa quotes a Buddhist tantric work to make the point clear: "Here, when the two organs unite, the semen that falls into the secret lotus and whatever blood is in the female organ (*bhaga*) are what are said to be seed." Then he reiterates the point himself: both the semen of the father and the blood of the mother are called "seed."[121]

Zurkharwa also elaborates how the menstrual cycle plays a crucial role in conception, drawing on *Aṣṭāṅgahṛdaya* and the Ayurvedic understanding of the period of fertility.[122] The uterus is only open for twelve days from the onset of menstruation, during which time the female can "hold the seed" of the male. After that, the mouth of the uterus closes, and she will not be able to hold his seed. In any event, the male's semen will not desire the inside of the uterus and will reverse its path and come out. Zurkharwa is equally clear that the presence of her own menstrual substances in the uterus during this same period is just as critical to the process. In short, one must wait for the appropriate moment in the menstrual cycle in order for the woman to get pregnant.[123]

Other sections of the *Four Treatises* support a view of joint contribution to the body of the fetus as well. It is one of the basic assumptions of the pulse chapters, a disparate segment of Tibetan medicine with likely Chinese rather than Indic sources. There the sex of the offspring is determined by how the mother's and father's pulses interact.[124] Only one passage in the *Four Treatises*, in the *rotsa* section, implies that the father alone contributes the substance that creates a fetus, but as we have seen, that statement is overdetermined by a set of social agendas around retaining control of sexuality and family lines.

SECOND SEX, THIRD SEX

A lot more could be said on the *Four Treatises*' conception of the female in physiological terms, including all the material on the red and white seminal substances, how they are sexually differentiated, how they are not, and how in Tibetan medicine the *samse'u* produces them both. But leaving that aside for another study (and in cognizance of the enormous work on related issues in Indian medicine already produced by Rahul Peter Das),[125] I turn to another principal ground for sexual identity: the genitals. These are often called "sign" (*mtshan ma*, Skt. *nimitta*) for both male and female, presumably in light of the genitals' indexical connotation of the person so marked. Here, in contrast to the physiology, knowledge of the female genitalia seems scarce, at least in the *Four Treatises.*

Other than the many references to the uterus, along with the door at its top and the mouth at the bottom, there are only a few scattered specifications relevant to female genital anatomy. The female genital chapter of the *Four Treatises* mentions the mouth of the urethra and perhaps a kind of mouth of the vagina that is bent.[126] The female pathology chapters mention a "mouth to the channel," which also seems to reference the vagina.[127] In one instance the locution "outer flesh-skin opening" is used, which probably refers to the vulva.[128] Usually just the general term "female sign" is used.[129] We saw in chapter 4 that Zurkharwa resorted to the prosaic functions of the tantric channels, in turn influenced by Indic *kāmaśāstra*, to describe her sexual plumbing.[130]

And yet the most salient metaphor for female identity, a term used broadly in medicine as well as most other Tibetan literature, is cast in terms of genitalia—albeit what she lacks. *Bud med* is an old and pervasive word for woman, or sometimes the female in general. (A more recent term for woman, *skye dman*, lit. "low birth," is already foreshadowed in the *Four Treatises*' female pathology opening.)[131] Apart from the most basic of all—*mo,* "female" and its opposite *pho,* "male"—*bud med* is used most frequently in the *Four Treatises*. Its etymology is somewhat of a mystery. *Bud* is probably the past tense of *'bud pa,* a verb that is usually transitive but can also be intransitive when it means to fall or be lost. The following gloss is found in the Desi's history of medicine:

> During the time of the first eon, when the male and female organs were close to emerging, at one point a protuberance sort of thing in a lump-like shape grew in some. It became the male sign and thus he is called "grown."[132] In some it fell off (*bud nas*) and so they became ones who possessed a hole that lacked it. Therefore they were known as "fell off and gone" (*bud med*).[133]

In brief, the *bud med* is the one with no penis. The Desi's etymology is likely fabricated, but it reflects the term's semantic resonance in common understanding. Here, then, the foundation of female identity—as in so many other places worldwide—is about a lack, an absence, now specifically in anatomic genital terms.[134]

A second term used in both the female pathology and virility/fertility contexts is harder yet to parse. *Za ma,* which I translated above as "deficient," often renders the Sanskrit term *ṣaṇḍha*, but not literally. Both terms have a complex semantic range, and they do not overlap. In Tibetan the more conventional sense of *za ma* is "food," but it is also a term for a female. One definition adds that it has to do with a lack of capacity to perform sexually.[135] Performance issues disqualify the female for primary status in the *rotsa* section. The reference to such dysfunction reminds us that both *za ma* and Sanskrit *ṣaṇḍha* are not only terms for the female but also names for sorts of eunuchs.[136]

In fact, the medical treatment of the eunuch—and the other sexual anomalies that I am calling the third sex—is especially generative for this study. Not only is it intimately connected with conceptions of the female, it also reveals some basic aspects of gender conception more generally in medicine. These become clear in the striking difference between Tibetan medical and Buddhist conceptions of the third sex.

Both Indian and Tibetan medicine clearly recognize the fact that the possible sexes of an infant are three, not two. Such a trio might seem to be a fixed trope of its own until we realize that the third sex is a very porous and elastic category that in a sense does not stand as a proper category at all. Nor it is singular by any means. Rather, the third sex stands for all of the aberrations in between the two normative poles of male and female. It represents a long-standing medical perception of the full range of human diversity with respect to genital anatomy.

As early as *Carakasaṃhitā* it is assumed that there are three possible sexes of a child: male, female, and neuter (lit., "not male," *napuṃsakam*).[137] The latter is perhaps the most common Indic term used for the third sex, but a closer look reveals a plethora of conditions that come under this heading, and a plethora of labels.[138] In one passage, for example, *Carakasaṃhitā* lists eight kinds of aberrational sexual identity.[139] *Ṣaṇḍha* is one of the words in Indic medical sources for a kind of abnormal or third sex.[140] Such conditions are largely ascribed to congenital causes.[141] In Tibetan the term of choice for the third sex is *ma ning*, another strange word of unclear etymology. I use this term in the following pages, alternating sometimes with the neutral phrase "third sex." Neither "neuter" nor "hermaphrodite" suffices. Note too that there are significant differences between the Tibetan and Sanskrit semantic ranges for any of the words used for the third sex. It would be misleading to gloss the Tibetan discussions on the *ma ning* as *napuṃsaka*, or *paṇḍaka*, to name another one of the common Sanskrit analogues. Each of these terms has different meanings in different contexts.

Sexual identity (*rten*, or "receptacle") is the first and most basic category by which the body is classified in the *Four Treatises*.[142] It reduces all of the anomalous third sex varieties to *ma ning*. This means there are three human sexes: male, female, and *ma ning*.[143] The same trio recurs in the embryology chapter, which speaks of three possible kinds of children. *Ma ning* is what will issue when the mother's and father's reproductive substances are in balance.[144] The impending birth of a *ma ning* instead of a boy or girl can be discerned when the mother's belly does not lean to the right or left but stays in the middle.[145]

Not all sexually relevant medical discourse references three sexes, however. The *Four Treatises* can also conceive only of the classic two, as is certainly the case in the exclusively heterosexual fertility/virility section and in the vision of sexual identity at the start of the female pathology section.[146] Even the discussion on the birth of children that so clearly anticipates the possibility that any one of the three sexes will be born goes on in a contiguous passage to count out the days of the menstrual cycle when a boy or girl is likely to be conceived, failing to specify when the third-sex child is conceived.[147] In the *Four Treatises*' basic passage on pulse type, which Tibetan medicine uniquely classifies as tripartite—male, female, or "bodhisattva" (here a euphemism for *ma ning*)—the third-sex person as such is not mentioned.[148]

And in yet another passage the *Four Treatises* classifies kinds of diseases in terms of men, women, children, and the aged.[149] While these examples might suggest that *ma ning* is not really considered a type of real person, one that medicine would treat, the failure to mention the third sex also has to do with its status as aberration from the male and female ideal types. In other words, it is often omitted because it does not really constitute a category on its own terms. The influence of long-standing categorical binaries, such as the red and white substances, light and dark, and many other gendered pairs, probably tipped the medical treatise back into a heterosexist conception of two sexes only. And yet when we get to one of the main terms that the *Four Treatises* coins to name sexual identity overall, "receptacle," the standard conception is three. We also know that Tibetan societies understood actual persons with anomalous genitalia to belong to this third sex.[150]

"Receptacle" is a very general concept in Tibetan. It often denotes something physical that is the basis or support for something more complex and culturally coded. In medicine the notion of sexual identity as receptacle has to do with the body, the basis for a much larger concept of the person that includes gender features as well as many other dimensions of human life.[151] When the third sex is defined in the context of the three kinds of receptacle, it has primarily to do with genital anatomy. Zurkharwa glosses the first two kinds of receptacle with the familiar terms. The male is the man (*skyes pa*); the female is the woman (*bud med*). But the third, *ma ning*, is the one who is "not definitely either of the first two."[152] He goes on to specify that there are three main kinds of *ma ning*: one who has two sexual organs, one who has no sexual organ, and one whose sexual organs change back and forth monthly.[153] In this scheme, then, male and female constitute stable identities recognizable through the shape of their genitalia; this stability is highlighted and bounded by the possibility that someone could have a genital identity that was indeterminate, or not definite (*ma nges pa*).

Being difficult to define endures as the basic nature of the third sex. Indeed, the *ma ning* or Sanskrit *paṇḍaka* is sometimes defined as someone who is not definable, or as already seen from Zurkharwa, as one who is not definitively either a male or a female.[154] As one early medical writer pertinently put it, the *ma ning* is the one who has no opposite.[155]

This category of noncategorizable sexual identity plays differently in medicine than it does in Buddhist paths of religious cultivation, where it has a consistently negative profile.[156] The third sex is highlighted early in Indian Buddhism within the context of monastic discipline. Kinds of third

6.7 The three kinds of sexes: male, female, and various types of *ma ning*.

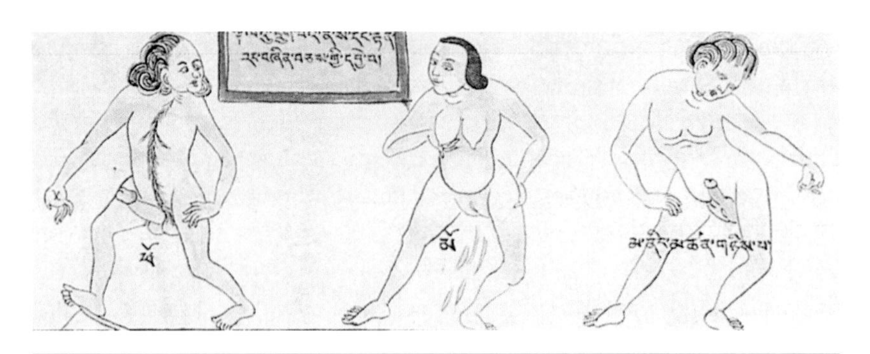

[a] Male, female, and the hermaphrodite *ma ning*. *Plate 16, detail*

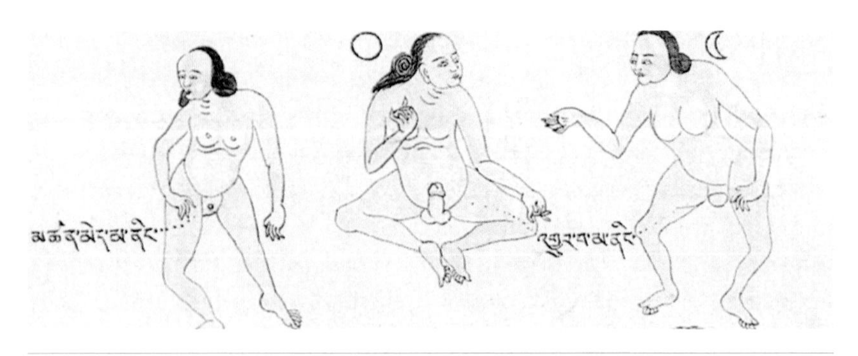

[b] The *ma ning* with no sexual organs, and the two alternatives for the *ma ning* who changes back and forth every month. *Plate 16, detail*

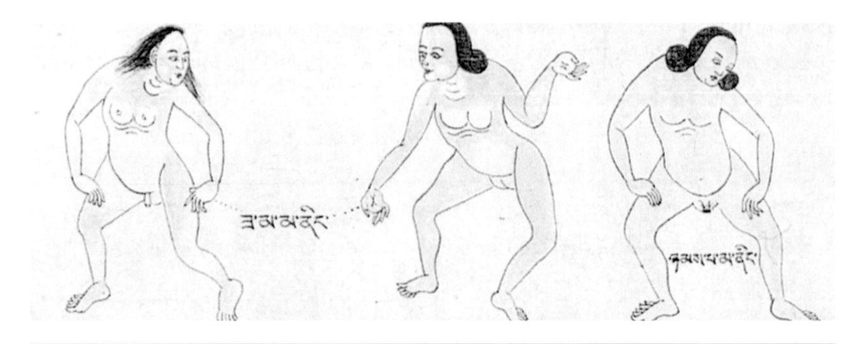

[c] Two figures illustrating deficient *ma ning*, one who appears to have a partial male organ and one who looks like a female, and finally the *ma ning* with compromised or injured organ. *Plate 16, detail*

sex conditions are included in the list of factors that disqualify a person from receiving monastic ordination.[157] Postulant ordinands are physically examined on just this matter. Passages in a variety of sūtras also warn against preaching the Dharma to the third sex, or even giving them donations, and they are said not to be able to meditate or do other kinds of religious practice.[158]

The term *paṇḍaka* emerges as the main Sanskrit word for the third sex in Buddhist texts.[159] By the early centuries C.E. a list of five kinds of *paṇḍaka* was consolidated. These can be roughly rendered as 1. a congenital neuter; 2. one whose sex changes from male to female and back again every month; 3. one who is only aroused by special kinds of sex; 4. a voyeur; and 5. a castrated eunuch.[160] Note that some of the members of this list are in the anomalous sex class on the grounds of the nature of their sexual desire and/or practice, in addition to anomalies regarding their organs as such.

The question of why the third sex was denied ordination in Buddhist monasticism is complicated, but I think the matter goes far beyond any aspersions that were cast about its sexual orientation, such as that the third sex person was homosexual. For one thing, homosexuality is attributed to men, women, and *paṇḍakas* in monastic literature, so it is not the defining feature of the *paṇḍaka*.[161] A better clue might have to do with the association that the third sex has with the second sex. This line of thought also leads to a notable intervention that Tibetan medicine makes into the notion of the third sex.

The Tibetan conception of the female as one who lacks a penis already suggests her affinity to at least some kinds of third sex. Ayurvedic sources, for their part, hint that there is an "unmanly" character of the third sex.[162] Some also recognize that there can be both male and female *ṣaṇḍhas*.[163] The *paṇḍaka* class as construed in early Buddhist monastic texts has a lot to do with problematic sexual functioning, including female varieties, such as problems with sexual organs or menstruation.[164] Particularly telling is the fact that the very list of female dysfunctions that render her sexually anomalous and thus prevent her ordination is almost exactly the same as an early list of insults that monks sometimes direct at any woman.[165] This suggests a close affinity between the third sex idea and misogyny. Women are said to have almost been excluded from ordination themselves in the early moments of Buddhist history, and were allowed in only after special rules guaranteeing patriarchal hierarchy were instituted.[166] Thus are there mounting parallels between the third and second sexes.

I have argued elsewhere that the undecidability, instability, and softness associated with both the female and the third sex were anathema generically to the "order" of monastic discipline. This set of features is germane to the connotations of the third sex category even beyond its defining genital specifications, and not only vis-à-vis Buddhist monasticism. We also see the metaphorical significance of the third sex as soft and unstable in Tibetan grammar, where the trio of male, female, and the third sex names important categories for verbal sounds (as they do in Sanskrit for noun gender). In Tibetan grammar *ma ning* sounds are singled out for their changeability, a crucial feature for phoneme construction, and obviously mirroring the anatomical anomaly of one of the kinds of genital *ma nings*.[167] Tibetan medicine also provides a number of cases where the trio of male, female, and third sex is used taxonomically. Here, in another key debate between Zurkharwa and Jangpa Tashi Namgyel, the slippery distinction between anatomical sex and its metaphorical or gendered resonances becomes a place to draw a line in the sand between the disciplinary and/or soteriological associations of the third sex and those of medical science.

OR GENDER? TIBETAN PULSE THEORY

The *ma ning* label was used to classify things that are not about sex at all, in both Tibetan medical and tantric physiologies. Early works list male, female, and *ma ning* as the three kinds of pulse, inhalation,[168] winds,[169] yogic channels,[170] and even digestive juices.[171] The three kinds of pulse, however, receive special attention from the medical commentators. I think that is because the seemingly metaphorical invocation of the three sexes in the pulse system is not neatly separable from the fact that these categories also reference sexual identity. But neither are the two senses collapsible. That undecidability poses interpretational challenges. In Zurkharwa's purview, it serves to elicit an explicit category of gender.

The relevant passage comes right at the beginning of the *Final Treatise*, when after a brief overview of kinds of illness and treatments, the text introduces its important section on pulse diagnosis. There is an initial discussion on how to prepare for the exam, and instructions on which fingers the doctor uses for which organ and on which spots on the patient's arms. The latter was at the center of the heart tip debate studied in chapter 5.

The text then introduces the first and most basic distinction in kinds of pulses:

The three types of common pulse flow in terms of receptacle [i.e., sex]
are male pulse, female pulse, and bodhisattva pulse, three.[172]
The male pulse is thick and throbs roughly.
The female pulse is fine and throbs quickly.
The bodhisattva pulse is long, soft, and pliable.

If a female pulse develops in a male, he will have a long life.
If a male pulse develops in a female, her sons will be greatly splendid.
If either gets a bodhisattva pulse,[173]
they will have a long life, infrequent illness,
their superiors will be kind to them,
their inferiors will dislike them,
their three [male] close relatives[174] will rise as enemies,
and in the long term, their line will be cut off.

When a male pulse meets another one there will usually be many sons.
When a female pulse meets another one usually there will be many
daughters.
Whether the bodhisattva pulse meets a male or female [pulse],
It will eventuate that one [child] will come, and then none.[175]

It is a basic feature of Tibetan medicine to divide people into three classes of pulse, those with male pulse, those with female pulse, and those with third

6.8 A male with a male pulse, a female with a female pulse, and a *ma ning* with a *ma ning* pulse. The squiggly columns indicate the quality of the pulse. *Plate 54, detail*

6.9 The destiny of a couple with *ma ning* pulse: they will have long life, they will be favored by those above them and hated by those beneath them, their family members will arise as enemies, and their family lineage will be cut. *Plate 54, detail*

sex pulse. In this passage from the *Four Treatises,* the concern, beyond just introducing this nomenclature, is largely how the pulses of the two members of a couple will affect the sex of their offspring. The passage also addresses the character and destiny of people who have these various pulses. Special detail is provided for those with the bodhisattva pulse.

It is pretty interesting to find the *Four Treatises* employing the very valorized Buddhist term "bodhisattva" as a gloss for the *ma ning* class.[176] Recall, the *ma ning* is excluded from taking Buddhist monastic ordination and is even deemed unsuitable for religious practice altogether—a harsh judgment indeed. Surprisingly, no medical commentator has anything to say about the anomalous medical usage of the term bodhisattva for *ma ning* pulse. I can add that the several Tibetan scholars whom I queried orally about the term all seemed uncomfortable with the question and brushed it away. It is clear that the gloss signals a very different valuation of the third sex category in medicine, in sharp contrast to its pariah status in Buddhist contexts.

But do the pulse categories refer to sexual identity, or are they only being used metaphorically as taxonomical devices? Although the passage introduces the taxonomy in terms of "receptacle," or sexual identity, the pulse types are not tied to the analogous sexed people.[177] That ambiguity yields important insights. Consider first the bodhisattva/*ma ning*'s overall profile in the passage. As elsewhere in medicine, the *ma ning* signals a middle point between two extremes. Here it denotes a balanced pulse between two more extreme options—a rough and thick pulse and a fine and quick one. The notion of balance is germane to the entire conception of health in Tibetan medicine. It is thus not surprising that medical tradition thinks that people with such a pulse will be healthy and have long lives. However, the passage also conceives a complex social destiny for people (or couples) with a bodhisattva pulse. They will be well regarded by the powerful, but will have problems with their family and inferiors. Now such a destiny would seem to pertain more to a person with *ma ning* genitals than pulse per se. If we can venture that the *ma ning*-sexed person is akin to a eunuch in Chinese society, or perhaps has a status like that of a homosexual, we can imagine scenarios in which indeed such a person did well in elite contexts but was resented by the common folk and relatives. The medical passage evinces nuanced appreciation of the vulnerable position of people with anomalous sexual status. Thus does it analogize anomalous pulse type with anomalous sex and gender type.

And yet that reading is largely based on inference. The *Four Treatises* passage explicitly disassociates genital sexual identity from pulse identity, since men can have female pulses and vice versa, and both can have *ma ning* pulses. Still, it suggests some connection between sex and pulse. The pulse identity of the parents has something to do with the sexual organs of their offspring, although it is not made clear what that connection would be. And those with a bodhisattva—i.e., third sex—pulse will have trouble bearing children and tend to have one at most. That seems to recall the idea of the *ma ning* as a person with dysfunctional sexual organs—although not fully. The *ma ning* pulse person has difficulty producing offspring, but it is not entirely impossible.

Such ambiguity drew Zurkharwa Lodrö Gyelpo's close attention to this passage, taking on what appears to him a badly confused comment by his predecessor Tashi Pelzang. Once again Zurkharwa drives home the point that a medical theorist must be able to distinguish nomenclature and heuristic categories from physical facts. And once again he is probably exaggerating his adversary's mistakes. But the point he ultimately makes is very significant.

Zurkharwa quotes Tashi Pelzang directly, excerpting several statements. As in the heart tip debate, we have the original passages that he is citing and can compare them with how he represents them.[178] But let us first read Zurkharwa's comment:

> In the commentary of Jangpa Tashi Pelzang it says, "The man is the male, the female is the woman, and that deficient one is the *ma ning;* all receptacles are subsumed into those three." And, "The male pulse is of the nature of wind, the female pulse is of the nature of bile, and that bodhisattva pulse is of the nature of phlegm: three." . . . And the one who has that which is known as the bodhisattva pulse, the pulse that corresponds to phlegm, "does not take the vows, is not an appropriate vessel for profound Secret Mantra, does not have the lot to practice Dharma and so on, has many faults." And, "When the parents have the same pulse they will give birth exclusively to males, or females. Most children will be in accordance [with the sex suggested by the parents' pulse]. When a couple with male and female pulse get together exogamously then both [male and female children] will be born. They will come in turns, depending on which [pulse] is strong and which is weak. When a *ma ning* meets another one there will not be family. If the mind stream changes, it is possible for it also to change."[179]

Note again the irony of the bodhisattva pulse person having no good lot to practice the Dharma. Zurkharwa never picks up on this grossly mixed metaphor, though he could have gone to town on Tashi Pelzang for it.[180] It is consistent with the widespread disparagement of the *ma ning* in Buddhist works. In any event, it would seem that Tashi Pelzang—or whoever is being quoted here, for it is possible again that Tashi Pelzang's statement is from someone else—is confusing his training in Buddhist monasticism and scripture with his medical training and errs in importing the former into the latter.[181] The blunder creates a striking categorical dissonance.

Zurkharwa takes Tashi Pelzang to task for what he sees as a simplistic equation between sexual identity and pulse identity. He first takes issue with the equation of the deficient *za ma* with the *ma ning*, maintaining that the root Vinaya scripture explains *za ma* as someone whose penis has been removed, i.e., a eunuch, whereas it lays out only three kinds of *ma ning*, the kind who changes monthly, the kind who has both male and female genitals, and the kind who has no genitals. He seems to be insisting on specificity; we should not gloss over the disparate conditions of differing types of sexual anomaly.[182]

But this is not Zurkharwa's main point. He goes on to add what would seem to be two hitherto unarticulated kinds of *ma ning* types to the foregoing standard list of three: one who has a male organ but cannot use it, and a female kind who is barren.[183] These two—impotence and barrenness—would be much more quotidian sexual aberrations and far more likely the kind that a physician might encounter. It is surprising that Zurkharwa would put them in the *ma ning* group at all, which by his day had become standardized as three types. This shows again his interest in empirical conditions rather than the surely rare occurrence, if not entirely mythological character of the standard three (especially the *ma ning* who switches back and forth between male and female monthly, a possibility already countenanced in a Buddhist *imaginaire* whereby bodhisattvas and others magically change their sex for salvific purposes[184]). But then he adds that it is unsure whether *any* of these kinds of *ma ning* will have a *ma ning* pulse (and now he uses that term explicitly). That is his main point.

It is with Tashi Pelzang's apparent equation of third sex pulse and aberrant sexual organs that Zurkharwa takes greatest issue. After dealing with what he sees as Tashi Pelzang's too simple identification of pulse types with humor imbalances, Zurkharwa returns to what he finds "really over the top": the connection between the *ma ning* pulse and the failure to take vows. Although he does not say it exactly in this way, I think he is so exercised because of the gross error of assuming that the term *ma ning* means the same thing in the very different contexts of monastic law and medical pulse. A good historian and empiricist, he insists that words can refer to different things in different contexts. He goes on to apply logic to back up this criticism.

> It is not the case that having a *ma ning* pulse renders one a *ma ning*. The *ma ning* pulse that is an examined pulse is not to be foisted onto the *ma ning* that is about receptacle. If it were, then the statement "It will eventuate that one will come, and then none" [would make no sense]. In that context, since you certainly posit that when a man with a male pulse and a woman with a female pulse can, through the power of that female pulse, have a female child, then how can the person who is able to have sex be the *ma ning* who can't take vows? Think about it![185]

Zurkharwa's language is elliptical, but his point is clear enough. If you understand the *Four Treatises*' statement about the sex of the children of couples with various pulses, then it should be obvious that in saying "one will come," the *Four Treatises* is predicting that the couple with *ma ning* pulse would have a child too—albeit probably only one. This should prove that the person who has a *ma ning* pulse is not the same as the *ma ning* who is forbidden to take monastic ordination, for the latter is understood as a *ma ning* by sexual identity, and therefore someone who is not capable—due to defective or aberrational sexual organs—of having sex at all, let alone bearing children. In the view of Buddhist monastic law, persons with genital aberrations can't have normal sexual intercourse.[186] This is the reason Zurkharwa cites Tashi Pelzang's statement that "When a *ma ning* meets another one there will not be family." The fact that Tashi Pelzang says this immediately after discussing the outcome of couples with various kinds of pulse combinations suggests that he is conflating the *ma ning*-pulsed person with the *ma ning*-sexed person.

Zurkharwa goes on to drive home the distinction between the sex of the genitals and the sex of the pulse category. He capitalizes on the notion of "mind stream" already brought up by Tashi Pelzang in the last sentence that Zurkharwa quoted. Zurkharwa develops that notion further, with a more radical implication than Tashi Pelzang suggests. The passage provides a prime example of a technical concept from Buddhist scholasticism serving the articulation of a medical question about the everyday body. Zurkharwa writes,

> When you say that "If the mind stream changes, it is possible for it also to change," are you thinking of the male and female organs, or the pulse on the arm? If you are thinking of the organ, then it is not the case that when the mind stream changes, the male or female organ changes. For example, there can occur a woman who has a man's mind stream, but how does that make her a man? If you are thinking of the pulse, why do you need a change of mind stream? Males with female pulse and females with male pulse occur.[187]

Tashi Pelzang seems to have added an unconnected point at the end of his discussion on *ma ning*. Its implication is not clear, but it has to do with some aspect of one's *ma ning* status changing if one's "mind stream" changes. Zurkharwa asks whether Tashi Pelzang means that the sexual organs would change, or the pulse identity. If the latter, he finds such a claim irrelevant or redundant, since it is already explicit in the *Four Treatises*' statement that people can have pulse identities that are at odds with their sexual identity. A change in pulse identity, let alone whatever is meant by mind stream, would not affect the genitalia.

However, the other possibility that Zurkharwa sees Tashi Pelzang suggesting, that the sexual organs themselves would change under the influence of the mind stream, deserves more consideration. And what is this mind stream (*sems rgyud,* Skt. *cittasantāna*) that Tashi Pelzang is invoking? *Cittasantāna* is an old term developed in Buddhist epistemology to denote a person's basic mental continuum over the course of a life. Given the absence of an essential soul, according to Buddhist doctrine, the mind stream serves as the mechanism whereby karmic deeds and experiences are registered and later reactivated. The notion was devised to provide continuity with the past but avoid the pitfalls of positing something eternal like a "self," or *ātman.*[188]

In the course of working through the nature of human pulse, apparently some medical theorists began to deploy this mind stream notion to name certain personal characteristics, including a sense of gender, not found in the Buddhist scholastic discussions. At present, Tashi Pelzang's work is our earliest evidence of the deployment of the idea for medical purposes, but this usage may well have a longer history. He talks about it even more than Zurkharwa does. Tashi Pelzang invokes the mind stream in an adjacent passage to what Zurkharwa quoted, where he tries to account for the many factors that affect a fetus *in utero.* One of these is the parent's mind stream, itself a product of good or bad karma. That in turn influences the child, including its pulse. Similarly, the predominance of certain humors in the body will also affect the kind of pulse one has.[189] Here the mind stream seems to function in a very general way to affect a person's overall disposition. In a later passage, Tashi Pelzang adds that someone whose mind stream is dominated by emotional obscuration and stupidity will be a *ma ning* and have a bodhisattva pulse. However, if that person collects merit and good karma, it is possible that the stupidity will lessen, the mind stream will change, and also the receptacle and pulse will change.[190] Tashi Pelzang believes that pulse, sexual identity, and mind stream are closely aligned and that the latter in some sense determines both of the others. We can also see that he regards the *ma ning* condition, in whatever that consists, as undesirable.

Why the extra category of mind stream was introduced to account for changes in either pulse or sexual identity is not entirely clear. Perhaps if we get access to other works from Tashi Pelzang's Jangpa predecessors we will have a better sense of the history of this idea. But even on the basis of what we have seen, the mind stream notion serves to separate out several factors. Zurkharwa goes further than Tashi Pelzang, and quite originally, as far as we now know, makes the mind stream tantamount to a category of gender, explicitly disassociated from sexual identity. A woman can have a man's mind stream, Zurkharwa opines, but that does not mean that she is a man. It is clear that he means by virtue of genitalia, since he already said he was talking about the sexual organs in his previous sentence. In other words, there is something called mind stream that carries gendered qualities, but it is not the same as genital identity. Moreover, Zurkharwa's final claim in the passage—that the mind stream need not determine what kind of pulse one has—implies that this mind stream, along with whatever gendered associations it has, is not the same as pulse, even though that also has gendered implications.

We have seen so far at least three categories that relate to sex and/or gender: pulse, mind stream, and receptacle. Receptacle, or genital anatomy, still seems to be the base line for sexual identity, although as this discussion proceeds, the more the gendered dimensions of that identity are displaced onto other, more flexible factors than anatomy itself. Pulse is one of these factors, constituting a gendered dimension of embodiment that has to do with the rhythms and speed of the body's metabolism. The notion of mind stream as introduced by the commentators suggests yet a further dimension of gender that, at least for Zurkharwa, seems to be free of physical reference. It appears to relate rather to personality type, or perhaps way of thinking, or style; indeed, it appears that "mind stream" in Tibetan today is a vernacular term for kinds of temperaments or personal styles.[191] Like pulse, mind stream drives a wedge between anatomy and destiny, although Tashi Pelzang sees it affecting anatomy under certain circumstances.

Tashi Pelzang also tries to line up the humors with gender and pulse in his comments, although that idea does not seem to receive further attention. But when Zurkharwa moves from his critique of wrong views to his own comment on the *Four Treatises*' passage he provides one further set of sexually associated personal characteristics relating to pulse. In the process, he makes a breakthrough on the entire question of how those characteristics map onto a person's genitally based sexual identity—or not—and what status such conventional categories have in the first place.

> In terms of the classification of examined pulse by type, if you condense and present them, there is none that is not included in the three: the

> so-called male pulse, female pulse, and bodhisattva pulse. This is explained by the fact that there is no one with a body who is not included in the three: method, primal awareness, or nonduality.
>
> If one were to analyze the three individually in terms of their characteristic mark *qua* pulse, that which is said to be male is thick and throbs roughly. Female pulse throbs finely and quickly. That which is said to be a bodhisattva pulse is of long duration and throbs softly and pliably—so it is said. For that reason, whether a person be male, female, or whatever, if method is predominant, they will have a male pulse. If primal awareness is predominant, they will have a female pulse. And if the two are equal or nondual, they will have what is said to be bodhisattva pulse. Such terms have been coined figuratively. For example, it is just like *kyangma* being said to be method, *roma* being said to be primal awareness, and the central channel being said to be *ma ning* or "discards all." Therefore, in most cases it occurs that a male has a male pulse and a female has a female pulse and a *ma ning* has a *ma ning* pulse. And yet from one perspective, it is said to be uncertain, and so it is said, "If a female pulse develops in a male he will have a long life."[192]

Once again Zurkharwa falls back on tantric categories to articulate a concept important to medicine. He lines up what are already gender tropes from the *Four Treatises*' original pulse doctrine with a further set of gender qualities enshrined in a classic tantric trio. We already saw Zurkharwa and his predecessors invoking the well-known tantric pair of method and primal awareness in the heart tip debate. Here he draws upon another very common tantric set, consisting in those two plus a third quality, nonduality.[193] Again, the male quality of method is associated with actively pursuing a goal through skillful means, while the female quality of primal awareness has to do with understanding the emptiness that underlies those skillful means. Nonduality would be a mediating factor whereby everything is fluid. By arguing that those in whom the male quality of skillful means is predominant will have a male pulse, and so on, Zurkharwa is tying the kind of pulse to these old tantric gender tropes for personal style.

Also worth noting in his discussion is the idea of predominance, another import adapted from Buddhist scholastic terminology. Predominance (Pali/Skt. *adhipati*) is one of the classic Buddhist kinds of conditionality, often invoked in epistemology and karma theory to address how the inevitably complex combination of disparate factors in human experience comes to have a defining character.[194] Its use now for a medically based notion of gender facilitates a discussion of sexual heterogeneity and the possibility of flexibility and deviation. In Zurkharwa's account of pulse, predominance implies that all three gendered qualities will be present together, and that people are not entirely governed by method or primal awareness or nonduality alone—or by maleness or femaleness or third sexness, as those notions are understood here, alone. It is only that one such feature rises to the top, to become the leader of the pack, as it were, and thereby affects or organizes the quality of the whole, which is reflected in the nature of the pulse.

Zurkharwa's reading of the pulse doctrine is of a piece with his solution to the heart tip debate. Both gesture to gendered styles and their analogues in the body. The pulse discussion sheds further light on what he was doing in the earlier debate, as well as how he views tantric categories altogether, particularly his specification that it is all a matter of "figurative" usage.[195] In fact, the foregoing passage brings up the tantric channels not to explain pulse but rather to consider another example of figurative usage. The same kind of association that obtains between the pulses and the gendered monikers with which they are labeled in the *Four Treatises* also pertains to the relationship of both the three kinds of pulse and the three tantric channels, and the qualities of method, primal awareness, and nonduality. Both the pulse monikers and the trio of qualities are cast as figures of speech, special kinds of suggestive designation. This tells us something important about Zurkharwa's understanding of the tantric channels: like the pulse types, they are associated with what we might call energy styles, or gendered orientations. In Zurkharwa's recourse to the tantric channels in explaining the heart opening too, the channels seemed to be a way of talking about personal tendencies that make for certain bodily patterns.

Mind you, such a reading still does not exactly explain why the heart opening faces in different directions in males and females. It just displaces the problem to a putative difference in directionality in the male and the female regarding the mind and the heart. But it does suggest that all these things—pulse types, tantric style types, gender associations, perhaps even the very disposition of the tantric channels themselves—are in some sense figures of speech, or ways of crystallizing or sedimenting what are really tendencies, orientations, and shifting and shiftable patterns. They are tendencies and patterns that readily lend themselves to metaphor. And that helps us to understand how tantric Buddhism, itself so rich in meaningful symbols and figures, becomes a useful handmaiden even for a medical theorist who is trying to make his theory match empirically verifiable data. It helps him describe aspects of human existence and experience that are not readily determined by bodily examination alone. Zurkharwa mobilizes the tantric vision to serve as a central explanation for human difference. Perhaps the more empirical medicine gets, the more it needs to look for conceptual tools to explain the significance of its findings.

One final aspect of Zurkharwa's foregoing pulse discussion deserves highlighting as well. Actually, his ending comment that "from one perspective, it is said to be uncertain." echoes Tashi Pelzang when he said, "Most children will be in accordance [with the sex suggested by the parents' pulse]." Both statements indicate that *most* people's sex will indeed line up with their analogues in pulse pattern, be that their parents' or their own. But like Kyempa did with respect to menstrual flow, both Zurkharwa and Tashi Pelzang open the door to deviation. Tashi Pelzang points to the fact that sex is not entirely predetermined by the pulse mix of a child's parents. He also allows sexual identity to change over the course of life if the mind stream and karma change sufficiently. In some ways he is less physically deterministic than Zurkharwa makes him out to be, although some of his statements betray a certain confusion and sometimes a collapse of gendered categories with sexual identity.[196] Zurkharwa himself is clearer and gestures robustly to the possibility of deviance from norms—of anatomical sex, pulse type, gender proclivity, or what have you.

Thus while both Tashi Pelzang and Zurkharwa show a general expectation that gender follows sex, both also make sure that there is room for individual difference. Most important, both suggest that the connection among the categories of sexual identity, pulse, and mind stream are variable and subject to many intersecting factors, among which human conception and culture play a significant part. Tashi Pelzang even admits that there is no such thing as a pure or singular type, either of person or pulse, other than in name.[197] It would seem to be an important point for medical science to realize.

GENDER AND MEDICAL SCIENCE

If medical theory in Tibet recognized the irreducibility of cultural construction to anatomical fact, it is hard to fathom how it spouted a virulent misogyny at the same time. A bewildering mix of positions on gender appears in the *Four Treatises*: the misogynist dicta, the androcentrism, the specialized female pathology, the straight descriptions of menstruation and pregnancy, the gender-bending implications of pulse theory.

Perhaps we should throw up our hands and say people—and texts—don't have to be consistent and leave it at that. Or perhaps we might borrow the image of the third sex as a trope for the entire intellectual inquiry into gender. Just as the *ma ning* is by its very nature undecidable and inconsistent, so is the conceptual space that makes for its articulation in the first place.

One of the points that has emerged in the foregoing is that male agendas, male prerogative, patriarchy—the war of the sexes, if you will—determined at least some of the treatment of the female in the *Four Treatises*. The social fact of patriliny—not often named so overtly elsewhere in Tibetan literature—along with the barely concealed claim to patriarchal privilege are used to legitimize and explain the male triumphalism in the *rotsa* section.[198] It is not so surprising that smuggled in with these social patterns is a tendency to blame women whenever reproduction goes wrong, and to ignore their investment in sex going right. Medical knowledge served as a discursive site for the advancement of male privilege.

What is unexpected in this environment is the flickering self-consciousness and occasional gesture toward egalitarianism.[199] It would be interesting to compare the record in Buddhist literary and social history on similar issues. The androcentrism, patriarchy, and misogyny seen in multiple Buddhist contexts have been well studied, along with a number of extraordinary scriptural moments when gender and enlightenment were intentionally decoupled, and even a few cases where the female gender was assigned superior virtues.[200] Such laudable efforts to respond to an otherwise widespread Buddhist androcentrism may have reflected ethical commitments, a sense of justice, a need to build community among religious virtuosi, a strategy for garnering lay patronage, or even the agency of strong and influential women. But I don't see evidence of any such factors in the medical case. Rather, when the physicians point out variations in menstrual patterns, question whether the male body should be considered the norm for all medicine, or notice gender deviation, these passages suggest more than anything a set of pragmatic needs: more healthy and predictable production of offspring, more accurate scientific knowledge.

The urge to get it right may well have inspired those moments in medical writing that seem modern and liberatory to our twenty-first-century eyes. Certainly to note dispassionately what is seen at the birth of an infant is a sign that doctors were looking closely, observing what is actually a wide range in genital appearance across human infants. It would also seem that the classification of gendered style in the pulse was based on observed differences, where the handy trope of male, female, and *ma ning* served to map associations in the actual feel of the pulse. Given the further stipulation that the gendered qualities of pulse do *not* always reflect the genital identity of the people who display them, we can say the *Four Treatises* had already gone a distance down the road toward noticing that secondary sexual characteristics are not necessarily tied to primary ones. Zurkharwa's confident assertion that there are women with men's mind streams takes another step yet. He moves beyond the particular matter of pulse to assert more broadly that sexually associated personal style does not always line up with sexual identity, and that this was common knowledge. Although a principled distinction between sex and gender may not have had a name outside the halls of medical scholastic debate, and its social ramifications may not ever have been discussed systematically, the medical commentators were beginning to make such things explicit.

Gender flexibility was a quotidian human fact for both Indian and Tibetan medicine, but the latter went further in seizing upon the image of the third sex as a model of health. Again, this did not issue out of an egalitarian agenda; rather, the person in the middle was an ideal figure to represent the propitious medical state of balance and harmony of the humors. Why that figure became glossed as the bodhisattva remains a mystery, though. It is tempting to read it as a direct affront to the massive bias against actual third sexed people in Buddhist monastic and sūtra literature (and I can say anecdotally that Tibetan scholars today are embarrassed by the evident monastic prejudice). Whether the bodhisattva designation might even have been a sly joke, we will probably never know. But certainly the principles of balance, equanimity, and flexibility are germane to the very definition of the bodhisattva in Mahāyāna Buddhist literature. Its adoption for the taxonomy of pulse is one more example where Buddhist rubrics helped to flesh out medical conception.

In the end, recognizing the disjuncture between sex and gender fits an attitude seen repeatedly in thinkers like Zurkharwa. The probative separation of construct from physical verity marks Tibetan medical thinking at its best. It reflects an urge to get things right, to be precise, and not to confuse values with medical science. And yet, having laid the ground for such a divorce, a careful scholar like Zurkharwa can also redeploy religious and cultural overlays to explain the medical facts more cogently. In the process he also betrays an appreciation of how fundamentally cultural/religious ideals and practices can affect the body. And what goes along with that is exactly to recognize that such effects are not always predictable or rational. Instead they are uncertain. It is often the case that gender mirrors sex, but not always. This recognition is itself a sign of good science.

So if we find in the medical treatment of women and gender a set of disparately tending stances, that may only be from the perspective of looking for gender justice. From the perspective of clients' needs to have boy children, rule the family, and stay alive and thrive, this medical picture of sexual and gendered states can make eminent sense. In the final chapter of this book I take up a fulsome statement on professional medical ethics, also from the period of the *Four Treatises,* that portrays the physician as fully beholden to the needs and expectations of his patients and patrons. It locates ethics not in an abstract conception but rather in the proximate, physical, and personal realities with which the physician must work.

7

THE ETHICS OF BEING HUMAN

THE DOCTOR'S FORMATION IN A MATERIAL REALM

If the teacher's mind is small, then also make your own thought small. If his is big, then make yours big. . . . If he has big desires, or likes farming, or likes to fight, or likes the Dharma, or likes to play and so on, in brief, whatever sort of behavior and orientation the teacher has . . . you should follow suit.[1]

When you are teamed up with a second doctor who is your equal, if you do the hard work to heal the patient but miss out on getting credit and the money, that's crazy.[2]

In the world of the *Four Treatises*, medical knowledge is embedded and intersubjective. Medical education forms whole persons, persons who are keenly aware of their material and social situations. Their teacher's own embodied example is at the heart of much of what they need to learn, as the first injunction, from an early commentary by Yutok Yönten Gönpo's leading disciple, urges here. Its ethos jettisons any judgment on the teacher's proclivities, whether oriented toward religion (the Dharma) or fighting. The virtue being advanced is rather about the student-teacher relationship itself.

The *Four Treatises* provides a full chapter on the training and character of the physician. Here the pragmatic and professional concerns distinctive to a medical career take center stage. The work of the physician requires astute observation of others, a well-honed "familiarity" in the realm of practice, great devotion in caring for those who are ill, and special kinds of artistry in dealing with patients and their families—all estimable virtues that the root text and its early commentary explore in detail. Medical learning on this account is grounded in the thick texture of bodily habits and social proficiencies, as much if not more than a conceptual repertoire or the study of texts. To be sure, texts and systematic knowledge are critical too. But the *Four Treatises*' physician's chapter pays special attention to the knowledge that grows out of clinical experience. Its shape is determined most of all by the contours of our earthly existence, with which the medical student is repeatedly asked to conform.

Such contours range from the disposition of instruments to the anxieties of patients to the very precariousness of life itself. Sometime the pursuit of livelihood and reputation make for a quite calculating agenda, including advice on how to "get ahead," to protect against malpractice suits, and even how to lure patients away from other doctors. These passages conjure an ethos that would have contributed to the competitiveness noticed throughout this book: in the Desi's relentless attack on his predecessor Zurkharwa; in Zurkharwa's own disingenuous failure to represent the contributions of his predecessors; in the patriarchal rhetoric and subordination of women in the *Four Treatises* itself. The chapter also provides a glimpse into the culture of prestige in Tibetan medicine at a formative moment, the inherited status of certain families dating to the royal period, and the economic realities of medical practice in twelfth- and thirteenth-century central Tibet. Much of this culture and its attendant sensibilities continued to inform what I have termed the medical mentality, down to the time of the Fifth Dalai Lama.

The bald self-assertion and concern to secure material rewards for one's services urged by the second epigraph, along with a certain tolerance for self-interested exaggeration, occasionally sound a distant echo of the manipulative ethos of Arthaśāstra, rather at odds with the circumspection advised by the Ayurvedic classics.[3] But for the most part the moral compass of the *Four Treatises*' physician chapter bears comparison to the worldly wisdom to be had in the Buddhist Nītiśāstra literature, along with many passages in sūtras and other works in Buddhist history that address the temporal rewards of the good life and the importance of looking after the welfare of self, family, and community. The chapter also pays considerable attention to the physician's compassion for the sick, even providing an elaborate meditation by which the doctor learns to model his vision of his vocation on the enlightened Medicine Buddha. In fact, we can recognize a number of familiar categories from

Buddhist scholastic and ethical discourse—from the ten nonvirtuous deeds to technical terms in psychology and epistemology and even a medical version of the "Middle View"—all marshaled to help elaborate an intricate account of the skills needed in the clinic.

And yet on the rhetorical level the chapter is a striking case where medicine deliberately distinguishes its main ethical profile from what it construes as a more selfless kind of medical practice associated with the teachings of the Buddha. The *Four Treatises* adopts an old Tibetan term for local cultural values—rendered here as "human dharma," or "the way of humans"—and distinguishes that from the Buddha's "True Dharma," or "True Way," whose relevance to medical practice the chapter also explores, but much more briefly. While our own reading will find that the medical way of humans is not without care for others and the True Way also evinces concern for the welfare of the self, the emphasis is different, and the *Four Treatises*' author found it important to make a distinction. In a similar move to others seen in this study, he developed a rubric under which matters particular to medicine could be freely explored.

In the end the *Four Treatises*' way of humans marshals a rich mix of resources—from practical experience in the clinic, to technologies of the self coming into Tibet under the name of the Buddha's dispensation, to local cultural and social values—all in service of laying out the ingredients for the formation of a physician. The medical way of being it describes is very much about how *both* care of the doctor's own needs *and* care of the patient turn on a lucid recognition that material reality and the course of human life remain beyond the doctor's ability to fully predict or forestall—much less transcend or magically control. Recognition of the fragility of life grounds the most urgent compulsion to protect it. This point is at the heart of the *Four Treatises*' physician's chapter. In medicine the welfare of the patient depends on the physician's ability to accord—in word, deed, and attitude—with the imperfectable human condition.

THE WAY OF HUMANS

The *Four Treatises* talks about human dharma on several occasions, but most of all in the chapter devoted to the physician, the thirty-first and final chapter of the *Explanatory Treatise*. This chapter takes up the many aspects of being a doctor: qualifications, education, what makes a good one and what makes a bad one, in sum constituting what we might characterize as a virtue ethics for the practice of medicine.[4] The chapter is not based on the *Aṣṭāṅgahṛdaya*, even though most of the rest of this second book of the *Four Treatises* is closely indebted to the *Aṣṭāṅga's Sūtrasthāna* and *Śārīrasthāna*.[5] In fact, there is no chapter devoted to the ethics of the physician in any Ayurvedic classic that compares to the *Four Treatises'* treatment.[6] Although several passages in Indic medical works do consider professional issues—income, family pedigree, general skills, the need for clinical experience, differences between good and bad doctors, issues around truth telling and other ethical questions—the terms, categories, and even the stance taken on these matters is different than the discussion in the *Four Treatises* (and considerably less calculating, I might add). Nor does the term "dharma of humans" (Skt. **manuṣyadharma*) figure in Indian medical ethics; it scarcely occurs in Sanskrit at all. All this tells us that while the broad issues that the *Four Treatises* raises in this chapter are well-nigh universal concerns for medicine, the way they are discussed reflects local and historically specific values and conditions from the area and period in which the text was written.

The early commentary on the physician's chapter that will closely inform our own analysis is *Small Myriad*, already encountered briefly in chapter 4.[7] This work was probably written by Yutok's student, Sumtön Yeshé Zung. It addresses the entirety of the *Explanatory Treatise,* not just the thirty-first chapter, but it deals in summary fashion with much of that, and seems to focus on what the author finds interesting. Its very lengthy comment on the physician's chapter elaborates the medical ethics sketched out in the *Four Treatises* in colloquial, direct terms. It says far more about the nitty-gritty of medical practice, including the often competitive atmosphere of human dharma, than the *Four Treatises* itself ventures. It may well be that its treatment of the topic reflects the situation on the ground more candidly than the root text was willing to do. In any event, *Small Myriad* had lasting influence, and later commentators such as Namgyel Drakzang, Kyempa, and Zurkharwa are largely dependent upon its discussion for their own remarks on the physician's chapter. Interestingly, these later commentators often eschew its more lurid details, especially those that seem most self-interested and manipulative.[8] More than any other medical work currently known to me, *Small Myriad* really fleshes out the category of human dharma. It also relates other sections of the physician's chapter to human dharma, where the *Four Treatises* did not do so. But it was still the *Four*

Treatises that introduced the category, and *Small Myriad*'s discussion closely follows the root text's verses.

The articulation of the way of humans in medical terms appears to be a unique contribution of the *Four Treatises.* The old Tibetan category of human dharma (*mi chos*) is already found in early Dunhuang manuscripts. While Geza Uray has shown that the sixteen *mi chos* rules attributed to the seventh-century king Songtsen Gampo were a later invention,[9] human dharma was one of the terms used to ease the transition to Buddhism at least by the time of the letter of Pelyang to the Tibetan king in around 800 C.E., which offered advice about social and political ethics.[10] Human dharma is also mentioned in the famous "Dharma That Came Down from Heaven" manuscript from Dunhuang, where it denotes the good customs of the people that set the stage for the establishment of Buddhism.[11] R. A. Stein noticed that these formulations seem to be citing the *Prajñāśataka* attributed to Nāgārjuna, but adds that this work may be apocryphal.[12] *Mi chos* is likely an old indigenous Tibetan term for everyday morals and lay etiquette. Stein noted that *mi chos* was regularly cast as the stepping-stone, or ladder, to the higher ethical vision of Buddhism, often cast as *lha chos*, "Dharma of the Gods."[13]

While Dunhuang manuscripts often make a pointed distinction between human dharma and the new Dharma of the Gods, or Buddhism, that the kings were promoting, at least one sustained passage shows *mi chos* occurring outside of the encounter with Buddhism altogether, which suggests that the term pre-dates it.[14] This is PT 1283, which records a discussion on morality between two brothers. Here *myi chos* denotes qualities like filial piety, compassion, and modesty, with no hint of Buddhist terminology or conception.[15] Rather, *m(y)i chos* is like a maxim.[16] As for the more general term *chos* that broadly translates the Indic *dharma*, Stein has pointed to its early history as an indigenous Tibetan term on the order of *gtsug lag*, which refers to a religious tradition or way.[17] Stein also noted its association with early Tibetan translations of Buddhist works from Chinese.[18] In PT 1283 *chos* can denote either good or bad action.[19] Both *chos* and *mi chos*, then, seem to precede the introduction of Buddhism into Tibet. Both came to serve as bridge terms (to wit, a ladder) between early Tibetan notions of morality and the transition to the Buddhist Dharma.

In the earliest documents, *m(y)i chos* and its simple opposite, *m(y)i chos ma yin pa*, denoted what is right and what is not. But when it came to be contrasted in Tibetan historiography and Treasure literature with the Way of the Gods, *mi chos* acquired a particular salience. Now the way of humans is clearly in a subordinate position to Buddhism. For example, the *Testament of Ba* quotes a minister saying to King Tri Songdetsen, "You have spread the Way of the Gods; now what will you give men as the way of humans? They were then given genealogies, forms of salutation, tales, stories, virtues, and diplomas."[20] In such narratives the dharma of humans is understood to fill out the rest of society's needs, on all kinds of legal, governmental, and social matters. It plays an important role but is clearly not the same as the transcendent virtues of the Buddha Dharma, whose label soon transitioned from the Way of the Gods to the True Way (*dam pa'i chos;* Skt. *saddharma*), a very common term for the exalted body of teachings of the Buddha throughout the history of Buddhism.[21] The distinction is comparable to the old Indic division between worldly and transcendent, or perhaps the famous Buddhist heuristic of conventional versus highest truth, although these lack the historical significance of the human/True Dharma pair for the introduction of Buddhism to Tibet.[22]

The history of the human dharma category helps us appreciate the significance of its deployment by the *Four Treatises.* Mirroring the term's early connotations, the way of humans is used to distinguish the text's main conception of medical ethics from a different approach, which it alternately labels the Dharma of the Gods and the True Dharma, i.e., the same terms that the early Tibetan documents use to refer to Buddhism. The text construes the distinction as the difference between looking after the physician's material rewards and eschewing all personal concerns to focus only on the patient's welfare. Its significance stands quite apart from whatever questions modern scholars might raise about its accuracy: certainly from the eightfold path onward, Buddhist scriptures and commentaries have paid attention to virtues that the laity should cultivate, virtues that sometimes facilitate quotidian success, temporal happiness, and material well-being in the present world.[23] However, Buddhist texts too make distinctions between temporal values and those that are fully salvific.[24] Moreover, accounts of the bodhisattva path recommending full renunciation of self-interest received a lot of attention in Tibetan Buddhism from an early period and were likely informing the *Four Treatises*' dyad as well.[25] In short, there is a complex history of efforts to separate—and to juxtapose—worldly and exalted values in Buddhist literature. However, no such case maps precisely onto the *Four Treatises*' human way/True Way distinction. What follows will be occupied with the particular import of that distinction, and what the text was trying to get at in drawing it out.

In the *Four Treatises'* rendition, the work of the distinction goes in the opposite direction from Tibetan Buddhist historiography, where *mi chos* was a ladder to a higher ethical tradition that eclipsed it. In contrast, the medical consideration of the way of humans largely eclipses the True Way. Now *mi chos* is not a handmaiden to *lha chos*, it is the main topic of interest. The putatively Buddhist variety seems added as an afterthought. Or perhaps more accurately, we can say it serves a watchful guardian, having some impact and giving some guidance and also legitimacy to an ethics that would ultimately be based in a scientistic and worldly mentality.

We might start with the place of human dharma in the overall structure of the *Four Treatises'* physician's chapter. As in so much Tibetan scholastic writing, the chapter is regulated by a well-articulated outline. This gives a useful overview of what topics were felt to need discussion, how they were positioned with respect to one another, and the relative weight and attention given to each.

One section of this outline is devoted to the way of humans. Human dharma is one of six subdivisions of the chapter's first general topic, what it calls the "causes," or requisites, for being a physician. These six requisite virtues are:

a. the kinds of mental capacity needed to practice medicine
b. white mind, i.e., the physician's dedication to the welfare of others
c. commitment to the precious value of medical practice, consisting in meditations on the Medicine Buddha
d. artistry, i.e., the bodily, linguistic, and mental kinds of agility needed to practice medicine
e. effort in work, i.e., how to study medicine and pursue a medical career
f. mastery of the way of humans.[26]

After considering these, the discussion goes on to five other general topics:[27]

2. the nature of a doctor, which amounts to a one-line definition of a healer
3. the definition of a doctor, another one-line gloss
4. the classification of doctors, a lengthy discussion of many kinds of good and bad physicians
5. the work of a doctor, a long discussion focusing on the art of giving prognoses, and then a variety of issues regarding attitude and professional ethics
6. the fruits of being a doctor, both material and spiritual, and the means to achieve success in one's medical career.

Actually the way of humans comes up in several parts of the chapter, in addition to the section so named. It is central to the discussion of the fruits of being a doctor and is also referenced when professional issues are explored, such as in the long section on the work of the doctor. Note too that the fact that the list of requisites names the sixth simply as human dharma suggests that it is primarily this human way that doctors must know and practice. Only when the text gets to its detailed discussion of the sixth requisite does it consider two variations thereof: that of the True Dharma and a mixture of the two.

In making this point, let us also make an analytic distinction of our own. On the one hand, the *Four Treatises*' physician's chapter endeavors to distinguish a distinctively Buddhist approach to healing from the quotidian *modus vivendus* of physicians. On the other hand, we can recognize a slew of heuristic devices that pertain to the human dharma doctor and, while not necessarily flagged as such, are clearly indebted to common forms of Buddhist thought and practice. Recognizably Buddhist concepts and categories figure even in exercises that help the physician gain material rewards and protect his reputation. Thus there are at least two registers on which things connected to the dispensation of the Buddha are at work in the chapter. One is at the rhetorical level and is deliberate on the part of the text itself: its overt invocations of the human dharma/True Dharma distinction. The other is discerned from our own knowledge of the history of ideas and practices and frequently reaches across the human/True divide. In the following I will be keeping track of both.

Despite the differences between human and True Dharma laid out by the physician's chapter and its commentary, it will become apparent in what follows that the distinction is less than airtight, with the human way also entailing careful concern for the patient's well-being. But it must also be noted that just because Buddhist terms and practices contribute to the cultivation of human dharma does not mean that their ethical and soteriological import remains unchanged in their medical guise. What is most accurate to say at this point is simply that the human way of medicine retains its own distinctive signature as a complexly ethical path. Even though the text can characterize it as that which accords with the "horrible world," human dharma entails sophisticated reflection on how medical learning unfolds and the full spectrum of preparation necessary both to thrive professionally and to serve patients effectively.[28] Then for the best of doctors, the most exceptional individuals, the way of humans can be perfected as the True Way. Everyone else is somewhere along the spectrum of human dharma, probably practicing some strategic mixture, as the chapter's discussion will come to suggest.

HOW TO GET THERE

As the main section on human dharma puts it, the human way requisite to medical practice involves three steps.

> First learn skillfully, then seek compliantly,
> and then control toughly.
> If you have these three, your wishes will be achieved.[29]

Small Myriad fleshes out the details. It reiterates the importance of skill and learning as the basis of a medical career. But unlike elsewhere in the chapter where learning is central to efficacy as a healer, the point here is all about career. Learning is part of the way of humans because the human truth is that if you are not skilled and learned, no one will seek you out to heal them, even if you carry your own supplies.[30]

The second step for mastering the way of humans, the cultivation of "compliance," or pliability, is even more instrumental. *Small Myriad* spells it out:

> Next, you need to proceed compliantly. Until your head stands out, you should act compliantly. However many people summon you, you should listen, saying you'll come. When you go, make your desires small. Be in accord with the people and act cooperatively. And as is said, it will come to pass that your head gets in.[31]

A key metaphor concerns the head and how it stands out, or has "emerged" from among the crowd and "gets in" or "established."[32] In contrast to an Indian medical injunction for the physician to keep his head lowered and to be inconspicuous, here the picture is of a head held high and evident to others as one attains autonomy as a practitioner and achieves eminence.[33] It has to do with the moment when one stands on one's own two feet, gets into the medical profession and has a place at the table, so to speak. One has gotten ahead. The advice is calculating—be nice, do whatever you are asked, make your desires small (the implication is that you should keep your charges low), and you will begin to gain a reputation and a following. Then you will be ready for the final step in this human dharma, acting "tough" (*gyong*). *Small Myriad* lays out its fruits in very concrete terms:

> Finally you need to control toughly. The head of your virtues has emerged. You have attained fame. At the point that you are being kind to all and they are respecting you, if you don't show toughness, they won't be moved to give you anything. They need to entice you with the transportation fee, the application pledge, the *lag dkar* fee, the *zhu khog*(?), the rack of mutton(?), the thank-you feast, the life-saving fee, and so on, all heaped up in accordance with custom.[34]

Now the picture is that you have proved yourself. Your patients, their families, and the larger community all respect you. Notably, you have benefited them. The worry is that even though they now owe something in return, perhaps they won't acknowledge that. It is a concern that will be repeated. The remedy is to make sure you get what you are owed. It will help if people are a bit afraid of you. You are tough, someone estimable, and they need to give you your due, at the same level as is the general custom: the way of humans.

The commentary then goes on to what it means to say "your wishes will be achieved" if you follow this advice about human dharma.

> Fame, things in hand, and everything that you need will come. It will come to pass that you won't have to worry about others. Rather they will worry about you. What's more, the desires of others will be fulfilled. If the patient dies, you can blame [someone else]. If she survives, then your head has emerged. [In either case] you need to be able to build their confidence.[35]

It is clear that being skilled in human dharma is about looking good and achieving a position of prestige.[36] *Small Myriad* adds later that this includes the acquisition of a significant amount of property. In addition to food and clothing, those who best achieve fame and esteem in these ways will be offered the amount of land a mounted rider can traverse in a day. Middling ones will receive the amount traversed on foot in a day. The least will gain power over the land traversed between breakfast and lunch—itself surely a sizeable plot.[37] *Small Myriad* assures its readers that in getting his due, the eminent doctor doesn't have to worry what others think of him. Rather, others will worry what he thinks of them. For the doctor who has the skills to come out looking good in all situations, all the material reward he desires will come. The key to that success is that others have confidence in him. But note too *Small Myriad*'s further point that the needs of others will be fulfilled in the process.

There can be no doubt that in turning next to a discussion of "the method of True Dharma," the *Four Treatises* is bringing into focus a distinction between the prosaic professional concerns just set out and a higher path. Both are organized in a sequence from start to middle to the ultimate ends of medical training, but what happens in those phases is painted differently. The root verses from the *Four Treatises* read:

> In the True Dharma method,
> if you are circumspect, easy to associate with, and satisfied,
> then benefit for self and others will ensue.[38]

Small Myriad characterizes this road as a "Dharma person's" way of acting.[39]

> The behavior of the Dharma person: with a mind of loving-kindness, you welcome [the patient or emissary] when you encounter them. You do not display your fame to people. In the middle, you are easy to associate with. Whatever food and service and so on your patrons provide, you do not resent it, and you act happy and fulfilled. In the end you are satisfied. Whatever feast and medicine costs they offer, you don't complain. Saying that their action is good, you show [gratitude].
>
> When you have those three, it will be beneficial to you. In this lifetime there will be great merit, and in the next one you will complete the accumulation of merit. It benefits others, you are easy to venerate, and it will come to pass that all praise you.

The commentary fleshes out the root verse with classic Buddhist ideals like loving-kindness and the collection of merit. The root verse itself is less explicit. It merely suggests a common-sense moral profile of kindness, humility, and friendliness without referencing distinctively Buddhist categories. But its final allusion to "the benefit of self and other" marks a departure from its characterization of the way of humans as the way to achieve one's own wishes. Although *Small Myriad*'s picture of human dharma also includes benefit to others as a result of acting tough, there is no question that the

True Way physician has a different profile. It seems that the doctor following the way of humans is not "satisfied" in the manner that the True Way physician is said to be. Nor is he exactly circumspect if he is acting tough; nor does he refrain from displaying his fame to others, as the True Way doctor will do. The commentary's point that the True Way doctor will accumulate "merit for his next life" closely echoes standard Buddhist language for the moral path of the laity, who are frequently said to focus on bettering their next life rather than aiming for the highest goal of buddhahood itself. The Buddhist layman is advised to make merit, which goes with the idea of not complaining about the thanksgiving feast and payment that the patient provides. Again, that contrasts with the way of humans, wherein the doctor cares very much about his payment and is explicitly instructed to complain.

And yet the commentary specifies that the True Way path is also beneficial to the doctor himself in terms of the merit he will gain and the reward of being easy to venerate and the recipient of much praise. And so position and esteem are also goals, or at least outcomes, for the True Way doctor. In other words, the physician following the True Way is still a physician, practicing his profession, just more honorably and less ruthlessly. The different tone and emphasis of the two approaches notwithstanding, the recognition of benefit to both self and other even for the True Way doctor is right in line with the pervasive picture that Buddhist scriptures paint of the esteem in which great masters are held, starting with the fame and exaltation of the Buddha in his time and continuing throughout the history of Buddhist hagiography. Even the self-abnegating path of the *Bodhicaryāvatāra,* which is trained upon advancing the welfare of others as assiduously as one would care for oneself, also eventuates in the welfare and religious cultivation of the self.[40]

FURTHER ON THE WAY OF HUMANS

A final option, the case where the physician pursues a path that is common to both approaches—in essence, a mixture—perhaps expresses the medical ethics best. The *Four Treatises* says that such an individual has compassion for the poor and takes care of his personal business with the upper classes.[41] The commentary makes clear that the former is in accordance with "true" behavior, i.e., that of the True Dharma. The worldly path, which entails the doctor's effort to treat lamas, members of the official class, the rich, the emperor, senior monks, and the rest of the nobility—and means that he will receive the right amount of payment and his own wishes will be achieved—refers to the dharma of humans.[42]

The possibility of mixing the worldly with the higher path may actually be a way of characterizing the optimal form of human dharma, what the *Four Treatises* most realistically envisions as a plausible model for doctors. It is a path on which instrumental virtues that promote one's career are mixed with a compassionate and selfless concern for the weak, making for a complex picture of a worldly, educated, and moral ideal. The fact that even the True Way is understood to bring respect and honor suggests that the *Four Treatises*' medical ethics is really about a spectrum.

A similar range is also evident in a different chapter, on everyday behavior. This chapter provides another sustained discussion of the way of humans.[43] The passage was noted by R. E. Emmerick for its distinctive meter that differs from the rest of the *Four Treatises.*[44] Emmerick also noted that the passage is independent from the *Aṣṭāṅgahṛdaya,* as is the physician's chapter.[45] It comes up in the context of describing the everyday lifestyle of a person who wants to live healthily and for a long time. The description has much in common with the doctor's professional ethics, although it is not focused upon career. All the everyday virtues this passage extols as imparting health—keeping one's word; repaying kindness; using prudence; being circumspect about personal matters in public but expressive and affectionate in private; being honest; being responsible; being able to admit wrongs; and especially the idea seen again and again in the physician's human dharma, making oneself accord with others, here cast as friends and family—are largely in sync with the general tenor of the doctor's ethics. There is also a line about not letting enemies go free but rather taking one's time and then subduing them skillfully. And there is a promise at the end of the passage that one will gain power over many through this path. Both of these last points recall the calculating side of the physician's human dharma.

Again, it seems that these more ambitious and self-directed aspects of the picture inspired the addition of a second lifestyle, called the True Dharma of the Gods. This is about being careful to practice the Dharma, which here clearly means Buddhist teachings, and not just pursue worldly happiness. One should obey the rules against the ten negative deeds so well known in Buddhism. One should also generate *bodhicitta*, be helpful and kind to those who are suffering—including insects and enemies—be disciplined, and consider the welfare of others. And so once more we find a distinction between a largely ethical path that includes ways to trounce one's competitors, and another version that lacks that component and is more selfless. Otherwise the True Dharma of the Gods as described here is not all that different from the dharma of humans; once again, it looks like an addendum to those basic ethics. It is an extra step that will confer further distinction and virtue.

If the overall portrait of the way of humans in the *Four Treatises* is readily recognizable as a moral path, what it sometimes entails distinguishes it from a higher kind of ethics. This emerges most clearly in a different section back in the physician's chapter, toward its conclusion, where the way of humans is explored in terms of the "fruits" of a medical career. There again the discussion is divided into two categories, "temporal" and "ultimate," clearly glossing the human/True dyad. But especially as unpacked in *Small Myriad*, the two profiles are conceived in more extreme terms than we have seen before, starting with the most manipulative side of the human way.

After an initial statement that the temporal fruits of being a doctor "are happiness in this life, power, enjoying possessions, and having joy," the *Four Treatises* goes on to lay out the means to achieve them.

> What makes you achieve all that is medicine.
> Brandish your own learning in front of people.
> Be good and circumspect, even with people who would harm you,
> as if they were your relatives.
> Give treatment according to your examination and assessment.
>
> As a result, you will come to have fame and merit.
> The food and possessions you desire will come naturally.
> At that time, maintain circumspection.
> When you have been kind, ask about your food and riches.
> If you delay, your kindness will be forgotten
> and a response will not be forthcoming.[46]

Here the *Four Treatises* is more explicit about the connection between reputation and material rewards than it was earlier. *Small Myriad* wastes no time in glossing the means to achieve these aims as human dharma, and sets out the strategy in detail.

> What produces fruit is knowledge of the medicines of medical science. To attain fame from medicine is to be skilled at human dharma.

Therefore, when you first learn to heal, you need to display your learning to people in order for your head to get established. How should you display it? Memorize words that are easy to understand from the summary instructions of medical science on types of medicine. Then, at a place where many people gather, or where two or three gather—the capital, or a village, or a great Dharma academy's intersection, and so on—slowly say to all the people, "I know the following." Then explain with sweet words.

Say a lot about your teacher's greatness, the greatness of the teaching, and how you performed austerities and so on. And then all the people will say, "This one has knowledge," and fame will come.

Then, wherever there is a patient, speak to those who take care of her. Smoothly dissimulate. Ask in detail about what kind of illness it is, who is the doctor, and what kind of remedy is being used. Then say this: "The name of the illness is such and such. I wonder if that doctor knows how to treat it. Maybe he knows. But if I were to treat her, I would heal her without difficulty. My teacher possessed the treatment for this disease and was especially skillful. I also have practiced this system a lot." Pour that into their ears and act nice.

Then that person will scrutinize you and will show you the patient. When they do, and you are considering whether to enter into treating her or not, take stock. If you decide not to treat her, say, "If I treat her, I will know [what to do] without difficulty; but if he treats her, she will also recover," and your own head will stand out.

If you can treat her, then match your behavior [with the other doctor's], and acting nice and friendly, treat her.

When in such a manner you treat one, and then a second, you will attain fame. Once you are healing many patients and attain fame, even if you don't want food and riches, they will come to you naturally. When first your head emerges, don't be intense (?) but just remain circumspect, take care, and attend to your virtues and knowledge of healing. But if you are always too small-minded, then your own head does not emerge. Think thus: "By acting in this way I will acquire [fame for having] special skill." Then when you gain certainty, act in the manner by which you know your head comes to stand out, and it will.

As for the moment for collecting your fee, when the patient has been cured of the illness and separated from death, and the illness has completely departed, then, while she is still largely dependent on doctors, it is time for your kindness to be repaid. That will be the time to take care

> of your wealth and material objects. If you delay beyond that, your earlier kindness will be forgotten, and later, since you are no longer needed, they will not be embarrassed about [not] repaying your kindness.[47]

There is a clear path here. It starts at the moment that the young doctor has become learned and skilled in medicine, as described earlier in *Small Myriad*'s discussion. The current passage recapitulates some of those strategies on how to be cautious and agreeable at first, even to enemies, until one's standing in the field is consolidated. But now the emphasis is on how to get the rewards. It gives particular support to the young doctor's efforts to gain attention and fame, advising him to memorize something easy for people to understand, then go out and recite it in public. It is very specific about where to talk oneself up in public places. Then it goes on to the quite bald-faced recommendation to move in on other doctors' patients. You show your knowledge and subtly (or not so subtly) suggest that you can do a better job. The text acknowledges that this involves some dissimulation, and it is clear that the young doctor is not certain it is true when he claims he could heal the patient "without difficulty." *Small Myriad* reminds him to be careful and assess cannily whether he can cure the patient when he finally gets to see her. It gives the doctor lines to say if he wants to back out and still keep his reputation and a semblance of generosity to his colleague. The passage closes with the strong injunction that once the doctor does take on a patient, he must be sure to get payment. In no uncertain terms, the text warns against being too "small minded"—often a virtue elsewhere in the Tibetan cultural universe, as it signals circumspection and honesty. But here the commentary makes clear that you can have too much of this good thing. You need to take care of yourself and be sure that you are given your due.

While Ayurvedic professional ethics makes a point of disapproving of boasting, acclamation of one's own learning and accomplishment is far from unknown even in the most exemplary Buddhist virtuosi, as in the famous "lion's roar" of the Buddha himself.[48] Closer in time and space to the culture of the *Four Treatises*, the same can be said of the robust self-assertion sometimes found in the autobiographies of Tibetan Buddhist teachers, even accompanied by the disparagement of rivals.[49] But *Small Myriad*'s advice on the self-promotion of human dharma seems cynical, especially for recommending a certain amount of deception with regard to the doctor's own abilities.[50] It is thus not a surprise to find once more a second kind of fruit described, the ultimate one.

The ultimate fruit is that having cast off all wish to deceive,
one enters into healing patients
and will progress to the unsurpassable level of the Buddha, it is said.
Thus explained King Healing Doctor.[51]

Note first the striking declaration, positioned at the very end of this last chapter of the *Explanatory Treatise.* The ultimate fruit of the True Dharma way of medicine is buddhahood. One has to wonder what weight to give this claim, especially when we recall the old question, still in people's minds, about whether medicine and the four outer *vidyāsthāna*s other than soteriology counted as salvific practice at all.[52] There is a similar pattern in classic works in *Dharmaśāstra* and *Arthaśāstra* traditions, where systems of practical ethics briefly gesture to a more optimal version at the edge of their horizon, thematized in soteriological terms.[53] How such a grand fruit issues out of medical practice is never fleshed out in the *Four Treatises*, in contrast with the fulsome account of its more materialistic strategies. For its part, *Small Myriad* makes the promised fruit of buddhahood seem distant: it may be the eventual result, but the focus is rather on the physician who just practices boilerplate Buddhist virtues. Avoid the ten nonmeritorious deeds, meditate on the four boundless minds, practice the perfections. This will cause a doctor always to be born as a physician either to gods or to humans. Either way, he will accumulate merit.[54] That sounds more reasonable as an ultimate fruit of medical practice.[55]

In any event, in specifying that the "ultimate" path eschews all wish to deceive, the root text and the commentary are surely contrasting it with the deception just seen in the path to temporal fruits. But there is no explicit critique of the temporal path as such. And indeed none of the various passages on the way of humans, even the most cunning of them, ever really tells the doctor to do anything terribly bad. They certainly advise ambition, but in the end it is all pegged to the doctor's real knowledge and capability. Human dharma in these passages thus serves as a guide to the physician on how to be a good public relations manager for himself—how to deal with humans, as it were.

I am quite convinced that even while they distinguish between temporal and ultimate modes of being a physician, both the *Four Treatises* and *Small Myriad* consider the human path to be upstanding and respectable on its own terms. The chapter does go on to make sharp distinctions between good and bad doctors, mostly on the basis of training and background. Moral transgressions and self-interested excess at the expense of the patient or even fellow doctors are condemned on several occasions, and I will touch on them below. But these constitute deviations. In contrast, human dharma as it is portrayed here does not jeopardize the patient's welfare or bring undeserved prosperity. Rather, it is part of the same list of basic prerequisites for becoming a physician as is "white mind," or compassion, and "mental capacity," whose perspicacity in discerning the nature of medical science, the right procedures, and the outcome of an illness is extolled by the root text as the most precious asset.[56] Apparently, along with the other virtues that come up in the chapter, it was expected that a medical practitioner would be competitive, assertive, and eager to enjoy the rewards of his hard work. Even what in our eyes might be rather scheming connivance is explained without embarrassment. Again, it is only that an alternate, "truer" path is added at the end, to say there is a way for a doctor to be even more virtuous.

We will return to further invocations of the way of humans in the physician's chapter, along with some larger reflection on its scope and ultimate horizons. But to appreciate the full range of the *Four Treatises'* medical ethics, we first need to consider a host of other rich passages in the chapter, some mediated by Buddhist categories, others seemingly inventing out of whole cloth the terms to describe the virtues and strategies that the student should master for clinical practice.

BUDDHIST TECHNOLOGIES OF THE SELF FOR MEDICAL ENDS

Quite different ends than with the human/True Way distinction are served by the devices the physician's chapter borrows from Buddhist symbolic and scholastic systems to elaborate the education of a doctor. The text does not flag these as Buddhist per se and makes no particular rhetorical point in adopting them; it merely seems to be drawing on resources close at hand that work for its purposes. It is our own historiography that prompts us to tag them as indebted to Buddhist traditions. Like the human way/True Way rubrics, devices of Buddhist pedigree are not found in the *Aṣṭāṅga* or other Ayurvedic treatments of the physician's training. They are not by any means the only components of the *Four Treatises'* account of the ethos of the doctor. But the fact that Buddhist categories of ethical formation, meditative practice, and scholastic analysis could translate into terms relevant not only to the education of the True Way doctor but also to the human way physician is germane to the entire problematic this book is attempting to address.

As elsewhere in Tibetan medical literature—and certainly at the heart of the dispute over the authorship of the *Four Treatises*—here too all moments that advert to the Medicine Buddha Bhaiṣajayaguru illustrate an *imaginaire* that sees a supreme enlightened buddha at the apogee of medical learning. Actually neither the Medicine Buddha nor the generic figure of the Buddha is mentioned often in the chapter on the physician's ethics. But when they are, important values are being created. *Small Myriad* warns, "If you accrue sin with respect to the patient, there is no difference than if you do it to the Buddha. And if you accumulate merit with respect to the patient, there is no difference from worshipping the Buddha."[57] Here the figure of the Buddha serves to indicate the deep worth and gravity of medical practice. The long-standing association of the Buddha with medicine also underlines the point that his enlightened compassion can serve as the optimal model for the physician's care of his patients.

The Buddha also symbolizes the pinnacle of learning and knowledge. Early in the *Four Treatises'* physician's chapter, the doctor is advised to regard his teacher as the Buddha and his medical teachings as the oral lineage of the Dharma. In fact, virtually everything involved in medical practice may be enhanced and elevated by likening it to its analogue in an exalted Buddhist domain, as when the implements of the medical "knowledge holders" are to be seen as the accoutrements of the deified Buddhist "Dharma protectors." The section of the physician's chapter on "commitment" where these analogies are drawn goes on to provide a comprehensive meditation on the Medicine Buddha that the physician can practice while he is making medical preparations. Thus does the *Four Treatises* make room for a close adaptation of Buddhist techniques of visualization, prayer, and mantra use. So, for example, while the physician is making medicine, he is advised to imagine himself in detail as the Medicine Buddha, his container as the Buddha's begging bowl, and so on. Then he goes on to dedicate the medicines he is producing to the well-being of sentient beings, recite a mantra, propitiate the medical "knowledge holders," and take up the resultant powers. This progression comes straight out of standard-issue Buddhist tantric *sādhana* formats.[58]

The physician's chapter translates and transforms other notions from Buddhist tantric traditions too. A perfect example is offered by the very section heading under which the meditations just described were presented. It is labeled with the key and suggestive notion of "commitment" (*dam tshig,* Skt. *samaya*). The semantic range of this term goes from vows to commitment to a teacher to a meditative perceptual practice whereby one consistently and always sees, in the Buddhist parlance, all phenomena as empty, or always remains in an experience of tantric bliss or other goal of Buddhist meditation.[59] The practice is akin to what is often called "pure vision" in Tibetan tantric Buddhism, whereby everything is to be viewed as "pure."[60] But in the medical context, *samaya,* or the ability to remain committed to and sustain a particular view or way of seeing, becomes a useful virtue of the physician. It is glossed here by the more prosaic phrase "placing the mind."[61] In effect the tantric technology provides a rubric for the physician to train and shape his perception of what he is doing, the instruments he is using, and most of all the project of healing itself. Doing the visualization helps him to see his medical activity as something precious and his medicines as functioning like a gem or powerful nectar. By virtue of imaginative processes borrowed from Buddhist ritual, medicine becomes something that contains blessings. The physician even becomes akin to the Medicine Buddha.

I am especially interested in tracking how such Buddhist practices and categories bolster skills and virtues that have little to do with soteriology and rather are trained upon professional and even scientific agendas. The most religious result that the *Four Treatises* posits as the upshot of doing the visualization it recommends is that blessings will rain down and auspiciousness and merit will ensue. Otherwise the effect is that the physician's own illnesses clear up, and he is able to save patients from death.[62] It would seem that really the visualization exercise functions to boost positive thinking. It enables the physician to imagine kinds of professional success. He might be able to do that anyway, if he is a positive-thinking kind of person, but the meditative structure focuses his imagination. It also gives it a patina of authority and importance. The latter is dependent upon the general position of Buddhist values and practices in Tibetan society. Given that status, Buddhist resources are available to be deployed widely: they spread their light to domains beyond the religious contexts in which visualization meditation is usually practiced.

Other kinds of Buddhist resources coming out of scholastic discourse help to organize thought. Ideas about education in the *Four Treatises*' physician's chapter are repeatedly expressed in terms of common Buddhist taxonomical devices like "body, speech, and mind"; the pedagogical heuristic of "hearing, thinking, and meditating"; or "view, meditation, and behavior." Buddhist scholastic styles of classification encourage the addition of detail and nuance to what might be an obvious medical virtue but has received less analysis than its analogue in the halls of Buddhist learning and spiritual cultivation. An example is the notion of white mind, the second of the six requisites for being a doctor. White mind can be just a way of expressing a general human virtue: white in Tibetan linguistic convention often implies good. But when the *Four Treatises* glosses white mind as *bodhicitta*, that opens the door to engage a slew of subcategories that have long been associated with Buddhist training techniques to cultivate compassion.[63] These help the physician imagine a rich picture of what compassion looks like in a specifically medical context.

In short, the Buddhist ethical apparatus serves to create fine-grained exercises for the doctor to go through. It becomes an instrument of his self-transformation. So, in the case of white mind, the use of the well-known distinction between arousing a preparatory wish for *bodhicitta* and actually entering into its practice creates an opportunity to follow in detail the train of perception and resolve that leads to full-fledged compassion for a patient. It begins in the pity felt when seeing the suffering of an ill person. Next comes the loving attitude that wishes to help the patient (*Small Myriad* is going through the classic Buddhist "four boundless minds"), the pleasure one feels in having such an intention, and the equanimity one must maintain in the resolve to heal the patient no matter what the conditions.[64] The fully developed *bodhicitta* that results from this progression, the resolve to help everyone who is sick, *Small Myriad* maintains, is then mobilized when a patient shows up and the doctor begins to treat him.[65] Thus do categorical precision and the education of the imagination become technologies of the self. The *Four Treatises* puts the upshot in professional terms: to have such well-developed compassion will make it easy to treat people, and many will recover.[66]

The hermeneutical facility in translating systems trained on salvific ends into medical terms is impressive. Regarding nonvirtuous sexual activity (one of the well-known "ten nonvirtuous deeds" from Buddhist ethics), the physician is sternly warned against fornicating with a patient or his wife. Lying, another of the ten (and here considered without reference to the dissimulation the text advised elsewhere), would consist in the doctor saying something that he hopes will delay having to work with a patient he doesn't want to treat. Wrong view is to be glad that someone died.[67] That is quite a long way from the usual definition in Buddhist texts, where wrong view has to do with a fundamental philosophical misconception.

The details that *Small Myriad* provides in these translations evince a well-developed imagination about the vicissitudes of the medical career in all of its conceivable permutations, salutary or not. Going through four detours, or "deviations"—a common trope in Buddhist ethical discourse for misconceptions and missteps on the path to enlightenment—the text speaks of things like pleasure taken in the fact that patients have many causes for illness, or pleasure if there is a chronic condition, or pleasure if the patient dies. These are all pleasure gone astray. To think that if the patient does not get well in good time, then I will take his food and wealth while he is still ill is to deviate from loving-kindness. To think that I will treat all good patients but not the bad ones is to deviate from equanimity.[68]

The medical skills that achieve a principal goal of the physician's chapter, fame and success, are also sometimes articulated by adapting Buddhist philosophical terms. The common Buddhist philosophical rubric of "avoiding extremes" is turned in the medical context into the precision needed to be on the mark with diagnosis. In the *Four Treatises'* rendition there are three kinds of extremes to be avoided: If you understand a serious illness to be minor that is to aim too low; to understand a minor illness to be serious is to be excessive; and if you don't know whether it is serious or minor, you are just wrong. Separate from all three, to understand an illness perfectly, exactly as it is, is the Middle View, i.e., the same term that names a highly valorized Buddhist philosophical realization. Or again, in the context of healing, if you don't defeat the illness, the treatment was pitched too low. If it causes one

7.1 The physician (left) as repository of compassion. *Plate 37, detail*

illness to end but another to start, something was in excess. To fix a heat illness with heat and a cold illness with cool substances is wrong. To treat in a way that is separate from those three extremes is the perfect healing, and the Middle View.[69] There is very little in these examples in common with the Buddhist philosophical notion of the Middle View, usually a considered meditation on the avoidance of essentialism. The term's use here seems an extraneous garnish on the common-sense notion that one needs to get one's diagnoses and treatments right. But then again, the textual rehearsal of the ways one can exceed the mark or undershoot it, the repetitive reminder to stay on target, serves as an exercise in remembering to be cautious and precise. The recurring sets of distinctions help the medical student develop the facility to distinguish finely.

The authors of the *Four Treatises* and *Small Myriad* sometimes show awareness that they are borrowing ideas. In a few cases they even betray caution about how far they can go in invoking capacities usually associated with religious transformation. One good example comes up when the *Four Treatises* speaks of "mental capacity" or "mind" (*blo*) as the first of the six requisites of medical practice. It proceeds to divide that into three: "big mind," which the text defines as understanding every medical treatment, in brief and in detail; "stable mind," which is about being confident and unhindered when performing medical procedures; and "fine mind," which involves examining with one's intellect, as a result of which the physician will have a "subtle prescience."[70] The commentary glosses the latter odd term as meaning that after the physician examines the texts and the key instructions of the teacher with incisive intelligence, something like prescience will shine forth regarding which medicines to use and which to avoid.[71] Both intellect (*rig pa;* Skt. *vidyā*) and intelligence (*shes rab*; Skt. *prajñā*) have worldly and enlightened guises in Buddhist scholastic discourse. But apparently the medical use of the term "prescience," which in Buddhist discourse is a power that comes only after intense meditative practice, sets off an alarm. Prescience (Skt. *abhijñā*) is an old technical term for supernormal abilities, including knowing the minds of others and having divine vision and hearing. The *Four Treatises* already tried to blur its claim by qualifying the kind of prescience the doctor will get as subtle, or slender. But when the commentator glosses this by saying that analyzing one's medical instructions with intelligence will lead to the dawning of something subtle *like* prescience, he is being cautious.[72] The qualification slightly but clearly shifts what the root text says, backing away from the suggestion that not only meditation but also sharp examination could lead to this special ability to know what is hidden to the senses. Apparently, this time the border between soteriology and medical method had been breached beyond the comfort level. The question continues to worry later commentators. Kyempa is driven to insist that the extrasensory perception that the physician has is "with outflows" (Skt. *āsrava*), another old Buddhist technical term for an epistemic state that is colored by emotional obscuration. It is not the more robust or "thick" prescience attained by virtue of doing meditation, which would lack outflows.[73]

As the discussion moves more deeply into the kinds of training and skills that a doctor needs, further structures of thought and practice coming out of Buddhist contexts prove useful and ready to hand. Still, the physician's chapter tailors its interests in marked ways. Some of the most remarkable aspects of its account have to do with the cultivation of manual and other bodily skills. The other distinctive feature of the *Four Treatises*' medical ethics has to do with the tenor and character of the doctor's interactions with patients and the broader community. These dimensions of the physician's formation contribute foundationally to the medical mentality and its considered notion of the way of humans.

BOOK LEARNING AS BASIC/BOOK LEARNING AS THIN

Kurtis Schaeffer has pointed to a series of statements from the premier early modern commentators on Tibetan medicine, including Zurkharwa, the Desi, and Darmo Menrampa, that insist upon the supreme importance of textual knowledge for medical training. Schaeffer deduces that there was a debate afoot in medical circles about this, and indicates the existence of another side that argued against the superior place of book learning.[74] Although we can question whether scholarship in medicine was actually on a par with Buddhist scholasticism in Tibet, I would concur that its promotion had considerable rhetorical cachet. By the sixteenth century, textual learning conveyed prestige and authority in Tibetan medicine. But it also was at odds with a fundamental requirement for the medical practitioner to have ample clinical experience.[75] We have already seen the very figures Schaeffer cites leaning heavily on experience and direct observation for their knowledge, and sometimes taking issue with what the texts say.

The *Four Treatises* and *Small Myriad* stand somewhere in the middle of the spectrum of views on the matter. But both arguably lean toward valuing embodied experience over textual study. This is not to say that textual work is not valorized too. For example, the discussion of kinds of effort positions basic literacy and an ability to write as the very root of a medical education and career.[76] *Small Myriad* elaborates that being able to read aids one's facility in unpacking examples and understanding fine points, while being able to write means one can copy manuscripts when needed.

> Reading and writing are requisites for you to know medicine. It is very important to gain proficiency at reading and writing. If you don't have proficiency at reading, then even if you gaze at deep key instructions you will be stupefied like a donkey who has been struck in the head with a stick. You won't figure out the examples, won't hold the words, won't get the meaning, and won't memorize it.
>
> If you don't gain proficiency at writing, then at the moment you have a text, there will be no copyist. And when you have a copyist, the text will not be there. And even if you have the text written out by the copyist, he won't return it or you won't feel like looking at it. In such a situation you won't investigate the teachings and won't learn masterfully.[77]

In emphasizing the inconvenience of having to locate a copyist, *Small Myriad* places considerable value on literacy and texts. Precious medical writings are hard to come by and even harder to have copied. But the picture painted here is not a scholastic environment. The passage values the instructions that texts can convey primarily for their practical use. It is especially interesting that reading is cast as a kind of mental exercise that aids in performing another kind of mental activity: the ability to "figure out the examples," which would be critical in the clinical setting. Skill in reading speaks to a capacity to unpack an elliptical or metaphorical textual statement and apply it in a particular clinical circumstance.

A slightly different issue that cut across the debate over reading versus clinical experience may be even more centrally at stake in the *Four Treatises*' conception of medical education. This concerns the form in which teachings are conveyed. A foundational and frequent distinction is made between extended discursive explanation and succinct statements that represent deep understanding. A similar issue is well attested in Tibetan Buddhist portrayals of how liberative teachings are conveyed. While Buddhist philosophy often critiques language altogether as incapable of representing reality, biography and other writings on transmission assume that some language is needed for communication to occur, and often valorize a special kind of teaching that consists in but a few, albeit deep and meaningful, words. This communication is usually understood to be conveyed orally, in an intimate utterance that a teacher specifically formulates for a particular student on a particular occasion. A common term for this kind of teaching is "key instruction" (*man ngag*).[78] The same term is used in the medical ethics discussion on learning. In the passage cited just above, the key instructions are apparently in writing, for the example speaks of gazing as the mode in which one would encounter them. This already indicates the kind of slippage that is possible for this very important notion, or "term of art," whose precise semantic range is not easy to delineate.[79] But even if a key instruction is written, it is understood to be a valued kind of statement that differs from other types of verbal expression.

Key instructions come up again in the next passage.[80] The student is advised that he must depend on a teacher. The *Four Treatises* defines one of the virtues that such a teacher should have as vast "mastery," or skill, which the commentary glosses as a kind of knowledge (*shes pa*). It is not specified what sort of knowledge that would be. The second virtue of the teacher is that he "possesses the key instructions." The commentary goes on to indicate just what these two kinds of educational credentials consist in, and differentiates them from conventional book learning:

> The knowledge of that teacher is vast, and he knows all sides of medical science. It is not necessary to search for and depend on a different teacher for each kind of key instruction, since [your own teacher] has penetrated them all. His mastery is that he has mastery of words, and therefore is hot at explaining. Mastery of meaning means that he can perform manual procedures. Being masterful in both words and meaning means that he is masterful at composing new key instructions. The one who possesses key instructions does not depend solely on books. Rather, he has the key instruction that has little complexity but big meaning, is easy to perform but hard to measure, and hits the essential point.[81]

Small Myriad is not devaluing learning as such. But it says explicitly that studying a book is not sufficient to gain the kind of knowledge and skill that is essential for medical practice. This passage provides a number of interesting ideas about what kind of learning, and especially what kind of verbal articulation of that learning, it *is* trying to foster. The teacher described here is certainly good with words—even "hot," or intense, in his explanations. The statement that one need not seek a different teacher for each key instruction has to do with particular medical procedures that the student endeavors to collect and master. Recall that the third of the *Four Treatises* is itself called *Key Instructional Treatise* (*Man ngag rgyud*); it provides many examples of what these instructions look like, including specific recipes for medicine, diagnostic tools, and therapeutic techniques and procedures. But in the passage above, we find a relatively rare attempt to specify the nature of key instruction generically, in contrast to other kinds of teachings. The ideal picture has something to do with the relationship between word and meaning. *Small Myriad* also specifies that meaning in this case is about procedure. A key instruction must be able to signify in words what one should do in the clinical setting. The passage goes on to assert that someone who is good at connecting words and their meanings in that sense can compose such instructions of his own.

We can set aside again the question of whether such composition is specifically in writing: the verb *rtsom* does usually point to writing, but it is not a stretch to use the term to talk about formulating an oral teaching. But the emphasis in the passage is not on whether something is written or not. Rather, what is important here is the way that teaching is articulated. It is simple in its formulation yet opens out to many dimensions of the world of practice. The procedure it describes is clear, but the understanding behind it is broad and far-reaching. It also gets to the heart of the matter. And it seems that this passage, at least, does not expect to find such key instructions primarily in books. It rather attaches such knowledge to the teacher with deep understanding, who is thereby the master and originator of key instructions.

I will explore below other, even less textual aspects of what—and how—*Small Myriad* says one should best learn from one's medical master. But we can notice already that *Small Myriad* sees a danger that what one learns from books might be superficial and insufficient. This is made clearer in an extended passage describing several classes of physicians who display kinds of book learning gone wrong.[82] One kind, who "thumbs through pages,"

> rushes to swallow what he has eaten. He does not stay at his teacher's place for more than one or two months. He takes up and learns a few

> medical recipes that are in the text but does not memorize them, nor does he become familiar with the manual procedures. He only looks at the text on the spot and then treats. Such a one will know how to treat those illnesses that accord with the text, but will not be able to treat those that do not accord. And as soon as he is separated from the text he falls back to the ordinary, and becomes a butcher.[83]

This is a sharp critique of superficial and instrumental use of texts. And yet in painting this disdainful picture, the commentary is not suggesting to remedy the syndrome by studying the texts more closely. Rather, the crucial problem is that the student has failed to spend sufficient time with the teacher. The issue has to do with time, or pace: swallowing too hastily. *Small Myriad* is also pointing out that written resources make shallow or instrumental use of medical instructions possible. One can hold the manual in one's hand, consulting it on the spot, like a cook consulting a recipe book as she makes food. But in such cases one may only know what the text explicitly denotes.

By pointing out that such a practice only works for ailments that correspond closely with their textual description, the commentary is noting a general problem about the relationship between systematic representation and the actual conditions of the material world. The medical texts only present ideal types of illnesses and cannot account for the variations one actually encounters in real life. What *Small Myriad* seems to recommend as the antidote to using texts in this way, then, is to take more time, becoming so intimate with the content as to memorize it. One also needs to spend a long period with the teacher and become "familiar" with the procedures. Familiarity itself is a very important virtue.

Small Myriad's next example of a poor medical practitioner furthers the impression that these are caricatures of doctors who are only educated by the book. This is "the doctor who amasses black letter grains," a jab at written scribbling devoid of meaning.

> He does not even go near the teacher, but depending only on the fact that he knows how to write, copies the old medical writings. Locating and borrowing a medical text, he looks at the black letters and then steals it, wandering off with the text that he has snatched from doctors and exchanging such writings [with others]. Then he stares intently at the black letters on how to practice and so on, and then he treats.[84] Even if he has memorized

> the text, since he did not receive key instructions from the teacher, he becomes a butcher.[85]

The picture might remind us of the classic story in the life of the yogi Milarepa, when he tried to meditate using instructions for which he had not received the key instructions—and ritualized permission—from his teacher. But it is also different on just that count. Nothing about religious ritual or lineal empowerment is pertinent in this rendition of medical education.[86] The point is more prosaic: written instructions need fleshing out. One can stare at them all one wants, but without a teacher one will never fully understand them. The statement can stand as a critique of textual study overall; something is missing, something that the written word cannot supply. Once again, that is characterized as the teacher's "key instructions," now most likely referring specifically to oral teachings that the teacher utters in the immediate context, precisely suited to the particular task at hand.

But there is even more that the medical student gets from the teacher. The passages we have examined hint at a larger set of skills, knowledge, and values when they speak of the need to stay with the teacher for a long time. What happens in that slow-paced process gets to the heart of the medical ethics envisioned here. This begins with a set of goods that are not conveyed primarily through verbal media of any kind, but rather through bodily means and by virtue of bodily kinds of knowledge.

FAMILIARITY

Although textual study is positioned at the beginning of the educational path, *Small Myriad* pairs it with another virtue, "familiarity," which is just as essential, if not more. Familiarity often has to do with what transpires between the teacher and the student. This goes beyond book learning.

The Tibetan term *goms pa* has to do with habituation and cultivation. It is etymologically related to *sgom pa*, to meditate, which is the causative form of the more basic act of getting repeated exposure to something.[87] Along with compassion, familiarity appears to be the most important virtue in *Small Myriad*'s vision of the doctor's training.

To be sure, not all instances of familiarity are valued without qualification. It is certainly not sufficient on its own to constitute the credentials to practice medicine. *Small Myriad* gives an interesting example of the elderly person who becomes a doctor based upon his experience of many years being ill himself. While such a person may indeed be able to treat others with the same illness, he will not know how to treat other illnesses. Here it would seem that the problem is not with the virtue of familiarity as such, but with its scope.[88]

Familiarity sometimes goes with textual education as a valued pair. In a list of the thirteen kinds of faults that mark a poor physician, the second is to fail to understand the medical books and the third is to lack familiarity.[89] Both point to gaping holes in the physician's preparation. If one does not understand the textbooks, one can neither recognize an illness nor perform a medical procedure. It is like being blind to what is in front of one. The failure to have familiarity is one step beyond that. Now one has entered the path of practicing medicine but has no idea what one will encounter. When attempting to diagnose an illness, such a physician will constantly be plagued with doubt.[90]

The familiarity that the latter kind of bad doctor lacks is glossed as "having seen." The object of such seeing and familiarizing is specified as the teacher's "style of acting" (*phyag bshes*).[91] Using the eyes to watch the teacher's way of acting comes up several times as a key part of the kind of familiarization that the *Four Treatises* wants the student to get, such as in a more general list of four kinds of "ordinary" doctors (as opposed to buddhas and other special ones with extrasensory perception). That list goes from the best

7.2 A doctor who has not used his eyes and cultivated familiarity sets out on a road about which he has no knowledge. *Plate 37, detail*

kind of doctor, whose family has royally bestowed medical status, to the kind who has studied in the footsteps of the great ones, to two who are less positive.[92] We already met the last of the four, the one who takes up physician's accoutrements on the spot and butchers lives; this is clearly not a good kind of doctor at all. But the third one deserves our attention here. This kind of doctor is described by the root text as a "friend of beings through having become familiar with action." The phrase "friend of beings" alone indicates that this is a path to respectable medical practice. The commentary makes clear just what it entails:

> As for the one who becomes a doctor by virtue of familiarity with conditioned action: even though he doesn't know the meaning of books, he depends upon the cycle of oral instructions. He is a doctor who has familiarity by virtue of seeing the style of acting of the teacher, and he has familiarity through his own experience. This is like Drangti Sibu or Nya Kongten or Tazhi Ngarpo are said to have been. Those three are said to have been the friend and mentor of all ill beings.[93]

This is a particular kind of familiarization that has to do with action (*las*). The commentary glosses that as "conditioned" (Skt. *saṃskṛta*; Tib. *'dus byas*), an old technical Buddhist term for the coming together of many factors that become falsely reified as a single whole. From the perspective of the life of beings, such reification amounts to the crystallization of habits and the collection of karma. For Buddhist soteriological purposes, conditionality is not a good thing; it is synonymous with the state of samsara. But for medical learning, habit formation and familiarization in performing manual therapeutic procedures is not only desirable but essential. This is a case where Buddhist scholastic categories had become part of a battery of analytic tools that could shed light on a very different, more secular project.

The fact that venerable figures of the past like Drangti are examples of this kind of doctor makes clear that even if the "friend of beings through having become familiar with action" is not the absolutely best kind of ordinary human doctor, he still is respectable and eminent. His familiarization through action comes from three sources, one of which has to do with a kind of study and discursive knowledge, but it is explicitly not textual. Rather, the "oral instructions" of the teacher represents something close to, if not synonymous with, the "key instructions" examined above.[94] They are seen to have value in their own right even if textual learning is entirely absent.

The other two sources of familiarity that *Small Myriad* is valorizing go beyond the discursive altogether. One is, again, the student's observation of the teacher's "style of acting." This revealing term refers to the manner in which the teacher performs various tasks in his medical practice. It may well be unique to him, like a particular way of inserting a needle or a way of blending medicines, but it is also eminently transmittable. How that happens will be fleshed out in a later passage. Clearly it is an important part of what the student learns, and begins to mimic and then become familiar with, through watching.

The last source of familiarization is the student's own experience. Although no more detail is given here, the picture is clear enough. To receive the teacher's oral instruction, and then begin to get familiar with medical knowledge and its applications by both watching the teacher's movements and performing these applications oneself, is a respectable path to becoming a physician.

Although the student's own experience is key, the discussion of familiarity seems to stress what the student gains from others, most of all the teacher. The value of intersubjectivity in medical education is given the most detailed attention in the discussion of effort, another of the six requisites for a career in medicine.[95] Here there is a direct progression from what is explicitly textual knowledge to that which will provide the critical familiarity. Having established a basic need for literacy along with the crucial role of the teacher, the text goes on to recommend a detailed method of relying on that teacher.

This involves complete trust and devotion. It bespeaks a dependence well known to us from Buddhist accounts of the teacher-student relationship. But in expanding on the root verse the commentary describes an unusual level of intimacy with the teacher that apparently was deemed especially apt for the medical student. After admonishing the student not to be mercenary about his educational needs and not to be greedy about receiving the all-important key instructions, it goes on to advise him in this way:

> Achieve everything that he commanded you when you were in front of him, and that you could not refuse to his face. It is not the case that privately you can not do what he commanded. Making no [distinction between] face and private, just do whatever he commands. In this way, all of your activities accord with what pleases the teacher. If his mind is small, then also make your own thought small. If his is big, then make yours big.[96] If he keeps medical materials, you should do likewise. . . .[97] If he has big desires,

> or likes farming, or likes to fight, or likes the Dharma, or likes to play and so on, in brief, whatever sort of behavior and orientation the teacher has, by whatever means there is to please him and in whatever order, you should follow suit and respect him.[98]

Part of this striking passage already appeared at the opening to this chapter. Its injunctions go beyond watching the teacher's style of performing medical procedures to fully assimilating his entire personality. It also requires full transparency, resolutely ruling out the possibility for the student to have a secret life. Rather, what transpires when the two are face to face becomes the guide for everything the student does when the teacher is not present. And this is not limited to whatever the teacher has explicitly instructed the student to do. It also includes the teacher's entire way of life, which the student must take up and make his own. It is apparently judgment-free. So if the teacher likes to fight, the student should too. Note that liking to fight and liking the Dharma (here referring to Buddhism) have been put on an even level, along with several other possible proclivities. These all are thought to constitute a kind of "orientation," (*bzhed gzhung*), an interesting term that indicates a general attitude or stance toward the world, where *bzhed (pa)* refers to both wishes and positions and *gzhung* implies a general order or set of possibilities. Everything about the teacher, from his stated injunctions and what pleases him to his way of behavior to his very stance toward the world and way of being, should be consciously taken up by the student as his own.

The intersubjectivity at the heart of medical education also extends to the student's compatriots—his friends, or fellow students. The *Four Treatises* sees them as both a source of information and as a spur to assiduousness. The commentary adds that these are the people who can help one with examples, explanations, how to do research, and how to understand difficult problems.[99]

Most of all, *Small Myriad* thematizes the very principle of familiarity itself. The final point of the section on effort maintains that this is what makes for excellence in medicine. In unpacking the abbreviated words of the root text the commentary returns to the teacher's style of practice and how deep is the imprint of the teacher's habits on the student.

> You become familiar through seeing and hearing the teacher's style of practice. Attaching to it, you become intimate with it. For that, there are the practices of earnest hearing, thinking, and meditating. Not thinking

> about anything else, integrate [the teacher's style] into all four activities—going, strolling, sleeping, and sitting. In all those times think about the main point without wandering, and you will become attached and intimate. And with no weariness whatsoever, you will attain confidence and get rid of all doubt.[100]

This passage sums up the main point of effort. The student takes the teacher's style of practice as an object and becomes intimate (*'dris pa*) and familiar with it. It is the focus of one's hearing, thinking, and meditating, a common Buddhist pedagogical trio, but it is also an integral part of everything one does, glossed by an everyday set of four kinds of bodily movement. It is notable, even surprising, that the text advises a certain attachment (*zhen pa*), usually anathema to normative Buddhist ethics. Here attachment instead makes for a useful fixation, trained on fully assimilating the teacher's mode of life.

ARTISTRY

The virtue of familiarity takes the medical student out of his isolated self and into imbibing the virtues of others. That not only makes interpersonal connection basic to medical training but also means that the kind of knowledge being transmitted is itself not confined to the private world of the subjective mind. Rather, what the student is becoming familiar *with* are propensities, orientations, habits, attachments, bodily skills, and patterns. All of this prepares the student doctor for the deep immersion in the material and social worlds in which his practice will perforce be situated.

Skill in manipulating physical things is most explicit in a passage in the physician's chapter on *bzo ba*, or dexterity.[101] Dexterity would of course be crucial for all the many things the doctor must do manually. Indeed, the first kind that the text names has to do with the hands. The commentary lists in summary fashion the various things the doctor must make, such as medicines and instruments, and then deploy with his hands, including kinds of bandages, splints, and other instruments connected with scraping, pressing, bleeding, and moxibustion. It is an obvious point. What is interesting is that the discussion goes on to a consideration of verbal and mental kinds of *bzo ba*. Given these latter dimensions, "dexterity" will not work as an overall rendering of the category. "Artistry" better captures the larger point that the *Four*

Treatises is extolling. Here is the commentary's slightly more elaborated discussion of the root text's comments on verbal and mental artistry.

> Verbal artistry is to know how to speak with soft, relaxed words that heal the sick person's mind, patiently sustain him, eliminate his grief, and can arouse pleasure in him.
>
> Mental artistry issues out of variegated intelligence. Not being mentally dull about all the body, speech, and mental artistries allows them to shine forth lucidly with scintillating clarity.
>
> If you know these kinds of full artistry, you will become a master of all artistries. They are all the artistry of human dharma. And since the possession of all the other virtues depends on that, you will become the master of all of them too.[102]

Once again, medical ethics has borrowed the Buddhist taxonomical device of body, speech, and mind, but aside from that bare resemblance, something different is at work. The mental kind of artistry is the most vague and the least compelling. We can say it is tautological—in effect, mental artistry enables mental artistry, although the notion of "variegated intelligence" is intriguing.[103] It may well be that "verbal artistry" captures the sense of the larger category best. This will become clearer when we turn to the topic of prognosis. Note for now how well verbal *bzo ba* parallels the manual kind. Verbal artistry has here to do with crafting one's words to effect a response in kind: one speaks softly and eases the patient's worry. Like dexterity, it requires one to artfully shape, craft, and work with what is there. In the doctor's work, what is there is often very difficult to deal with and always requires ingenuity and skill.

In its final statement summing up the section, *Small Myriad* takes the liberty of making a fundamental connection between artistry and human dharma. This suggests once again the larger sense of the medical way of humans already detected above, quite beyond its self-interested dimensions. For *Small Myriad,* the way of humans is at the heart of all artistry. Artistry conceived in this way, the commentary goes on to maintain, governs all of the other virtues being considered in the physician's chapter. I think *Small Myriad* says this because the doctor's need to deftly manipulate physical and social situations reflects the deep imbrication of medicine in the concrete realities of the world. The physician's chapter suggests repeatedly that such manipulation is a two-way street. Be it an interaction between the student and the teacher, the doctor and the patient, the doctor and the patient's family and larger community, or even the doctor's hands and the patient's body, a medical compound, or a surgical instrument, neither the doctor's initial training nor his actual work can proceed purely in a spiritual or mental register. While textual knowledge and ideal schema are extremely important too, they must be implemented in material and social milieus. And these are realms that the physician cannot completely control. The only way for him to move forward is through careful observation, receptivity, and negotiation.

I believe that such interplay is central to the ways human dharma is thought to govern the career of the physician. It is especially germane to the dimension of human dharma that is concerned with reputation and the opinions of others, which no physician can eschew.

FAME AND FORTUNE

The fact that a physician must craft his practice in response to the physical and social exigencies of the world around him adumbrates some very basic features of human dharma that shape the contours of medicine's epistemic horizon. But the point can also be approached more prosaically, from a side of the way of humans that we have already observed: its unabashed concern with fame and fortune.

The quest for fame is part and parcel of a larger constellation of concerns that say a lot about how the authors of the *Four Treatises* and *Small Myriad* saw the medical profession. Some of these issues become apparent in their classification of doctors, both good and bad, along with a variety of subtypes. It is telling that at the head of the list "ordinary" kinds of physician is the one who is descended from a "family (whose status) was bestowed."[104] *Small Myriad* explains this as a family to whom the kings of the imperial period had granted permission to practice medicine.[105] It is a prestigious class to which all the medical writers refer in admiration in the centuries that follow. The family with royal appointment would seem to be a class designation comparable to medieval European crafts groups with inherited or royally bestowed rights. The implications of class in Tibetan medicine are elaborated further in lists of thirteen features of a superlative doctor and thirteen features of the lowest kind of doctor. Family prestige figures in the first of each.

The first virtue of the superlative doctor is that he has "venerable bones."[106] He is the son of a lineage of the "knowledge holders" and sages.[107] The *Four Treatises* is referring to family pedigree. Further elaboration is supplied when the text gets to the opposite case, the first of the qualities of the lowest doctor, who lacks such a family line.[108] The *Four Treatises* glosses the latter as being "like a fox who has seized the capital: such a doctor will not be venerated on the head of all."[109] Here *Small Myriad* explains:

> If your paternal ancestors lack good family lineage, even if the extent of your mastery is known, you will not assume leadership.[110] No one will want to respect you. It is as if a smart fox were to emerge at the top in the capital of wild animals. Even the weakest of the tigers and lions and so on would not want to venerate him, since he is not in the wild animal family lineage.[111]

The wild animal category includes lions, tigers, and bears; foxes are not part of it. But rather than extolling the intrinsic virtues of this group, the passage is concerned with pragmatic issues. Coming from a venerable patrilineage—and here the text explicitly names the normative gender of the practitioner of Sowa Rikpa—means first and foremost not that you are good, but that

7.3 The physician without good family lineage is like a fox who would seize the throne. *Plate 37, detail*

people will respect you. The commentary appreciates the irony that the physician who lacks such a pedigree might be very learned and capable, and might have achieved some recognition for that reason. But it sees a social reality: without the requisite good family, people simply won't accord that doctor the highest regard, justifiably or not.

The physician's chapter does hold out other categories of good doctors, based on the mastery of their teachers and the depth of their own study, diligence, and skills. But fame and status emerge repeatedly as key desiderata. The text unabashedly advises the young medical student on how to get ahead. Yet what goes along with the valorization of competitiveness, which I want to stress here, is the conviction that it is vital that a doctor's community respects and trusts him. As one passage concludes, "you need to be able to build their confidence."[112] A similar point is fundamental to the *Four Treatises*' very definition of what it means to be a physician: since he benefits beings and protects them like a father, "all kings venerate him as a lord." *Small Myriad* goes on to tie this prestige to the old title for a noble physician, "divine lord" (*lha rje*), which was granted by kings.[113]

This is not to say that there is no ambivalence about self-promotion or ruthless advancement at the expense of colleagues. On several occasions *Small Myriad* indicates some moral pause, and though such admonishments only come up outside the explicit discussion of human dharma, they do suggest a concern that ambitious aims and principles might be misunderstood or misused. The commentary criticizes selfish thinking about one's own fame and fortune at the expense of the patient.[114] Or again, "To roughly examine the patient and think that whether she is ill or dies I should work to get as much of my own fame and reward as I can is loving-kindness gone wrong."[115] "To make one who has absolutely no knowledge famous is senseless, because it closes off their merit."[116] And again: "To scorn a doctor who is one's equal in status is senseless, for it diminishes one's own life force."[117] Several of the pitfalls of competitiveness are cast in terms of compromising common Buddhist values such as merit or loving-kindness; another has to do with the old Tibetan concern for care of one's life force, or vitality (*srog*). All evince a sense that excessive ambition and antagonism bring harm to the doctor himself. This can be quite calculating. As *Small Myriad* explains, here in familiar Buddhist terms, to favor the immediate returns of fame and fortune in this life over the negative results that one will experience as a result of such behavior in the next lifetime is crazy, for it privileges the present over something that will last much longer, i.e., the future.[118]

Most of the discomfort about excessive self-promotion is framed in terms of its impact on the doctor's reputation. The commentary insists that to perform little tricks like pretending to extract a stone or do eye surgery or take out a tumor through magical means just to collect wealth, while meanwhile failing to do the hard work of treating chronic conditions, is blatant misbehavior. The upshot is that others will fail to accord one the appropriate regard.[119] We also see in this statement a concern to rein in certain dubious medical practices, along with a sense that there is a larger moral community with a governing set of values. This is evident in yet another warning about why too much ambition will actually be counterproductive: "When you are teamed up with a second doctor who is more skillful than yourself, to think that you should pretend to be virtuous and to arouse conceit so that you won't be put under him is crazy. It is highly embarrassing to the masters."[120] The jealousy and competitiveness the statement describes are clearly ridiculous. They also violate deep-seated expectations and the highest values in the profession.

And yet, similar to the human dharma and fruits sections of the physician's chapter, we still find here, alongside the cautionary notes, a strong sense that rivalry with other physicians is not only inevitable but also important. One should in fact try to assert one's own merits. One has to win over patients and make a living; one should earn regard for one's medical proficiency. The statement immediately following the one just cited suggests a way to theorize the defining principle. There is a big difference between competing with physicians whose knowledge is clearly superior to one's own and being sure not to be swindled by one's peers. This is how *Small Myriad* works out the other half of the formula: "When you are teamed up with a second doctor who is your equal, if you do the hard work to heal the patient but miss out on getting credit and the money, that's crazy. He will have cheated you, and you won't have achieved your aim."[121]

The sentiment is repeated many times. There can be no question that we are seeing here the professional concerns of a medical guild with scarce resources and a felt need to aggressively recruit patients. There are also ample signs that medical techniques were jealously held as exclusive patrimony. In one well-known collection of medical instructions based on a master's clinical experience, each teaching is bequeathed to a particular student from a particular clan, with the implication that that student would have exclusive access to the technique.[122] Such rivalry seems to make for a fundamental anxiety. As *Small Myriad* says at one point, "If you don't have medicines in your own bag, it is bad. It means you need to borrow medicines and instruments from others, and therefore be dependent upon them."[123]

But such bald counsel notwithstanding, the competitiveness does not merely represent a quotidian ambition to do better than one's peers. It also reflects real anxieties about the efficacy of medical practice overall. These worries contribute substantially to the need to maintain the respect of clientele, and have to be factored into the picture we are painting of medical ethics. Medicine is very difficult to practice well. It is exceedingly hard, in fact, to achieve a reputation of being able to cure people, as the *Four Treatises* advises. The reasons have to do with the real conditions of illness, and those are outside of the physician's control—no matter how wise and skilled he may be.

LIFE AND DEATH: REALITY PUSHES BACK

The *Four Treatises* describes the basic situation that the physician faces lucidly:

> As for all the [issues regarding whether the patient] lives or dies
> or whether the danger is great or small:
>
> By virtue of the power of wind horse, karma, merit, and circumstances,
> perhaps he will survive or perhaps he will die.
> Therefore, it is not appropriate to pronounce a very clear prognosis.
> When the danger seems great, say nonetheless that it is possible
> to heal him.
> When the danger seems smaller, say nonetheless that care is necessary.
> In any case, it is important to be in accord with the way of humans, that of
> the horrible world.[124]

The passage comes up in a longer section on how to give prognoses, to which we will return. It lists a number of *kinds* of factors that make for life's uncertainties. The old Tibetan notion of wind horse is set side by side with the Buddhist categories of karma and merit, and both are juxtaposed with the general disposition of circumstances.[125] If one of these doesn't get you, another will. There is a wide array of variables coming from many sides, which means that it is not possible to predict anything for sure. Thus does the text advise the physician how to hedge his bets.

Note especially the invocation of human dharma again, that which accords with the horrible world of samsara. In connecting its strategies for prognosis with the way of humans, the text is not only referring to the cautious and clever language that it instructs the doctor to use. The way of humans also has to do with the very condition of being subject to unpredictable forces in the first place. It is this basic condition that makes the caution—and the humanness—necessary.

Small Myriad adds that the unpredictable forces mentioned in the passage refer to both the patient and the physician.[126] Not only do the patient's karma and luck and so on contribute to the mix; the physician's own karma and so on will also affect the outcome of an illness. That multiplies the number of variables by two. The upshot is that almost anything can happen, no matter what the doctor's medical expertise tells him: "Even if the danger is great and you understand she will die, it is possible that she will survive. Similarly, even if the danger is small and you understand she will survive, it is possible that she will die."[127]

The unpredictability of the situation also means that the physician must not let down his guard for a moment: "No matter how small the danger and even if she seems to be surviving, there is the danger of the error of thinking that if I let it be, nothing will happen. A little bit of laziness can make for a full fathom of a lawsuit."[128] This becomes the occasion to insist once again on the necessity of being competitive, blaming one's rivals for mistakes, and taking credit for victories oneself. The comment ends by restating the *Four Treatises'* point: whatever happens, it is very important to understand the way of humans.[129] The fundamental situation is that there are real dangers afoot for anyone who would attempt to predict the course of an illness.

The point that the doctor must never let down his guard or become too confident is also explored in another passage that makes very clear what is at stake in medical practice. The root text's statement is brief, but *Small Myriad* and other commentaries know it refers to a story.[130] It has to do with the need for the doctor to apply effort not only to enhance his own career but also for the sake of others. The story is about a king who once asked his preceptor about a performance going on in town. The preceptor replied that he was not aware of it. When the king was surprised that someone could be so focused as to utterly miss this raucous event, the preceptor performed a stunt to demonstrate his powers of concentration. He stood on a high wall and one of the ministers handed him a bowl filled to the top with ghee. All sorts of interesting shows and plays were being performed down on the ground, but the preceptor demonstrated his ability to walk along the wall and pay exclusive attention to the bowl of ghee so that he did not spill a drop. He didn't even see the performances below. In the same way, the physician should give undistracted attention to his patients.[131]

Small Myriad's final lines in this passage drive home a very fundamental point about medical ethics. Life is easy to lose; life is very precious.

> Just like in that example, when working to heal the sick, one needs to be tightly focused and not interrupted by beer and amusements and happy conversation and work and so on. If the life force is cut but once, it is impossible to restore it. Therefore it is very precious.[132]

The stakes of medical practice are as high as the attention required to walk along a narrow wall without spilling a bowl of butter is fierce. The life force is an all-or-nothing affair. Cut it off once and it is gone forever; there is no redressing such a slip.[133]

The first and foremost goal of medicine is to keep the patient alive. The special significance of that life, and the survival—or demise—of the patient is foundational to the mindset of the physician. And yet as we have seen, it is difficult to ensure survival. Many factors contribute to an illness, and the physician can neither control them nor confidently predict how they will play out. Both of these incapacities render the physician's reputation in the world—which directly affects his ability to practice his profession—vulnerable to forces beyond his power.[134]

We have seen how important is the community's estimation of a doctor's skills and knowledge. Not unlike the bottom line of death itself, the doctor's reputation is a real force in the world that he cannot completely command, only try to manage. The perceptions and reactions of the people around him are the deciding factor of his very ability to practice medicine. He can never ignore them or consider his reputation as mere vanity. If no one respects him, he can't function as a doctor. Like death, reputation can't be brushed aside by the physician.

TELLING THE FUTURE

One of the most sustained passages in the physician's chapter where reputation and death are closely intertwined is its advice on how to formulate prognoses.[135] Here the recurrent specter of death and a repeated worry about the perceptions of others triangulate with a third factor, the relationship between words and their meaning.

The text calls on the physician to do several things with his prognoses at once. He needs simultaneously to protect his reputation, protect the patient's own feelings and optimism, and actually figure out what the real condition is in order to cure it. It's a complex negotiation. Recall Zurkharwa's efforts to allow language about constructs like Tanaduk or pulse gender to play an important figurative role, while still holding apart a space where the empirical truth of the matter can be ascertained on its own terms. The prognosis instructions in the *Four Treatises* make one wonder if Zurkharwa's facility in making such distinctions might have been cultivated through his own training in this aspect of the doctor's work. The *Four Treatises*' conception of prognosis is all about wrestling with words, the kinds of reality they may not actually reference, and most of all, other kinds of realities that they do affect.

The discussion of prognosis immediately precedes the passage considered above on the many factors that make for uncertainty. It is all part of the larger section on "the work of the doctor." Prognosis is the aspect of that work that has to do with speech, and it is given special attention. The speech work of the doctor is to make a definitive pronouncement (*kha dmar*), a prediction about the patient's chances of survival, hard as that might be. The *Four Treatises* provides two sets of discussion about how to deal with this challenging expectation. The commentary understands these as representing, first, a set of kinds of "double prognosis," and second, a set of kinds of "triple prognosis."[136] Exactly where *Small Myriad* sees doubleness or tripleness is never exactly specified, but in trying to figure that out for ourselves we are brought to important insights about the medical ethics of human dharma.[137]

DOUBLESPEAK

Despite how hard it is to discern all the factors in the course of an illness, the *Four Treatises* holds out optimism that at least sometimes one can be certain. This is its first statement on what the commentary classes as a double prognosis:

> If you realize what [the illness] is,
> then announce it in the marketplace as if blowing a conch.

If he is going to survive, then make a pledge.
If he is going to die, then it is important to specify when.[138]

Small Myriad unpacks the plan of action:

If you realize which way [will come to pass], whether he will live or die, you can announce the prognosis clearly in the presence of all of the patient's circle. If the patient is going to live, then say, "This one definitely will survive, I guarantee it without a shred of doubt," and pledge that he will survive. If he is not going to survive, then say, "As for this one, even if the Buddha Bhaiṣajyaguru were to look down on him, it is definite he would

7.4 When you know the diagnosis, announce it publicly, as if blowing a conch. *Plate 37, detail*

> not survive. You must understand this!" and then specify the time. For that, one needs to draw together the key instructions of the lama, one's own experience, and the situation of the merit. This takes courage.[139]

With the first kind of prognosis, the doctor makes a public and definitive proclamation. If he is really sure what will happen, then he stands to gain credit and respect. But even here the commentary points out that although one may be certain that the patient will die, it is especially hard to predict exactly *when.* That requires all the knowledge gained from the teacher's crucial key instructions, along with one's own experience and an assessment of how much merit and/or bad karma is at work. Even then, one will still need to be bold.

How exactly this case illustrates a double prognosis is not clear. It may simply refer to the fact that two options are laid out, to say either that the patient will live or that the patient will die. Or the classification may be serving to set up the second kind of prognosis, for there indeed a kind of doublespeak is recommended. Now subtlety and thought are required. If doctors always knew exactly what would happen, they would simply say so without further ado. But for this second kind of prognosis, the doctor is looking at something not seen before: "If it is not in your experience, then name it as if with the tongue of a snake. Ride two and then whichever is good, flee on that one."[140]

With this short verse, the root text provides two metaphors about doublespeak, the proverbial serpent's forked tongue and what would be a technique familiar to people in equestrian societies: riding two horses at once. Again, *Small Myriad* explains the technique:

> If you don't know whether the patient will live or die, then like a snake's tongue, you need to name it while holding two. To the inside people say something like she's going to recover. To the outer people say that there is great danger. If she does recover, then it will transpire that the outer people will say, "At the time the doctor himself was alarmed, but now since she has recovered, his skill is clear." And the inner people will all say, "Since last year he promised that she would recover. His skill is clear."
>
> But if on the other hand the patient dies, all the outer people will say, "Since last year he's been saying there is danger. He probably didn't dare say it to you inner people." The inner people will say, "The doctor at the time said to the inner people that she would recover, but this was to console us. To all the outer people he gave a straight diagnosis." So they will think. There will be no bad ridiculing of the doctor, that he didn't know.[141]

There are actually several doubles in play. Uncertain about the outcome, the doctor is advised to say two different things to two different audiences: the patient and her attendants, and other people in the community. Then two possible outcomes are considered: the patient either survives or dies. In both cases, the two groups will interpret whichever of the physician's original two prognoses they heard in different ways. No matter what the outcome, the physician ends up looking good.

Better than the snake's forked tongue, the metaphor of riding two horses at once is most apropos. One rides two horses until one determines which is better, and then continues on that one alone. The doctor has said two things,

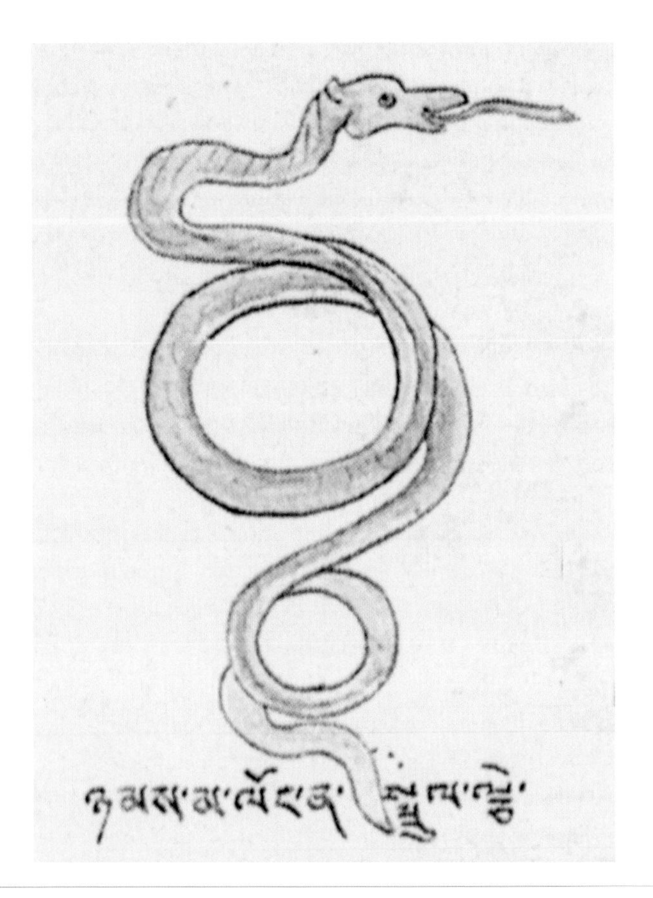

7.5 If it does not match your experience, speak with the forked tongue of a snake. *Plate 37, detail*

that the patient will live and that she will die. But the meaning of both utterances will go with the outcome that turns out to be true. Amazingly, the two audiences both make sense of what he said in accordance with that outcome, even though in each case, one of them heard the opposite of what actually came to pass.

In the context of an inquiry into ethics, it is interesting how intersubjective inference, coupled with a basic disposition toward trust, works to make the strategy possible. If the patient recovers, the outer people understand that the doctor's original warning showed that the danger was real and that the recovery was only due to his exceptional skill; meanwhile, the inner people are impressed that he knew well what would happen. In any event, the worry is not very great if the patient survives. The problem is if she dies. Then both the outer and inner people will understand that the physician told an untruth in order to preserve the patient's feelings and optimism. There is also the expectation that the two groups will compare notes.

Social assumptions and processes are a key part of the doctor's strategy. They function to make his sentences always name the truth, no matter what their semantic content. Even when what he says is false, people understand it as pointing to a true concern for the patient, which is at the heart of being a good doctor.

But even more than the reception of his words, what really drives the entire strategy is what actually happens. That is truly the unknown factor, but when it does become known, eventually, all of the talking and thinking have to accommodate it. No matter what is said in the double prognosis, its ultimate meaning is determined by what in fact ends up happening.

TRIPLESPEAK

I think the idea of a triple prognosis actually thematizes this third element, the reality of what actually happens. That outcome appears to be counted as a factor on its own in the triple prognosis section. In other words, while the double prognosis allows the doctor to say two things, one of which will be correct in the end, the triple prognosis seems to triangulate what the patient wants to hear about the illness, what the doctor says about it, and what the illness actually is.

The first example of a triple prognosis is a more straight-ahead situation in which both the doctor's suspicion of what the disease is and the patient's suspicion are the same. If that is the case, the doctor can simply state the prognosis. The commentary adds that due to their accord, the patient will have confidence in the doctor. The doctor can call the illness as it is because there is agreement with both the patient's idea and the reality of the situation. The second example, wherein there is no such accord and the doctor will say something different from what is actually the case, helps us see that there are three factors in play even in the first example: what the patient thinks, what the doctor says, and what actually is.

Here is the second case. The *Four Treatises* says,

> With the patient suspecting it is poison
> it is possible for you to say, "No one has come up with anything else."
> Thereby name it in a way that makes your assessments accord.
> [But] give treatment on the basis of your identification of
> what the illness is.[142]

The text is recommending a skillful way to say something ambiguous without telling an outright lie. As before, there is concern about accord between doctor and patient. In this case, the goal is to make the patient think you are agreeing with his assessment of his condition. The commentator adds that this means the patient will trust the doctor:

> The patient says, "My illness here really is poison" and has come to a unilateral assessment. The physician understands it to be something like *smug po*,[143] but if he says, "This is *smug po*," he won't be coming up with something that confirms [the patient's] conception. With [the patient thinking], "I really have poison," if the physician says that it should be identified differently, then it will come to pass that [the patient] doesn't trust [the physician]. Therefore, say, "Yes, it seems like *smug po,* but it is really poison, just as you suspect." By saying that, you will accord with his assessment and he will believe in you. But as for your method of treatment—medicine, examination, food, behavior, all that—treat in accordance with what the illness actually is.[144]

Once again skillful speech is required, and once again great care is taken with what people think and whom they trust. The doctor's ambiguous statement serves to form a social bond and generate trust. But a third level of reality has come into high relief, beyond whatever the doctor or the patient says. Strategic use of language keeps the patient going with the attitude the doctor wants him to have, and mollifies him sufficiently to allow the doctor to turn to what he needs to do: heal the patient. He will attend to that guided by his own true—if private—assessment of what the condition really is. That knowledge, and that condition, are in the physician's back pocket all the while, and will determine the treatment regime.

The passage gives one more example of a threefold prognosis, called "running to the fortress."[145] *Small Myriad* never quite decodes this metaphor, but its larger import is clear. We already broached this passage above in the context of the root text's point that many forces affect an illness. Now there is more to say about how the commentary explains the specific scenario imagined here.

> Due to the flourishing and/or decline of the merit of both the patient and the doctor, and through the power of common conditions either coming together or not coming together, even if the danger is great and you understand she will die, it is possible that she will survive. Similarly, even if the danger is small and you understand she will survive, it is possible that she will die. Therefore, you may have your own assessment, but it is not appropriate to say it. No matter how great the danger and how much you see death coming, say, "There is actually a way to treat this. I am changing your treatment. Until man, horse, or dog dies, they are to be treated. You must not die." In this way console her and treat her.
>
> Also, no matter how small the danger and even if she seems to be surviving, there is the danger of the error of thinking that if I let it be, nothing will happen. A little bit of laziness can make for a full fathom of a lawsuit, so study medicine and behavior without being cocksure.[146]

As in the previous example, the doctor is unable to pronounce a clear prognosis, but now it is because he himself is not sure. The salient implication is that the doctor is not to be blamed for this failure. The problem is really the situation itself: many factors contribute to it, and no one can quite tell what will happen. It is inscrutable. Still, the doctor is given similar advice, to the extent that he is told to encourage the patient and to appear optimistic, although he is not sure if optimism is merited.

In fact, *Small Myriad* suggests that the physician sees that death is likely. This becomes key to discerning the tripleness of the prognosis. Let us speculate that the patient thinks he is dying, the physician thinks he is dying, but no one is sure. The passage tells us that the doctor should work very hard and not give up until the patient actually dies. This work ethic is enshrined in an aphorism. Apparently dogs and horses are entitled to unflagging effort too.

The third factor that the prognosis is taking into account seems to be the actual situation, even though the doctor does not know what that is. The emphasis is upon doing everything one can anyway. There is no magic bullet. The attitude is realistic and humble. It is also compassionate, for it seeks to protect the patient's feelings. Perhaps the fortress that one runs to is just this realistic, humble, and compassionate bottom line: no matter what, the doctor will just keep trying. Such an idea is reflected in the way the *Four Treatises* glosses the fortress metaphor: it is the kind of prognosis that one gives when the door to samsaric existence cannot be blocked.[147] In other words, the situation looks bad, and no possibilities can be ruled out.[148] Although the text does not say so, perhaps the idea is that when the patient actually does die, adhering to this bottom line, i.e., the safe haven of the fortress, means that everyone will realize that at least the doctor did everything he could; at least he cannot be charged with laziness.

The text also gestures to the possibility of a lawsuit if its advice is not followed. What such a lawsuit would entail in this time period is unclear, but it is evidently something to worry about. That would be another eventuality looming out there that the doctor must consider in making prognoses and treating patients. Not only must he work with the feelings and conceptions of both the patient and the larger community, he must also face up to the possibility of legal action. That serves as a checking device to keep him honest and working hard. Yet again, the physician is shown to be subject to external forces that he is forced to reckon with.

BEING HUMAN AND THE HORIZON OF DEATH

What stands powerfully on the horizon for all of these prognosis scenarios is the survival or death of the patient, the limit case that will determine once and for all the accuracy of the doctor's prognosis. That is the final reality with which the doctor must reckon. His livelihood depends on his skill with respect to it.

The *Four Treatises*' medical ethics, I suggest, is most fundamentally about managing realities and forces outside the physician's individual power. These include not only the ultimacy of death but also the multiple causes of illness, the recalcitrance of bodies, the scarcity of materials and implements, and basic social facts like standing in the community, common customs, and expectations. Such realities are what the text is referencing when it admonishes the doctor to practice "in accord with the way of humans, that of the horrible world." The sense we get from reading the physician's chapter is of a concern for the many difficult things in the world that can be neither controlled nor denied. I am calling that sense a limiting horizon: it both defines the domain in which the physician must function and marks the fact that there are forces at the edge of that domain beyond his ken.

That the physician's human dharma must take account of hard realities has everything to do with the very nature of medicine. While the *Four Treatises* recognizes many forces that contribute to illness, including karma, luck, behavior, state of mind, and even demons, it is most concerned with imbalances of the humors in the body and physical injuries. The preponderance of its therapies proceed through material means: the administering of pills and concoctions and the physical therapies of moxibustion, bloodletting, bodily manipulations, purgatives, and restorative baths. Even though a few passages in the *Four Treatises* mention ritual therapies, the chapter on the physician does not take them up in its detailed consideration of the professional tools or skills to which he can take recourse.[149] While his deployment of visualization techniques to imagine himself as the Medicine Buddha when he prepares medical compounds serves to boost his self-conception and his commitment to the values of medical practice, the chapter does not instruct the physician how to become adept at healing rites or magic. Nor does it take comfort in their possible efficacy; in fact, as we saw above, it betrays some concern about such practices. More than anything else, the physician's chapter points to the ultimate inability of all treatments—including ritual and even the power of the Buddha—to certainly prevent death. *Small Myriad* advises the physician to warn his terminal patient's family, "Even if the Buddha Bhaiṣajyaguru were to look down on him, it is definite he would not survive."[150]

Paramount concern with the realities of the material world is a central feature of what I have posited as the medical mentality of Sowa Rikpa. But concerns about all of the realities with which the physician must work, not just those that are strictly physical, inform the *Four Treatises*' medical ethics. Even things like the style of the teacher or the expectations of patients require long experience, bodily familiarity, and most of all artistry—just as working with veins and orifices and bones and needles and plants and concoctions demands. None can be willed to suit the physician's desires. All demand rigorous attention, ingenuity, flexibility, and personal, embodied engagement. And yet none of these qualities or implements guarantees success. Neither will whatever moral or spiritual perfection the physician may have achieved.

The complex task of the physician draws on a wide range of capacities and types of knowledge. We have seen that Buddhist resources, from meditation to structures of knowledge to practices of learning, contribute to the cultivation of the medical practitioner's virtues. But the *Four Treatises*' understanding of medical ethics is not reducible to those resources, nor does their appropriation for medical purposes necessarily bring along the same spiritual or salvific horizons toward which they might reach in other contexts. Medicine's immersion in material and everyday social realities necessarily affects the terms and scope of what is possible. While the ascetic hermit in his cave can visualize all of his accoutrements and even his offerings to the deities, a physician with nothing but his imagination simply cannot practice medicine.

This signature immersion in material and social realities is all too evident in the insistence that the human way physician must achieve fame and promote his own standing in his community. If he were to eschew all social relations and stay humbly by himself mixing medicines and studying texts, never exhibiting his knowledge or skill, he would likely never be called upon to practice medicine. The text does add a more optimal True Dharma version of the physician's behavior in which he does not promote himself, in this reflecting a long-standing cultural valorization of circumspection. Indeed, even in the most self-assertive cases of Tibetan autobiography we can note a certain hesitation and apology—sometimes amounting to false modesty, at other times a real concern that one might not actually be entitled to a "lion's roar," but always evincing a sense of the virtue of humility. Yet it is also worth remembering what Richard Gombrich cannily noted about monks and their lay donors in Theravāda Buddhist societies: "Indifference to comforts thus causes them to be provided."[151] The expectation that modesty and austerity will bring social and material rewards is evidenced repeatedly in the Tibetan world as well, where high regard is accorded to the indigent virtuoso.[152] But the *Four Treatises*' physician's chapter is concerned nonetheless to distinguish that economy from the everyday practice of medicine, and worried about too much "small mind," reminding the human way doctor not to trust that humility will bring the requisite patronage and material fortune on its own. This line of thought also tells us something about the chapter's vision of the True Dharma doctor. In acknowledging the possibility that the exceptional physician might indeed trust that virtue will bring him patients, even without self-promotion, the chapter still does not conceive of the True Dharma doctor as a renunciate ascetic. He may be modest, but he is still dependent upon social processes, and by force a creature of interpersonal relations. Even the True Way doctor hones his craft in the clinic, in the thick world of bodies, substances, and human demands—not in his imagination, however lofty or detailed.

For me, the medical mentality's ineluctably social and material center of gravity becomes most evident in the way that death figures in the *Four Treatises*' physician's chapter. Death delivers a double whammy to the career of a physician. When the patient dies, the doctor has a public relations problem on his hands. His reputation is at stake. And yet there is nothing he can do to ameliorate the situation. Hence his care to hedge his bets in what he says before this ultimate disaster takes place. Death is notably an all-or-nothing situation. No talk of the patient's next life enters into the doctor's consideration, nor are there even hints in the physician's chapter of the patient's after-death experience, or miraculous reappearance, or transcendent signs from beyond the grave, or imminent reincarnation, as are all so regularly seen in Tibetan religious narratives of death.

Death in the medical context figures as irreversible. And much as the physician might like to prevent the discovery of the patient's death by saddened or angry relatives, he cannot do that either. Short of creating some elaborate ruse, the physician's failure to save the patient cannot be concealed. That is because death is empirically evident in a patently definitive way. Debates about when exactly the moment of death occurs notwithstanding, no one can sustain an argument that a patient is still alive when everyone can see she is dead. The physician's chapter gives the doctor no resources to maintain that the patient's passing is a matter of interpretation, or an illusion, to use the Buddhist parlance.

The place of death in the practice of medicine is one grounds upon which I would draw a critical distinction between the horizons of the human way of medicine and the soteriological aspirations of religion. This is not to say that the paths of cultivation that feed into Buddhist notions of enlightenment are not themselves oriented in foundational ways around the inevitability of death and the virtue of accepting impermanence.[153] But the distinctive undeniability of death in the clinic makes for a telling difference in how the accomplishment of ultimate goals is determined. If the ultimate goal of the physician is to keep his patients alive and well and the ultimate goal for the Buddhist virtuoso is some sort of liberation, there is a critical difference in the way their achievement is ascertained. No one (save perhaps an omniscient buddha) can assess for sure the degree of spiritual realization in another person. In fact, what the exalted achievement of enlightenment means and who is judged to have achieved it have changed throughout Buddhist history and are in any case always subject to local and socially contingent processes of evaluation.[154] But there is no question if a doctor's patient has died.

I submit that the basic human dharma principle defined in the physician's chapter means that there is always something that will forever remain outside the doctor's control. When and how it will come into play is unknown, but its eventual intrusion is certain. There is no other recourse, no higher truth by virtue of which the physician could transcend the material and social facts of the way of humans, despite his best efforts at verbal ambiguity. It all makes the doctor anxious, cautious, and careful—just as the *Four Treatises*' instructions on human dharma want him to be. Conceptual and practical disciplines from across the spectrum of his cultural heritage give him rich resources to manage his practice and his efforts to ameliorate human suffering. But the particular circumstances of medicine render him subject, at the most defining and fundamental level, to an empirically discernible standard that has a powerful logic of its own. This standard makes for a type of accountability that marks a basic difference between matter and the life of the mind, and, I dare say, between science and religion. It also marked the character of academic Tibetan medicine, into its heyday under the Ganden Podrang state.

CONCLUSION

WAYS AND MEANS FOR MEDICINE

Medical knowledge in Tibet faced at least two challenges, both having to do with ideals. One is very basic, and shared across the history of medicine: the incongruity between the need to catalogue information and the need to heal individuals. Standard, generic rubrics are essential to structure knowledge. But their very ideality can stand in the way when the physician is trying to pinpoint the unique combination of factors in a particular pathological condition. It is the problem of the discrepancy between rational system and nature.

A second challenge has to do with a proclivity to favor ideal bodies—now in the sense of optimal or perfected—over ordinary ones. This is the problem of how the perfect and divine relate to the human body with which medicine deals. It too obtains far beyond Tibet. Jean-Pierre Vernant puts the conundrum well for the ancient Greeks: how to think of the human and the divine under "the double figures of the same and the other," especially given "all the signs that mark the human body with the seal of limitation, deficiency, incompleteness"?[1]

These issues were inflected in particular ways in the Tibetan cultural sphere. Old conceptions, both local and spreading onto the plateau from West, South, and East Asia, contributed to the complex picture of human vitality and embodiment found in the classic medical formulation by Yutok Yontan Gonpo. The root text's diagnostics, pharmacologies, and therapies were also received from across Asia, while its theoretical structures were indebted primarily to Indian medicine. As for the resources to challenge and change medical knowledge, the primary deliberative model available was Buddhist intellectual practice. But that massive presence in the Tibetan world posed problems of its own with regard to both of the challenges just noted. Indeed, the entire edifice of classical Buddhist scholasticism was taken up with enumerating its own standardized elements of mind and body, foundational building blocks that early on had achieved canonical status. Although debated and ultimately deconstructed, these left an indelible mark on habits of thought. More foundational yet to Buddhist discourse was the wide-ranging invective against the impurities of the ordinary human body, the folly of our attachment to it, and its incommensurabilities with the truer, far more spiritualized embodiment that constitutes buddhahood. This is not to say that Buddhist literature had nothing to say about the ordinary body. Buddhist sūtras and their commentaries adopted Ayurvedic anatomies for exercises to develop mindfulness of the body, and Abhidharma had long described the bodily bases for sense perception and the elements of the material world. Buddhist traditions of self-cultivation also used multiple devices—in intellectual, ritual, and somatic spheres alike—to mediate between quotidian human existence and its enlightened counterpart. But with few exceptions over the history of Buddhist discourse, the emphasis of valuation remained on the ideal part of the equation, and the emphasis of scholastic attention on the cognitive and the spiritual.

Yet the material realities and deficiencies of the everyday human body are what the physician has to know, in all their idiosyncratic detail, in order to heal patients effectively, not to mention to survive professionally in a competitive environment. Like physicians everywhere, Tibetan doctors handled those realities as best as they could in the clinic. But as soon as they began to write anything to expand on the *Four Treatises*' statements, to reconcile the occasional discord between medical notions adopted from East or West Asian medicine with Ayurvedic ones, or especially to incorporate new empirical data or therapies or shift existing conceptions, they ran into the legacy of two well-codified knowledge systems, Ayurveda and Indian Buddhism, inherited and translated from Indian literary culture. And that is not to mention the textual authority with which both the Tibetan translation of the *Aṣṭāṅgahṛdaya* and all the scriptural resources that the medical commentators marshaled from the Buddhist canon and its exegesis were imbued.

How medicine handled ideal system and authoritative tradition became exponentially more fraught in the seventeenth century, with its close association with the newly centralized Tibetan Buddhist state. But the implications of the alternate episteme suggested by medicine's emphasis on the ordinary body and ordinary empirical realities seems to have been sensed for centuries. The intimate imbrication of Buddhism in the trappings of political power ever since its introduction into Tibet thus stood as a third challenge facing the development of medical knowledge. Certainly by the onset of the commentarial disputes followed in this book, the political capital of Buddhist symbols was central to the careers of the Sakya and Karmapa hierarchs under whose patronage medical scholarship flourished. Much the same could be said of the Changpa and Rinpungpa lords for whom leading medical writers served as court physicians. Although the parameters of medicine's own political import prior to the seventeenth century remain to be tracked, if we are to judge from the way things came to a head at the time of the Ganden Phodrang, there was charged awareness throughout the history of Tibetan medicine of what its distinctive mentality might mean for a Buddhist hegemon. Unlike the Indian case, where medicine flourished as a separate field, even if sometimes studied in Buddhist monasteries, and also unlike anything that ever obtained in Buddhist China or Japan, the religious-political landscape of Yutok's day already was such that he was moved to frame his synthetic and on many counts original medical treatise as entirely the teaching of the Buddha.

We have seen that the *Four Treatises*' bid for a religious authority of its own did not solve all of the problems that medicine encountered as it evolved in Tibet, not least because the empirical gaze that medicine was fostering led both medical theorists and other critically minded members of the Tibetan intelligentsia to see the ruse. What then ensued was a host of occasions when outstanding medical thinkers tried to face down the issues of medicine and the evidence of the body and the material world on their own terms. They did so through the lens of classic Ayurvedic and other medical systems, sometimes with the help of a Buddhist intellectual or even yogic apparatus, and sometimes in creative ways independent of those systems and tailored instead to the local conditions of practicing medicine. The latter entailed honing their insights into legible formulations, knowing the limits of flexibility, and knowing as well the parameters of etiquette, credibility, and rhetorical impact.

In addressing the bodily and social realities of being human, medical knowledge in Tibet evolved in both content and its approach to the problem of representation. At an early formative moment, medical theorists could summarily change one of the classical eight branches of medicine to gynecology, simply out of a conviction that this would facilitate more specific and inclusive coverage. Other shifts were more subtle but sometimes even bolder, as commentators found ways to clarify, redefine, or even question the very status of hallowed categories in medical theory. Understanding categories to be but provisional heuristics that stand for a far more complex situation on the ground, medical thinkers came to recognize that opposing taxonomical pairs do not reference polar opposites in reality. Rather, they are better understood as markers along axes that can often be asymmetrical and that in any case admit a wide array of permutations. Medicine in both South and East Asia had long suggested that ideal types are far from iron-clad by enumerating many hybrid variations in the humors and the pulses and the sexes. In exploring a few commentarial exchanges in Tibet when such a point was underlined and its implications further unpacked, this book has only scratched the surface of how this important insight informed both medical theory and practice.

The commentators studied here were also exercised to question artificially normative standards for human functioning that the root medical text sometimes seemed to posit. They endeavored to correct certain assertions in order to account for human variation and were comfortable with approximation. The "tip of the heart" does not necessarily refer to the actual tip, but something just *near* the tip. Gender is like sexual identity, but not exactly. The number of bones in the body depends on how you count them. The amount of menstrual discharge depends on the individual. Unpredictability has an essential place in medical theory, and not only as an expedient to be invoked in order to protect the physician's reputation. More fundamentally, unpredictability marks the way the material world is for medicine: far too varied and complex to ever be pinned down with absolute certainty. And yet, given the enduring presence of system and structure in classical medical doctrine, it seems that such a point had to be reiterated and instated ever more minutely, particularly in the context of growing aspirations for commentarial acuity.

The unavailability of material realities to strict systematicity also pertains to conceptual or linguistic formulations themselves. Once realized, this principle also served the development of medical knowledge, especially given the delicacy of challenging authoritative textual doctrine. Stretching a canonized rubric can enable empirical accountability without having to jettison the classical statement. In one case, homophony between a set of verbs facilitates a certain slippage whereby an assertion about a "facing" tip could be read to say "looking to," or perhaps "turned to face," in order to remove the implication that facing implies leaning. In other cases, slippage in the meaning of a single lexeme allowed for important innovation or clarification. "Dharma" did double duty in the root text's physician's chapter to both join and distinguish two profiles of medical ethics. Zurkharwa blurs the distinction between the "connecting" function of the channels to link organs and vaguer kinds of linkages, in order to reconcile the problematic misfit of yogic anatomies with empirical ones. He trades in the ambiguity between the embryonic "vital channel" and the mature white and black "vital channels" to similar ends.

We have seen how clinical practice itself occasions linguistic flexibility. A critical skill for doctors to learn at the outset of their training is how to say things with double or even triple meanings, so as to manage the expectations of a patient, her family, and the larger community while still pursuing the medically prudent course of action. Perhaps the doctor's everyday encounters with actual fleshly bodies or botanical specimens, and the discrepancy between these and their ideal descriptors, served as a further reminder of the difference between rubrics and their referents. Perhaps this clinical experience enhanced commentarial agility as well.

One way that the commentators capitalized on the differing meanings of words in order to clarify medical doctrine was by pleading the variability of context. Zurkharwa vociferously chastises his Changpa predecessor for mixing up the negative associations of the third sex for monastic discipline with its symbolic import in pulse examination, thereby preserving the label's value for medical diagnostics. A considered distinction between the field of medicine and the tradition of Secret Mantra allowed the commentators to account for the absence of the latter in the former, even while taking liberties to introduce tantric elements when they could extend the reach of medicine. Or again, the notion of context helped dispel confusion about when the term "body" serves as a rubric of medical learning and when it merely refers to human bodies, thereby eliminating a conundrum that might have derailed the creation of a gynecology branch of medical learning.

Semantic fungibility enabled a resolution of thorny questions about the author of the *Four Treatises* and the place where it was first taught, when with stunning boldness Zurkharwa could subvert a common taxonomy to make the "secret" level of identity the most prosaic and human of all. This move presupposes audience expectation, albeit cynically, and participates in the pervasive medical preoccupation with reception. The same can be said of Zurkharwa's take on why the *Four Treatises* had to be cast as Buddha Word: without the sacred rubric, people would not take the text seriously.

In several important cases, the flexibility in what words and categories denote allows crucial distinctions to be made. When Zurkharwa insists that the root text's reference to the mountains around Danaduk is figurative and does not map onto the actual mountains with the same names, he forges a critical distance between metaphorical language and empirical fact. Again and again this is what the medical theorists endeavor to do, preserve a space apart for material facts—the evident channels in the body, places on planet Earth, what will happen to a patient, what treatment the physician will actually perform. Language is understood to be an expedient to accomplish any number of pedagogical aims, but there is also something behind the discourse that has its own robust reality, a kind of reality that receives little attention in Buddhist skillful means theory, where everything, ultimately, is empty. In fetching the leaf out of the field to check the accuracy of its taxonomical description, the Desi has his colleagues peer below the rubric to a decidedly physical and ordinary state of affairs that must be allowed to speak for itself. The same for the bones of Darmo Menrampa's boiled corpses, even if the way we count them can vary.

The disambiguation of categories, mentalities, and spaces had the virtue of allowing medicine to elaborate different kinds of knowledge—and to manage audience expectations and values—without necessarily having them infect each other. Genre conventions sometimes produce statements that can be respected on their own terms even if they don't point to historical realities. But domains that have been separated can also remain porous, and we have seen several important instances of having things both ways. Zurkharwa respects the tantric channels' nature and logic of operation, making sure they are not confused with the empirical channels of the body, but then allows the two systems somehow to relate. Medicine's human way of practicing medicine holds apart a space for the competitive exigencies of a professional life, but the root text also knows that all medical practice is founded on altruism and envisions a mixed approach whereby the superior True Way can also inform the doctor's mode of operation, at least some of the time.

The rapprochement in these last two cases worked differently, however. For one thing, the issue of the body's channels is more controversial, and there is a reluctance to really spell things out. The case of the tantric channels also illustrates another, and I think very important, kind of move that can follow on attempts to disambiguate. This involves jumping levels, something not seen in the case of the medical ethics. I have floated the term "postempirical" to name Zurkharwa's redeployment of tantric categories like the channels as potent devices to organize knowledge about aspects of embodiment that are not entirely available to empirical analysis—gender difference is a prime example—even as he notes that such categories are but figures of speech. Recall that the tantric channel problem was the issue around which the commentators were the most self-conscious, feeling compelled to acknowledge explicitly the liberties they were taking in intentionally juxtaposing otherwise disparate knowledge systems. Apparently once such an acknowledgement is in place and the logic of the juxtaposition is defended, the door is open for the learned physician to make use of culturally legible tropes to enhance and expand the range of medical understanding, as the channels did for both the heart tip question and the relationship between pulse and gender. Or in another case, once it could be demonstrated conclusively that the root medical text was not really composed by a buddha in a Pure Land and that the language used to describe the preaching scene is not literal, the commentator could point out that the distinction between mythological imagination and human reality is not so momentous after all. There is value in realizing that a historical person's insights, especially if properly motivated and cultivated, can be almost as good as a buddha's revelation.

Methodological self-consciousness is one sign that helps us distinguish between an uncritical repetition of past tradition and an intentional redeployment of such inheritance for new purposes. In the cases studied in this book, this line is not always easy to draw. But as medicine moved into the early modern period, its appropriation of Buddhist nomenclatures and practices became increasingly deliberate and its awareness of the stakes heightened. From the perspective of medicine, too much rapprochement threatened medicine's autonomy and attention to prosaic realities. Too little, on the other hand, might undermine medicine's standing and credibility in a world pervaded by Buddhism. From the perspective of religion, the availability of sacred states to bear upon quotidian realities could dilute their sacred status. And yet the same appropriability extended the reach of Buddhist knowledge systems into the lives of people. A good example is that the old Buddhist understanding of extrasensory perception could shed light on the diagnostic acuity of the physician, and yet with an accompanying caution, distinguish the medical variety as mundane. Buddhist scholastic tradition provided a number of categories and hermeneutical strategies that focused and refined medical knowledge and practice, intentionally and with caveats or not. This shows the extent to which these resources were available for purposes quite beyond the salvific domains of religion as such.

◎ ◎ ◎

This book has had the complex task of pointing both to disparities between the Tibetan medical mentality and religious soteriology and to the many registers on which Buddhist thought and practice positively affected the ways and means of Tibetan medicine. Indeed, despite the robustly materialist and worldly orientation of medicine, the impact of Buddhist scholarly and meditative traditions cannot be gainsaid. We see this in ingrained modes of discourse and interpretation; in yogic insights about the body; and in a toolbox of particular categories, strategies, and practices. The first is too enormous and fundamental to fully characterize, but it includes virtually everything about writing texts and commentaries and parsing categories and negotiating scriptural authority. It also includes fundamental dispositions in critical disputation. And yet in medicine, principles like the disassociability of linguistic labels from what they would name or the profound dependence of meaning upon context never point to an underlying metaphysical principle as they do in Buddhist thought. Certainly medicine never countenances a transcendence of language. When medicine invokes skillful means or notices the variability of linguistic reference, it is to serve a particular instrumental end. Pointing to the skillful means of the Buddha in teaching different kinds of medicine in different parts of Asia worked as a sop to allow a more accurate account of the history of medicine. In another context, it allowed certain apologists to *avoid* confronting historical evidence regarding the *Four Treatises*' Tibetan provenance. Realizing that sex and gender categories are relative designations does not amount to critiquing them as altogether illusory, but rather inspires ways to name their variability in order to better account for the shifting gender configurations of pulse and personality and even sexual organs of real, material bodies.

Chapters 4 through 6 noted several instances where medicine directly appropriated knowledge produced by tantric contemplative communities. One had to do with details on the sexual organs; another with matrices of channels that focused kinds of feelings and cognitive capacities in particular parts of the body; and another included fine specifications on the channels around the human heart, coming out of a long tradition of observing the interaction between mind and body in meditation. There was discomfort in certain medical circles when such details were not visibly identifiable in the clinic, but this material was too interesting or useful to reject. Future research may find that yogic insights on the bodily winds and their relation to breath and to health had significant repercussions for Tibetan medical theory too. I also expect that there is influence to discover with respect to the humors, refined sexual substances, and the substances underlying human vitality.

We have also seen a host of particular conceptual tools, practices, and sentiments from scholastic and/or soteriological Buddhism that lent themselves to medical adaptation. The Buddhist analytical model that posits all things to be constituted by multiple parts whose predominance shifts, making for shifts in character, proved useful in understanding the heterogeneous nature of gender and its impact on pulse. Buddhist taxonomical tools and categories such as the ubiquitous body/speech/mind distinction; the ten non/virtuous deeds; the differences among view, contemplation, and behavior; or the long interest in the nature of conditionality and habit helped articulate fine-grained aspects of the doctor's education. In the last case, even if habit and the conditionality of karma work against salvation from the Buddhist perspective, they are still eminently helpful, if not critical, to becoming a doctor. In some places the translation shows only superficial adjustment, such as the straight-ahead appropriation of Buddhist visualization techniques to enhance the physician's self-conception and sense of his medical calling; another example would be the Buddhist pedagogical category of "key instruction" that can also name the specially tailored guidance that a teacher of medicine can give a close student. But in other cases the concept or practice morphed significantly. The old Abhidharmic notion of "mental continuum" was put to an altogether new use in undercutting a perceived essentialism in medical theories of pulse and sexual identity. The moral category of karma was used to account for human variation really having nothing to do with moral issues, such as differences in menstrual cycles (although in other medical contexts karma's invocation directly mirrored Buddhist misogyny). The teacher-student relationship in medical education was deeply informed by the enormous culture of guru devotion in Tibetan Buddhism, but its moral compass was oriented differently. And Buddhist distinctions between types of love and compassion helped elaborate the concern to heal and to help in medicine—as it seems they did in the Ayurvedic classics as well—even while building the requisite reputation to actually heal and help was deemed to entail self-interested contrivance.[2]

Distinguishing between Buddhism as a civilizational force and Buddhism as a religion might help us better construe what it is tempting to characterize as Buddhist elements in Tibetan medicine. Charles Hallisey and Frank Reynolds understand the civilizational force of Buddhism to have issued out of the close relations between royal courts and elite, internationally mobile clerics who, we might add, were instrumental in the spread of international forms of medicine in medieval Asia.[3] On a smaller scale in Tibet, the propinquity of Buddhist personages and political power facilitated the naturalization and wide incorporation of Buddhist symbols, concepts, and practices, such that these could stretch well beyond themselves and inflect metaphors and templates for many things in the lives of people apart from the central activities and concerns of religion. An outstanding example would be the formations of governance and prestige in the Ganden Phodrang bureaucracy. The civilizational dimensions of Buddhism are particularly evident in modernity, especially in the increasing laicization of practices (the transformation of *vispassanā* meditation into an exercise to reduce stress is a good case in point) and the adaptation of Buddhist principles to respond to colonialism in Asia.[4] Medicine provides one suggestive case of a related pattern in early modern Tibet. We would certainly expect to see such a phenomenon elsewhere in the history of religion and science too: the adaptation of the Vedic initiation of the twice-born for the Ayurvedic initiation of the medical student is one such instance.[5]

Attending to the civilizational dimensions of Buddhism helps use see how medicine was influenced by Buddhist values and ways of knowing, and under what circumstances it adapted Buddhist resources for new purposes in disparate domains. But this book has also shown the multiple ways medicine worked to distinguish itself from Buddhist values and ways of knowing, sometimes explicitly and sometimes relatively silently, evident only by dint of our own efforts to pick out differing mentalities and epistemic grounds. In the end we have to follow the lead of commentators like Kyempa and Zurkharwa, and indeed the root text itself, in recognizing that Sowa Rikpa was an intentionally heterogeneous endeavor. It could relativize virtually any of its constituent components, from its imputed revelatory origins to its most particular diagnostic precepts, depending on the exegetical or cultural or even political agenda at hand. It was primarily concerned with pragmatics: what works best to heal human illness. It drew variously upon several medical traditions, ongoing adjustments coming out of clinical practice, and an array of goods from Buddhism, not to mention another array of local conceptions of human life, and by the seventeenth century, the cachet of state-level patronage to boot.

How that mix continued to percolate in the centuries following the Desi's rule will become clearer as we identify and examine other relevant documents. Whether interpretive innovation reached its climax in the work of the Desi and Darmo Menrampa and colleagues or there were new insights and directions in the Qing period remains to be tracked.[6] We can expect that medical theory had new cultural and political implications in the complicated circumstances after the Desi's execution, not only in the capital but also in medical schools and other contexts across the Tibetan plateau. It may also be the case that once medicine achieved state patronage under the Ganden Podrang, its distinctive mentalities and discursive practices—along with those of other evolving pragmatic traditions, such as administrative record-keeping and geography—contributed more broadly to the fostering of new kinds of social mores, autobiographical self-reflection, registers of national self-conception, and even new uses of the visual arts. Hopefully the picture here proposed of medicine's ways of knowing in its Tibetan heyday will help us appreciate the full range of means by which it could enhance human flourishing in a Buddhist world—and elsewhere too.

NOTES

INTRODUCTION

1. The parameters of the "early modern" are variously understood by historians. For Tibet the category has barely been invoked, but for our purposes can roughly correspond to the development of Tibetan self-consciousness of its political and cultural position vis-à-vis other powers in the region. On the category's purchase on South Asian history see Subrahmanyam 1997 and 1998; Kaviraj 2005a and 2005b.
2. On tantra's hoary interpretive challenges, see Wedemeyer 2013; Gray 2007; and Dalton 2011. On the spiritual sides of embodiment see White 1996 and 2003. R. P. Das 2003 considers some of the impact of tantric ideas on Indian medicine.
3. Cullen 2001. The Tibetan *nyams yig* is not a case study but does reflect individual physicians' experiences: J. Gyatso 2004.
4. On Tibetan healing rituals see Schrempf 2007, part 3; Adams, Schrempf, and Craig 2011; and Garrett 2009. On medicine in Bön, see Millard and Samuel forthcoming.
5. I do retain the common phrase "Medicine Buddha" to refer to Bhaiṣajyaguru, who mostly had to do with ritual healing.
6. For a preliminary survey of the primary "external" therapies and procedures of Sowa Rikpa, based primarily on the *Instructional Treatise,* which has yet to be translated into a Western language but contains the large majority of the practices in the *Four Treatises,* see Yonten 2014. And that is not to mention the many "internal" therapies based in pharmacological remedies.
7. See chapter 2, nn. 95 and 102, and pp. 361–62. Even the chapters on demon-produced illness are not dominated by ritual healing methods and primarily recommend material therapies.

8. Garrett 2008, 55 argues that medical and ritual therapies are not easily distinguished, citing ritual therapies to be found in several texts in the *Eighteen Pieces from Yutok* (*G.yu thog cha lag bco brgyad* 1976). The textual histories of that complex collection and the *Heart Sphere of Yutok* (*G.yu thog sñiṅ thig* 1981) remain to be worked out. On the former, see p. 119; on the latter see Garrett 2009.
9. J. Gyatso 1998. Even the most prosaic passages of Tibetan autobiography are usually framed in larger presumptions about making merit, or advancing on the path toward enlightenment, or the flourishing of religious institutions. But there may be exceptions in the autobiographical writings of lay aristocrats in the last two centuries.
10. It would be interesting to compare the intellectual issues of Tibetan writings on astrology, law, and grammar with those of medicine. Cf. n. 19 and chapter 2, n. 57.
11. I discuss the term "mentality" on p. 91.
12. Medicine's retreat from boldness after the late seventeenth century in some ways parallels what Pollock (1998, 2001b, 2005, 2011) has found in South Asian Sanskrit culture in the same period. But the reasons modern science developed first in Europe and not elsewhere remain elusive: see, e.g., Pollock 2005, 84 seq.
13. For the larger issues of the following paragraphs, useful sources include Toulmin 1990; Jameson 2002, 6 ff; Israel 2001; Pocock 1999–2005; and Shapin 1996. For the relevance of these issues to South Asia see Subrahmanyam 1997 and 1998; Kaviraj 2005a and 2005b.
14. Schuh 1973c, 20 and 1974.
15. See Elman 2005.
16. *Rgya dkar gyi mu stegs pa'i rtsis pa.* Tuttle 2006, 83–84. Tuttle 2006, 6 points out that the Dalai Lama might only have known of Jesuits coming into Asia through Goa.
17. On missionaries in Tibet see Petech 1944–46. Cf. Tucci 1949, 78.
18. Pomplun 2010.
19. A Tibetan translation of a Mongolian translation of the *Shixianli,* a Chinese translation of a European work on astronomy in the early eighteenth century, is studied by Huang and Chen 1987. Thanks to Sokhyo Jo for bringing this study to my attention. On eighteenth- and nineteenth-century Tibetan knowledge of the world and European influence on the latter, see Aris 1994 and 1995; Kapstein 2011; Wylie 1962 and 1970; Yongdan 2011; and Tuttle 2011.
20. Eisenstadt 2000; Gaonkar 2001, 1; Kaviraj 2005b. Cf. Chakrabarty 2000. Dussel 2006 argues for a "trans-modernity" in many parts of the world. See also Subrahmanyam 1997, 737, 745, et passim. For the relevance of modernity theory to Tibet since the eighteenth century see J. Gyatso 2011b.
21. Subrahmanyam 1997 and 1998.
22. Pollock 1998, 2001b, 2005, and 2011; Kaviraj 2005a; Dominik Wujastyk 2005; Bronner 2002. See also Bronkhorst 2006 and Pingree 1996.
23. Goble 2011.
24. Elman 2005, 25, 227.
25. Ali 2004, concluding chapter. The terms "indigenous modernities" and "practices that modernized" are from Kaviraj 2005a, 131 and 138–39; see also Kaviraj 2005b.

26. See, e.g., Adams and Li 2008; Adams 2007; Adams 2001; and Craig 2012.
27. On the varying agendas behind such definitions see Almond 1998; Masuzawa 2005; and Abeysekara 2002.
28. On the former, see Eberhardt 2006, 8–11.
29. See p. 156.
30. Garrett 2008, chapter 1. My difference in emphasis from Garrett may have a lot to do with the fact that embryology is one of the topics least concerned with physical realities in all of Tibetan medicine, and one of the most indebted to Buddhist narratives.
31. For a critical reading see Lopez 2008. On early twentieth-century apologists see McMahon 2008; Braun 2013; and Pittman 2001. Examples of the more recent Buddhist encounter with neuroscience and physics include B. A. Wallace 2003 and 2007. On Buddhism as science or religion, see Garrett 2008, chapter 1.
32. "If science proves some belief of Buddhism wrong, then Buddhism will have to change." T. Gyatso 2005a and 2005b, 3; cf. Lopez 2008, 135. See also pp. 197–98.
33. Dreyfus 2003, chapters 13 and 14.
34. In this I am taking issue with Garrett 2008, 4, although I fully embrace a "contextualist" approach in what follows.

1. READING PAINTINGS, PAINTING THE MEDICAL, MEDICALIZING THE STATE

1. The following refers to the set reproduced in Parfionovitch 1992. I will also refer to plate numbers according to that version, even though the original set of the Desi included two other plates, numbered 6 and 7, that are absent in the Ulan Ude set. Specific images will be referred to by plate and image number according to the system used in Parfionovitch vol. 2. See also nn. 31 and 51.
2. Haskell 1993; Barthes 1977, 32–51.
3. In addition to Barthes 1977, Schapiro 1996 has been suggestive for some of the issues in this chapter.
4. See Pollock 2006, especially the analysis of modifications in rhetorical style in *praśastis* at chapter 3.2.
5. See Berger 2003.
6. Elias 1994, 168–78.
7. Wang le and Byams pa 'phrin las 2004, a reproduction that may include parts of the original set, has been consulted for comparison for all the images discussed in what follows. See also nn. 10, 45, and 136.
8. The captions are translated in Parfionovitch 1992, vol. 2. They are also rendered in Wang le and Byams pa 'phrin las 2004 as well as Williamson and Young 2009.
9. Cf. [G.yu thog] 1992, 57 and Sde srid 1973a, vol. 1, 349 seq.
10. One precedent for the snow lion's impudence is found at Dratang, where the two traditional lions holding up the throne are irreverently scratching their necks. Reproduced in Finnegan 2007. Note that the yak in the analogous plate in Wang le and Byams pa 'phrin las 2004 (plate 23) is not licking his mate, but there are many other

ways the animals on this and other plates in that version exceed their taxonomical specifications idiosyncratically, such as by stretching their necks up, sticking out their tongues, making calls, sleeping, and so on. I do not have permission to reproduce the images in Wang le and Byams pa 'phrin las 2004; in any event, we do not know their age or original provenance. In general the plates in Wang le and Byams pa 'phrin las 2004 and Parfionovitch 1992 are identical, but in each case the artists take liberties with small details on occasion.

11. [G.yu thog] 1992, 54–55.
12. Cf. James Wood's point that the addition of extraneous detail itself conveys the message that there is a realist impulse at work. Wood 2005, commenting on Barthes 1989.
13. Teiser 2006.
14. One more case is to be found in the murals on the north wall of the Lukhang in Lhasa, painted in the early eighteenth century. They appear to illustrate the source of life, along with an embryology according to a tantric tradition. See Baker 2000, plates after 64. Thanks to Michael Sheehy for drawing my attention to it.
15. Cf. [G.yu thog] 1992, 92; Sde srid 1973a, vol. 1, 601. "Newlyweds" is expressed as *khyim gsar,* lit., "those with a new house," hence the image of the couple on top of a house.
16. The special sexual methods described in Buddhist tantric sources are not relevant to the everyday sex decried in much Buddhist literature or depicted in the medical paintings.
17. [G.yu thog] 1992, 53; Zur mkhar ba 1989, vol. 1, 353–54.
18. This plate was missing in the Ulan Ude set and was copied only recently from the plates kept in Lhasa: Parfionovitch 1992, vol. 1, 8. The images illustrate [G.yu thog] 1992, 567, which only mentions sex (*nyal po*); Sde srid 1973a, vol. 4, 85 elaborates by glossing *nyal po* as "engaging in sex together with a female, or otherwise the types [of activities] that become [the occasion for] the depletion of vital fluids."
19. See J. Gyatso 2014 for a fuller examination of the overtly Buddhist images in the set.
20. Ruch 1990, 512.
21. [G.yu thog] 1992, 85. Regarding the teachers on whom the patient has turned his back, *Four Treatises* only mentions the "spiritual friend" (*dge bshes*), but Sde srid 1973a, vol. 1, 581, divides that into two types, but does not name them. The image illustrates the two types as a monk and a tantric practitioner; once again, the illustrations elaborate on the written text.
22. Cf. p. 348.
23. From a section of *Blue Beryl* that supplements the *Four Treatises:* Sde srid 1973a, vol. 1, 536.
24. Indeed, the group of seven disparate individuals immediately preceding this one has no monk in it.
25. A recent biographical sketch is Byams pa 'phrin las 1996b. See also Skal bzang legs bshad 1994, 213–27. For lists of the Desi's writings see Lange 1976 and Schaeffer 2006. For studies of his life, see chapter 2, n. 54.
26. On the production of art during the period, see Lo Bue 2003. See also n. 28.
27. At the same time that he began to write the *Blue Beryl.* Sde srid 1973a, vol. 4, 492 specifies it was the twenty-sixth day of the *sa ga* month, when he was 35.

28. See pp. 42–44. On the Desi's patronage of artists, see Jackson 1996, 206–14.
29. 'Byams pa 'phrin las 1996a, 372 and 1996, 424 both cite Sde srid 1973b, f.281.5 for a reference to a set of 50 medical *tangkas*, but I do not see such a statement on that folio, despite the Desi's mention there of his production of the medical paintings: Sde srid 1973b, vol. 1, f. 281a.6. In Sde srid 1973a, vol. 4, 494 the Desi mentions 60 completed plates. In Sde srid n.d.a., f. 203b he reports offering a set of 62 medical paintings to the Sixth Dalai Lama when he ascended the throne, which would have been in 1697. The Desi lists the contents of 79 plates in Sde srid 1982, 386–94. Wang le and Byams pa 'phrin las 1986, 2 give 1703 as the date when the original set was completed. For a history of the set, see Byams pa 'phrin las 1996a and Meyer 1992, 5–8.
30. It is possible that scholars at Mentsikhang or the Tibetan Medical College in Lhasa do know. 'Byams pa 'phrin las 1996b, 426–29, transcribes the verses inscribed on the back of the main painting; 'Byams pa 'phrin las 1997, 374 adds that the main painting with the lineage of lamas has a seal on it. See also nn. 41 and 42.
31. The set lacks plates 6 and 7 in the other sets. These are the channel and moxibustion illustrations prepared by Lhünding Namgyel Dorjé (see n. 51); they were not copied with the rest of the set. See Parfionovitch 1992, vol. 1, 8 and Bolsokhoyeva 2007, 347–67.
32. Cf. 'Byams pa 'phrin las 1996a, 374–75.
33. 'Byams pa 'phrin las 1996a, 374, states that the ones that were at Chakpori were the real originals (*ma phyi ngo ma*). On the building of Chakpori, see chapter 2, n. 168.
34. 'Byams pa 'phrin las 1996a, 374, citing what a scholar of antiquities told the author about the medical paintings at the *mdzod sbug* of the Chensel Podrang at Norbulingkha, which number 164.
35. Ferdinand Lessing had a copy made of twelve medical paintings from there and brought them back to the East Asiatic Library of University of California at Berkeley, which are reproduced in Veith 1960. See also Emmerick 1993, 56–78 and Emmerick 1995.
36. Katherina Sabernig delivered a paper on this topic entitled "Tibetan Medical Paintings Illustrating the *Bshad rgyud*" at the 12th seminar of the International Association of Tibetan Studies held in Vancouver in August 2010.
37. The Field Museum in Chicago has several anatomical paintings, obtained by Berthold Laufer, that Meyer thinks are from Yonghe Gong: Parfionovitch 1992, vol. 1, 5. There is also a small set of the mnemonic diagrams and anatomy in the private collection of Thomas Pritzker that do not seem to be from the Desi's set, but are closely related in subject matter and style; I am grateful to Amy Heller for sending me photos of these images. Cf. Pal 2008. Modern copies of the set include one by a Nepalese artist. See Williamson and Young 2009. There are also two new copies at the Tibetan Medical College in Lhasa.
38. Lange 1964. Several examples of single xylographs based on but not directly copying the anatomical illustrations of the Desi's painting are on display in the National Museum of Mongolia in Ulaan Baator.
39. Some of these are on display today in the Museum of Medicine in Ulan Baator. Others are in a variety of private collections. They are being studied as part of a doctoral dissertation project by Stacey van Vleet at Columbia University.

40. ’Byams pa ’phrin las 1996a, 374–75; ’Byams pa ’phrin las 1996b, 429–30. On Khyenrap Norbu, see Rechung Rinpoche 1976, 22 seq.
41. One set put together from holdings in the Mentsikhang is published in Wang le and ’Byams pa ’phrin las 2004. See also Massin 1982. Emmerick 1993 studies the mixed provenance of the paintings at Mentsikhang. Meyer opines that among others at Mentsikhang, plates 55 and 79 may be from the original set: Parfionovitch 1992, vol. 1, 7. Yang Ga is of the opinion that the originals at Chakpori were moved to Mentsikhang during the time of Thirteenth Dalai Lama (personal communication 2008). Theresia Hofer told me that she was informed by Jampa Trinlė that he negotiated the transfer of the remaining medical paintings at Chakpori to the Mentsikhang in March 1959, before Chagpori was shelled (personal communication 2013).
42. Some of the plates at the Norbulingkha were published in *Bod kyi thang ga* 1985, plates 130–138, and in *Trésors du Tibet* 1987, plates 86–93.
43. Sde srid 1982, 386–94.
44. Meyer makes the same point: Parfionovitch 1992, vol. 1, 8. I have found a few small differences, e.g., as discussed in nn. 10, 45, and 136.
45. There are some differences in whether nipples are noted or not in the small vignette figures, and also in a few cases regarding whether genitals are depicted or not. In one case the small female figure on plate 8 in the Ulan Ude set is positioned in a different corner of the plate than the one pictured in Wang le and ’Byams pa ’phrin las 2004. I have also noted some minor differences in the rendering of hair. See J. Gyatso 2011a, e.g., 273–74.
46. Sde srid 1973a, vol. 4, 495.
47. Sde srid 1982, 388.
48. Reproduced in Lha sding 2000, 137. I am grateful to Rae Dachille for bringing this book to my attention.
49. Sde srid 1973a, vol. 4, 487. He states that the paintings would render the *Blue Beryl*’s meaning in its entirety at Sde srid 1982, 386.
50. Tib. *cha bsdur* or *go bsdur*. Based on what he mentions in Sde srid 1982, 388–94 and the colophons of the paintings themselves, the works that he consulted on how to render the anatomy and medical botany would include *Zla zer* (*Padārthacandrikāprabhāsa*) of Candranandana, *Somarāja*, various texts from the Zur tradition, and the teachings of Jinamitra (also mentioned in Sde srid 1982, 148 as being from Oḍḍiyāna and the author of *Gso stong dgu bcu rtsa gcig pa*). Sde srid 1973a, vol. 4, 480 also discusses sections that he drew from the *’khrungs dpe* traditions of medicinal plant classification and recognition, including the work *’Khrung dpe bstan pa* in 120 chapters translated by the Indian scholar Śāntigarbha “and the 7 or 4 scholars,” as well as the *Sgrol ma sngo ’bum*, and the *sngo ’bum* of Jamyang and of Yutok. The latter have been published in G.yu thog Yon tan mgon po sogs 2005.
51. Plates 6 and 7 in the original set; mentioned in Sde srid 1973a, vol. 4, 486 and Sde srid 1982, 388. These were not included in the Ulan Ude set, as noted above. On Lhünding Namgyel Dorjé see Hofer 2012. For further discussion of the Desi’s relationship with him, see pp. 46 and 86–88, and n. 106.

52. Sde srid 1973a, vol. 4, 494. This comment refers to the set at an initial stage when it only included 60 plates.
53. The former would be Lhodrak Tendzin Norbu; see n. 106.
54. As when he invokes his own understanding (*rang lugs*) of the channels of the body: Sde srid 1973a, vol. 4, 492.
55. Barthes 1977 poses virtually the same question about the advertising image he analyzes: is the illustration redundant or does it add fresh information to what is already communicated in the captions?
56. Sde srid 1973a, vol. 4, 487.
57. Sde srid 1973a, vol. 4, 491.
58. Sde srid 1982, 386.
59. This point is made compelling by Stein 1972, 195; see also J. Gyatso 1998, 117.
60. The Tibetan Treasure (*gter ma*) tradition would be one major case in point: see J. Gyatso 1993 and 1998.
61. For one example see Sde srid 1973a, vol. 4, 490.
62. On various levels of newness see Kaviraj 2005. I propose a different but related set for the Tibetan context in J. Gyatso 2011b.
63. See n. 84.
64. 'Byams pa 'phrin las 1996a, 370–71, mentions the *Ro bkra 'khrul gyi me long* by Biji Tsenpashilaha in 26 chapters, along with the *Ro bkra tha gu dgu sbyor*, *gSon thig*, *Ro thig*, and other "explanations with illustrations" (*rnam bshad ri mo yod tshul*) of the thoracic cavity. He also mentions *Byang khog gi rtsa pra* and *Deb chings sgyog gsum*, composed during the time of Yutok Yönten Gönpo.
65. Sde srid 1982, 294, 345.
66. Bai ro tsa na? 2007. The date and authorship of this work are not clear. I am grateful to Michael Sheehy for bringing it to my attention.
67. Sde srid 1973a, vol. 4, 486; Sde srid 1982, 326. On these Jangpa scholars, see Hofer 2012.
68. See Sde srid 1973a, vol. 4, 486; Sde srid 1973b, f. 281a–b. See also 'Byams pa 'phrin las 1996a, 371 and 'Byams pa 'phrin las 1996b, 429.
69. Cited in the colophon to plate 49; Parfionovitch 1992, vol. 2, n. 154 states that this dates from the fifteenth century and is associated with Lhodrak Menlha Döndrup.
70. Parfionovitch 1992, vol. 1, 8.
71. The *Tashrīḥ-i Manṣūrī* by Manṣūr ibn Muḥammad ibn Ilyās is based in the Galenic anatomical tradition. See Savage-Smith 1997.
72. Elman 2005, chapter 1.
73. Sterckx 2008. See also Sterckx 2002.
74. For example, Wang Qi's *Assembled Diagrams of the Three Powers* (*Sancai tuhui*) of 1609. See Hanson 2003. On the *Golden Mirror* see also Wu 2008.
75. Although there is influence from Chinese medicine in the pulse and urinalysis sections of the *Four Treatises* and elsewhere.
76. See Egmond et al. 2007.
77. Entitled *Ajā'ib al-makhlūqāt wa-gharā'ib al-mawjūdāt* (*Marvels of Things Created and Miraculous Aspects of Things Existing*). It is manuscript P2 at the U.S. National Library

of Medicine. See http://www.nlm.nih.gov/hmd/arabic/natural_hist4.html and also http://www.nlm.nih.gov/hmd/arabic/p18.html (accessed 2013), which states, "No anatomical illustrations of the entire human body are preserved from the Islamic world before those which accompany the Persian treatise composed by Manṣūr ibn Muḥammad ibn Aḥmad ibn Yūsuf ibn Ilyās," from the late fourteenth or early fifteenth century.

78. Cf. p. 8.
79. On secular scenes at the Anxi Yulin grotto in Gansu see Dunhuang 1997. Scenes of cooking, commerce, and lovemaking were painted on the sides of coffins found at Dulan in contemporary Qinghai province. See Xu 2006.
80. For overviews of this history see Heller 1999 and Jackson 1996.
81. See, e.g., Kossak and Singer, plate 1; Heller 2006; Heller 2009, 88–89; and Klimburg-Salter 1988 figs. 2–5, 48–50, 139–42, 151.
82. I saw this mural in 1986.
83. I am grateful to Amy Heller for showing me her photos of the main "foundation fresco" on the east wall of the Red Temple. See also Tucci 1936, plates CIII–CIX.
84. Examples include the murals of the life of the Buddha and the story of Sudhana from the *Gaṇḍavyūhasūtra* at Tabo: see Klimburg-Salter 1988, figs. 120–32. See also the scenes from the life of the Buddha on a book cover in plate 34 of Kossak and Singer 1998. On early fourteenth-century murals of the life of the Buddha at Zhalu and Jonang, see Vitali 1990 and Ziegler 2010. On the *Lalitavistara* murals at the Red Temple at Tsaparang, from the sixteenth century, see Heller 1999, 185. An important early instance of the life of the Milarepa in painting is from the late fifteenth century: Dollfuss 1991.
85. A beautiful and large example is to be found in the Zanabazar Museum in Ulaan Baatar, although that would postdate the seventeenth century.
86. These are in the East Sishi Püntsok Hall and are dated by the authors to 1645–48. See Lha sdings 2000, 52–82.
87. Lha sdings 2000, 84–139.
88. Heller 1999, 85, plate 63. Earlier attempts to render historical specificity would include the foreign monks around the Buddha on a mural at Dratang from the eleventh century (Vitali 1990, plates 30, 33), and the lamas depicted in Kossak and Singer 1998, plates 11 and 18.
89. As in *Bod kyi thang ga* 1985, plates 77 and 78; see also 170. Cf. Jackson 1996, plates 32 and 33.
90. Jackson 1996, plate 63; Heller 1999, 198, plate 114.
91. A later copy published in Jackson 1996, plates 52, 268. See also Lin 2011 on the painting of Rinchen Mingyur Gyaltsen.
92. Yang Ga 2010, 32–34 suggests that *'khrungs dpe* represents a local and indigenous Tibetan medical tradition. De'u dmar dge bshes Bstan 'dzin phun tshogs (b. 1672), produced several important studies of the *'khrungs dpe*: De'u dmar 1970. For inklings of the earlier history of *'khrungs dpe* see van der Kuijp 2010, 31 n. 2. See also nn. 50, 66, and 95 infra.
93. Kusukawa 2009, 464.

94. Cuevas 2011. He points to *dpe ris* in Sangs rgyas gling pa 1981–84, vol. 18. The original cycle dates from the fourteenth century, but the manuscript in which the illustrations appear is in the last volume of an undated collection, and it is not clear when they were included.
95. See n. 66. Another *'khrungs dpe* manuscript is shown at http://www.alessandroboesi.eu/Books.html. It was collected by Giuseppi Tucci in the mid-twentieth century, but its date has not been ascertained.
96. I saw one such example at Zhalu Monastery in 1987, but I do not know its date.
97. Karmay 1988b.
98. Many of the frescoes published in Baker 2000.
99. An early example would be the Cakrasaṃvara maṇḍala in Kossak and Singer 1998, plate 2, which they date to around 1100.
100. *Bod kyi thang ga*, plates 2, 3, and 6.
101. *Bod kyi thang ga*, plates 17, 23, and others.
102. Or in two occasions, the physician's visualization of deities in Buddhist tradition is connected to the rejuvenation substances he is making (plates 52 and 53). See J. Gyatso 2013.
103. Tsering 2005, 172, refers to one Jomo Namo, listed by the Desi, who might have lived before the fifteenth century.
104. J. Gyatso 2011a.
105. Barthes 1977.
106. A few images on plate 49 are labeled *phung po kha shas la lho brag bstan 'dzin nor bu can gyis bltas pa ltar*; another is labeled *le'u 'di'i sha rus don snod kyi gnas lugs.* On Lhodrak Tendzin Norbu, see Jackson 1996, 208, 213–14. The Desi also credits Lhünding Namgyel Dorjé with teaching him a tradition of anatomical iconometry and other anatomical points associated with Lhünding Dütsi Gyurmé, and also with creating "side illustrations" (*bris cha zur*), which would be plates 6 and 7 in the original series: Sde srid 1973a, vol. 4, 486. Cf. Sde srid 1982, 326 and 388. For further significance of the corpse for anatomical questions see p. 205.
107. Sde srid 1973a, vol. 3, 218.
108. On this point see Daston and Galison 2007 and Kusukawa 2009, 357–58. There has also been considerable criticism of realism for its frequently conventional nature: see James Wood's review of *Realist Vision* by Peter Brooks: Wood 2005.
109. Sde srid 1973a, vol. 4, 489.
110. Sde srid 1973a, vol. 4, 496–97. It is not clear to what text of Mendrongpa the Desi refers.
111. Kusukawa 2009.
112. See n. 49.
113. Sde srid 1973a, vol. 1, 282–83.
114. Discussed at length in chapter 32 of the *Instructional Treatise*. See chapter 7.
115. Meyer 1992, 12.
116. Sterckx 2008.
117. Walsh 1910.
118. Personal communication, 2004.

119. A classic study of this practice is Yates 1966.
120. Hanson 2003.
121. Thub bstan tshe ring. 1986, 165; see also 156–57 on the didactic value of the paintings and the use of corpses for the anatomical depictions.
122. Along with books in twenty sizes of the five sciences. Sde srid n.d.a, f. 203b. Cf. n. 29.
123. Kusukawa 2009.
124. Sde srid 1973a, vol. 4, 494.
125. Hanson 2003.
126. LaCapra 1983.
127. [G.yu thog] 1992, 243. The Desi's *Blue Beryl* talks of these things in more detail, bemoaning the sadness of the degenerate age and the decline of morality, but then moves seamlessly to discuss, also in detail, the other actions that also cause illness. Sde srid 1973a, vol. 3, 337.
128. As for example in the third chapter of *Abhidharmakośa*.
129. The connection between health and social life is in much evidence in the *Explanatory Treatise*'s chapter 6 and elsewhere; the relation of pulse to human issues is discussed in detail in the first chapter of the *Final Treatise*.
130. Sde srid 1973a, vol. 4, 487.
131. See Sterckx 2002.
132. A study on the relation of the production of the Degé canon to the ruler's prestige is Scheier-Dolberg 2005. See also Schaeffer 2009, chapter 5.
133. See Gene Smith 2001a and 2001b. The former concerns the remarkable encyclopedia, with much material of a secular nature, by Don dam smra ba'i seng ge, entitled *Bshad mdzod yid bzhin nor bu*, held at the Library of Tibetan Works and Archives. Compare also the five-volume encyclopedia *Mdo rgyud mdzod* (Klong chen chos dbyings 2000) with illustrations in the original block print, especially in the last volume. We also know of an illustrated manuscript from Mongolia, entitled *Khang sa shing sa dur sa s[o]gs kun la gces pa'i spang sa dan brtag tshul 'dra ba'i dpe'i ri mo kun gsal me long*, by Agvan-lodoi. See http://echo.mpiwg-berlin.mpg.de/ECHOdocuView?url=/mpiwg/online/permanent//echo/buddhism/mongol_book2/pageimg&pn=1&mode=imagepath. It would seem to be based on the Desi's *White Beryl*, and probably dates from the eighteenth or nineteenth century (Petra Maurer, personal communication, 2013).
134. For examples of this genre in translation see Köppl 2008 and Kongtrul 1996.
135. Cf. the phrase *bskyed rim rdzogs rim ci dang cir bsgom yang* ("whatever creation phase or completion phase meditation practice you do") in Kongtrul 1996, 66 (translation on 32).
136. This example provides one more case of a difference from the plate reproduced in Wang le and 'Byams pa 'phrin las 2004, plate 20, where the person reading is a layman with long hair.
137. E.g., Kossak and Singer 1998, plate 55. The figure is not identified, but the authors speculate it might be Atiśa.
138. Vincent van Gogh's *Still Life with Bible*, painted in 1885, is in the Van Gogh Museum, Amsterdam. We might compare the earlier masterwork of Rembrandt, portraying his mother reading an open page of the Bible on which some Hebrew letters seem to be

visible, but the words are impressionistically portrayed, and yet again it is specific in its referent. Rembrandt's *The Prophetess Anna* (also known as *Rembrandt's Mother*) was painted in 1631 and is in the Rijksmuseum.

139. Hanson 2003, 143.
140. Meyer 1992, 12.
141. Barthes 1989, 141–48.
142. Sde srid 1973a, vol. 3, 150–51. The discussion of head shape has to do with humor imbalances and a variety of issues in the brain and skull bones, but nowhere is hair discussed. Note too that the figures at the bottom of plate 48 (48.68–74) illustrate the same head shapes from plate 47, and the two sets of people have completely different appearances and hairstyles.
143. [G.yu thog] 1992, 362 and Sde srid 1973a, vol. 3, 3 say nothing of the dress style or region of mothers who mishandle their children in these ways.
144. Another good example of several dress styles together is the group of women milking animals on plate 22.
145. Toulmin 1990.

2. ANATOMY OF AN ATTITUDE: MEDICINE COMES OF AGE

1. Pollock 2005 and 2001; Bronner 2002.
2. Sde srid 1982, 349–55. The Desi's account seems to be based on Zur mkhar ba 2003a. See p. 145.
3. Cf. Zur mkhar ba 2003a, 10. Zurkharwa's death date is not clear. As the Desi notes, the last age to which Zurkharwa refers is 63: cf. Zur mkhar ba 2003a, 73. Modern studies of Zurkharwa's life include Byams pa 'phrin las 2000, 226–29; Pa sangs Yon tan 1988, 11; Gerke and Bolsokhoeva 1999; Gerke 1999; and Czaja 2005b.
4. See Czaja 2005a.
5. *G.yu thog pa'i phyag dreg ma'i rgyud bzhi gser mchan* (corrected from *mtshan*): Sde srid 1982, 350. See also pp. 122 and 123.
6. This would have been in the 1540s. See n. 117 in chapter 3.
7. The Desi titles this *Rang 'grel gser gyi phra tshom;* it does not seem to be extant.
8. This would be Zur mkhar ba 2003a; see pp. 145–46. At least one other response is extant: see Bod mkhas pa 1986. Both works and the questions and answers themselves are studied in Czaja 2008.
9. *Zur za 'am bsting tshig gsungs.*
10. Cf. [G.yu thog] 2005a, 700–703.
11. He is referring to Zur mkhar ba 1986. See p. 182 seq.
12. Czaja 2005b, 139 seq. and n. 54, misconstrues the passage to be the words of Zurkharwa himself. Rather, it represents the opinion of the Desi. Cf. D. S. Gyatso 2010, 313.
13. Sde srid 1982, 352, citing the *Sa skya legs bshad*: cf. Gyaltsen 2000, 118.
14. *'Phral skad gso dpyad rang lugs:* Sde srid 1982, 352–53.
15. The metaphor evokes *Raghuvaṃśa* 1.4: Kālidāsa 1928, 1. Thanks to Robert Goldman for this reference.

16. Sde srid 1982, 353: cf. [G.yu thog] 1992, 665.
17. Cf. Sde srid 1973a, vol. 4, 483.6. He also had smallpox, probably in 1547, and he complains of being unwell in the course of trying to complete his writing projects: see Zur mkhar ba 2003a, 12 and 74.
18. Sde srid 1982, 354–55.
19. See Coda, n. 69.
20. As one example, Jetsünpa Chökyi Gyeltsen (1469–1544) criticizes the philosophical views of his own teachers: Ary 2007, 142. On fierce loyalties and the culture of scholastic disputation see Hopkins 2002, chapter 1. On the criticism of other lineages in the writing of autobiography see Gyatso 1998.
21. Compare Tukwan's apologies in differing with the views of his teacher Sumpa Khenpo: Kapstein 1989, 226.
22. Sde srid 1973a, vol. 4, 485–86. Cf. Sde srid 1982, 394–95.
23. See also chapter 1, n. 67.
24. Sde srid 1982, 325–27.
25. Sde srid 1973a, vol. 4, 487.
26. Sde srid 1982, 326.
27. Sde srid 1973a, vol. 4, 495–96.
28. Sde srid 1982, 381. Cf. Thondup 1986, 168.
29. Sde srid 1973a, vol. 4, 487.
30. Sde srid 1973a, vol. 4, 495. See also n. 34.
31. See n. 4 in chapter 4.
32. Sde srid 1982, 379.
33. Sde srid 1982, 378.
34. He is referring to Lingtö Chöjé Lozang Gyatso and Mengom Drangyepa: Sde srid 1982, 378.
35. Not to mention scoping out the site for Chakpori: see p. 114. Cf. Sde srid 1973a, vol. 4, 494.
36. Sde srid 1982, 327.
37. Sde srid 1982, 395. Mipam Gelek was famously at odds with the Fifth Dalai Lama and the Desi on poetic style.
38. Cf. Weintraub 1978.
39. Petech 1966, 271–73; Petech 1972, 11–12; Sperling 2003, 130; and Tsyrempilov 2006, 56–57.
40. Tucci 1949, 74.
41. J. Gyatso 1992 and 1998.
42. Among other things, the Desi is shown (usually as a short-haired man with darker skin and larger size than those around him) supervising the building of the Potala, presiding over celebrations, writing his main works, issuing proclamations, and even practicing archery as a child, all over a door in the *bar khyams* section of the Potala: Lha sding 2000, 127. According to Samten Karmay, personal communication, the autobiographical paintings are located on the ninth floor of the Potala's Red Palace. His autobiographical passages include Sde srid 1982, 372–96; 547–56; 569–73; Sde srid 1973a, vol. 4, 480–81 et passim and 493 et passim; Sde srid 1973b, vol. 2, 399 seq.;

and scattered self-referential statements throughout his biographies of the Fifth and Sixth Dalai Lamas. For biographical studies of the Desi, see n. 54.

43. For an example of the Desi's rich writing style see Tucci 1949, 71–72. Among other works, the Desi added three volumes to the journalistic autobiography of the Fifth Dalai Lama (Ngag dbang blo bzang rgya mtsho 1989–91; Sde srid n.d.b.)
44. Vovelle 1990, Introduction.
45. Cf. p. 160, and chapter 3, n. 72; see also Introduction, nn. 14 and 19 regarding possible comparisons with Tibetan astronomical writing.
46. Two outstanding examples would be the doubts on the *Four Treatises*' Indic origins (see p. 160) and the empirical existence of the tantric channels (see p. 205 seq.).
47. See Nattier 1991.
48. J. Gyatso 1992; 1998.
49. See Dreyfus 2003; Lopez 1996, 217–28; Kapstein 2002; and J. Gyatso 1999.
50. Shapin 1998, 119–23.
51. Dasgupta 1952, 373–402; Prets 2000. For an English translation see Sharma and Dash 1976.
52. Sde srid 1973a, vol. 4, 474–76. The critique extends to physicians in India and Nepal: Sde srid 1973a, vol. 4, 488–89.
53. Sde srid 1973a, 488.
54. Modern Tibetan biographical sketches of the Desi include Nor brang o rgyan 2006; Byams pa 'phrin las 1996b and Skal bzang legs bshad 1994, 213–27. See also Lange 1976; Yamaguchi 1999; Schaeffer 2006; and Cüppers et al. 2012, Introduction. On the Desi's medical career see Czaja 2007. For the Desi's own autobiographical writings, see n. 42.
55. Sde srid 1973a, vol. 4, 493. Cf. Sde srid 1982, 383, 569.
56. See Sde srid 1982, 376 and rGya-mTSHo 1999, 304 seq. on the huge amount of administration that the Desi oversaw starting from 1679. For the *vidyāsthānas*, see p. 102.
57. On the *White Beryl* (*Baiḍūrya dkar po*) and related astrological and calendrical systems see Schuh 1970; 1973c; 1974.
58. He began the *Blue Beryl* in 1687 and finished it in 1688: Sde srid 1973a, vol. 4, 492.
59. The Desi is famed for having concealed the death of the Great Fifth for sixteen years but it was not unknown to all parties; see Petech 1972, 9 and Shakabpa 1967, 125–26.
60. Sde srid 1982, 571–72.
61. Sde srid 1982, 372–73; cf. Sde srid 1973b, vol. 2, 400.
62. Sde srid 1982, 378–80. See also Byams pa 'phrin las 1996b.
63. Sde srid 1982, 380. See also Sde srid 1982, 383–85; and Sde srid 1973a, vol. 4, 488 seq.
64. Sde srid 1982, 380.
65. E.g., Sde srid 1982, 396–99.
66. *Rje 'dzin gyi rig pa*: Sde srid 1973a, vol 4, 494.
67. Sde srid 1973a, vol. 4, 491.
68. Sde srid 1982, 368–72 et passim.
69. Sde srid 1982, 382–83; Meyer 2003, 99–117.
70. See Sde srid 1973b, vol. 2, 400–402; Richardson 1998; and Petech 1972, 218.
71. Sde srid 1982, 373; see also Sde srid 1973b, vol. 2, 400–401.

72. Sde srid 1982, 373; 571; Sde srid 1973b, vol. 2, 399–400.
73. Sde srid 1973b, vol. 2, 399. He reports his father's name as Asuk, and supplies details on his Drongmé family and ancestors. His mother was sent away from his father for a while, but she became pregnant with the future Desi when she returned. See also Byams pa 'phrin las 1996b, 401–42. Cf. Richardson 1998, 454–55, and Tucci 1949, n. 145; Petech 1972 assumes that the Desi was the Dalai Lama's son.
74. E.g., rGya-mTSHo, 264, 266, 274.
75. *G.yu thog sñiṅ thig* 1981. See n. 140.
76. Sde srid 1982, 382–83; Meyer 2003, 106.
77. Cf. p. 111 seq.
78. Sde srid 1973a, vol. 4, 494.
79. Sde srid 1973a, vol. 4, 495.
80. Schopen 1978. See also Birnbaum 1979.
81. Demiéville 1985 (first published in 1937).
82. E.g., *Khog 'bugs khyung chen lding ba* 1976, 15.
83. *Suvarṇaprabhāsottama*, chapter 16; see Emmerick 1996. The physician's knowledge is characterized in terms of the Ayurvedic "eight branches" (*aṣṭāṅga*). See also Salguero 2013, which argues that the sūtra does not represent an independent tradition of Buddhist medicine but rather summarizes general Indic medical knowledge.
84. See Kritzer 1998a, 1998b, and 2009.
85. Demiéville 1985; Zysk 1991; Haldar 1977; Mitra 1985. See Birnbaum 1979 for a summary of the *materia medica* in the *Mahāvagga*.
86. Demiéville 1985, 31–35; Zysk 1991, 44–46. Cf. Bapat and Hirakawa 1970, 329–31.
87. *Mahāsatipaṭṭhāna* 4.5–6 (Walshe 1987, 337–38); *Visuddhimagga* VIII, 83–138 (Buddhaghosa 1999, 244–58).
88. See chapter 4. On the intersection between tantrism and late Indian Ayurveda see R. P. Das 2003.
89. Dutt 1962, 134.
90. Hsüan-tsang 1884, 78; Beal 1973, 112. On its later fortunes in India see McKeown 2010, 32, 405, 410, 448. On the five sciences in Chinese Buddhism see Overbey 2010, 34–35. See also n. 107.
91. Lévi 1886, 480–82. Meulenbeld 1999–2002, vol. 1A, 105–15 dates *Carakasaṃhitā* between 100 B.C. and 150–200 A.D. See also Meulenbeld 1974, 403–406; Filliozat 1975, 12–20; Dasgupta 1952, vol. 2, 393–99; and Dagmar Wujastyk 2012, 16–17 for current thinking on the dating of *Caraka* and *Suśruta*.
92. Meulenbeld 1999–2002, vol. 1A, 597–612.
93. Hoernle 1983.
94. Demiéville 1985, 94 seq. See also Salguero 2010; and Capitanio 2010.
95. The most significant use of Buddhist *sādhana* technique in the *Four Treatises*, where the physician is advised to visualize himself as the Medicine Buddha when he prepares medicine, is discussed below in chapter 7. On several other occasions the text advises the physician to use rituals or mantras while preparing medicine, as in the *bcud len* substances in the gerontology section ([G.yu thog] 1992, 549–50), but not always involving Buddhist figures, e.g., [G.yu thog] 1992, 125 (for *mkhris pa* ailments)

and 269 for disease caused by *gnyan* spirits. The work also mentions a few mantric spells that help heal disease, as on 292 for heart problems; 322 for vomiting; and 329 for urinary tract problems. The mantra of Vajrapāṇi is advised for diseases caused by nāgas: 396. Most of the ritual means of healing have to do with illnesses caused by demons and spirits, as in chapters 77–81, also mentioned in passing in a few other cases. See also 280–81 for references to ritual means that might help eye disease; 532 and 536 with regard to poison, reflecting Indic Ayurvedic lore, and 655 seq. with respect to plague. A rite to magically change the sex of a fetus, provided by [G.yu thog] 1992, 18, is directly indebted to Ayurveda; see also chapter 6, n. 63. But all of these statements on ritual means of healing taken together occupy less than 10 pages, scattered through a work of more than 660 pages.

96. When Buddhist scriptures speak specifically of physical health and illness, they often refer to the balance of the four standard elements of the physical world (*mahābhūta*, i.e., earth, water, fire, and wind). But they also sometimes speak instead of four humors, which consist in the usual medical three plus a fourth that represents the three combined. Buddhist texts also discuss three humors on occasion, in the Pāli canon and elsewhere: Demiéville 1985, 65–76. There is anyway significant blurring of the distinctions between the humors and the elements in the pathogenesis of Ayurveda: see Haldar 1977, 5; Mitra 1985, chapter 2.
97. Demiéville 1985, 70. Demiéville also points to a slippage between *doṣa* as humor and *doṣa* as hatred: 78.
98. Haldar 1977, 4–10. Cf. *Carakasaṃhitā* I.25.18–19; 26–29; IV.2–3. See also Lyssenko 2004.
99. E.g., *Abhidharmakośabhāṣya* 4.5. Note too the attribution of the "temperaments" not only to karma but also the prominence of certain elements and humors in the physical body, as in *Visuddhimagga* III.80–81: Buddhaghosa 1999, 102–103. Cf. the eight causes of experience listed in *Samuttanikāya* 26.21, i.e., bile, phlegm, wind, imbalance, climate, carelessness, assault, and kamma: Bodhi 2000, 1278–79.
100. Halbfass 1980, 272.
101. [G.yu thog] 1992, 16–17; 22; 27.12 seq.; 552.6; 35.1–2. See also pp. 301 et passim, 319. Cf. Zur mkhar ba 1989, vol. 1, 264 seq. See also Garrett 2008; and Goble 2011, chapter 4.
102. For example, the *Four Treatises* advises its readers to work for auspiciousness and good merit and to drive out demons at the birth of a child, or to make offerings and do recitation and purification rituals when beset with an illness caused by *'byung po* spirits. [G.yu thog] 1992, 360–62; 386–87; see also 389. However, these instructions are usually very general. As the text's chapter on pulse examination points out, once demons are recognized to be causing a condition, one should do whatever rituals are appropriate and match the area in which one lives. [G.yu thog] 1992, 566.
103. Halbfass 1980, 296. Cf. the *Upāyakaulśalyasūtra*'s attribution of a thorn on which the enlightened Buddha injured his toe to his own karmic retribution: Chang 1983, 457–58.
104. A good overview of the range of physical external therapies in Sowa Rikpa is to be had from Yonten 2014. And that is not to mention the huge array of the internal therapies in the *Four Treatises* and its commentaries.
105. *Brahmajāla* 12. 27: T. W. Rhys Davids 1977, vol. 1, 25–26; Norman 1969, vol. 1, 88 (Pārāpariya 939). Cf. Birnbaum 1979, 5–7.

106. Demieville 1985, 38–39.
107. *Mahāyānasūtrālaṅkāra* 11.60: Asaṅga 1907, 70–71. Sakya Paṇḍita glosses *adhyātmavidyā* as "the authoritative Dharma" (*lung gi chos*) in *Mkhas 'jug*: Gold 2007, 189, n. 49. Rhys Davids and Stede, eds. 1997, 618, note eighteen *vijjaṭṭhānāni* in *Jātaka* I.259. There is also a common list of ten.
108. See Ruegg 1995, 102–105.
109. Gold 2007, chapter 1.
110. Schaeffer 2003 and 2011.
111. See Ishihama 1993 regarding the massive number of Avalokiteśvara rituals carried out by the Fifth Dalai Lama.
112. Sde srid 1982, 396–97; 549–53.
113. Thondup 1986; J. Gyatso 1993 and 1998. On the Desi's use of Treasure prophecy for the birth of the Fifth Dalai Lama, see rGya-mTSHo, 235–42.
114. Sde srid 1982, 550.
115. Karmay 2003, 66; Shakabpa 1967, 82; neither cites their primary sources.
116. Kapstein 1992; Dreyfus 1994; J. Gyatso 1998.
117. See Lo Bue 2003, 182–84; Townsend 2012.
118. See also pp. 378–80.
119. It is mentioned repeatedly, for example Zur mkhar ba 2001, 261, 288, 290, 292, 299; Sde srid 1982, 156; 213.16–17.
120. See Khro ru 1996, 4, on *sman pa rab 'byams pa*, *sman pa bka' bcu pa*, and *sman pa bsdus ra ba*.
121. E.g., Sde srid 1982, 170.13. Somarāja also discusses this term: Yang Ga 2010, 179. See also chapter 7, n. 113.
122. The following is based on Zur mkhar ba 2001, 287–321; Sde srid 1982, 148–83; Dkon mchog rin chen 1994; and Khro ru 1996. See also Beckwith 1979 and Meyer 1981, 77–94.
123. Garrett 2007.
124. *Ngag rgyun rim par brgyud pa*. Zur mkhar ba 2001, 287.
125. Zur mkhar ba 2001, 299; Sde srid 1982, 150.
126. Zur mkhar ba 2001, 253–55; 287–89; Sde srid 1982, 150–51; Beckwith 1987, 20.
127. Yoeli-Tlalim 2010a, 3.
128. See Beckwith 1979, 300 and n. 24.
129. E.g. PT 127, 1044, 1057 and 1058. See Richardson 1987, 10–11; Schaeffer, Kapstein, and Tuttle 2013, 115–17; and Yoeli-Tlalim 2012a and 2012b.
130. Jinamitra is one of the translators, and he is probably the same as the well-known Indic scholar/translator in Tibet in the early part of the ninth century: Emmerick 1980–82, vol. 1, 2.
131. Beckwith 1979, 303; Yang Ga 2010, 39 seq. and 60 seq. On the Drangti lineage see Zur mkhar ba 2001, 255, 285, 291; Sde srid 1982, 155 seq.; De'u dmar 1994, 703–708; Yang Ga 2010, 28–51, 129–36; and van der Kuijp 2010. On Greek and/or Arabic medicine in Tibet see Yoeli-Tlalim 2010b and Martin 2010.
132. Zur mkhar ba 2001, 295–97. Cf. Sde srid 1982, 176 and Garrett 2007.
133. The college, called Mengi Drongkhyer Tanaduk, is said to have been located at Kongpo Menlung: Khro-ru 1996, 3. On Tsenpashilaha see Beckwith 1979 and Yang Ga 2010, 36–42 et passim; 61.

134. Zur mkhar ba 2001, 289–90.
135. *Sman dpyad zla ba'i rgyal po* 1985. There are several versions of the text and much debate about its provenance. Sum ston reports that Yutok studied a text by a very similar name at Varanasi: Ye shes gzungs 1981, 32. Zur mkhar ba 2001, 289, implies it was translated from Chinese; cf. Sde srid 1982, 152. Yang Ga 2010, 69 cites Kongtrul Lodro Taye's opinion that the work was composed originally in Tibetan. It is currently the topic of doctoral research by William McGrath of the University of Virginia.
136. Toh. 4310 and 4312. The Sanskrit is Das and Emmerick 1998; for an English translation see Murthy 1991. On Vāgbhaṭa and Candranandana's *Zla zer*, i.e., *Padārthacandrikā Aṣṭāṅgahṛdayavṛtti*, see Meulenbeld 1999–2002, vol. IA, Part 5, and vol. IB, 732, n. 158. On the Tibetan translation of the *Aṣṭāṅgahṛdaya* see Martin 2007, 311. On the relation of the *Four Treatises* to *Aṣṭāṅgahṛdaya*, see Yang Ga 2010. Cf. Emmerick 1977, 1990, and 1991; Taube 1980; and Vogel 1963. For the Tibetan translation of the root text and the *Padārthacandrikā* see Kha che Zla ba mngon dga' 2006. Other commentaries were also translated into Tibetan and may be found in vols. 198–202 of the Degé version of the Tengyur. See also Ruegg 1995, 110.
137. Yang Ga 2010, chapter 1 touches briefly on Gampopa's medical writings, and treats the Biji tradition at length.
138. Yutok's dates have been disputed. The best source on Yutok's life and on Sumtön is Yang Ga 2010, chapter 1; see also Yang Ga 2014. Cf. n. 196 and chapter 3, n. 51.
139. See p. 119.
140. *G.yu thog sñiṅ thig* 1981. The core of this collection may have been compiled by Yutok's disciple Sumtön Yeshé Zung, but the chronology of the entire collection awaits detailed study. Cf. n. 50 in chapter 3. Other Old School Treasure ritual healing scriptures are found in the *Bka' brgyad* section of the Old Tantras, especially *'Chi med bdud rtsi yon tan gyi rgyud* (Kaneko 299–314) as well as in *Bima snying thig*, and in the Treasure cycles of Nyangrel and Guru Chöwang. An initial study of some of these materials is Garrett 2009. By the Desi's time, the Treasure-related *sman grub* teachings are central to the myth of the life of Yutok. The *Bdud rtsi 'bum pa* lineage is depicted at the top of Parfionovitch et al. 1992, plates 7 and 8, and related teachings are discussed in detail in Sde srid 1982, 183–201 (cf. D. S. Gyatso 2010, 175–89.) For Zurkharwa's critical assessment of Old School influence see chapter 3, nn. 17 and 29. For other signs of Old School teachings in Tibetan medicine see p. 210 and chapter 5, n. 22.
141. See Coda, nn. 37–40.
142. Karmay 2003; Shakapba 1967, chapters 6–8; and Ishihama 1993.
143. Schuh 2004.
144. Sde srid 1979.
145. See Schaeffer 2006.
146. rGya-mTSHo, 305–53; Tucci 1949, vol. 1, 69.
147. Ngag dbang blo bzang rgya mtsho 1989–91.
148. Schuh 1978, 1981a, and 1981b.
149. His conflicted feelings about that are evident in his "secret autobiography" *Rgya can*: Karmay 1988b, 15, 17–18.

150. This first appears in the biography of the First Dalai Lama: Ishihama 1993. Cf. rGya-mTSHo, 136 et passim. On the larger background see van der Kuijp 2005, 24; and Kapstein 1992.
151. Sde srid 1982, 373–74, although on 375 he displays modesty: *yon tan thams cad rdzogs pa'i sangs rgyas dngos yin 'dug pa de lta bu'i re ba ci.* On Muné Tsenpo, see Sde srid 1982, 571 and Sde srid 1973b, vol. 2, 402.
152. Sperling 2003. See also Tucci 1949, 74; and Schaeffer 2011.
153. See Petech 1952 and Toscano 1981. Cf. Tucci 1949, 78. See also Pomplun 2010.
154. Sde srid 1982, 368, 381. Cf. rGya-mTSHo, 260 and Dkon mchog rin chen 1994, 93–94.
155. Sde srid 1982, 306–64; Pa sangs yon tan 1988; Dkon mchog rin chen 1994; Khro ru 1996, 6–7; Taube 1981, 51–73; Gerke 1999; Hofer 2012; Czaja 2005b.
156. Dkon mchog rin chen 1994, 93–94, also mentions Tratsang Lodrö Chokgi Dorjé; Panchen Lozang Chögyan; Zurpa Chöying Rangdröl; Zhalu Rinchen Sönam Chokdrup; and Terdak Lingpa.
157. Sde srid 1982, 368.
158. Dkon mchog rin chen 1994, 98–99.
159. Gya-mTSHo, 328. Cf. Dkon mchog rin chen 1994, 104.
160. Dkon mchog rin chen 1994, 94–95; rGya-mTSHo, 320.
161. Sde srid 1982, 369.
162. Sde srid 1982, 369.
163. Dkon mchog rin chen 1994, 95.
164. Dkon mchog rin chen 1994, 99.
165. This might have had to do with the Dalai Lama's participation in war magic: see Karmay 1988b, 15 and 22.
166. See pp. 175 and 281.
167. Sde srid 1982, 395–96. See also Meyer 2003, 99–101.
168. On the building of Chakpori, see Thub bstan tshe ring 1986 and 'Byams pa 'phrin las 1996b, 417 seq. See also Khro ru 1996, 7.
169. The Desi calls it a *grwa tshang*: Sde srid 1982, 547. See Thub bstan tshe ring 1986. Meyer 2003, 111 mentions the possibility of lay students, but does not give his source.
170. Sde srid 1978.
171. Sde srid 1982, 547.
172. Khro ru 1996, 7.
173. Sde srid 1982, 369–71.
174. Ngag dbang blo bzang rgya mtsho 1989–91, vol. 2, 460. cf. Sde srid 1982, 371; Dkon mchog rin chen 1994, 99.
175. Ngag dbang blo bzang rgya mtsho 1989–91, vol. 3, 64–65. Cf. Sde srid 1982, 222 attributing the curing of the Dalai Lama's eye problems to a statue brought to Tibet by Yutok from India.
176. Sde srid 1982, 371. See also Dkon mchog rin chen 1994, 101–103 and van der Kuijp 2010.
177. Dkon mchog rin chen 1994, 100.
178. *Tshe'i rig byed mtha' dag gi snying po bsdus pa* 1982–85, ff. 333b–334b. Studied by Schaeffer 2011. See also Dkon mchog rin chen 1994, 100–101.

179. Sde srid 1982, 371. Some of these works were included in volume *No* of the Degé Tengyur.
180. Some portion of the *Caraka* seems to have been available in Tibetan: Skyem pa Tshe dbang 2000, 7, 8.
181. Dominik Wujastyk 2005. Cf. Pollock 2000, 2001a, and 2001b.
182. Including Jonang, although see Karmay 2014, 9. Some of the monastic libraries that were confiscated and sealed under the Fifth's reign have recently been found at the Nechu Lhakhang library at Drepung.
183. Sde srid 1982, 371. These works may be found at TBRC W15888; TBRC W00EGS1017628 or W2DB13635; and TBRC W17713; the last is Byang pa 2001. See also van der Kuijp 2010.
184. Dkon mchog rin chen 1994, 102–104.
185. Hanson 2003.
186. The original blockprint is *G.yu thog cha lag bco brgyad* 1976. Discussed by Taube 1981, chapter 4; and Gerke 2001.
187. Sde srid 1973a, vol. 4, 481.
188. *Cha lag bco brgyad kyi them yig* 1976, 5; Sde srid 1982, 368; Ngag dbang blo bzang rgya mtsho 1989–91, vol. 2, 239.
189. *Cha lag bco brgyad kyi them yig* 1976, 5 provides a publisher's colophon by the Fifth Dalai Lama. See also Dkon mchog rin chen 1994, 97–98.
190. See Sde srid 1982, 277–78.
191. Zur mkhar ba 2001, 321–22; see also 279. Cf. dPa' bo 1986, vol. 2, 1523–24.
192. *Cha lag bco brgyad kyi them yig* 1976, 4–5 says that the *them yig* was arranged by Yutok Gönpo, from whom it was transmitted to Yutok Wönga, then to Yutok Bumseng, then to Yutok Pelzang, then to Yutok Gönrin, and then to Remen Gönpo. See Yang Ga 2010 on the descendents of Yutok.
193. Sde srid 1973a, vol. 4, 481–82; Sde srid 1982, 277–88. See also Taube 1981; and Gerke 2001.
194. See Yang Ga 2010 and 2014.
195. Jo bo and Dar mo 2005. See TBRC W23640 for a recent print from the original Zhöl blocks, signed in the printers' colophon by the Dalai Lama. Sde srid 1982, 372 says the Dalai Lama commissioned Larawa to produce the biographies of the Younger and Elder Yutok based on old documents that Larawa had collected. However, both works are attributed in their colophon to Darmo Menrampa, with his students Mermowa Lodrö Chömpel and Larawa Lozang Dönden acting as scribes. See also Ngag dbang blo bzang rgya mtsho 1989–91, vol. 3, 340 and 384, and Dkon mchog rin chen 1994, 104.
196. The dates of Yutok Yönten Gönpo are still in question. Yang Ga 2010, 97–99 places him in the twelfth century; van der Kuijp 2010, 24–26 proposes that he flourished around 1100.
197. Yang Ga 2010, 96 notes that the "Yutokpa" mentioned by Sumtön is characterized as a physician "who had to study," rather than from a family with royal appointment. See Ye shes gzungs 1976, 280. But cf. Jo bo and Dar mo 2005, 63. Sde srid 1973a, vol. 4, 461 refers to the Elder's status as *bla sman lha rje*. It is not clear to me that Sumtön's comment is necessarily about the Elder Yutok.

198. The colophon to the biography of the Elder states it was based on old and fragmentary documents which were copied by Mermowa Lodrö Chömpel and Larawa Losang Dönden: Jo bo and Dar mo 2005, 302. According to the editors of Jo bo and Dar mo 2005, 1–2 the work is based on autobiographical statements of the Elder Yutok transcribed by Kongpo Degyal, and was later edited by Jowo Lhündrup Tashi. Both of the latter are mentioned in the biography itself: see Jo bo and Dar mo 2005, 122, 284, 301. A biography of Yutok associated with Kongpo Degyal is mentioned by Sog bzlog pa 1975, 224 but he never uses the appellation Elder Yutok. Sog bzlog pa 1975, 234 also refers to an extended biography of Yutok and a *mgur bka' rgya ma*. Karmay 1998c, n. 5, refers to a biography of Yutok the Elder called *Rnam thar bka' rgya ma* by Jowo Lhündrup Tashi but warns that it should not be confused with a biography of the Younger Yutok by the same name (see chapter 3, nn. 34 and 183 for more on a work with this or similar title). Karmay 1998c, 230 and n. 13 also states that the Elder Yutok is a fictitious character. Cf. Taube 1981, 48–49. Yang Ga 2010, 94–96 notes a reference to an early biography of the Elder Yutok by Lungmar Gönpo Rinchen in Dbang 'dus 2004, 5 but has not been able to confirm its existence. He also notes that our current biography of the Elder Yutok mentions the fourteenth-century *Gser bre*, and suggests that an Elder Yutok biography was first composed in the fourteenth century.
199. In his colophon to the biography of the Younger Yutok Darmo Menrampa says that it was based on the *Lo rgyus nges shes 'dren byed* and the songs (*gsung mgur*) of this master. The former is the same as the *Heart Sphere of Yutok Story* (see p. 154 seq.). Exactly what is meant by "songs" is uncertain but may refer to the *Shog dril skor gsum*. See Ehrhard 2007, nn. 7, 9, 15. On the use of the biographical fragments of the Younger Yutok for the story of the Elder, Sde srid 1982, 219 cites the *Rnam thar bka' rgya ma* (see n. 198).
200. Sde srid 1982, 225 speaks of fools who tell a story of the Younger Yutok's grant of powers to Drejé from a corpse in India, an episode that actually belongs in the biography of the Elder Yutok. On 227 he notes that as in the case of the former (*snga phyir*), again the Younger is said to have gone to India six times. On 222 he mentions the predictions of the medicine goddess Dütsima; cf. Sde srid 1973a, vol. 4, 459–60 where the father of the Elder predicts the life of the Younger with the same name in thirteen generations. This and other statements about the Younger can also be found throughout the biography of the Elder.
201. Although mention of the Elder Yutok is oddly missing in his colophon to his edition to the *Four Treatises*: [G.yu thog] 1978, *Phyi rgyud*, f. 63 seq. His account of the life of the Elder in Sde srid 1982 goes from 206–25. The Younger's life goes from 225–29 and again from 275–84. See Coda for further discussion.
202. TBRC dates him to 790 and attributes many works to him. See Yang Ga 2010 for the assumption in western scholarship that the Elder Yutok is a historical figure.
203. Yang Ga 2010 and 2014. Among the works that betray no knowledge of a Yutok the Elder are Ye shes gzungs and Gzhon nu ye shes, 1976 (thirteenth century*); Sman gyi byung tshul khog dbubs rgyal mtshan rtse mo 'bar ba* (thirteenth century; see Martin 2007); Brang ti n.d. (fourteenth century); Sog bzlog pa 1975 (sixteenth century), which states on 231 that Yutok was a contemporary of Drakpa Gyaltsen (twelfth to thirteenth centuries centuries); the sketch of Yutok in Dpa' bo 1976, vol. 2,

1521–23 (sixteenth century; cf. Taube 1981, 42–45); and Zur mkhar ba 2001 (sixteenth century).

204. Sde srid 1982, 372.
205. Sde srid 1982, 368–69.
206. The colophon to both the *Instructional Treatise* and *Final Treatise* commentaries attributes their composition to Darmo alone: Dar mo 1989, vol. 2, 527; 778–82. In Sde srid 1982, 372 the Desi writes that Namling and Darmo wrote the commentary to the *Final Treatise* together, and that Darmo, Larawa, and Mermopa did the notes on the *Instructional Treatise*. Cf. Sde srid 1973a, vol. 4, 484.1–2.
207. Sde srid 1973a, vol. 4, 482–84. Cf. Sde srid 1982, 372. One principal *Instructional Treatise* commentary that was in existence was the *Skyem 'brel* (Skyem pa 2000), which the Desi certainly knew: see p. 85.
208. Dar mo 1989, vol. 2, 779; 782. Cf. Ngag dbang blo bzang rgya mtsho 1989–91, vol. 3, 152.
209. One example would be the anatomy of the "mind-entrance opening" in the heart: see chapter 5, n. 28.
210. For an overview of the editions of the *Four Treatises* see Bstan 'dzin don grub 2005–8; and Yang Ga 2010, 120–25. Sde srid 1982 mentions several of the earlier editions of the *Four Treatises* on 381, 384–85. Cf. [G.yu thog] 1978, *Phyi rgyud*, ff.61a–66b.1. See also Czaja 2007, 357–58. An earlier version of the *Four Treatises* by Nyamnyi Dorjé is mentioned in Zur mkhar ba 2003, 99; cf. [G.yu thog] 2005, 701.
211. Sde srid 1982, 368, 381 cf. [G.yu thog] 1978, *Phyi rgyud*, f. 65a.
212. A print is extant and I have a photocopy of some chapters. The modern edition is [G.yu thog] 2005. The blockprint seems to have been completed in 1546. See Zur mkhar ba 2003a, 12. Although Dratang was connected to Drapa Ngönshé, a major figure in the Treasure version of the *Four Treatises*' history, I have no evidence that this connection was significant for the location of the first blockprint edition of the text. See pp. 153–54, 178, and 278.
213. For Zurkharwa's own comments on the range of sources he consulted and continuing doubts about accuracy and errors, see [G.yu thog] 2005, 700–703.
214. Sde srid 1982, 381–82; cf. Sde srid 1973a, vol. 4, 481.1.
215. Sde srid 1982, 381. His similar account in the colophon as preserved in the Chakpori edition is more diplomatic in its characterization of Jangopa: [G.yu thog] 1978.
216. *Cung zad thal*. Sde srid 1982, 381–82.
217. Shapin 1994, chapter 1.
218. Sde srid 1982, 382.
219. The blocks were carved by 1694. See [G.yu thog] 1978, *Phyi rgyud*, ff. 64b and 66b.
220. Sde srid 1982, 384.
221. Sde srid 1982, 385. Cf. [G.yu thog] 1978, *Phyi rgyud*, ff. 63b–66b.
222. Sde srid 1982, 385. See also Schaeffer 2003 and 2009, 76–78.
223. See Kaviraj 2005 and J. Gyatso 2011b.
224. See p. 127 and n. 249.
225. E.g., Sde srid 1973a, vol. 1, 519, 527, 529. On the history of *'khrungs dpe* in Tibet see Yang Ga 2010, 32–34. Cf. chapter 1, nn. 50, 66, and 92.
226. Czaja 2007 studies the composition and sources for this work.

227. On the possibility of creativity in South Asian commentary see Bronkhorst 2006.
228. The *Four Treatises* commentary by Kyempa Tsewang, which precedes the *Blue Beryl* by more than a century, takes up the tasks and virtues in writing commentary in more impersonal terms: Skyem pa 2000, 4–15.
229. Sde srid 1973a, vol. 4, 492. Cf. Cüppers et al. 2012, 2.
230. *Baiḍūrya g.ya' sel* and *Snyan sgron nyis brgya brgyad pa*.
231. Cf. his comments on his own originality in creating the conception of Lhasa, cited by Schaeffer 2006, 14.
232. Sde srid 1982, 386. Cf. Sde srid 1973a, vol. 4, 491: "I cannot bear for [medicine] to decline."
233. Sde srid 1973a, vol. 4, 490. The word for exposition is *bstan bcos*, or Skt. *śāstra*, but here this does not have the same significance that it does in the Buddha Word debate considered in chapter 3.
234. Although he only says Zur, it is clear that he is referring to *Ancestors' Advice*. At Sde srid 1973a, vol. 1, 151 he claims to have fleshed out what was just a rough account by Zurkharwa when in fact he has copied Zurkharwa's words exactly: see p. 271. He does point to the fact that others copied *Ancestors' Advice* directly, such as Nyitang Tulku: Sde srid 1973a, vol 4, 476–77.
235. Sde srid 1982, 386.
236. Sde srid 1973a, vol. 4, 492.
237. Sde srid 1973a, vol. 4, 490.
238. He is citing *Bodhicaryāvatāra* I.2 (Śāntideva 1995, 5).
239. See Sde srid 1973a, vol. 4, 488–90. He is talking about the *rkyang sel* chapter. Extensive additions were made to the information in the *Four Treatises* on medicinal plants and other *materia medica*, and portrayed in plates 29–31.
240. Sde srid 1973a, vol. 4, 490–91.
241. Sde srid 1982, 381.
242. Sde srid 1982, 384. Cf. pp. 94–95.
243. J. Gyatso 2011a. The Desi's *nyams yig* is Sde srid 1978.
244. Sde srid 1982, 395.
245. Sde srid 1973a, vol. 4, 495.
246. Sde srid 1973a, vol. 4, 483.
247. Sde srid 1973a, vol. 4, 495.
248. Sde srid 1973a, vol. 4, 496.
249. Sde srid 1973a, vol. 4, 496–97.
250. Sde srid 1973a, vol. 4, 491.
251. Sde srid 1973a, vol. 4, 495.
252. Sde srid 1982, 558–59.
253. Sde srid 1973a, vol. 4, 492 (reading *ngo char* as *ngo tshar*).
254. One example of the former is Butön Rinchen Drup who sent his famous history of Buddhism to colleagues before finalizing it: Bu ston 2000, 191–92 (ff. 22b–23a). Thanks to Leonard van der Kuijp for this reference.
255. J. Gyatso 1992.
256. Sde srid 1982, 562–69. Cf. Zur mkhar ba 2001, 246–87.

257. Stein 1972, 195. See also J. Gyatso 1998, 116–18. Among the most common historical genres are *chos 'byung* and *rgyal rabs*. Clan histories are sometimes called *rus mdzod*.
258. See n. 28 in chapter 3.
259. Garrett 2006, 216–17.
260. See pp. 151–52.
261. See n. 34 in chapter 3.
262. *Sman gyi byung tshul khog dbubs*. Martin 2007.
263. Sde srid 1982, 568; cf. 178–79; Zur mkhar ba 2001, 252.
264. Ye shes gzungs (?) 1976, 136 calls itself *don gyi khog dbub stong thun*.
265. Zur mkhar ba 2001, 251; Sde srid 1982, 562. Another *khokbuk* is from the sixteenth century: Czaja 2005a.
266. Occasionally *khokbup* and its variants can actually be used for histories of Buddhism, and so this implication does not hold in all instances, and as already noted, *khogbup/k* sometimes name other kinds of texts as well.
267. Other variants in Tibetan include *khog 'bug*, *khog phub*, *khog 'bubs*, *khog dbub*, *khog dbubs*, and *khog 'bub*.
268. Czaja 2005a, 156–58; Schaeffer 2003, 624, n. 10; Martin 2007, 307; Karmay 1998c, 231.
269. Krang dbyi sun 1993, 241 defines *khog 'bubs* as something that can cover the general meaning of a text, or also an outline thereof: *khog sgrig gi za bcad*; *khog sgrig ste go rim khog 'bubs pa*. The term is also used to describe the act of putting up a tent or an umbrella. Sde srid 1982 seems to prefer *khog 'bugs* (although see 567). In other contexts the term can refer to a handbook of practices; cf. the common genre term *sgrub khog*.
270. *Khog 'bugs khyung chen lding ba* 1976, 11–12. Cf. Sde srid 1982, 7–8. Note that our current version of this work was edited by the Desi's colleagues; Zur mkhar ba 2001, 246 calls it *khog dbub*.
271. *Khog 'bugs khyung chen lding ba* 1976, 11–12 refers to the act of *khog dbug pa*. Zur mkhar ba 2001, 276 concludes his discussion of this work by saying, "thus did [this tradition] lay out the inner space (*khog dbub par byed*) [of the history of medicine]."
272. Brang ti n.d. See Yang Ga 2010, 28–30. There might have been another copy at Lhasa Mentsikhang. Considered in the class of *khokbup/k* by Zur mkhar ba 2001, 252 and Sde srid 1982, 562. For Drangti Pelden Tsoché's dates, see van der Kuijp 2010, 33 n. 2.
273. Zur mkhar ba 2001. See p. 177. Sde srid 1982 provides two different titles for the work, *Khog 'bugs gtan pa med pa'i mchod sbyin gyi sgo 'phar yangs po*, and *Drang srong kun tu dga' ba'i zlos gar*.
274. Zur mkhar ba 2001, 29–30.
275. Sde srid 1982, 566. *Rang nyid kyis gang mthong gso dpyad chos 'byung du ngom pa* (here using *chos 'byung* in the general sense of historical account).
276. Sde srid 1982, 567. The title is *Spyi don legs bshad rgya mtsho drang srong kun tu dga'i ba'i zlos gar*. It is not extant as far as I know.
277. E.g., Drangti's *Shes bya rab gsal* (=Brang ti n.d.); or Jarpo Panchen Dorjé Palam's *Khog 'bugs legs bshad gser gyi snye ma*; or Rinchending Lozang Gyatso Rinchen's *Chos 'byung drang srong dgongs rgyan*: Sde srid 1982, 562–66.

278. Sde srid 1982, 567.
279. Sde srid 1982, 567–68.
280. Cf. Schaeffer 2003.

3. THE WORD OF THE BUDDHA

1. Zur mkhar ba 1986, 64.
2. Bronner 2002, 458.
3. J. Gyatso 1993; 1996.
4. The Buddhist affiliation of *nag rtsis* was later disputed: Schuh 2004. Other branches of Tibetan astronomical conception had long been based on the *Kālacakra*: Schuh 1973c and 1974.
5. We could compare here certain early Buddhist attempts to advert to some realist criteria for what was actually taught by the Buddha or appointed surrogates. See Davidson 1990; Kapstein 1989, nn. 10, 13, citing *Kathāvatthu* and comments of Buddhaghosa. On critical standards in Tibetan historiography, see Bjerken 2001, 83. See also n. 94 and pp. 189–90.
6. Karmay 1998c.
7. Israel 2001.
8. Zur mkhar ba 2003a, 13.
9. Zur mkhar ba 2003a, 13–15.
10. Contrary to Czaja 2005a, 142, Zurkharwa is well known in Tibetan medical circles today.
11. He is the same as Lhatsün Tashi Pelzang, described in Sde srid 1982, 319–21; see also 328. Note that he was the son of the great Jangpa physician Mi'i Nyima Tongden, not just his disciple, as Karmay 1998c, 23 has it. See also Hofer 2012 and 2007. Hofer places his birth date between 1440 and 1450. Cf. Gerke 1999, who dates his birth to approximately 1459, based on Pa sangs yon tan 1988, 100.
12. Sde srid 1982, 328.
13. Sde srid 1982, 350.
14. Samuel 1993 and J. Gyatso 1998 both assume that the Treasure modality served innovation and creativity.
15. Cf. *Carakasaṃhitā* I.1.3–40 and *Suśrutasaṃhitā* I.1. 5–9. Human sages (probably not historical) credited with recording medical science (*cikitsam*) include Ātreya Punarvasu and Agniveśa. See Meulenbeld 1999–2002, vol. IA, 105–15 and 333–57.
16. On canonical Buddhist works that discuss some medical topics, see p. 98. *Aṣṭāṅgahṛdayayavṛtti*, or Tibetan *Zla zer* (see chapter 2, n. 136) begins by making obeisance to the Buddhist *triratna* but is explicitly human-authored. So is the *Siddhasāra*, a medical work from around the time of Vāgbhaṭa, whose author, Ravigupta, was probably Buddhist: Emmerick 1980–82. According to Pierce Salguero (personal communication 2013), the Chinese works *Sūtra on the Buddha, King of Physicians* (*Foshuo Foyi Jing*), and *Sūtra on the Saunas and Baths of the Sangha* (*Foshuo wenshi ziyu zhongseng jing*) are cast as translations of Indic Buddhist sūtras, but their Indic provenance is

questionable, and in any case they only cover a small range of medical topics. The *Sūtra on Saunas* begins with a version of the classic opening "Thus have I heard," but the *Sūtra on the Buddha* does not. See also Salguero 2014.

17. There is also the possibility that some works on healing in the *Rnying ma'i rgyud 'bum* are attributed to the Buddha. Cf. chapter 2, n. 140; Zur mkhar ba 2001, 310; Garrett 2009.
18. See chapter 2, n. 135.
19. *Bi ci'i pu ti kha ser*, 2005. It does provide a Sanskritized version of its title. A third early Tibetan medical work, *Byang khog dmar byang gsal ba'i sgron me*, described by Yang Ga 2010, is also attributed to a human author.
20. These earlier "drafts" of the *Four Treatises* merely pay homage to the Medicine Buddha. *Rgyud chung bdud rtsi snying po* ([G.yu thog] 1976b) is said in its colophon to have been written by Candranandana, but Tibetan scholarship attributes it to Yutok Yönten Gönpo. *Bu don ma* (G.yu thog 2005b) repeatedly attributes its authorship to Yutok. *Nor bu'i 'phreng ba* (G.yu thog 2003) is attributed to Yutok on 255. For more on these earlier drafts see Yang Ga 2010; Yang Ga 2014. See also n. 160.
21. Most notably Sumtön Yeshé Zung. See also Yang Ga 2010.
22. The *Root Treatise* commented upon by Drangti Pelden Tsoché reads "Thus have I heard": Brang ti 1977, 24. Cf. nn. 68 and 110. According to Bstan 'dzin don grub 2005–8, vol. 1, 4, n. 1, the first two printed editions of the *Four Treatises*, the Dratang ([G.yu thog] 2005a) and the edition prepared by Gampopa, read "Thus have I heard." Also, the first edition prepared under the Fifth Dalai Lama, the Potala edition edited by the Desi, and a later edition printed at Kumbum Monastery all read "heard." Seven other editions of the *Four Treatises* compared in Bstan 'dzin don grub 2005–8 read "Thus have I explained." Certain other Tibetan compositions begin with the same opening line as a kind of literary flourish but without implication of canonical scriptural status; the life story of Milarepa would be one example: Heruka 2010, 11.
23. [G.yu thog] 1992, 1–2.
24. Das 1981, 734. Often expanded as *nang pa sangs rgyas pa*.
25. I.e., Avalokiteśvara, Mañjuśrī, and Vajrapāṇi. Caraka here refers to *Carakasaṃhitā*. On works related to *Gso dpyad 'bum pa*, see chapter 2, n. 140.
26. [G.yu thog] 1992, 660.
27. The work even includes a prophecy of future conditions and the person entrusted with its propagation in the degenerate age: [G.yu thog] 1992, 665–67. Cf. J. Gyatso 1998.
28. *Khog 'bugs khyung chen lding ba* 1976. It is not clear who the author is, although it is signed by Yutok Yönten Gönpo. See also Taube 1981, 39–50; Garrett 2006.
29. Zurkharwa couches his critique in terms of its probable Bönpo or Old School affiliation, and it does indeed invoke the primordial Buddha and self-reflexive ground of awareness: *Khog 'bugs khyung chen lding ba* 1976, 12. See Zur mkhar ba 2001, 321; see also 279. Cf. chapter 2, n. 140, and p. 210.
30. *Khog 'bugs khyung chen lding ba* 1976, 20. The third kind, *rjes su gnang ba'i bka'*, is equivalent to Skt. *anujñā*. Cf. the use of *anubhāva* (Tib. *mthu*) as in the *Bhaiṣajyaguru*: Schopen 1978, 80.

31. See Schopen 1978, 166–72; MacQueen 1981 and 1982; Davidson 1990. An early Tibetan discussion would be Bsod nams tse mo 1992–93, 533–34. Thanks to Ian MacCormack for locating it.
32. *Khog 'bugs khyung chen lding ba* 1976, 20–21.
33. *Khog 'bugs khyung chen lding ba* 1976, 26.
34. *Brgyud pa'i rnam thar med thabs med pa* (Ye shes gzungs and Gzhon nu ye shes [?] 1976). It appears to be a composite work, some of it extracted from an apparently longer biography, *Sku lnga lhun grub*, cited in the first line. The latter also appears to be a biographical work by Sumtön Yeshe Zung and Zhönnu Yeshé. *Sku lnga lhun grub* is cited by Brang ti 1977, ff. 36 b; 37a. Karmay 1998c, 229, n. 5 seems to equate *rNam thar med thabs med pa* and *Sku lnga lhun grub*; cf. Ehrhard 2007, n. 17. Karmay 1998c, 229 gives the main title of this work as *Rnam thar bka' rgya can*; cf. Sde srid 1973a, vol. 4, 464. Zurkharwa also mentions a *Gsung mgur bka' rgya ma rnam thar*: Zur mkhar ba 2001, 320; cf. Zur mkhar ba 1986, 71. See also n. 183 and chapter 2, n. 198. Zur mkhar ba 2001, 321; Zur mkhar ba 2003a,11; and Dpa' bo 1986, 1523 all refer to *Med thabs med pa rnam gsum*. There appear to have been several autobiographical fragments. Ye shes gzungs 1981, 31, 33; Zur mkhar ba 2001, 315–20; and Dpa' bo 1986, 1521 all cite verses from Yutok's composition *Shog dril skor gsum*; see also n. 50. Its verses are gathered in 'Ju Mi pham 1992. There is also a *Gsang ba'i rnam thar*: Taube 1981, 48, n. 199. Ehrhard 2007 refers to this material as "the spiritual songs" of Yutok. See also Yang Ga 2010.
35. J. Gyatso 1993.
36. Bkra shis dpal bzang 1986, 102 maintains it is equivalent to 150 human years.
37. Ye shes gzungs and Gzhon nu ye shes (?) 1976, 136. He also calls himself Jñānadhāra, the Sanskrit equivalent. See also n. 51.
38. Ye shes gzungs and Gzhon nu ye shes (?) 1976, 138, 140.
39. Ye shes gzungs and Gzhon nu ye shes (?) 1976, 137; cf. 140. See also n. 183 as well as Coda, n. 44.
40. Ye shes gzungs 1981. Usually abbreviated as *G.yu thog lo rgyus* or *Lo rgyus dge ba'i lcags kyu*. Cited by Skyem pa 2000, 14 and 17; Zur mkhar ba 2001, 320; Sog bzlog pa 1975, vol. 2, 232–33; and Sde srid 1982, 280. See also Ehrhard 2007.
41. See chapter 2, n. 140.
42. Ye shes gzungs 1981, 9.
43. Ye shes gzungs 1981, 31, 33. This text also incorporates long quotes that the author attributes to Yutok's oral lectures.
44. E.g., Ye shes gzungs 1981, 6.
45. Ye shes gzungs 1981, 12 predicts that he will go to the land of the Medicine Buddha after he dies; on 10 the text claims that Yutok would appear to others as a deity when he performed rituals that made him into a *nirmāṇakāya*, implying that this status is dependent on human ritual participation.
46. *Gso dpyad kyi bstan bcos*: Ye shes gzungs 1981, 14.
47. Ye shes gzungs 1981, 14, 18.
48. Ye shes gzungs 1981, 14.
49. This sentiment too is attributed to Yutok's own utterances: see, e.g., Ye shes gzungs 1981, 14, 26–27.

50. Perhaps close scrutiny could reveal which of the two was written first. Ehrhard 2007, nn. 17 and 18, takes the statement *phyi mo zin bris dpe la bshus* to refer to the writing of *Rnam thar med thabs med pa* itself, and hence datable to 1198, but the statement is likely talking about the codification of either the *Heart Sphere of Yutok* cycle or the *Four Treatises*, or both. 1198 is given as the date when the seal on the *Heart Sphere of Yutok* was released (Ye shes gzungs 1981, 40), making that the *terminus post quem* for both biographies. Ehrhard 2007, n. 9 suggests 1201 as the date for *Shog dril skor gsum*, which would make that the *terminus ad quem* for the *Heart Sphere of Yutok Story*, but that reckoning depends on the overall dates for Yutok. Yang Ga 2010 postulates that the *Heart Sphere of Yutok* was already in existence by 1188, thus pushing back the dates of Yutok's birth and death.
51. Yang Ga, personal communication, 2010. Two different disciples of Yutok by the name of Yeshé Zung are listed by Zur mkhar ba 2001, 322: Dzayé Yeshé Zung and Sumtön Yeshé Zung. See also Brang ti n.d., ff. 37b–38a.
52. *Stong thun mdzes pa'i 'ja' ris* 1976, 43. It also labels *Gso dpyad 'bum pa* as Word. This work provides an interesting overview of medical works from India, China, and Tibet. It is signed "Mkhas pa G.yu thog mgon po" in its colophon, but see Zur mkhar ba 2001, 279.
53. E.g., *Sa dpyad sgag mos rngam thabs* 1976, 78. *Ṭīkka mun sel sgron me* also provides a brief discussion of kinds of translation methods: G.yu thog sman pa dge bshes mañju (?) 1976, 122–24. On this work, see Zur mkhar ba 2001, 321.
54. Bronkhorst 2006.
55. See chapter 2, nn. 131 and 272. His *Four Treatises* commentary is Brang ti 1977.
56. Brang ti 1977, 27.
57. See Brang ti 1977, 28. The full discussion of Tanaduk goes from 27.3 to 30.1.
58. The field of the *Bhaiṣajyagurusūtra* is Vaiḍūryanirbhāsa (Tib. Baiḍūrya'i snang ba): Schopen 1978, 37 and 80; see also 235–42.
59. In addition to other places with this name, Mount Sudarśana is one of the golden mountains around Meru: *Abhidharmakośa* 3.48–49; Sasaki 1965, vol. 1, 280 (4139). The name is listed in a variety of contexts in *Mahāvyutpatti*, where it is also rendered Legs mthong or Shin tu mthong; see La Vallée Poussin 1923–31, vol. 2, 141, n. 2.
60. He addresses the Buddha Word controversy at Brang ti n.d., ff. 21a.5–24a.4, and the transmission of medicine in Tibet at ff. 27a.1–33a.7 and 36a.5–41b.1.
61. The section on *śāstra* in Brang ti n.d is at ff. 24a.4–26b.7.
62. On his life see Hofer 2012, 84–92. He summarizes his position in Byang pa 2001; on 9 he also refers to a larger discussion in his *Rtsa rgyud* commentary, but the version we have does not address the introductory section of the *Four Treatises*: Byang pa 2008. However, in a brief comment on 3 he seems to take a different position from Byang pa 2001, characterizing the *Four Treatises* as "made" (*mdzad pa*) by Yutok, and praising the *Four Treatises* for being keyed to the particular conditions of Tibet, unlike the *Aṣṭāṅga*. He also states that although some show that the *Four Treatises* are Buddha Word, there are problems with that position, and he does not rule out of hand that the *Four Treatises* is a *śāstra*. This statement is later contradicted by a reference to the translation of the *Four Treatises* by Vairocana and its subsequent burial as Treasure: 252–53 (cf. n. 65).

63. Byang pa 2001, 7.
64. Here he trades on the ambiguity of the Tibetan term *rgyud*, which I rendered above as "scripture." *Rgyud* usually translates *tantra*, which names many Buddha Word scriptures, but it also names the *Four Treatises*, i.e., *Rgyud bzhi*, which I translate as *Four Treatises* since the work is not like a tantra as we generally know it in Buddhist literature. Cf. n. 126. Note that the Indian medical work *Siddhasāra* seems to refer to itself as a tantra in its introduction and in its first and also other chapters: Emmerick 1980–82, vol. 2, 11–13 passim.
65. Byang pa 2001, 8; 710.
66. See n. 11.
67. At least three versions are available: Bkra shis dpal bzang 1977; 1981; and 1986. Karmay 1998c (orig. 1989) introduced this work to modern scholarship but did not have access to Zurkharwa's writings. On 232, n. 21, Karmay refers to a reference in Sde srid 1982 to another work by Tashi Pelzang, but this is would actually be the same work under discussion. See also n. 81.
68. The Taktenma edition, produced under the direction of Tāranātha (1575–1634), and the third printed version of the *Four Treatises* according to Bstan 'dzin don grub 2005–8, vol. 1, 2, begins, "Thus have I explained at one time." While further versions maintained the "heard" reading, the influential Degé blocks of 1733 ([G.yu thog] 1999?) read "explained" (*bshad pa*). So does the Chakpori print of 1888 ([G.yu thog] 1978) and the Zungchusé edition ([G.yu thog] 1975). See also n. 22.
69. Several Old School Buddhist scriptures also read "explained" in their opening lines; Tashi Pelzang would have known this well. Cf. n. 82.
70. Karmay 1998c, 232.
71. Chapter 6 gives an example of such implications when Tashi Pelzang confuses system with the empirical realities of sex difference (see p. 331 seq.).
72. The specific issue of whether there was an original Indic text is not entirely unprecedented in Tibet; compare the old debate about the apocryphal status of the *Vajrakīlatantra*, which the *Blue Annals* reports was resolved when Sakya Paṇḍita discovered a Sanskrit manuscript of the work. Roerich 1976, 103–106.
73. Bkra shis dpal bzang 1986, 92–97. He also considers the possibility that it was handed down orally.
74. Bkra shis dpal bzang 1986, 104.
75. See p. 180.
76. TBRC mentions a medical *lo rgyus*, but it does not seem to be available. Two versions of his *gSo dpyad nyer mkho gces bsdus* have been published in China, one of which is Stag tshang 2004. Cf. Sde srid 1982, 569 and p. 180.
77. On Shakya Chokden's work on the authorship of the *Four Treatises* see Czaja 2005a, 147, n. 43. Tashi Pelzang is said to have received vows from Shakya Chokden (Hofer 2012, 98), but he must have differed with this teacher on the *Four Treatises*' authorship.
78. Zur mkhar ba 2001, 311–12. I am not confident that Karmay 1998c is correct that Tashi Pelzang is responding to a particular disputational work (*rtsod pa*). Tashi Pelzang's points of discussion start by saying "someone says" and don't appear to be direct quotes.

79. See Shākya mchog ldan 1975, 325–26. Czaja 2005a, n. 43 provides a translation but misconstrues several key points.
80. A detailed account of the story of those origins and its variations in a Tibetan source is Zur mkhar ba 2001, 153–73.
81. There are more than the sixteen points listed by Karmay 1998c. Among others, Karmay fails to mention the debate on Tanaduk and on "one teaching, two teachers" (Bkra shis dpal bzang 1986, 77).
82. Bkra shis dpal bzang 1986, 75–76. Tashi Pelzang shows affinity with Old School doctrines in identifying the teacher of Buddha Word with its addressees; cf. J. Gyatso 1986. Mind-Born is never called "compiler" (*sdus ba po*) in the *Four Treatises* itself, but rather the interlocutor. In a further sign of indebtedness to Old School tradition, Tashi Pelzang invokes the tantra *Guhyagarbha*, which also used the word "explained" in its opening lines. Cf. Zur mkhar ba 1989, vol. 1, 20. See also chapter 5, n. 22.
83. Bkra shis dpal bzang 1986, 76–77. See also n. 169.
84. See pp. 290-99.
85. Bkra shis dpal bzang 1986, 78. While the *Suvarṇa* mentions the eight branches of medicine several times in its chapter 16, it does not seem ever to list them: Emmerick 1996.
86. Bkra shis dpal bzang 1986, 85–87.
87. Bkra shis dpal bzang 1986, 87–88.
88. Bkra shis dpal bzang 1986, 79–80.
89. On *nag rtsis* in Tibet, see Schuh 1972; 1973a; 1973b; and 2004.
90. Bkra shis dpal bzang 1986, 80–82.
91. Bkra shis dpal bzang 1986, 89–91.
92. Bkra shis dpal bzang 1986, 88–89.
93. Bkra shis dpal bzang 1986, 91.
94. Note however Sumpa Khenpo's (1704–87) recourse to historical philological criteria in the course of questioning the Buddha Word status of certain *dhāraṇī* scriptures, briefly mentioned by Kapstein 1989, 238. I use the term "buddhalogical" analogously to the category of "Christology," that aspect of theology concerned with the nature of Christ.
95. Shapin 2010.
96. Zurkharwa cites Kyempa, i.e., Tsoché Tsewang Gyelpo, as a requestor of his *Root Treatise* commentary: Zur mkhar ba 1989, vol. 1, 88. Skyem pa 2000, 39, cites Zurkharwa on the Tanaduk mountain controversy. Kyempa is also one of the responders to Zurkharwa's posted questions: Zur mkhar ba 2003a, e.g., 41 and 43. On Kyempa's life, see Sde srid 1982, 347–48; no dates are provided. The colophon to Kyempa's *Explanatory Treatise* commentary (Skyem pa 2000, 504) names the earth female pig year as its date of completion, which Byams pa 'phrin las 2000, 220, places in the eighth *rab byung*, or 1479. This does not seem possible since Kyempa knows the work of Zurkharwa and was his contemporary. The commentary to the *Instructional Treatise* was completed in his fifty-fourth year: Skyem pa 2000, 996. The commentary to the *Final Treatise* was completed in the sheep year: Skyem pa 2000, 1245. If we consider these years to be in the ninth *rab byung*, it would make the completion date of the work 1547, and his

birth date approximately 1488. Kyempa's birth is dated to 1514 by Yang Ga 2010, 137. See also van der Kuijp 2010, 27, nn. 1 and 2.

97. Skyem pa 2000.
98. Skyem pa 2000, 16; 22–23. *Rtsom gzhi lung gyi brgyud pa; rnam bshad 'chad nyan gyi brgyud pa; sprul pa don gyi brgyud pa.* I have not seen this list before. Regarding the second, he adds that this is like the *śāstras* of the *ḍākinīs*.
99. Skyem pa 2000 16; 17. Cf. [G.yu thog] 1992, 660.
100. Skyem pa 2000, 17–18.
101. Cf. n. 81.
102. Skyem pa 2000, 16–17.
103. Skyem pa 2000, 39.
104. Skyem pa 2000, 18; he also cites the *Aṣṭāṅgahṛdaya* as referring to its own composite nature.
105. Skyem pa 2000, 17.
106. Some attention to temporal issues is in evidence in Indic Buddhist defenses (and presumably critiques) of the authenticity of the Mahāyāna and tantric scriptures (Davidson 1990, 309 and 113) but in far less historical terms than in the *Four Treatises* debate. None of this is to say, of course, that Kyempa and the others were determining historical dates and so on in the way that we do in contemporary scholarship, but only that the basic concept of history was on the table.
107. Skyem pa 2000, 17.
108. Skyem pa 2000, 23.We might recall the famous "indeed" (*kila*) with which Vasubandu punctuated his reproduction of Sarvāstivāda system in *Abhidharmakośa* as a subtle way to indicate his own doubts.
109. In other contexts Tibetan authors sometimes provide a Sanskrit title for a composition as a kind of literary flourish and sign of erudition without any connotation that the entire work was originally written in Sanskrit (cf., e.g., n. 22), but in the case of the *Four Treatises* Buddha Word debate, that implication is clearly at stake.
110. Skyem pa 2000, 30.
111. Skyem pa 2000, 29–30.
112. Skyem pa 2000, 39.
113. Yang Ga concurred with this reading: personal communication 2010.
114. Paul Ricoeur makes a distinction in similar terms but on more nuanced hermeneutical grounds than is intended here: Blundell 38–39.
115. Sde srid 1982, 352 indicates Zurkharwa also has another work on this topic, entitled *Byang ba'i bka' sgrub kyi lan dkar po chig thub bam dbang po'i lag nyal.*
116. See Czaja 2008, 130–34.
117. Zur mkhar ba 1989, vol. 1, 88 dates the completion of the *Root Treatise* commentary under the patronage of Gyelwang Dorjé Drakpa to 1542, and on 664 dates the completion of the *Explanatory Treatise* to 1545. Zur mkhar ba 2003a, 12 puts the beginning of his four-year period under the patronage of Dorjé Drakpa in which he wrote the *Ancestors' Advice* to 1544, but *Old Man's Testament* was written much later and perhaps confuses dates. According to Taube 1981, 63, *Ancestors' Advice* originally had a different arrangement. Czaja dates Zurkharwa's comment on the pulse chapter to 1566:

Czaja 2005a, n. 19. Mipam Gelek, contemporary of the Desi, implies that Zurkharwa first edited and published the *Four Treatises*, then wrote the *Ancestors' Advice* commentary and his medical history *Pitched Interior*: Bod mkhas pa 1986, 4.

118. Zur mkhar ba 1989 vol. 1, 4.
119. Ibid.
120. Zur mkhar ba 1989 vol. 1, 18–22. See n. 22: it seems that the Dratang blocks produced by Zurkharwa do read "heard."
121. Zur mkhar ba 1989 vol. 1, 20.
122. Zur mkhar ba 1989 vol. 1, 21.
123. Ibid.
124. Zurkharwa cites the *sNying rje pad ma dkar po*, which can either refer to Toh. 111 or Toh. 112, i.e., Skt. *Mahākaruṇāpuṇḍarīkasūtra* or *Karuṇāpuṇḍarīkasūtra*. The former is cited frequently by Tibetan authors, but I have not been able to locate the lines that Zurkharwa quotes in any version of the Tibetan canon. However, similar lines, but with no citation, are to be found in Bsod nams tse mo 1992–93, 534. The *Karuṇāpuṇḍarīkasūtra* studied by Yamada 1968 also discusses at length what will happen in the future after the Buddha's demise, but does not include the passage in question either.
125. Zur mkhar ba 1989 vol. 1, 21–22.
126. He rejects the mistaken view that since the work is called *Rgyud bzhi* it is a tantra, and therefore, like other tantras, it must have been preached in Oḍḍiyāna: Zur mkhar ba 1989 vol. 1, 22. He had already distinguished the term *tantra* in the work's title from that of the scriptural Vajrayāna tantras: Zur mkhar ba 1989, vol. 1, 14; 22.
127. Zur mkhar ba 1989, vol. 1, 25.
128. Zur mkhar ba 1989, vol. 1, 24.
129. Zur mkhar ba 1989, vol. 1, 22–25. Cf. pp. 189–90.
130. Zur mkhar ba 1989, vol. 1, 25–26.
131. Zur mkhar ba 1989, vol. 1, 25: *da lta 'dzam gling*. Cf. n. 168.
132. Zur mkhar ba 1989, vol. 1, 24: *don la snyeg pa'i ngo bo mi 'dug pas gyi na ba yin no.*
133. Zur mkhar ba 1989, vol. 1, 26.
134. Ibid., referring to *sngon rabs pa kun.*
135. [G.yu thog] 1992, 1.
136. Zur mkhar ba 1989, vol. 1, 22.
137. Zur mkhar ba 1989, vol. 1, 25–26, citing *Bye brag tu bshad pa*. Zur mkhar ba 1986, 67 attributes the work to Vasubandhu. The *Vibhāṣa* is not thought to have been translated into Tibetan. The passage that Zurkharwa quotes is also quoted in the history of Buddhism by Butön, but without the attribution to Vasubandhu: Bu ston Rin chen grub 1988, 115–16 (cf. Obermiller 1931, pt. 2, 70). It is likely that Zurkharwa got the passage from this text, which he cites elsewhere.
138. Zur mkhar ba 1989, vol. 1, 26.
139. Especially evident in Zurkharwa's detailed treatment of the four mountains: Zur mkhar ba 1989, vol. 1, 28–45.
140. Cf. Davidson 1990 on the shifting of sites for the preaching of Buddha Word in Mahāyāna scriptural apologetics as one means to avoid historicist critique.

141. Lopez 2008, chapter 1. It is not clear that Tominaga, the first to articulate such distinctions, had knowledge of modern geography: Tominaga 1990, 24–30.
142. Zur mkhar ba 2001. It is not clear when this work was written. Czaja 2005a, n. 23 maintains that it was written at the end of Zurkharwa's life but does not give his evidence. Czaja speculates that Zurkharwa changed his mind and only believed that the *Four Treatises* are a Tibetan composition after writing Zur mkhar ba 1986. This is at odds with my own reading, as argued below.
143. Zur mkhar ba 2001, 276–87.
144. Zur mkhar ba 2001, 277–78, citing *Zla zer*.
145. The entire Treasure tradition has the same vulnerability, which arguably accounts for its elaborate theory of transmission. Such a point was noted by the Treasure tradition's critics: Kapstein 1989.
146. Zur mkhar ba 2001, 283–84.
147. Zur mkhar ba 2001, 284. On 310 he repeats the quote, specifying that it is from the *Padma bka' thang* of Orgyen Lingpa.
148. Zur mkhar ba 2001, 285.
149. The six lies of the owl proverb is mentioned in Sangpo 1974, 60.
150. Zur mkhar ba 2001, 287–300.
151. Zur mkhar ba 2001, 300–309. His main biography of Yutok is at 313–22.
152. Zur mkhar ba 2001 309–22.
153. Zur mkhar ba 2001, 310.
154. Zur mkhar ba 2001, 311. He gives it a different title than our current version, and his passage contains the additional words *dpal ldan rgyud bzhi* after *gso dpyad kyi bstan bcos*.
155. Zur mkhar ba 2001, 311–12.
156. Zur mkhar ba 2001, 312.
157. Zur mkhar ba 2001, 312. Cf. *Theg pa chen po rgyud bla ma'i bstan bcos* 1982–85, 144.
158. Cf. Davidson 1990; Kapstein 1989.
159. Zur mkhar ba 2001, 313.
160. One other indication of Zurkharwa's views on Yutok's authorship in his history of medicine is at Zurkharwa 2001, 320–21. He recounts that at the end of his life, after depending only on the *Aṣṭāṅga*, Yutok started writing, after collecting all the medical works in Tibet: first he composed a work on pulse diagnosis; then he wrote *Lag len pod chung*; then *Rgyud chung bdud rtsi snying po*, "made to seem as if it was authored by Khaché Daga (=Candranandana)"; and then the longer *bDud rtsi snying po gsang ba man ngag gi rgyud* in 33 chapters, which, in order to arouse faith, he composed as if it were Buddha Word. Probably the third work Zurkharwa refers to here is the *Four Treatises* itself. Our current edition of *Rgyud chung* has 100 chapters, but is indeed attributed to Candranandana (see n. 20); our current *Rgyud bzhi* has 156 chapters. We don't know of a third work with a similar title in 33 chapters. Cf. Yang Ga 2010, 112–15.
161. Zur mkhar ba 1986, 71. The colophon does not mention the year; Tibetans consider a person to be one year old at birth.
162. Zur mkhar ba 1986, 64.
163. Sde srid 1982, 352 quotes these lines in outrage.

164. Zur mkhar ba 1986, 64.
165. Czaja seems to miss this point and proceeds to misread the entire argument in this text as arguing that the *Four Treatises* are indeed Buddha Word: Czaja 2005a, 134–39.
166. The common taxonomy of three kinds of biography—outer, inner, and secret—is a representative example: J. Gyatso 1998.
167. Zur mkhar ba 1986, 64–66.
168. Zur mkhar ba 1986, 67.
169. Cf. p. 173. The suggestion that Tanaduk as described by the *Four Treatises* did exist during the time of the Buddha is close to the proposal of Tashi Pelzang (see p. 161) with the difference that Tashi Pelzang maintains that the city may well still be in existence but just not perceptible to everyday firsthand witnesses. In any event, Zurkharwa's own statement here is of a piece with his larger strategy to keep Tanaduk out of the realm of the empirical, albeit here granting the magical show some historical specificity. I would only add that if I am right that a scholar such as Czaja was misled by the convoluted rhetoric, he is hardly to be blamed.
170. Zur mkhar ba 1986, 68–69.
171. He largely repeats here his discussion in his history of medicine regarding other scholars who reject the *Four Treatises* as Word, adding now a reference to Shakya Chokden.
172. Zur mkhar ba 1986, 70.
173. Zur mkhar ba 1986. 70. Czaja 2005a, 138 misreads this key sentence and the following argument about the foundational idea (*dgongs gzhi*).
174. Compare the comments on Tibetan credulousness by Tāranātha (1575–1634), a generation after Zurkharwa: "In this land of Tibet, a story irrespective of being true or untrue, if once circulated among the people, [then] nobody will listen to anything else, even though that be a firm truth" (Bjerken 2001, 83).
175. *Dngos la gnod byed.*
176. Zur mkhar ba 1986, 71.
177. Dpa' bo 1986, 1522–23. I am grateful to Yangga for drawing my attention to this passage. Tsuklak Trengwa, like Zurkharwa, is primarily associated with the Karma Kagyü; he also highlights Nyamnyi Dorjé in his survey of Tibetan medicine on 1525, and seems to lean toward the Zurkharwa lineage. See also Taube 1981, 33.
178. According to TBRC (accessed 2012) he was born in 1456, his main title was Pelkhang Lotsawa, and he was a famous grammarian.
179. Mentioned in *Abhidharmakośa* 3.57.
180. "*su mdzad*" at the end of the statement governs all the elements in the sentence. Tibetan historiography locates the birthplace of Yutok Yönten Gönpo in the neighborhood of Gozhi Retang in Tsang: *Zur mkhar ba* 2001, 314; cf. Byams pa 'phrin las, 2000, 128 n. 1. Zur mkhar ba 2003a, 11 refers to Yutok's birthplace as Nyang Tömé. See also Ferrari 1958, 58.
181. Gling sman, 7–8.
182. Zur mkhar ba 1986, 71. On the *Rnam thar bka' rgya ma*, see chapter 2, n. 198.
183. See p. 153. Zurkharwa's version of the *Rnam thar bka' rgya ma* reads differently from the lines from *Crucial Lineage Biography*, which only said that Yutok is like (*'dra*)—rather than the real (*dngos*)—Intelligent Gnosis.

184. Dpa' bo 1986, 1522.
185. Once again, critical assessments of the *Four Treatises*' status come very close to old Buddhist hermeneutics on what is at the heart of authentic Buddha Word, such as kinds of enlightened gnosis (Davidson 1990), but in the medical context the move serves to shift authorship to historical persons (their personal realizations of gnosis notwithstanding).
186. As in Jé Trinlé Zhap himself: Dpa' bo 1986, 1522.5.
187. Dpa' bo 1986, 1522–23. The statement is similar in content but worded differently than Zur mkhar ba 2001, 312.
188. *Bag chags snga ma.*
189. Zur mkhar ba 1986, 71
190. Compare the use of *sngon rabs pa kun* at Zur mkhar ba 1989, vol. 1, 26. *Snga rabs pa* often refers to those from whom Zurkharwa distinguishes himself, e.g., Zur mkhar ba 1989, vol. 1, 150; 166. See also pp. 233–34. Skyem pa 2000 uses the same phrase, e.g., at 1005–6, where the issue also concerns discrepancies between what the text says and what is discovered in practice. For an example of Zukharwa participating in the more conventional rhetoric berating contemporary scholarship see Schaeffer 2009, 79–80.
191. Nattier 1991. On the dark times before the advent of Buddhism in Tibet, see Bjerken, 76 seq.
192. Kapstein 2011. See also Bjerken 2001, 83.
193. Huber 1990.
194. I.e. *dngos po'i gnas lugs* vs. *skyon yon sngags pa*. See Sa skya Paṇḍita 1992–93, ff. 32a–33b seq. Cf. Gyaltshen 2002, 139–41.
195. Newman 1996.
196. Newman 1996, 492. This sentiment continues into the nineteenth century, according to Newman, in Nomonhan Trülku's early nineteenth-century *'Dzam gling rgyas bshad*, which says that Shambhala is an emanated village. Bernbaum 1980, 36–37 reports on a conversation with the Dalai Lama in the twentieth century that attempts to locate Shambhala in the empirical world. See also Kollmar-Paulenz 1992–93; Yoeli-Tlalim 2004.

4. THE EVIDENCE OF THE BODY: MEDICAL CHANNELS, TANTRIC KNOWING

1. Its simultaneity with the public anatomy lessons of the Amsterdam Guild of Surgeons, albeit entirely coincidental, may explain why the episode has received considerable attention in modern scholarship, e.g., Meyer 2003, 110, and in a Tibetan journal on medicine: Dar mo 1996, 22–24. The passage is excerpted from Dar mo 2006.
2. *Dmar khrid du phab pa*: Dar mo 2006, 98.
3. Dar mo 2006 presents this statement as an interlinear note and moves its placement back a few syllables from where it is in Dar mo 1996. This means that the latter reads that they found 365 bones, whereas Dar mo 2006 suggests that they found 360 but would have come up with 365 if they had counted the skull fissures differently.

Compare Dar mo 1996, 24 with Dar mo 2006, 97–98. I do not have access to the original edition of the text and it is not clear where this caveat is positioned, if it is indeed an interlinear note, and if so, who wrote it.

4. A detailed biographical sketch is offered by Dkon mchog rin chen 1994, 107–15. For a bibliography of his writings, see http://www.tbrc.org.ezp-prod1.hul.harvard.edu/#library_person_Object-P1791 (accessed 2012).
5. *Sman bsod.* Sde srid 1982, 378–79. Cf. Sde srid 1973a, vol. 4, 476, where the Desi singles Darmo out as one of the few worthy medical scholars.
6. Sde srid 1982, 555.
7. Sde srid 1973a, vol. 4, 486.
8. See chapter 2, nn. 175, 195, and 206.
9. Dkon mchog rin chen 1994, 105–106.
10. *Suśruta* Śā. 5.17, *Atharvaveda*, and *Caraka* Śā. 30.20 count the bones as 360. See Hoernle 1907; Dasgupta 1952, 278–79. *Visuddhimagga* VIII.101 (Buddhaghosa 1999, 248) also counts only 300 bones. Robert Kritzer, who is engaged in a comprehensive study of the *Garbhāvakrāntisūtra,* informs me that each of our available versions supplies different lists of bones and different total numbers (personal communication, July 2011).
11. *Aṣṭāṅgahṛdaya* Śā. 5 (see Hoernle 1907, 92 seq.); [G.yu thog] 1992, 22; V. Wallace 2010, 13–14; Rang byung rdo rje 1970?, f. 5a. Darmo cites the latter and several unnamed sources at Dar mo 2006, 98.
12. Dar mo 2006, 98.
13. Dar mo 2006, 100.
14. Glossed at Dar mo 2006, 98 as *zhib rtsis*, "detailed count." The term is also used to denote meditation instruction based on personal experience. See J. Gyatso 1981, 104.
15. Dar mo 2006, 98.
16. Dar mo 2006, 100.
17. In particular, Dar mo 2005 is being taken up for close study by Katharina Sabernig with regard to just these questions.
18. See the discussion of the channels of the body in Dar mo 2006, 106–28. Cf. p. 272.
19. See n. 25.
20. See Lopez 2008, 131–52, and T. Gyatso 2005a; cf. Introduction, n. 32.
21. *Des na lus rtsa yi gnas lugs rgyud don zab mo'i dgongs pa mkhas pa'i bzhed tshul 'di bzhin khas ma blangs par 'di las gzhan du bshad na lung rigs gyi gnod pa zhig mi 'grub pas rang bzo spangs nas mkhas pa'i rjes su 'brang ba legs so.* Tshul khrims rgyal mtshan n.d., 88 (my translation). I am grateful to Frances Garrett for sending me this essay. Cf. Garrett and Adams 2008, 110. Note that Tsültrim Gyeltsen argues at some length that just because something is invisible does not mean that it does not exist: Tshul khrims rgyal mtshan n.d., 88–89; Garrett and Adams 2008, 111–12. Cf. p. 161.
22. Garrett and Adams 2008, 94–95.
23. Skyem pa 2000, 128, specifies that the discussion of the black channel begins here, and also identifies where the white vital channel discussion begins, as indicated below.
24. [G.yu thog] 1992, 22–23. In the last two lines there is a play on words between *rtsa ba*, "root," and *rtsa*, which in this context refers to a channel.

25. Yang Ga 2010, 158 fails to find any source for the system of channels represented in the *Four Treatises* in either *Aṣṭāṅgahṛdaya* or earlier Tibetan medical works. The principal discussion of channels in *Aṣṭāṅgahṛdaya* is in Śā. 3. See also Dasgupta 1952, vol. 2, 342. Buddhist lists of body elements such as *Mahāsatipaṭṭhānasutta* 4.5–6 (Walshe 1987, 337–38) and *Visuddhimagga* VIII 83–138 (Buddhaghosa 1999, 244–58) fail to provide a category covering veins or other channels of the body.
26. Cf. the embryology itself: [G.yu thog] 1992, 18.
27. See chapter 2, nn. 96, 97, and 101.
28. *Srogs rtsa dkar po* and *srog rtsa nag po*.
29. [G.yu thog] 1992, 19.
30. See [G.yu thog] 1992, 341 seq.; 434–36; 448; and 471–72 seq.
31. Veins are often termed *khrag rtsa*; arteries are either *'phar rtsa* or *gnyis 'dom*. Thub bstan phun tshogs 1999, 86 provides the term *sdod rtsa* for vein, but its origin is not clear. He also equates *roma* with the main *sdod rtsa sdong po*, which would be the vena cava, and the tantric channel as the *'phar rtsa'i sdong po*, i.e., the aorta, both illustrated with modern biomedical figures on 89 and 90.
32. *Lus 'gul skyod kyi bya ba byed pa*: [G.yu thog] 1992, 23; Skyem pa 2000, 884.
33. See nn. 114 and 115.
34. Cf. Skyem pa 2000, 850–51, citing Somarāja on the "artery trunk" (*'phar rtsa sdong po*) at the thirteenth digit of the spine. For more on the channel trunks see pp. 215–19, 245. Contemporary Tibetan studies of the channels include Thub bstan phun tshogs 1999, 80–112; and G.yang dga' 2002, 81–114.
35. White 1996, 30–31. The proximate source of this idea in Tibet would be Buddhist tantras, especially *Kālacakra*; see V. Wallace 2010, 115–23. Cf. Tsuda 1974, chapter 5. Cf. also Klong chen rab 'byams pa 1983, 210–11.
36. See Tucci 1980, 190–94; Karmay 1998a; Heller 1985.
37. Classically described by Eliade 1958; see also Dasgupta 1952, vol. 2, 352–57, and White 1996. A seminal source in Buddhist tantra would be the first chapter of *Hevajratantra* (Farrow and Menon 1992).
38. The central channel is often *avadhūtī* or *suṣumnā* in Buddhist sources, but most often just called "the central one" (*dbu ma*) in Tibetan.
39. The Indic notion of the subtle body has an extensive history: Dasgupta 1952, vol. 2, 302–12.
40. Most explicit in Zur mkhar ba 1989, vol. 1, 132.
41. Jordens 1978, 28–29.
42. *Suśrutasaṃhitā* Śā. 5.49–56.
43. Yang dgon pa 1976. There are also other manuscript versions extant. It was also published in Dor zhi 1991, 1–108, but incorrectly attributed to Sakyapa Gyeltsen Pelzang in the table of contents. Miller 2013 studies the work in detail.
44. See Garrett 2008.
45. Interlinear notes in this work vary among the various manuscripts. I provide here the rendition from the Pajoding manuscript (Yang dgon pa 1976).
46. These are listed just below, in the quote from *Mahāyogatantra*. It is not clear what text that refers to.

47. *Srog gi dbyug pa.*
48. Yang dgon pa 1976, 434–35. I have deleted a few interlinear notes in my translation; their provenance is uncertain and they differ in the various manuscript versions of the work.
49. It is possible that for Yangönpa the vital channel is either the aorta or the spinal column. Cf. pp. 210, 217. The category of vital channel continues to receive attention in the Tibetan tantric tradition, such as where it is distinguished by Rangjung Dorjé from the central channnel in the third chapter of his own commentary to *Profound Inner Meaning*. Thanks to Willa Miller for this reference.
50. On the principle of *dngos po'i gnas lugs* and Yangonpa's larger arguments, see Miller 2013.
51. As in *Hevajra* I.I.14 seq. Farrow and Menon 1992, 12–23.
52. Yang dgon pa 1976, 425; Zur mkhar ba 1989, vol. 1, 132.
53. E.g., [G.yu thog] 1992, 16.
54. The term *chags pa'i lus* for the embryo occurs in *Sems nyid ngal gso* of Longchen Rapjampa. (Klong chen pa 2000?, 16). *Grub pa'i lus* also refers to the developing body, e.g., Zur mkhar ba 1989, vol. 1, 134, quoting *Zab mo nang don.*
55. Ye shes gzungs 1998; Ye shes gzungs 1976, i.e., *'Bum nag* and *'Bum chung*. "Myriad" means 10,000 but a secondary meaning is simply a large number, which *'bum*, lit. 100,000, also connotes.
56. On Sumtön, see chapter 2, n. 138.
57. *Dwangs ma.*
58. Ye shes gzungs 1998, 27. It is also said to be surrounded by channels that are passageways for the five senses, the emotional afflictions, and mental consciousness. Cf. Ye shes gzungs 1976, 169.
59. I.e., the practices of *thod rgal*. Cf. n. 29 in chapter 3.
60. See, e.g., Klong chen rab 'byams pa 1983, 247.
61. Indic works like *Caraka* are far more liberal in incorporating the metaphysics of the mind in their medical systems. Cf. Dasgupta 1952, vol. 3, 366–73.
62. Byang pa 2001, 77.
63. Byang pa 2001, 78.
64. Byang pa 2001, 84 seq.
65. Byang pa 2001, 90.
66. See Sde srid 1982, 321–25; see also Hofer 2012, 98–99. He was court physician to a Rinpung lord and appears to have been killed during a siege.
67. Bsod nams ye shes rgyal mtshan 2002. A photocopy of a manuscript version of this work in cursive script was given to me in Lhasa in 2000. In J. Gyatso 2004 I incorrectly identified it as the composition of his father, Tashi Palzeng.
68. Including the *Explanatory Treatise* commentary of his father, Tashi Pelzang, According to Sde srid 1982, 321, this work would have been entitled *Rgyud bzhi'i rnam nges dpag bsam ljon shing*. I think I saw it on display in the Tibet Museum in Lhasa in 2000. Cf. http://www.pbase.com/bmcmorrow/image/105568921 (accessed 2013).
69. Bsod nams ye shes rgyal mtshan 2002, vol. 1, 81: *chags gzhi'i ro rkyang dbu ma'o.*
70. Bsod nams ye shes rgyal mtshan 2002, vol. 1 82.

71. Bsod nams ye shes rgyal mtshan 2002, vol. 1, 84. We can confirm that this second kind of channel is the white vital channel, as when it is discussed at Bsod nams ye shes rgyal mtshan 2002, vol. 1, 94.
72. He frequently quotes an authoritative source for this discussion, which is not named. That might indicate that the original equation of the tantric channels with the connecting channels is not his, but that has yet to be determined.
73. See V. Wallace 2010, 71–75.
74. See nn. 31, 34, and 86.
75. Bsod nams ye shes rgyal mtshan 2002, vol. 1, 97.
76. I.e., Rang byung rdo rje 1970?
77. Skyem pa 2000, 126.
78. Skyem pa 2000, 127.
79. Skyem pa 2000, 851.
80. Skyem pa 2000, 127.
81. Skyem pa 2000, 128.
82. Implicitly is *shugs bstan du*.
83. Skyem pa 2000, 128.
84. Ibid.
85. Cf. Skyem pa 2000, 731 specifying that mind, channels, winds and seminal drops are what the tantras discuss with respect to the "vajra body."
86. The quote is from [G.yu thog] 1992, 434. The referent of *de* is unclear, including whether it refers to a vital channel trunk in the singular or plural. The same can be said of *nang rtsa srog rtsa dkar nag sdong po ste*.
87. *Rgyungs pa* often means spinal cord, but since the locative is used, it seems here to mean the spinal cavity. Ye shes gzungs 1998, 28 reads *rgyus pa ltar song*, in contrast to Skyem pa 2000's rendition, *rgyungs par mar song*.
88. Skyem pa 2000, 128–29.
89. Skyem pa 2000, 850. See also Zur mkhar ba 1989, vol. 1, 159.
90. As at [G.yu thog] 1992, 341.
91. Skyem pa 2000, 731–32. See n. 115.
92. *Sdong po*. Zurkharwa too seems to focus on the black channel as the one he will call "trunk." Zur mkhar ba 1989, vol. 1, 165: *srog rtsa nag po ni khrag rtsa'i yal ga thams cad bskyed pa'i gzhi sdong po lta bu yin pa*. This might correspond to the superior and inferior vena cava. Zurkharwa also depicts the black and white channels on 162, where he does not use the term *sdong po* for either of them, and locates both as running inside the spine, albeit using slightly different terms: he puts the black vital channel in the *sgal pa'i nang gzhung*, which picks up the wording of [G.yu thog] 1992, 434, while he locates the white vital channel as *sgal tshigs thams cad kyi nang du zhugs*. A second description of the white vital channel may be found at Zur mkhar ba 1989, vol. 1, 166.
93. Skyem pa 2000, 850: *de gnyis ni bshad rgyud du/ drug par lte ba la rten srog rtsa chags/ zhes pa de'o/*
94. Skyem pa 2000, 129: *srog rtsa dkar po ni gsang sngags nas gsungs pa'i rtsa dbu ma dang mthun pa/chags srid tshe'i rtsa sogs thams cad kyi rten gzhi yin la/nag po ni ro ma dang*

mthun par khrag rgyu bar mngon te. Yang Ga (personal communication 2010) agreed that *mngon* refers to the *Four Treatises* statement, not a channel that is "appearing" in the body.

95. These name specific veins identified in the *Four Treatises*' anatomy. The quote is found at [G.yu thog] 1992, 448.
96. Skyem pa 2000, 129.
97. But see Skyem pa 2000, 135–36, citing *Kālacakra* on the life-related subtle substances of the body.
98. For some of the Tibetan tantric sources with such doctrines, see Garrett 2008, 96–102 and Miller 2013 passim. See also n. 126.
99. We have no reason to think that prenatal infants were the object of medical treatment in Tibetan medicine. The *Four Treatises* only considers how to perform difficult deliveries or to abort the fetus: see p. 314.
100. See e.g., Spyan snga 1982, 3–4. I am grateful to Willa Miller for this reference.
101. See pp. 246–47, 258 seq., and 337 seq.
102. Zur mkhar ba 1989, vol. 1, 154.
103. Zur mkhar ba 1989, vol. 1, 132.
104. *'Grub tu rung ba ma yin*.
105. *Mngon du 'gyur ba*. Zur mkhar ba 1989, vol. 1, 132.
106. See n. 151.
107. Zur mkhar ba 1989, vol. 1, 132.
108. *Commentary of Avalokiteśvara* is a euphemism for *Vimalaprabhā*, the famous commentarial work on *Kālacakratantra*. Thanks to Matthew Kapstein for pointing this out to me (personal communication 2012). Toh. 1347 lists the author as Spyan ras gzigs dbang phyug; cf. Toh. 845. *Kun tu kha sbyor* is probably Toh. 1197, *Kha sbyor thig le*, Skt. *Sampuṭatilaka*; cf. Toh 1199. Like Yangönpa above, Zurkharwa does not specify which *Mahāyogatantra* he is citing.
109. *Tshu rol mthong ba*.
110. Zur mkhar ba 1989, vol. 1, 133.
111. Ibid.
112. *Sbubs*. It is not clear if he saying that the central channel can't be inside of the vital channel because it is the vital channel in total, a point that he makes in a different context: see p. 246. Even there he provides no explanation.
113. Cf. Yang dgon pa 1976, 427 and Zur mkhar ba 1989, vol. 1, 167–68.
114. [G.yu thog] 1992, 23: *bya ba byed pa'i chu rtsa*.
115. See [G.yu thog] 1992, 471, and 341 where four main kinds of wind-bearing white channels are mentioned. But Zurkharwa quotes the *Four Treatises* selectively; see Zur mkhar ba 1989, vol. 1, 151, where he cites the *Instructional Treatise* to gloss the white channels as water channels (*dkar po rgyas par bshad pa la/gzhan dang 'brel ba'i chu rtsa phyi nang gnyis*), but I have not been able to locate this quote in the text. Zurkharwa also states, puzzlingly, that the *Four Treatises* never call the white channels "wind channels" (Zur mkhar ba 1989, vol. 1, 151). Skyem pa 2000, 884, commenting on [G.yu thog] 1992, 471 maintains that the term "water channel," or *chu rtsa*, is a moniker given to the white channels because they look like ligaments (or tendons: *chus pa*).

Just a few lines above, Kyempa states that the white vital channels conduct wind, and that wind makes for the ability of the body to move.

116. Zur mkhar ba 1989, vol. 1, 133.
117. *Don du; nyid.*
118. Zur mkhar ba 1989, vol. 1, 133–34.
119. Skyem pa 2000, 127. See also n. 126. Zurkharwa also cites other tantric works that Kyempa marshaled in the same context.
120. See n. 115.
121. Zur mkhar ba 1989, vol. 1, 134–35.
122. Cf. p. 218 seq. on Kyempa's view on this.
123. See also Spyan snga 1982, 3–4. Yangönpa's idea would seem to be contravened by several of the tantric sources that Zurkharwa cites here, which gloss the central channel as "vital channel."
124. Zur mkhar ba 1989, vol. 1, 135.
125. [G.yu thog] 1992, 18–19.
126. Zurkharwa's quote is from Rang byung rdo rje 1970?, f. 5b seq., which is on the channels in the mature body. Cf. Rang byung rdo rje 1970?, f. 4b, which is in the embryology chapter itself. See also 'Jam mgon kong sprul 1970?, f. 43a.
127. [G.yu thog] 1992, 19. But as per the previous note, *Profound Inner Meaning* itself makes this error, although it is not as beholden to the *Four Treatises* as Zurkharwa's commentary would be.
128. See Garrett 2008, 98–102; cf. Yang dgon pa 1976, 428 seq.
129. Zur mkhar ba 1989, vol. 1, 135.
130. Zur mkhar ba 1989, vol. 1, 150–57.
131. Zur mkhar ba 1989, vol. 1, 150–52.
132. Zur mkhar ba 1989, vol. 1, 151.9–10.
133. Zur mkhar ba 1989, vol. 1, 152: *Zhes gsungs pa ltar sgal pa'i nang du 'byar ba'i tshul gyis sbom phra yang mthe'u chung tsam 'byung ba dngos su mthong zhing/ro ma ni shin tu phra ba la lus kyi g.yas phyogs su gang la'ang ma 'byar ba'i tshul gyis gsungs pa sogs gnod byed shin tu mang ba.*
134. Zur mkhar ba 1989, vol. 1, 152.
135. I.e., at Zur mkhar ba 1989, vol. 1, 133. In that context he was discussing the central channel as the embryonic vital channel in total, which indeed ramifies in the direction of the brain and heart, i.e., above the navel, in addition to below the navel.
136. It is not clear why he makes only the *roma* channels plural.
137. Zur mkhar ba 1989, vol. 1, 153.
138. Zur mkhar ba 1989, vol. 1, 153–54.
139. See, e.g., Zur mkhar ba 1989, vol. 1, 184.
140. Zur mkhar ba 1989, vol. 1, 154. The passage he cites is at [G.yu thog] 1992, 435, but that specifies the ninth digit of the spine, while Zurkharwa has the passage addressing the eighth digit. Cf. Zur mkhar ba 1989, vol 1, 162.
141. [G.yu thog] 1992, 19.
142. Zur mkhar ba 1989, vol. 1, 154–55.

143. Cf. [G.yu thog] 1992, 22.
144. Zur mkhar ba 1989, vol. 1, 155.
145. See chapter 5, n. 28. See also Zur mkhar ba 1989, vol. 1, 135, on the movement of wind and vitality at the heart.
146. Zur mkhar ba 1989, vol. 1, 155–57.
147. *Ye shes kyi khams.*
148. Zur mkhar ba 1989, vol. 1, 155–56.
149. Zur mkhar ba 1989, vol. 1, 156.
150. See chapter 6, n. 96.
151. Zur mkhar ba 1989, vol. 1, 156–57. The last line of this passage is dropped in Zur mkhar ba 1989 vol 1, 157.1 (following *dmar rgyu ba'i*), but the Lhasa Zhöl blockprint provides the missing words: Zur mkhar ba 1980–85, vol. 1, Bshad rgyud f. 58a: *mtshams 'di nas phran bu zhig chad 'dug pa slar brtag/*
152. Neither the *mo'i gsang mtshan* nor the *mo nad* chapter in *Man ngag rgyud* provides anatomical detail on the female genitalia. See pp. 313–14.
153. Zur mkhar ba 1989, vol. 1, 110–11; 125. See p. 316 seq. A century later, Darmo and the Desi do add a few tantric images to their description of female menstrual physiology, e.g., at Dar mo 1989 vol. 2, 314.
154. Zur mkhar ba 1989, vol. 1, 135. See p. 230.
155. Zur mkhar ba 1989, vol. 1, 157. The entire discussion of the channels of being runs from 157–59.
156. Zur mkhar ba 1989, vol. 1, 157: *srid ces pa de nyid dngos po 'ga' zhig la brten nas gnas pa'i don.*
157. Zur mkhar ba 1989, vol. 1, 157. He is citing *Aṣṭāṅgahṛdaya* Śā.3. 39–40. Cf. *Yan lag brgyad pa* (Arura ed.) vol. 1, 714. The *Four Treatises* itself uses the phrase *rtsa bo che* in its description of the channels of being: [G.yu thog] 1992, 22.
158. Zur mkhar ba 1989, vol. 1, 157.
159. A mainstream Buddhist account of the *indriyas* is found in the second chapter of *Abhidharmakośa.*
160. Zur mkhar ba 1989, vol. 1, 158; cf. [G.yu thog] 1992, 434.
161. Zur mkhar ba 1989, vol. 1, 157. But see chapter 5, n. 28, for *Instructional Treatise* commentators who did add tantric language here.
162. *'khor lo bzhi 'am drug.* He also uses the phrase *'khor lo bzhi'am drug* on 133.
163. Zur mkhar ba 1989, vol. 1, 158–59.
164. Zur mkhar ba 1989, vol. 1, 158.
165. Zur mkhar ba 1989, vol. 1, 166–67.
166. Zur mkhar ba 1989, vol. 1, 168.
167. Zur mkhar ba 1989, vol. 1, 166–67 argues that even though the black vitality channel system does pervade the whole body, it can't be considered the life channel since there are also other channels, namely the arteries (*'phar rtsa*), apparently not considered part of the black vitality channel system, that would also have to be considered the life channel. This is hardly a compelling point.
168. Zur mkhar ba 1989, vol. 1, 167; see also 168.

169. Zur mkhar ba 1989, vol. 1,167.
170. Zur mkhar ba 1989, vol. 1, 168–72.
171. Zur mkhar ba 1989, vol. 1, 170.
172. Zur mkhar ba 1989, vol. 1, 173.
173. Zur mkhar ba 1989, vol, 1, 173; cf. [G.yu thog] 1992, 558 seq. On the anatomy of *bla rtsa* see Dar mo vol. 2, 1989, 406–407.
174. Zur mkhar ba 1989, vol. 1, 173: *'dir kha cig na re/khyod kyis bshad pa de ltar na rtsa'i dbye bar mi 'gyur te/ srog gi cha shes kho nar bshad pa'i phyir zhe na bden mod/'on kyang de rnams la rtsa'i ming gis btags pa tsam ma gtogs rtsa dngos ni ma yin kyang/de rnams kyi rten rtsa kho na yin pa'i dbang gis 'dir bshad pa yin la.*
175. The Desi's own rhetoric betrays an expectation that physicians readily know the connecting channels, but have less knowledge about the other three kinds of channels: Sde srid 1973a, vol. 4, 487.
176. Zur mkhar ba 1989, vol. 1, 159–66.
177. Hence the detailed anatomy of the white and black vital channels in chapter 60 on the maladies of the white channels, chapter 85 on wounds to the torso, and chapter 86 on wounds to the limbs in the *Instructional Treatise*.
178. Zur mkhar ba 1989, vol. 1, 159.
179. *Lung rigs mngon sum*. Zur mkhar ba 1989, vol. 1, 160.
180. Here he cites the *Vajramālā*, a tantric scripture, to disqualify the equation with the medical channels. Zur mkhar ba 1989, vol. 1, 160.
181. Zur mkhar ba 1989, vol. 1, 165; see p. 230.
182. Actually all four kinds of channels are glossed by the Four Treatises as "connecting channels" (*'brel rtsa*) in a very general sense: [G.yu thog] 1992, 22: *'brel pa rtsa yi gnas lugs bstan pa ni/chags srid 'brel pa tshe yi rtsa dang bzhi.*
183. Zur mkhar ba 1989, vol. 1, 165.
184. Zur mkhar ba 1989, vol. 1, 166.
185. Zur mkhar ba 1989, vol. 1, 162.
186. [G.yu thog] 1992, 25.
187. This was already anticipated at the end of the *Four Treatises*' main statement on the four kinds of channels: [G.yu thog] 1992, 23.
188. Zur mkhar ba 1989, vol. 1, 184.
189. Skyem pa 2000, 143, and even though he is citing the tantric *Profound Inner Meaning* here.
190. Zur mkhar ba 1989, vol. 1, 162, commenting on the black vital channel: *sha khrag 'phel bar byed pa'i nyer len gyi rtsa yin pa'i don gyis srog pa rtsa shes pa.*
191. It is commenting on a statement in [G.yu thog] 1992, 559 having to do with the relation of a category of light and shade (*gdags sribs*) to the breath patterns of humans. On the latter, a category influenced by Chinese medicine, see pp. 255, 267, and chapter 6, n. 3.
192. Zur mkhar ba 1989, vol. 1, 690.
193. In another passage, regarding the spots on the arm where the physician reads the pulse of various organs, Zurkwarwa rejects again the idea that the tantric central channel is being read in the pulse diagnosis: Zur mkhar ba 1989, vol. 1, 693.
194. Zur mkhar ba 1989, vol. 1, 690.

5. TANGLED UP IN SYSTEM: THE HEART, IN THE TEXT AND IN THE HAND

1. The term *rtsa*, normally "channel," can also be translated as "pulse." For an overview, see Rabgay 1981a and 1981b. Comparative studies of pulse diagnosis include Hsu 2008 and Kuriyama 2002.
2. See Kuriyama 2002, 26–27. Cf. Liu 1988, 214.
3. [G.yu thog] 1992, 559–60.
4. Skyem pa 2000, 1028, merely glossing the gender difference as "method and primal awareness and so on."
5. The tip of the heart is already said to face downward in *Suśruta* Śā. 4.32.
6. Tashi Pelzang argues that the heart does not lean at all: see Coda, n. 5. The Desi and Lingmen Tashi maintain that the heart does lean, but in the same direction for both sexes: see n. 32 and Coda, n. 10.
7. *Zur mkhar ba* 1989, vol. 1, 696–98; the verses attributed to Tashi Pelzang are cited at 696.
8. Cf. Bkra shis dpal bzang n.d., 293–98. The verses cited by Zurkharwa are split between 296 and 298.
9. See chapter 6, n. 96.
10. See [G.yu thog] 1992, 564–65. Part of this passage was discussed by Emmerick et al. 1988. The pairing of light and shade also pertains to astrology, as well as to kinds of breath; see [G.yu thog] 1992, 559. See also p. 267 and n. 15.
11. See [G.yu thog] 1992, 559.
12. Zur mkhar ba 1989, vol. 1, 696. Zurkharwa references the "Jangpa master's" middle composition" (*rtsom pa bar pa*) and his *Bka' phreng mun sel*. We do not seem to have access to either of these works now. Cf. Hofer 2012, 89.
13. Zur mkhar ba 1989, vol. 1, 696: *ngo mtshar kyi gnas yin*.
14. Zur mkhar ba 1989, vol. 1, 696: *de gnyis mi mthun phyogs 'gal ba yin. . . .*
15. See also Zur mkhar ba 1989, vol. 1, 688–91, commenting on [G.yu thog] 1992, 559, where Zurkharwa characterizes the *gdags sribs* distinction as a linguistic heuristic (*tha snyad btags pa*).
16. Zur mkhar ba 1989, vol. 1, 722. Zurkharwa also references *Somarāja* on this idea: Zur mkhar ba 1989, vol. 1, 693.
17. This reading of the passage was suggested to me by Tupten Püntsok (personal communication 2000). Cf. Zurkharwa's further point on 696 that if the hollow organs follow on the solid ones (i.e., in Tashi Pelzang's way of understanding that) then there would be no need to explain each of the six hollow organ channels, since they already follow (ineluctably) on the solid ones. In other words, either collapsing the solid and hollow organs since one entirely entails the other or separating them as polar opposites fails to appreciate their true nature, which is complexly related, but neither the same nor totally different.
18. *Cho rol tu mi 'byung zhing*.
19. Zur mkhar ba 1989, 722.

20. Cf. Coda, n. 5.
21. In the wrong view of Tashi Pelzang, as characterized by Zurkharwa, the sides to which male and female hearts would face has switched. See p. 267 Cf. Lingmen's critique of a similar wrong view: see Gling sman 1988, 477–78.
22. Zur mkhar ba 1989, 697. Again, Tashi Pelzang's assumptions align him with the Old School, where there is a tradition that *roma* and *kyangma* are on reverse sides of the body in males and females: see, e.g., 'Jigs med gling pa 1985, 66.
23. Zur mkhar ba 1989, 698.
24. The association between the two tantric channels, primal awareness and means, and gender is already clear in *Hevajra* 1.15–16 (Farrow and Menon 1992, 13). Note, however, that *Hevajra* here associates *rasanā*, bearer of female substances, with means, and *lalanā*, the bearer of semen, with primal awareness. Elsewhere *Hevajra* invokes the conventional associations, as at 8.26.
25. Zur mkhar ba 1989, 698. Just what "at" means is critical to the subtle strategy here. Zurkharwa's phrase is *snying gi rtse mo na. . . .* In order to do the work that is necessary, the aperture could not be centered on the very tip as such but would have to be at one side or the other, even if very close to the tip. On the Desi's clarification see Coda, n. 10.
26. [G.yu thog] 1992, 31; 387.
27. Kha che Zla ba mngon dga' 2006, vol. 1, 809. The Desi also cites another passage from *Aṣṭāṅgahṛdaya*: Sde srid 1973a, vol. 3, 62.
28. See Skyem pa 2000, 775–76: the lowermost channel at the heart is the mind consciousness' path, blue in color. Cf. Ye shes gzungs 1998, 27 and Ye shes gzungs 1976, 169. Zurkharwa never commented on this section of *Four Treatises*. But see Dar mo 1989 vol. 2, 328; Sde srid 1973a, vol. 3, 62; Dbang 'dus 1983, 549–50 and illustration 20; and Gling sman 1988, 478.
29. See pp. 210 and 236–37.
30. As far as I know; perhaps it is to be found in a tantric work, but none is cited here.
31. Gling sman 1988, 477: *snying gi rtse mo de ltar bsten zhes pa ma nges te*. On this substitution of *bsten* for *bstan* see p. 262.
32. Gling sman 1988, 478: "*bdag gis ni pho mo mang po'i ro bshas pa mthong / rang gis kyang gri snying blangs pas pho mo thams cad snying rtse cung zad g.yon phyogs brang ngos la bsten pa mthong*."
33. http://library.thinkquest.org/J0112390/circulatory_system.htm; http://www.naturalhealthschool.com/8_1.html (accessed 2011): "The apex, or bottom of the heart, is tilted to the left side."
34. Gling sman 1988, 477–78. The text is a modern edition and may inaccurately represent Gling sman's original spelling.
35. That is, at least in the editions that I have consulted and in Tashi Pelzang and Zurkharwa's representation of the passage.
36. Or perhaps the edition of the *Four Treatises* that he was using had already emended the line as an intervention in the ongoing consternation over its meaning.
37. See pp. 202 and 246.
38. See n. 28.
39. Gling sman 1988, 478.

CODA: INFLUENCE, RHETORIC, AND RIDING TWO HORSES AT ONCE

1. Even Darmo Menrampa, who may have added many new specifications to the *Four Treatises'* anatomy based on extensive empirical research (personal communication, Kathrina Sabernig), seems to have demurred before questioning the material existence of the tantric channels. See p. 272.
2. Bkra shis dpal bzang n.d., 294.5–297.2.
3. Bkra shis dpal bzang n.d., 296.6–297.2.
4. Bkra shis dpal bzang n.d., 296.2.
5. Bkra shis dpal bzang n.d., 297.5–298.1: *spyir snying gi stod sa lus kyi dbus drang po na gnas kyang, snying rtse mo'i ston lugs de ltar du go zlog du stan pa'i phyir te. Skyes pa rnams la snying rtse bu kha [sic] g.yon gyis phyogs bsegs stod la stan/bud med rnams la bu kha g.yas kyis phyogs bsegs stod la stan mod pas/snying rtsa pho mo gnyis bo zlog 'byung bas de dang de'i phyir ro.*
6. Bkra shis dpal bzang n.d., 298.1.
7. *Ancestors' Advice* precisely quotes the verses cited in Tashi Palzeng's *Final Treatise* commentary (see chapter 5, n. 7) and represents other points from Tashi Palzeng's work in prose.
8. See p. 388.
9. Sde srid 1973a, vol. 4, 39 (cf. Zur mkhar ba 1989, vol. 1, 697). He makes the important clarification that the heart's mind opening is on the side of the tip when he says it is close (*nye ba*) to the left of the heart tip in males and the opposite in females.
10. Sde srid 1973a, vol. 3, 218.
11. Cf. p. 55.
12. Sde srid 1973a, vol. 1, 150–75. He omits most of Zurkharwa's topic headings and the critique sections. Cf. Zur mkhar ba 1989, vol. 1, 152–66. See also nn. 14 and 16, and chapter 5, n. 69.
13. Sde srid 1973a, vol. 1, 151. This might conceivably be read to refer to Zurkharwa Nyamnyi Dorjé, but that would only be more disingenuous, given his close dependence on Lodro Gyelpo.
14. In some cases the Desi's language departs somewhat from Zurkharwa's, e.g., at Sde srid 1973a, vol. 1, 158 seq.
15. Sde srid 1973a, vol. 1, especially 159–63.
16. Sde srid 1973a, vol. 1, 128. The Desi deletes Zurkharwa's critiques at Zur mkhar ba 1989, vol. 1, 132–33, 152, 159–60 and 167, all discussed above in chapter 4.
17. See chapter 5, n. 28.
18. Sde srid 1973a, vol. 3, 219–20.
19. Dar mo 2006, 106–28. We can also find a few more references to tantric channels in Darmo's commentary on the *Instructional Treatise* (Dar mo 1989) than are evident in the Desi's *Blue Beryl*, but here Darmo is circumspect and largely sticks to the root text's distinctly nontantric formulations, as does the Desi himself. According to Katharina Sabernig (personal communication 2013), elsewehere Darmo provides an expanded account of the medical cardiovascular and nervous systems based on his examination of human corpses. His entire *oeuvre* deserves close study.

20. The tantric channels of the *Kālacakra* system had already been portrayed in a pair of images attributed to the eighth Karmapa, Mikyö Dorje (1507–1554), recently acquired by the Rubin Museum in New York City. See http://kalacakra.org/phyinang/phyin5.htm and http://kalacakra.org/phyinang/phyin6.htm (accessed 2013). These appear unrelated to the medical systems of Sowa Rikpa. I am grateful to Michael Sheehy for bringing them to my attention.
21. This count includes the two extra plates, numbered 6 and 7 in Wang le and Byams pa 'phrin las 2004, which are missing in the Ulan Ude set.
22. This colophon, which continues on plate 10, breaks off and is incomplete in both the Ulan Ude and Lhasa Mentsikhang versions.
23. The statement may refer to the side image on plate 10, but I cannot locate the indicative letters *pa* or *na* on plate 9 or 10, or any image illustrating vital channels mixing with the tantric central channel.
24. *Ro rkyang gi rang rtags kyi dkar dmar mdog. . . .*
25. The caption at the top reads *ro rkyang dbu gsum mtshon byed las cha dpyad med*. The colophon says *'di la'ang dbu rkyang ro gsum rags tsam mtshon byed du bkod pa.*
26. Actually plate 9 does double duty, showing primarily the bloodletting vessels that are part of the connecting channels system. But it also references the growth channels and the channels of being.
27. There is at least one other, seemingly random tantric intervention on plate 47, which covers the iconometric conventions for much of the frontal anatomy. On the right figure is a channel that is specified to come from Great Perfection tradition on the path of primal consciousness from the heart to the eye. It appears to have nothing to do with the rest of the plate.
28. Gling sman 1988, 44–45. He attributes such a distinction to the teachings of an unspecified "Khyenzig Rinpoché."
29. Gling sman 1988, 46. Interestingly, Lingmen includes Rangchung Dorjé's *Profound Inner Meaning* in the category of material conceptions of the body, along with medicine, or Sowa Rikpa, citing again the teachings of Khyenzig Rinpoché.
30. 'Jam mgon kong sprul 1970?, ff. 49b–51a.
31. Thub bstan phun tshogs 1999, 86; illustration on 89.
32. Adams 2002; 2007; Adams and Li 2008.
33. The special efforts of the Desi in promoting the *Four Treatises* as Buddha Word is also noted by Schuh 2004, 21.
34. Sde srid 1973a, vol. 1, 13 seq. The only thing that tips his hand is a brief salvo on the troublesome word "explained" in the deviant opening line. The Desi opines that the fact that the compiler of the text is a "speech manifestation" of the Buddha and no different than its original teacher (which justifies the word "explained") makes the *Four Treatises* better Word than most.
35. Sde srid 1982.
36. The earlier account begins at Sde srid 1973a, vol. 4, 409; see especially 457 seq. His discussion in Sde srid 1982 goes from 40 to 86. See also Taube 1981, 32–33.
37. For example, the Desi copies Zur mkhar ba 2001, 287–94 almost verbatim, but then adds his own information on Vairocana, the translation of the *Four Treatises*, and its

burial as Treasure, such as at Sde srid 1982, 160–71, that is missing in Zurkharwa's history. There are other examples as well.

38. Sde srid 1982, 160–67. For the larger legend of Vairocana, see Karmay 1988a, chapter 1.
39. Sde srid 1982, 202–206. We can compare the less detailed biographical sketch from the Treasure tradition in its twentieth-century iteration: Dudjom Rinpoche 1991, vol. 1, 753–54.
40. The full history of the development of these legends remains to be traced. References to Vairocana and Drapa Ngönshé may be found in Ye shes gzungs and gZhon nu ye shes (?) 1976, 135, and Brang ti n.d., ff. 32b and 36a.
41. See chapter 2, nn. 200 and 201.
42. His main sections on Yutok the Younger are at Sde srid 1973a, vol. 4, 459–65; Sde srid 1982, 225–29; 275–84.
43. Sde srid 1973a, vol. 4, 463–64.
44. Sde srid 1973a, vol. 4, 464. Like Zurkharwa, the Desi is actually citing *Rnam thar bka' rgya ma*, whose statement about Yutok reads *rig pa'i ye shes dngos*. Cf. chapter 3, n. 183.
45. Sde srid 1973a, vol. 4, 464; cf. Ye shes gzungs 1981, 14.
46. Sde srid 1973a, vol. 4, 461; 463.
47. Sde srid 1973a, vol. 4, 464.
48. Sde srid 1982, 275. Elsewhere the Desi indicates that Oḍḍiyāna is only one of the places with which Tanaduk has been identified. See pp. 280–82.
49. Sde srid 1973a, vol. 4, 464.
50. Sde srid 1973a, vol 4, 461–62.
51. Sde srid 1973a, vol. 4, 465.
52. Yang Ga opined that this statement implies that the Desi considered the entire *Root Treatise* to be Yutok's invention (personal communication 2010).
53. Sde srid 1973a, vol. 4, 462; Sde srid 1982, 275. On *Somarāja* see chapter 2, n. 135.
54. Sde srid 1973a, vol. 4, 461.
55. Sde srid 1982, 229: *'gyur gsar du bcos zhing. . . .*
56. Sde srid 1973a, vol. 4, 462: *thams cad bka'i tshul ni . . . thams cad bka' ltar mdzad pa*. Cf. Zur mkhar ba 1986, 70: *bka' yin pa ltar ma byas na*; and 70–71: *bka' yin pa ltar mdzad pa*. See also Sde srid 1982, 275, where he says that since he had established himself as the body of the Medicine Buddha in reality, Yutok had "completed all the conditions for being a composer."
57. For the further fortunes of the debate see Taube 1981, 33 seq. I have not been able to consider the very complex arguments in Sog bzlog pa 1975, which followed on Zurkharwa and was known to the Desi (see Sde srid 1982, 52).
58. Sde srid 1973a, vol. 1, 16- 29; Sde srid 1982, 51–58.
59. Sde srid 1982, 51.
60. Sde srid 1982, 53–54. This is *De bzhin gshegs pa'i stobs bskyed pa baiḍūrya'i 'od ces bya ba'i gzungs*. I have not been able to identify this work. Cf. chapter 3, n. 137.
61. Sde srid 1973a, vol. 1, 18. He does take issue with Zurkharwa's comments on Oḍḍiyāna: Sde srid 1982, 57–58; Sde srid 1973a, vol. 1, 17. Cf. Zur mkhar ba 1989, vol. 1, 22.
62. Sde srid 1973a, vol. 1, 19 seq.

63. See pp. 84–85 and 182–83.
64. Sde srid 1973a, vol. 1, 16–17. Cf. Sde srid 1982, 275. The verses that he cites are found in the biography at Jo bo and Dar mo 2005, 162–63. See also Sde srid 1982, 52 and 565–66.
65. But the Desi rejects the most radical version of Tanaduk, which he attributes to Sokdokpa, namely that the land of the Medicine Buddha is emanated by Yutok in the realm of the samsaric world. Sde srid 1982, 52. Cf. Sog bzlog pa 1975.
66. Cf. one more use of the heuristic at Sde srid 1973a, vol. 4, 461–63.
67. Sde srid 1982, 274–75; see also 220–21.
68. Sde srid 1982, 229–75. The life story, which began on 225, picks up again at the end of this detour.
69. The Desi highlights Zurkharwa's membership in the Karma Kagyü lineage in the first line of his biographical sketch: Sde srid 1982, 349. Yang Ga, one of several Tibetan scholars who have suggested orally to me the political background to the Desi's dislike of Zurkharwa Lodrö Gyelpo, also pointed out that in the *Blue Beryl*'s section on embryology, the Desi reproduces all of Zurkharwa's quotes, but oddly omits the one that Zurkharwa attributes to Rangjung (i.e., Karmapa Rangjung Dorjé): Sde srid 1973a, vol. 1, 129; cf. Zur mkhar ba 1989, vol. 1, 133. And yet a few lines later, he does include the Karmapa's *Nang don rang 'brel* that Zurkharwa also cites, but referenced by title instead of by author.

6. WOMEN AND GENDER

1. "Gender" has been used in this sense since the medieval period. Ann Oakley (1972) is sometimes credited with articulating the modern sense of the term. Judith Butler's *Gender Trouble* (1990) influentially challenged the sex-gender distinction.
2. Yang Ga 2010.
3. One case of an issue of national pedigree may be found at Zur mkhar ba 1989, vol. 1, 693 where the Chinese origin of the light/shade dyad (*gdags sribs*) has significance.
4. *Yan lag;* Skt. *aṅga.*
5. Skt. *kāya*; *bāla*, *graha*; *ūrdhvāṅga*; *śalya*; *daṃṣṭrā*; *jarā*; and *vṛṣa*. The list of eight varies: Vogel 1963, 290–95. Vogel impugns Butön undeservedly, based on Vogel's own ignorance of the Tibetan introduction of a female pathology branch and several other misconceptions. See also n. 24.
6. Yang Ga 2010 suggests that Sumtön Yeshé Zung might have been involved in the codification of the *Four Treatises.*
7. *Lus*; *byis pa*; *mo nad*; *gdon*; *mtshon*; *dug*; *rgas*; *ro tsa*: [G.yu thog] 1992, 5.
8. G.yu thog sman pa dge bshes mañju (?) 1976 and 'Tsho byed dkon skyabs 1976. The former (*Ṭīkka mun sel*) is implied in its colophon to have been composed by Yutok but Zur mkhar ba 2001, 321 doubts that.
9. *Gso dpyad spyi.*
10. 'Tsho byed dkon skyabs 1976, 307–8.
11. See e.g., G.yu thog sman pa dge bshes mañju (?) 1976, 126: *skyes pa dar ma'i lus gtso bor byas pa.* [G.yu thog] 1992, 8, counts seventy chapters for the body branch; three each

for the pediatrics, female pathology, and poison branches; five each for the demons and wounds branches; one for geriatrics; and two for the fertility/virility branch. All save the body branch are found only in the *Instructional Treatise.*

12. G.yu thog sman pa dge bshes mañju (?) 1976, 127.
13. 'Tsho byed dkon skyabs 1976, 306–7.
14. G.yu thog sman pa dge bshes mañju (?) 1976, 126–27.
15. G.yu thog sman pa dge bshes mañju (?) 1976, 128.
16. 'Tsho byed dkon skyabs 1976, 307; G.yu thog sman pa dge bshes mañju (?) 1976, 128–29. *Bu* can mean either son or child; see discussion below. On the *samse'u*, see n. 96.
17. The uterus is alternately called *bu snod* and *mngal* in Tibetan medicine. I have not discerned a clear distinction between the two terms in medical writings.
18. 'Tsho byed dkon skyabs 1976, 308.
19. This point is confusing since diseases of the uterus are covered in the *Instructional Treatise*, not the *Final Treatise.* In addition, I don't see where twenty kinds of uterus problems are listed anywhere in the work. The passage may reference an earlier version of the *Four Treatises.*
20. 'Tsho byed dkon skyabs 1976, 306.
21. See 'Tsho byed dkon skyabs 1976, 307–8.
22. However, some chapters seem to leave aside female pathologies intentionally. [G.yu thog] 1992, 39–40 discusses blood imbalances, but does not mention female disorders, the large majority of which result from excesses of blood.
23. The one exception is information on pregnancy. See Selby 2005, 256–75 on the sections of *Suśrutasaṃhitā* where female disorders are scattered. *Caraka* provides at least one chapter devoted to "Disorders of the Female Genitalia" (*yoni-vyāpat*).
24. Zur mkhar ba 1989, vol. 1, 62–64. The latter idea is attributed to one Tsangtö Dargön.
25. *Skabs.* See Zur mkhar ba 1989, vol. 1, 62–63.
26. See Scott 1993.
27. Scott 1993, 188. See also Minow 1990 and Lacquer 1990.
28. J. Gyatso 2011a.
29. Za ma mo. See p. 323 and J. Gyatso 2003, nn. 75 and 76.
30. Here *khu ba* names both the male and female sexual fluids. In the next line the text refers to her fluid as *zla mtshan.* In the next line *khu ba* would seem refer specifically to the male sexual fluid, as it usually does.
31. The Degé edition and others say twelve but most say thirteen: Bstan 'dzin don grub 2005–8, vol. 2, 413.
32. [G.yu thog] 1992, 375.
33. Actually men are regularly recognized to have breasts (*nu ma*), from which certain anatomical measures are often specificed (e.g., Skyem pa 2000, 847). But the idea of the breast as a fleshly organ only applies to women. On occasion nipples are recognized in males. Zur mkhar ba 1989, vol. 1, 185 records an objection that men have apertures at the breasts too. For Ayurvedic references to female anatomy as extra see *Suśruta* Śā. 5.9 and 5.42. *Aṣṭāṅgahṛdaya* Śā. 3.41 specifies that the female has three extra orifices.
34. See [G.yu thog] 1992, 21, 22, and 25; Zur mkhar ba 1989, vol. 1, 146–47; and J. Gyatso 2011a.

35. [G.yu thog] 1992, 42.
36. Also, the exact count varies. G.yu thog sman pa dge bshes mañju (?) 1976, 127, counts forty-two special illnesses in females. Darmo discusses other ways to count them: Dar mo 1989, vol. 2, 310.
37. Three for men, two for women, as well as five for children and two for the elderly. Skyem pa 2000, 1006.
38. Skyem pa 2000, 760. Dar mo 1989, vol. 2, 310 adds that she is thereby said to be the best illusion.
39. See, e.g., *Hevajratantra* II.4.33–35; 43–50: Farrow and Menon 1992, 214–15; 217–19.
40. *Skabs thob*. Skyem pa 2000, 760. Cf. Dar mo 1989, vol. 2, 310.
41. Skyem pa 2000, 988, glosses *ro tsa* as if it had a Tibetan etymology: "*Ro* is the aspect that is the experience (*myong ba*) of the relish of taste (*ro*); *tsa* means to multiply (*'phel ba*) or to propagate (*spel ba*). Together they make the term *ro tsa*, or "desire-multiply" (*'dod 'phel*)." Colleagues have suggested that *ro tsa* might be connected to the Sanskrit root *ruc*, which can mean "to like," but I have yet to find evidence of such a Sanskrit word to denote sexual stimulation or fertility.
42. These are the last two chapters of the *Instructional Treatise*, 90 and 91. The *Aṣṭāṅga* chapter is the very last in the work, chapter 40 of *Uttarasthāna*. *Suśruta* has separate chapters on male and female reproductive maladies, at Ci. 26 and Ut. 38 respectively, but they are not close to the *Four Treatises*' chapters in content or structure.
43. *Rigs brgyud bu tsha*. *Bu tsha*, "offspring," is gender-neutral but more frequently male. The same can be said of *Rigs brgyud*, "family lineage." See n. 63 and chapter 7, n. 110. Cf. also the term "generational lineage" (*rabs brgyud*), which occurs several lines down.
44. *Gtso (bo)* and *yan lag*. I render *yan lag* here differently than when it translates Skt. *aṅga* as branch of medical learning.
45. *Bu tsha*. I am correcting the modern edition's *'phil* to *'phel*, as in the Degé and most other xylograph editions. See J. Gyatso 2009. Note that n. 19 in that essay mistakenly refers to its n. 25; that should read instead n. 24.
46. *Za ma bud med*. See below on both terms.
47. *Pha yi rabs brgyud*.
48. That this line refers to her karma, etc. is implied by the logic of the statement, and is also specified by the commentators: see n. 65.
49. At least two versions of the *Four Treatises* read *bud med* instead of *bu med* here, the Chakpori edition ([G.yu thog] 1978), and the Loro edition, printed in Lhokha in the sixteenth or seventeenth century: Bstan 'dzin don grub 2005–8, vol. 2, 684. See J. Gyatso 2009, nn. 21–24.
50. [G.yu thog] 1992, 551–52.
51. Cf. Skyem pa 2000, 988.
52. Sde srid 1973a, vol. 3, 512; cf. Skyem pa 2000, 989.11, commenting on a similar phrase later in the same passage, viz. *de la bu[d] med btsal thabs yan lag gces*; see also 989.14 and 994.5.
53. *Bud med btsal ba'i thabs*. I have discussed the following point with more precision in J. Gyatso 2009. At the time of writing that essay I did not have access to Bstan 'dzin don

grub 2005–8, but all of my findings are supported by the latter with the exception of the history of the title of the second *ro tsa* chapter, regarding which see n. 55.

54. All versions are unanimous in this spelling: Bstan 'dzin don grub 2005–8, vol. 1, 9.
55. Bstan 'dzin don grub 2005–8, vol. 2, 688 shows that the three earliest blockprints, Dratang, Gampo and Takten, read *bu med btsal ba'i thabs*. So does the Lhodruk version ([Gyu thog] 199?, f. 208b).
56. See Skyem pa 2000, 990.12; also 994.3; cf. Sde srid 1973a, vol. 3, 521.6.
57. That is, save one statement at the end of the second *ro tsa* chapter, where it asserts that if one's female partner is barren one should look for a friend (*grogs*, a euphemism for the female consort) with the "right marks." [G.yu thog] 1992, 556.
58. See previous note. Sde srid 1973a, vol. 3, 526 adds lines from *Aṣṭāṅgahṛdaya* on obtaining an attractive consort.
59. [G.yu thog] 1992, 16–17 characterizes both *khu* and *khrag*, i.e., semen and female reproductive blood, as *sa bon*, or "seed" at 17.1. Ayurveda has long referred to both male and female reproductive fluids as seeds (*bīja*), although more commonly for the male. The term also denotes the fertilized combination of the two right after conception: R. P. Das 2003, 19–29; cf. *Aṣṭāṅgahṛdaya* Śā. 1.2.
60. The commentators also add that the female has the function of growing (*'phel ba*) the seed as well as holding it: Skyem pa 2000, 989; Dar mo 1989, vol. 2, 517; Sde srid 1973a, vol. 3, 512.
61. *Gang gis kyang*. Sde srid 1973a, vol. 3, 512.
62. *Pha yi rabs brgyud*.
63. Zurkharwa's comment on the *Four Treatises*' instructions for how to ritually ensure the sex of a fetus as male ([G.yu thog] 1992, 18) explains its importance in terms of family line (*rigs brgyud*) as well (Zur mkhar ba 1989, vol. 1, 127). The *Aṣṭāṅgahṛdaya* version of the rite has it changing the sex of the embryo to either male or female: *Aṣṭāṅgahṛdaya* Śā. 1.41a. Elsewhere the *Aṣṭāṅgahṛdaya* shows preference for sons, e.g., Śā. 1.27b–30. See also chapter 7, n. 110.
64. Indicated explicitly in *Small Myriad*. See p. 379.
65. [G.yu thog] 1992, 552. Kyempa makes clear that the line refers to the woman: Skyem pa 2000, 989: *bud med de bsod nams dman pa la bu mi 'byung bas*. Sde srid 1973a, vol. 3, 512 adds the gloss *las mi mthun pa'i dbang*.
66. Cf. nn. 49 and 56.
67. 9.31–42 (Olivelle 2005).
68. Skyem pa 2000, 989; Dar mo 1989, vol. 2, 518.
69. The Desi introduces the small difference that the male's white elements face downward and red face upward in the *samse'u*, whereas it is the reverse in the female. Sde srid 1973a, vol. 3, 513. See n. 96.
70. [G.yu thog] 1992, 552. This verse echoes images in *Aṣṭāṅgahṛdaya* Ut. 40.9. Cf. Kha che Zla ba mngon dga' 2006, vol. 2, 1996 and 1998.
71. The following lines are partly drawn from *Aṣṭāṅga* Ut. 40.39–43; cf. Kha che Zla ba mngon dga' 2006, vol. 2, 2002–5.
72. Yang Ga commented that these enemas must be administered to a man by a physician (personal communication 2011).

73. The *Four Treatises* verse is concise; in its parallel section the *Aṣṭāṅga* refers to the *Kāmasūtra*: Ut. 40.41. cf. Kha che Zla ba mngon dga' 2006, vol. 2, 2003.
74. Yang Ga 2010, 239–40.
75. [G.yu thog] 1992, 553; vaguely similar points in *Aṣṭāṅga* Ut. 40.41. Cf. Kha che Zla ba mngon dga' 2006, vol. 2, 2003–4.
76. [G.yu thog] 1992, 533; *Aṣṭāṅga* Ut. 40.25 and 34; cf. Kha che Zla ba mngon dga' 2006, vol. 2, 2002.
77. [G.yu thog] 1992, 554; *Aṣṭāṅga* Ut. 40.27–28a; cf. Kha che Zla ba mngon dga' 2006, vol. 2, 2001.
78. [G.yu thog] 1992, 554; *Aṣṭāṅga* Ut. 40.33; cf. Kha che Zla ba mngon dga' 2006, vol. 2, 2002.
79. [G.yu thog] 1992, 554; *Aṣṭāṅga* Ut. 40.1–3; cf. Kha che Zla ba mngon dga' 2006, vol. 2, 1997. This statement is placed at the end of the first *ro tsa* chapter, while in *Aṣṭāṅga* the parallel verses introduce the chapter.
80. Primarily *Aṣṭāṅga* Śā. chapters 1 and 2. Some comparison with the *Four Treatises* on embryology is provided by Garrett 2008. I am not convinced that more attention to female experience and care of the mother is provided by *Aṣṭāṅga* (Garrett 2008, 82–83); one would need to check throughout the *Four Treatises*, beyond the embryology chapter.
81. See n. 17.
82. E.g. [G.yu thog] 1992, 349; 352. Cf. Fenner 1996.
83. See R. P. Das 2003. I have consistently translated *khrag* as "blood" despite its multiple connotations, and *zla mtshan* as "menses" or "menstruation."
84. [G.yu thog] 1992, 375–79.
85. Selby 2005, 272–73 et passim.
86. [G.yu thog] 1992, 376–78. See also 380–82 regarding treatments for tumors and parasites.
87. [G.yu thog] 1992, 379; 382.
88. Cf. *Aṣṭāṅga* Ut. 33 and 34; Kha che Zla ba mngon dga' 2006, vol. 2, chapters 33 and 34.
89. [G.yu thog] 1992, 311–13. The entire pathology in this chapter is summed up as *mngal nad*: [G.yu thog] 1992, 313.7.
90. *Bu rtsi*. [G.yu thog] 1992, 555–56.
91. [G.yu thog] 1992, 17.
92. *Aṣṭāṅga* Śā. 1.7; 20b–21a; 22b–23a. Cf. Kha che Zla ba mngon dga' 2006, vol. 1, 655–660.
93. See *Suśrutasaṃhitā* Śā. 3.10. R. P. Das 2003, 100 is not convinced that these are the fallopian tubes.
94. Ye shes gzungs 1998, 12; cf. Zur mkhar ba 1989, vol. 1, 109.
95. [G.yu thog] 1992, 18.
96. The *Four Treatises* notion that the seminal substances are spread all over the body and yet also situated at the heart seems to have caused some confusion about the *samse'u*: [G.yu thog] 1992, 27. [G.yu thog] 1992, 21 and 552 also says the *samse'u* is the support of the refined distillates of the red and white sexual substances. See also [G.yu thog] 1976b, 377–79. Skyem pa 2000, 112 describes the *samse'u* as located at the thirteenth digit of the spine and as a kind of channel knot or fleshly protuberance (*sha rmen*)

the size of a bird's egg. For the history of this idea in Tibet see Dbang 'dus 2004, 664–67, who maintains the term is a faulty rendering of a Chinese term phoneticized in Tibetan as *gsan rtsa'u*, but that the Tibetan understanding of the *samse'u* is unique to Yutok. Zurkharwa provides an account of the *samse'u* in terms of the tantric channels: Zur mkhar ba 1989, vol. 1, 156–57; cf. Sde srid 1973a, vol. 3, 513. See also Meyer 1981, 69; 110–13; 155–56. Zurkharwa knows that the Ayurvedic sources never mention such an organ but states that in the context of embryology, the heart mentioned by the *Aṣṭāṅga* is the same as the *samse'u*: Zur mkhar ba 1989, vol. 1, 124.

97. Skyem pa 2000, 108. He is citing *Aṣṭāṅgahṛdayavaiḍūryakabhāṣya*. See Pha gol 2006, 317.
98. [G.yu thog] 1992, 18. Clark 1995, 49 translates *lte ba* as placenta, but Skyem pa 2000, 112, specifies that it refers to the navel of the infant. Cf. Zur mkhar ba 1989, vol. 1, 124–25. The placenta is often called *rog ma*, e.g., [G.yu thog] 1992, 382–84.
99. Skyem pa 2000, 112.
100. The exact denotation of Skt. *yonimukha* has long been uncertain: see R. P. Das 2003 11.11–12; 572–75.
101. That detail is added later in his comment, at Zur mkhar ba 1989, vol. 1, 125.
102. Zur mkhar ba 1989, vol. 1, 110.
103. Zur mkhar ba 1989, vol. 1, 124.
104. Neither Kyempa's commentary nor the *Four Treatises* itself make such a point.
105. Zur mkhar ba 1989, vol. 1, 110. *Zla mtshan* refers both to the menstrual cycle and the menstrual fluids.
106. Zur mkhar ba 1989, vol. 1, 110. *de yang tshes gcig nas bco lnga'i bar du zla ba'i khams lhag par 'phel zhing nyi ma'i khams 'bri ba'i dus yin pas dkar phyogs la 'dzag pa shes che* (corrected from Zur mkhar ba n.d., vol. 1, f.18a) *zhing zla ba re re bzhin gyi de dang de'i dus su 'byung bas na zla mtshan zhes bya'o.*
107. See n. 31.
108. Skyem pa 2000, 109.
109. Ibid.
110. Zur mkhar ba 1989, vol. 1, 110. Zurkharwa never commented on the female pathology chapters.
111. Cf. Goble 2011, chapter 4, on the Japanese medical use of karma to account for incurable illness.
112. Actually the *Four Treatises* uses the opposite wording but the meaning turns out to be the same. See discussion below. Ayurvedic works vary on this system: see R. P. Das 2003, 447–52. Cf. Selby 2005, 259.
113. Cf. *Aṣṭāṅgahṛdaya* Śā. 1.5; 23–27; Kha che Zla ba mngon dga' 2006, vol. 1, 654–55; 661.
114. [G.yu thog] 1992, 17.
115. Skyem pa 2000, 109–10.
116. *Aṣṭāṅgahṛdaya* Śā. 1.26b–27a. Cf. Kha che Zla ba mngon dga' 2006, vol. 1, 661.
117. Zur mkhar ba 1989, vol. 1, 111–16. Tibetan tantric tradition has its own ideas on the matter, as in Klong chen pa 1975, 274–75.
118. Khro ru 2001, vol. 1, 96–97, moves the "opening of the uterus door" to the fourteenth day from the onset of menstruation, after which there is a twelve-day period of fertility.

119. See n. 59.
120. The contribution of both parents' reproductive substances to the sex of the child is clear, e.g., at *Carakasaṃhitā* Śā. 2.12–16; *Suśrutasaṃhitā* Śā. 5.2.
121. *Sa bon.* Zur mkhar ba 1989, vol. 1, 116; he is citing Vimalaprabhā. See also Zur mkhar ba 1989, vol. 1, 120–21 for further discussion and citations from Indic sources. He also uses the term here to refer to the combination of the father and mother's substances and the consciousness of the reincarnating being. See also n. 59.
122. Zur mkhar ba 1989, vol. 1, 112–13.
123. Zur mkhar ba 1989, vol. 1, 113–14.
124. [G.yu thog] 1992, 560.
125. R. P. Das 2003; on *samse'u* see n. 96.
126. *Chu kha* and *kha 'chus*, respectively. [G.yu thog] 1992, 312.2; 312.3.
127. *Rtsa kha.* [G.yu thog] 1992, 383.1.
128. *Phyi yi sha lpags mig.* [G.yu thog] 1992, 383.9.
129. *Mtshan ma* or *mo mtshan*: [G.yu thog] 1992, 379.15; 379.16; 382.10.
130. Zur mkhar ba 1989, vol. 1, 156. See p. 238.
131. Ra se 2003 coins the more egalitarian term *skyes ma* to refer to women.
132. *Skyes pa*, i.e., the common Tibetan term for man (Skt. *jana*).
133. Sde srid 1982, 462. Cf. S. C. Das 1981, 872 citing an undocumented definition that "a *bud med* is one who is not put out (*bud pa*) at night outside."
134. These points were famously made by Irigaray 1981 and 1991.
135. Dbang 'dus 2004, 529 cites De'u dmar dge bshes Bstan 'dzin phun tshogs (b. 1672) to say, "The one who does not have the capacity for those village practices (*grong pa'i chos dag*, a common euphemism for sex) is the *za ma mo*." Dbang 'dus himself attempts to join the senses of female, the third sex, and food: "*Za ma* refers to one whose testicles have been removed, or who has diminished desire, and so forth. But in this context it is a word for the woman (*bud med*) who, when having sex, other than 'eating' for herself the experience of taste and pleasure of desire [created] by another, does not have the capacity to have an aroused organ of desire and do it to another."
136. In the monastic context *za ma* is sometimes distinguished from *ma ning*, as in Dge 'dun grub 1999, 403. Elsewhere the two are often synonomous: Krang dbyi sun 1993, 2443; Btsan lha 1997, 792. Zur mkhar ba 1989, vol. 1, 698 resists the equation and defines the *za ma* as a castrated person. In Sanskrit Vinaya and Abhidharma literature the *ṣaṇḍha/ṣaṇḍa* is frequently mentioned alongside the *paṇḍaka*, suggesting a distinction between the terms: See J. Gyatso 2003, n. 11.
137. *Carakasaṃhitā* Śā. 4.10; 14. *Aṣṭāṅgahṛdaya* Śā. 1.5 uses the term *klība*, which was translated into Tibetan as *ma ning*.
138. R. P. Das 2003 provides lengthy entries on *klība* (538–41); *dviretas* (551–53); *napuṃsaka* (558–60); *paṇḍaka* (560–62) and *ṣaṇḍ/h/a/ka* (581–84).
139. Including hermaphroditism, various conditions making for weak sexual desire, disfunctioning or misshaped sexual organs, voyeurism, and absence of testicles. *Carakasaṃhitā* Śā. 2.17–21.
140. Alt. *ṣaṇḍa/ka*, *ṣaṇḍha*/ka, and *ṣāṇḍya*: *Carakasaṃhitā* Śā. 2.21; *Suśrutasaṃhitā* Śā. 2.41–42, 44.

141. See especially *Suśrutasaṃhitā* Śā. 2.38–41.
142. As in the sixth chapter of the *Explanatory Treatise*, on the classifications of the body.
143. [G.yu thog] 1992, 29.
144. [G.yu thog] 1992, 17.
145. [G.yu thog] 1992, 20. cf. *Aṣṭāṅgahṛdaya* Śā. 1.69–72.
146. But Darmo Menrampa expands the first statement in the female pathology section to apply to the *ma ning*: Dar mo 1989, vol. 2, 309.
147. [G.yu thog] 1992, 17.
148. [G.yu thog] 1992, 560. But the *ma ning* person is pictured in the Desi's paintings on the passage. The caption to plate 54.62 reads *ma ning la ma ning gi rang bzhin*. See figure 6.8.
149. [G.yu thog] 1992, 42.
150. A Tibetan colloquial term, *pho lus mo lus* (sp?), names a range of sexually anomalous persons. Oral reports suggest that out of fear that the penis of a child born male will "split" and the child will become either a female or a *pho lus mo lus*, parents sometimes wrap the infant's organ.
151. Cf. Tib. *sku rten*, "body receptacle," which names a painting or statue of a Buddhist deity, a medium through which the power and knowledge of the deity is localized and made perceptible.
152. Zur mkhar ba 1989, vol. 1, 219.
153. Zur mkhar ba 1989, vol. 1, 219; see also 698, and Skyem pa 2000, 162. An almost parallel trio occurs in the Tibetan translation of *Vinayavastu*: *mtshan med* (= Skt. *animitta*), *mtshan gnyis* (= ?Skt. *dvinimitta* or *ubhatovyañjanaka*), and *gle gdams pa*, which can refer either to a castrated man or to a woman whose vagina and and anus are joined (= *sambhinnavyañjana*): *'Dul ba gzhi* 1934, vol. 1, f. 69b.
154. So defined in the *Vinayavastu*'s description of the *jātyāpaṇḍaka* (*'Dul ba gzhi* 1934 vol. 1, f. 133a) and in Yaśomitra's gloss of the class of *paṇḍaka* and *ṣaṇḍha* in general, in *Abhidharmakośavyākhyā*: Law 1949, 94.
155. 'Tsho byed dkon skyabs 1976, 327.
156. J. Gyatso 2003, 102–3 finds one anomalous tradition, rare and seemingly unique to Tibetan tantric Buddhism, where the third sex is deified as a form of Mahākāla; otherwise the third sex is a pariah in Indo-Tibetan Buddhist literature.
157. Pāli *antarāyikā dhammā*; Tib. *bar chad*. The list varies, but it often includes some of the *anantarika* crimes such as murder of parents or an *arhat*. The list for women includes more specifications of sexual ambiguity. See J. Gyatso 2003.
158. J. Gyatso 2003.
159. "Eunuch," a common translation for *paṇḍaka*, is inadequate to cover its semantic range. Many other labels for sexual anomalies which merit exclusion from ordination are to be found in Vinaya texts as well.
160. The earliest source I have found for the list of five is *Vinayavastu*: *'Dul ba gzhi* 1934, vol. 1, ff. 132b–133a. The list is absent in the Pali Vinaya and seems to emerge first in Pali in Buddhaghosa's *Samantapāsādikā* (Takakusu 1999, 1016), where it diverges on some of the types from *Vinayavastu*'s list. A later source is Yaśomitra's *Sphuṭārtha* II.1 (Law 1949: 94–95). See also dBang 'dus 2004, 410.

161. See J. Gyatso 2003 responding in part to Zwilling 1992.
162. *Suśrutasaṃhitā* Śā. 2.41–42 suggests that the *ṣaṇḍha* is like a female; 2.43 refers to the female who acts like a man.
163. As in a comment on *Suśrutasaṃhitā* Śā. 2.45: Sharma 2000, 137.
164. *Vinayavibhaṅga* regularly distinguishes the category of male *paṇḍaka* (Tib. *ma ning pho*), e.g., *'Dul ba rnam par 'byed pa* 1934, vol. 5, f. 49a. *Cullavagga* X.17's list of female sexual anomalies disqualifying her for ordination include: lacking sexual organs, having defective sexual organs, being without (menstrual) blood; having stagnant blood; being always dressed (*dhuva colā*); being dripping and deformed (*sikhariṇī*); being a woman *paṇḍaka*; being a manlike woman (*vepurisikā*); having genitals that are joined (*sambhinna*); being a hermaphrodite (*ubhatovyañjana*): Horner, 1997–2000, vol. 5, 375. Cf. Takakusu 1999, 548.
165. *Suttavibhanga*, Saṅghādisesa III.3.1 (Horner, 1997–2000, vol. 1, 217). A list of those from whom the monk may not beg for alms (prostitutes, widows, unmarried women, nuns and *paṇḍakas*) also suggests that the *paṇḍaka* is like a potential female sexual partner: *Mahāvagga* I.38.5 (Horner, 1997–2000, vol. 4, 87).
166. J. Gyatso 2003.
167. Thon-mi Sambhoṭa 1973, 9.
168. Yang dgon pa 1976, 457.
169. Sa chen 1983, 4; 266–67.
170. Yang dgon pa 1976, 454.
171. [G.yu thog] 1976a, 344. On the authorship of this work see Zur mkhar pa 2001, 321.
172. The line reads *byang chub sems* for metric reasons; the full phrase, *byang chub sems dpa'i rtsa*, occurs three lines below.
173. The Tibetan reads *gnyis kar*. Skyem pa 2000, 1029 and Zur mkhar ba 1989, vol. 1, 700 (see also the Desi's medical paintings, plate 54.68–73) understand this line to refer to "both" members of a couple, but while combined pulses of a couple are considered a few lines later, the present line would seem to follow the lines before it to discuss the possible disjunctions for an individual between sex and pulse.
174. *Drung*. Skyem pa 2000, 1029 specifies these are paternal uncle, male descendent, and maternal uncle.
175. [G.yu thog] 1992, 560.
176. The *Four Treatises* passage itself does not gloss bodhisattva pulse as *ma ning*, but the larger context and the commentators make the equation clear: Skyem pa 2000, 1028; Zur mkhar ba 1989, vol. 1, 699: *byang chub sems rtsa'am ma ning gi rtsa*.
177. But see n. 148.
178. I.e., Bkra shis dpal bzang n.d. The four passages Zurkharwa cites, with minor variations, are at 300–1; 301; 311; and 313.
179. Zur mkhar ba 1989, vol. 1, 698.
180. Bkra shis dpal bzang n.d. also states that a bodhisattva pulse results from a predominance of emotional obscuration and stupidity: see p. 335.
181. Bkra shis dpal bzang n.d., 311.5 says *bshad do* and later *zhes so*, but it is not clear where the quote starts.
182. Zur mkhar ba 1989, vol. 1, 698.

183. Ibid.
184. This topic was explored in detail by Jan Nattier in an unpublished essay entitled "Gender and Awakening: Sexual Transformation in Mahāyāna Sūtras."
185. Zur mkhar ba 1989, vol. 1, 699.
186. See J. Gyatso 2003, 99 and J. Gyatso 2005. Having the ability to have sex is a defining feature of what it means to swear off it in Buddhist monastic law.
187. Zur mkhar ba 1989, vol. 1, 699.
188. Skt. *cittasantāna* or *cittasantati* or *cittaprabandha*. See, e.g., Lamotte 1988, 21; 28; Tucci 1971, 9.
189. Bkra shis dpal bzang n.d., 308–309.
190. Bkra shis dpal bzang n.d., 312–13.
191. Krang dbyi sun 1993, 2939.
192. Zur mkhar ba 1989, vol. 1, 699–700.
193. Cf. *Hevajratantra* I.1.15; see also I.8.26 (Farrow and Menon 1992, 13; 94).
194. One of the 24 types of conditions in the Pali Abhidhamma: Narada 1969, 1–3. Discussed in terms of karma theory in *Abhidharmakośa* 2.56b; 2.58c–d.
195. *Brda btags pa.*
196. The manuscript is illegible in spots but Bkra shis dpal bzang n.d., 301–10 seems to build a case for a variety of combinations of humors and heterogeneous pulse types, including the possibility that one's sex will be at odds with one's pulse type.
197. *(R)kyang pa'i mi rigs dang rtsa rigs . . . ming tsam zad do.* Bkra shis dpal bzang n.d., 309–10.
198. On the preference for sons, see Schaeffer 2004, 133; and Havnevik 1999, e.g., vol. II, 149–153.
199. The Desi's medical paintings also display an egalitarian view of gender in everyday life despite their androcentrism and misogyny in the anatomical plates. See J. Gyatso 2011a.
200. Most scholarship in Buddhist Studies has focused upon gender prejudice rather than corrections thereof. Examples of the latter include J. Gyatso 1998, 248–49; Ruch 2002; Jacoby 2007; and Meeks 2010.

7. THE ETHICS OF BEING HUMAN: THE DOCTOR'S FORMATION IN A MATERIAL REALM

1. Ye shes gzungs 1976, 273–74.
2. Ye shes gzungs 1976, 293.
3. See nn. 33 and 48.
4. In what follows, the notion of ethics refers to both concern for others and the care of the self, connected by Charles Hallisey to the thought of Ricoeur and Foucault respectively: Hallisey 2010. In this chapter I use the term "moral" for a more general and culturally embedded sense of right and wrong. But the distinction between the two terms is hard to maintain. Cf. V. Das 2012, 134.
5. A related sign of the anomalous character of this material is the fact that chapter 13 of the *Explanatory Treatise*, which also treats the dharma of humans, stands out from the entire work for its lines of eleven syllables: Emmerick 1981. See pp. 355–56.

6. The *Aṣṭāṅga* has only scattered, short statements on the qualities of the physician. Cf. *Caraka* Sū. 9 on the "four pillars of medicine," physician, patient, medicine, and the assistant; Sū. 10.7–8 on a physician's bad reputation and possible punishment; Sū. 29 on bad doctors who try to win over the patients of others and cover up their own ignorance; Vi. 7 on diagnosis; and Vi. 8 on medical education and debate. See also *Suśruta* Sū. 3, and Dagmar Wujastyk 2012.
7. Ye shes gzungs 1976.
8. Kyempa and Zurkharwa, followed closely by the Desi, provide substantive comments on the physician's chapter. I do not have the space to compare how they domesticate or elide *Small Myriad*'s more challenging passages. Jangpa Namgyel Drakzang also drew on *Small Myriad*'s discussion, not noted by Schaeffer 2003.
9. Uray 1972, 58–59; Stein 1972, 100; Stein 2010, 215–18.
10. Stein 2010, 128 n. 23; Tucci 1986, part 2, 142. Pelyang (Śrīghoṣa) was South Asian, but *manuṣyadharma* is virtually unknown in Indic sources.
11. ITJ 370.5, as listed in La Vallée Poussin 1962, 122. See Richardson 1977. Stein 2010, 216 estimates the date of the manuscript to be 800–820. See also Stein 1972, 192.
12. Toh 4328 and Toh 4501. This work does not exist in Sanskrit. See Stein 2010, 128 n. 23; 46–47; and 57. A *nītiśāstra* attributed to Nāgārjuna and a work of Aśvaghoṣa also mention *myi yi chos lugs*: Stein 2010, 128 n. 23. None of these works is extant in Sanskrit. Brunnhölzl 2007 reviews the writings of Nāgārjuna, with no mention of *Prajñāśataka*.
13. Stein corrected Richardson's reading of *'dzem* in ITJ 370.5, arguing that it is a version of *'dzegs/bzegs*, to climb, and that this work is thus citing *Prajñāśataka*: Stein 2010, 198–200. Stein 1972, 192 notes that *Blon po bka' thang* says that basing oneself in the field of *mi chos* will make *lha chos* (Buddhism) grow there as a fruit.
14. This thesis would be at odds with Stein 1972, 192. Stein connects the concept of *mi chos* to the introduction of the "two great organized religions," i.e., Bön and Buddhism.
15. Stein 2010, 128 n. 23; see also 88. The text is reproduced at Old Tibetan Documents Online (http://otdo.aa.tufs.ac.jp/). Also studied by Bacot 1956 and others. Reproduced in Tibetan script in Chab spel 1997.
16. See also PT 2111. Maxim texts called *mi chos* continued to be produced later. Stein 1972, 270 notes that *Rlangs po ti bse ru* contains a series of adages called *mi chos*, many in verse. Cf. Gung thang Bstan pa'i sgron me 2006.
17. See Stein 2010, 117–90. On 189 he notes that *gtsug lag* could be synonymous with *mi chos* by the time of *Mahāvyutpatti*. On 22 he seems to affirm the use of *chos lugs* in non-Buddhist texts.
18. See Stein 2010, especially 43–49 and 30–33 on the Tibetan terms *drang po* vs. *yon po* or *g.yon*, "left." Stein also suggests Chinese analogues, and says that PT 1283 may be either a Tibetan composition or a translation. But his discussion on 37–38 and at n. 48 suggests that the dialogue between the two brothers is an indigenous set of maxims. On 88 he indicates the work combines Chinese-influenced vocabulary with ancient or idiomatic Tibetan.
19. As in terms like *rgyal po'i chos khrims, drang po'i chos*, and *ya rabs chos* on the one hand, and *ma rabs gyi chos* and *rtsing chos* on the other.
20. Stein 1972, 193; Stein 1961, 53.

21. As, to cite one example, in *Cullavagga* X.1.6 (Horner 1997–2000, vol. 5, 356) where the Buddha is talking about the quicker demise of his dispensation with the admission of women into the monastic order.
22. Skt. *Laukika* and *lokottara*; *saṃvṛtti* and *paramārtha satya*. [G.yu thog] 1992, 101 makes a parallel distinction between temporal and ultimate (*gnas skabs* and *mthar thugs*) that maps directly onto the human/True Dharma pair. See p. 356.
23. A number of such passages are found, for example, in *Aṅguttaranikāya*: see Bodhi 2005, 124–26; 127–28. A classic Tibetan Buddhist instance from close to the period of the *Four Treatises* is Gampopa's *Thar rgyan*, which discusses the gift of material objects, but distinguishes that from the gifts of fearlessness and the Dharma: Guenther 1971, 153–57.
24. As in Guenther 1971, 160–61.
25. Even if such works also evince care for the self. See n. 40.
26. [G.yu thog] 1992, 95–96: *blo ldan; bsam pa dkar ba; dam tshig dang ldan pa; rnam pa bzo ba; bya ba la brtson pa; mi'i chos lugs mkhas pa*. In this chapter I alternate between translating *mkhas pa* as "skill" and as "mastery."
27. [G.yu thog] 1992, 95.
28. [G.yu thog] 1992, 100. The association of human dharma with *'jig rten* is repeated in Buddhist works down to the contemporary period, e.g., Rdza dpal sprul 2003, 34.
29. [G.yu thog] 1999? (Degé edition), f.47b reads *gnyan gyis btsa'*. [G.yu thog] 1992 (Chakpori) f.41b reads *gnyen gyis btsa'*. Bstan 'dzin don grub, vol. 1, 144 has all versions reading *gnyen gyis btsa'*. Ye shes gzungs 1976, 276 reads *mnyen gyi btsal*. The best reading would be *mnyen gyis btsal*, as at 277.
30. Ye shes gzungs 1976, 277, *'chol* should be emended to *'tshol*. The language here is colloquial and difficult; I am grateful to Yang Ga for his help.
31. Ye shes gzungs 1976, 277.
32. *Mgo thon; mgo tshud; mgo 'dzugs*. The sense of the metaphor in this context may be different than what contemporary lexicons indicate for *mgo thun*, which emphasize the competence that goes along with the head's emergence.
33. Cf. Dagmar Wujastyk 2012, 90; 93.
34. Ye shes gzungs 1976, 277.
35. Ibid.
36. See also the 25th chapter of the *Explanatory Treatise* and Schaeffer 2007.
37. Ye shes gzungs 1976, 296.
38. [G.yu thog] 1992, 98.
39. *Chos pa*. Ye shes gzungs 1976, 277–78.
40. E.g., *Bodhicaryāvatāra* 1.2; 1.17; 1.19; chapter 8 famously emphasizes how compassion for others is modeled on the concern normally lavished upon the self, e.g., 8.94. Śāntideva 1995, 5–7; 96.
41. [G.yu thog] 1992, 98.
42. Ye shes gzungs 1976, 278.
43. [G.yu thog] 1992, 48.13–49.17.
44. Emmerick 1981.
45. Note, however ,the brief invocation of Dharma as a general guide to daily human behavior at *Aṣṭāṅgahṛdaya* Sū. 2.20 and 2.30.

46. [G.yu thog] 1992, 101.
47. Ye shes gzungs 1976, 297–99.
48. As in *Mahāsīhanādasutta*: Walshe 1987, 155–56. On Ayurvedic disapproval of physicians who boast see Dagmar Wujastyk 2012, 47; 89; 91.
49. J. Gyatso 1998. But see my caveat below, p. 394.
50. Deception is a complex topic for Buddhist ethics, especially in "skillful means" discourse, and is famously addressed in the *Lotus Sutra*. It takes on yet further metaphysical implications in Tibetan Buddhist visualization tradition: J. Gyatso 1998, 214–18. But these are issues of a different order than the doctor's behavior described here.
51. [G.yu thog] 1992, 101.
52. See p. 102.
53. E.g., Olivelle 2005, the last section of chapter 12.
54. Ye shes gzungs 1976, 299.
55. Zurkharwa Lodro Gyëlpo also had doubts that practicing medicine led to full enlightment. See Czaja 2008, 120–30.
56. [G.yu thog] 1992, 96.
57. Ye shes gzungs 1976, 293.
58. [G.yu thog] 1992, 96–97; Ye shes gzungs 1976, 266–70. Other parts of these visualization formulae refer more specifically to medical tradition, as when the speech of the medical teacher is to be equated with the words of the *ṛṣis*.
59. See Rangdrol 1993, especially the chapter on *samaya*.
60. See J. Gyatso 1998, chapter 4.
61. *Blo bzhag*. [G.yu thog] 1992, 96.
62. [G.yu thog] 1992, 97.
63. [G.yu thog] 1992, 96.
64. Cf. Conclusion, n. 2.
65. Ye shes gzungs 1976, 264–66.
66. [G.yu thog] 1992, 96.
67. Ye shes gzungs 1976, 292.
68. Ye shes gzungs 1976, 291–92.
69. Ye shes gzungs 1976, 290–91.
70. [G.yu thog] 1992, 96. Here "big mind" (*blo che*) is not the opposite of the "small mind" (*sems chung*) mentioned below (see n. 96), but is a different notion altogether.
71. Ye shes gzungs 1976, 265.
72. *Mngon shes lta bu'i phra mo*: Ye shes gzungs 1976, 265.
73. Skyem pa 2000, 469. Cf. Zur mkhar ba 1989, vol. 1, 633–34.
74. Schaeffer 2003; 2009, chapter 4.
75. Cf. *Caraka* Sū. 9.24, where the physician is said to require both textual knowledge and his own intelligence (*buddhi*).
76. [G.yu thog] 1992, 97–98.
77. Ye shes gzungs 1976, 271–72. In this quotation I have translated *dpe* as "text," but the overall discussion seems to use *dpe* and *gzhung* interchangeably, and I myself vary among "text," "work," and "book" in rendering both terms.
78. See Kapstein 1996, 284, n. 1. *Man ngag* is often connected to Sanskrit *upadeśa*.

79. For the notion of a term of art as applied in Buddhist Studies, see Wedemeyer 2011.
80. [G.yu thog] 1992, 98.
81. Ye shes gzungs 1976, 272–73.
82. The entire discussion of kinds of doctors is found at [G.yu thog] 1992, 98–100. Ye shes gzungs 1976, 280–81, adds subtypes to the *Four Treatises*' general category of *'dod pas 'phral bslabs*, the doctor who learns on the spot and in an ad hoc manner based on his desires.
83. Ye shes gzungs 1976, 281. A later rendering of this passage by Jangpa Namgyel Drakzang is mentioned by Schaeffer 2003, 629.
84. Or makes medicines: *'chos* is ambiguous here.
85. Ye shes gzungs 1976, 281.
86. There is no indication in the *Four Treatises*' physician's chapter of the tantric transmission of medical knowledge, such as is found in the *Heart Sphere of Yutok* tradition.
87. One could usefully compare here the Buddhist notion of *bhāvanāmārga* (Tib. *sgom lam*), which has to do with the repeated meditative practice that both sets up and then internalizes a doctrinal insight (Skt. *darśana*). These ideas are developed in detail in the sixth chapter of *Abhidharmakośa.*
88. Ye shes gzungs 1976, 281.
89. [G.yu thog] 1992, 99.
90. Ye shes gzungs 1976, 283–84.
91. Ye shes gzungs 1976, 284.
92. [G.yu thog] 1992, 99.
93. Ye shes gzungs 1976, 280.
94. Like *man ngag*, *gdams ngag* can also be either oral or written, but in this case it is explicitly the former.
95. [G.yu thog] 1992, 97–98.
96. In general a small mind (*thugs* or *sems*) is well regarded for its care and circumspection, a large mind less so. It is not clear how these terms are meant here.
97. I have skipped one line since its meaning is unclear; it is another injunction to imitate the attitudes and actions of the teacher. Ye shes gzungs 1976, 273: *khong phyi phyogs che na de dang mthun par byed.*
98. Ye shes gzungs 1976, 273–74.
99. [G.yu thog] 1992, 98; Ye shes gzungs 1976, 274.
100. Ye shes gzungs 1976, 274–75.
101. [G.yu thog] 1992, 97. *Caraka* Sū. 9.6 briefly mentions dexterity (*dākṣyam*) as one of the qualities of the physician.
102. Ye shes gzungs 1976, 271.
103. *Rig pa bkra ba.*
104. *Gnang rigs.* [G.yu thog] 1992, 99; Ye shes gzungs 1976, 280 renders it *snang rig.*
105. Ye shes gzungs 1976, 280.
106. *Gdung btsun.* [G.yu thog] 1992, 99. Skyem pa 2000, 489 provides details on the virtues of such physicians.
107. [G.yu thog] 1992, 99.

108. *Rigs brgyud*, correcting [G.yu thog] 1992, 99 (Bstan 'dzin don grub 2005–8, vol. 1, 144).
109. [G.yu thog] 1992, 99.
110. The gloss *pha mes bzang po* that is given here for the *Four Treatises'* term *rigs brgyud* is another example where the latter term specifically denotes patriliny. Cf. chapter 6, nn. 43 and 63.
111. Ye shes gzungs 1976, 283.
112. Ye shes gzungs 1976, 277.
113. [G.yu thog] 1992, 98; Ye shes gzungs 1976, 279 explains that the kings were called gods (*lha*), and since they venerated doctors, the term refers to the fact that doctors were lords to the gods. See also Skyem pa 2000, 487.
114. Ye shes gzungs 1976, 291–92.
115. Ye shes gzungs 1976, 292.
116. Ye shes gzungs 1976, 294.
117. Ye shes gzungs 1976, 294.
118. Ye shes gzungs 1976, 293.
119. Ye shes gzungs 1976, 294.
120. Ye shes gzungs 1976, 293.
121. Ye shes gzungs 1976, 293.
122. Gong sman 1969.
123. Ye shes gzungs 1976, 295.
124. [G.yu thog] 1992, 100; in the last line I follow the slightly different rendering of the passage in Ye shes gzungs 1976, 289: *lar na 'jig rten mi chos bstun pa gces.*
125. On wind horse see Karmay 1998b.
126. Ye shes gzungs 1976, 289–90.
127. Ye shes gzungs 1976, 290.
128. Ye shes gzungs 1976, 290.
129. Ibid.
130. [G.yu thog] 1992, 98. Kyempa associates this story with a narrative about Katyana (i.e., Kātyāyana) and King Rabnang from an unidentified sūtra, which includes a tale in which the famous Buddhist philosopher Dharmakīrti was searching for a patron and went before King Metok Küntu Gyepa. The king proceeds through an elaborate check of Dharmakīrti's credentials and finally puts him through a ghee-balancing test: Skyem pa 2000, 483–84. Zurkharwa is uncertain about the source of this story and whether it is pure or not, but refers his readers to the biography of Dharmakīrti in the *Bu ston chos 'byung*: Zur mkhar ba 1989, vol. 1, 645. The latter does indeed tell a story of Dharmakīrti being tested by the same king, but says nothing about carrying ghee along a wall: Obermiller 1931, part 2, 153–54.
131. [G.yu thog] 1992, 98; Ye shes gzungs 1976, 275–76.
132. Ye shes gzungs 1976, 276.
133. The term for life force is *srog*, the same word that I am translating as "vital/ity" in chapter 4 in the context of the *srog rtsa.*
134. A critical connection between reputation and the survival of a patient is made in Indian medical writing: Dagmar Wujastyk 2012, 114.

135. Similar topics are also addressed in the 25th chapter of the *Explanatory Treatise*, on how the physician uses ruses to avoid censure, and the 26th chapter, which advises the physician how to avoid patients who are beyond treatment. See Schaeffer 2007.
136. [G.yu thog] 1992, 100; Ye shes gzungs 1976, 287.
137. Kyempa and Zurkharwa know of these classifications but do not explain the doubleness or tripleness. Skyem pa 2000, 492–94; Zur mkhar ba 1989, vol. 1, 653.
138. [G.yu thog] 1992, 100.
139. Rendered loosely. Ye shes gzungs 1976, 287.
140. [G.yu thog] 1992, 100.
141. Rendered loosely. Ye shes gzungs 1976, 288.
142. [G.yu thog] 1992, 100.
143. According to Yang Ga, this is an illness with symptoms similar to poison (personal communication 2011).
144. Ye shes gzungs 1976, 289.
145. *Rdzongs la bros*. [G.yu thog] 1992, 100; see also 84, where the same metaphor refers to a technique in which the doctor is not sure of the diagnosis but mentions exotic and impressive names of diseases and medicines. No one understands what the doctor is saying, but he comes off looking smart. Skyem pa 2000, 417–18 comments on the practice but does not explain the metaphor here. See n. 148.
146. Ye shes gzungs 1976, 289–90.
147. *Srid pa'i sgo mi bgag par rdzong la bros*: [G.yu thog] 1992, 100.
148. Kyempa unpacks the metaphor to say that among the many doors of existence, one holds one of them as what will happen but cannot shut out the other possibilities, so the freed prisoner runs to the fortress: Skyem pa 2000, 493.
149. On ritual in the *Four Treatises*, see chapter 2, nn. 95 and 102.
150. We could compare here PT 1057, advising recourse to religion when all medical procedures fail: Schaeffer, Kapstein, and Tuttle 2013, 116.
151. Gombrich 1988, 95.
152. L. Gyatso 1998, 82–83.
153. For a typical example of the rhetorical force in Tibetan Buddhist path teachings of invoking the uncertainty of when we will die but the certainty that we will, see Patrul Rinpoche 1998, 41. But in contrast to the medical injunction to protect the life of the patient, there the import is to prepare for death and one's future lives.
154. On the many theoretical questions that have been raised historically about nirvana see Collins 2010.

CONCLUSION

1. Vernant 1989, 23.
2. The physician's kindness toward the patient is discussed in terms close to the Buddhist four *brahmavihāras* in *Caraka* and *Aṣṭāngasaṁgraha:* Dagmar Wujastyk 2012, 31 and 219 n. 482.

3. Hallisey and Reynolds 1987 only see this aspect of Buddhism as operative at the international and multilinguistic level, which according to them comes to an end by the ninth or tenth century C.E. On the role of Buddhist monks in the pan-Asian spread of medicine see Goble 2011.
4. For example, in colonial Cambodia, where the old notion of *satisampajñña* served to articulate kinds of prudence in daily life: Hansen 2007, 155–56. On the modern incarnation of insight meditation in colonial Burma, see Braun 2013.
5. Dagmar Wujastyk 2012, 82–84; 99–101.
6. It will be worth keeping track of the new research of Stacey von Vleet, who is completing a dissertation at Columbia University on the role of Tibetan medicine in the Manchu empire, and Katrina Sabernig, who is studying the empirical anatomical writings of Darmo Menrampa, on these questions.

BIBLIOGRAPHIES

TIBETAN AND TIBETAN-CHINESE BILINGUAL PUBLICATIONS

Krang dbyi sun, ed. 1993. *Bod rgya tshig mdzod chen mo*. Beijing: Mi rigs dpe skrun khang.

Klong chen chos dbyings stob ldan rdo rje. 2000. *Mdo rgyud rin po che mdzod*, 5 vols. Chengdu: Si khron mi rigs dpe skrun khang.

Klong chen rab 'byams pa, Dri med 'od zer. 1975. *Thig le dangs snyigs 'byed pa dang/Rigs rgyud gzhag thabs lag len*. In *Snying thig ya bzhi*, 13 vols., vol. 11, 273–80. From an A 'dzom chos sgar blockprint. Delhi: Sherab Gyaltsen Lama.

——. 1983. *Gsang ba bla na med pa 'od gsal rdo rje snying po'i gnas gsum gsal bar byed pa'i tshig don rin po che'i mdzod*. In *Mdzod bdun*, 7 vols., vol. 5, 157–519. From a Sde dge blockprint. Gangtok: Sherab Gyaltsen and Khentse Labrang.

——. 2000?. *Rdzogs chen sems nyid ngal gso*. In *Gsung 'bum*, 5 vols. Blockprint from the nineteenth-century Sde dge blocks ordered by the fifth Rdzogs chen rig 'dzin chen mo, vol. 4, 1–79. Sde dge: Sde dge par khang. Available at TBRC W00EGS1016299.

Dkon mchog rin chen. *Bod kyi gso rig chos 'byung baiḍūrya'i 'phreng ba*, Lanzhou: Kan su'u mi rigs dpe skrun khang, 1994.

Bkra shis dpal bzang, Byang pa. N.d. *dPal ldan phyi ma brgyud* [sic] *kyi 'grel pa rin po che'i bang mdzod dgos 'dod 'byung ba*. Photocopy of incomplete manuscript, given to me in Lhasa in 2000; the pages have been numbered recently.

——. 1977. *Rgyud bzhi'i rnam nges dpag bsam ljon shing*. In *Commentaries on the Rtsa rgyud, Bśad rgyud, and Phyi ma rgyud*, by Gong sman Dkon mchog bde legs, 479–533. From a manuscript. Leh: T. Sonam Tashigang.

——. 1981. *Theg pa kun dang thun mong du byas pa gso ba rig pa'i rtsod spong*. In *Bzo rig kha śas kyi pa tra lag len ma and Other Texts on the Minor Sciences of the Tibetan Scholastic Tradition*, 111–52. Manuscript. Dharamsala: Library of Tibetan Works & Archives.

——. 1986. *Rgyud kyi bka' bsgrub drang srong bkra shis dpal bzang gi mdzad pa.* In *Bod kyi sman rtsis ched rtsom phyogs bdus*, ed. Bod rang skyong ljongs sman rtsis khang, 72–116. Lhasa: Bod ljongs mi dmangs dpe skrun khang.

Skal bzang legs bshad. 1994. *Rje btsun byams pa mthu stobs kun dga' rgyal mtshan gyi rnam thar.* Xining: Khrung go'i bod kyi shes rig dpe skrun khang.

Skyem pa Tshe dbang. 2000. *Rgyud bzhi'i rnam bshad.* Xining: Mtsho sngon mi rigs dpe skrun khang.

Kha che Zla ba mngon dga'. 2006. *Yan lag brgyad pa'i snying po'i rnam 'grel tshig don zla zer*, 2 vols. Beijing: Mi rigs dpe skrun khang.

Khog 'bugs khyung chen lding ba. 1976. In *G.yu thog cha lag bco brgyad* 1976, vol. 1, 7–37.

Khro ru Tshe rnam. 1996. "Bod lugs gso rig slob grva rim byung gi lo rgyus gsal ba'i gtam dngul dkar me long." In *Bod sman slob gso dang zhib 'jug* 1:1–11.

——. 2001. *Gso rig rgyud bzhi'i 'grel chen drang srong zhal lung.* 6 vols. Chengdu: Si khron mi rigs dpe skrun khang.

Gung thang Bstan pa'i sgron me. 2006. *Bstan bcos kyi skor.* Xining: Mtsho sngon mi rigs dpe skrun khang.

Gong sman Dkon mchog phan dar. 1969. *Nyams yig rgya rtsa: The Smallest Collection of Gong-sman Dkon-mchog-phan-dar's Medical Instructions to the Students.* From a manuscript from Ladakh. Leh: Lharje Tashi Yangphel Tashigang.

Gling sman Bkra shis 'bum. 1988. *Gso ba rig pa'i gzhung rgyud bzhi' i dka' 'grel.* Chengdu: Si khron mi rigs dpe skrun khang.

Dge 'dun grub. 1999. *Legs par gsungs pa'i dam pa'i chos 'dul ba mtha' dag gi snying po'i don legs par bshad pa rin po che'i 'phreng ba.* Beijing: Mi rigs dpe skrun khang.

Ngag dbang blo bzang rgya mtsho. 1989–91. *Za hor gyi ban de ngag dbang blo bzang rgya mtsho'i 'di snang 'khrul ba'i rol rtsed rtogs brjod kyi tshul du bkod pa du kū la'i gos bzang.* 3 vols. Lhasa: Bod jongs mi dmangs dpe skrun khang.

Cha lag bco brgyad kyi them yig dkar chag me long 'phreng ba. 1976. In *G.yu thog cha lag bco brgyad* 1976, vol. 1, 1–5.

Chab spel Tshe brtan phun tshogs. 1997. *Tun hong gi gna' yig phu bos nu bor btams shing bstan pa'i mdo zhes pa'i rtsa 'grel.* Beijing: Mi rig dpe skrun khang.

Jo bo Lhun grub bkra shis and Dar mo sman rams pa sogs. 2005. *G.yu thog gsar rnying gi rnam thar.* Beijing: Mi rigs dpe skrun khang.

'Jam mgon kong sprul Blo gros mtha' yas. 1970? *Rnal 'byor bla na med pa'i rgyud sde rgya mtsho'i snying po bsdus pa zab mo nang gi don nyung ngu'i tshig gis rnam par 'grol ba zab don snang byed.* From a blockprint. Gangtok, Sikkim: Rum btegs chos sgar.

'Jigs med gling pa. 1985. *Yon tan rin po che'i mdzod dga' ba'i cha.* In *The Collected Works of 'Jigs-med-gliṅ-pa Ran-byuṅ-rdo-rje Mkhyen-brtse'i-'od-zer*, 9 vols., vol. 1, 1–101. From a Sde dge blockprint. Gangtok, Sikkim: Pema Thinle for Ven. Dodrupchen Rinpoche.

'Ju Mi pham rnam rgyal rgya mtsho. 1992. *G.yu thog shog dril skor gsum gyi ma bu don bsdeb tu bkod pa.* In *Sman yig phyogs bsgrigs* by 'Ju Mi pham, 569–668. Chengdu: Si khron mi rigs dpe skrun khang.

Stag tshang lo tsā ba Shes rab rin chen. 2004. *Gso dpyad nyer mkho gces bsdus.* In *Gso rig*, ed. Kun dga' bzang po, 1–32. *Dpal ldan sa skya pa'i gsung rab*, vol. 9. Xining: Mtsho sngon mi rigs dpe skrun khang.

Stong thun mdzes pa'i 'ja' ris. 1976. In *G.yu thog cha lag bco brgyad* 1976, vol. 1, 39–75.

Bstan 'dzin don grub. 2005–8. *Dpal ldan rgyud bzhi*, 2 vols. Beijing: Krung go'i bod rig pa dpe skrun khang.

Thub bstan phun tshogs. 1999. *Gso bya lus kyi rnam bshad*. Beijing: Mi rigs dpe skrun khang.

Thub bstan tshe ring. 1986. "Gangs ljongs sman pa'i grong khyer lcags ri 'gro phan rig byed gling gi byung rabs brjod pa gsal ba'i sgron me." In *Bod kyi sman rtsis ched rtsom phyogs bsdus*, 148–81. Lhasa: Bod ljongs mi dmangs dpe skrun khang.

Theg pa chen po rgyud bla ma'i bstan bcos. 1982–85. In *Sde dge bstan 'gyur*, 213 vols., vol. Phi (= TBRC vol. 123), 107–45. From the Sde dge blocks carved 1737–44. Delhi: Delhi Karmapae Choedhey.

Dar mo sman rams pa Blo bzang chos grags. 1989. *Rgyud bzhi'i 'grel pa mes po'i zhal lung* by Zur mkhar ba Blo gros rgyal po, vol. 2. Beijing: Krung go'i bod kyi shes rig dpe skrun khang.

——. 1996. "Rus pa'i dum bu sum brgya drug cu'i skor bshad pa." In *Bod lugs gso rig sman rtsis ched rtsom phyogs bsdus*, 22–24. Lhasa: Sman rtsis khang.

——. 2005. *Bshad rgyud kyi sdong 'grems legs bshad gser gyi thur ma*. In *Legs bshad gser gyi thur ma*, 1–143. Beijing: Mi rigs dpe skrun khang.

——. 2006. *Gser mchan rnam bkra gan mdzod*. Beijing: Mi rigs dpe skrun khang.

De'u dmar Bstan 'dzin phun tshogs. 1970. *Principles of Lamaist Pharmacognosy: Being the Texts of the Dri med shel gong, Dri med shel phreng, and the Lag len gces bsdus*. From a Lcags po ri blockprint. Leh: S. W. Tashigang.

——. 1994. *Gso ba rig pa'i chos 'byung rnam thar rgya mtsho'i rba rlabs drang srong dgyes pa'i 'dzum phreng*. In *Gso rig gces btus rin chen phreng ba*, 632–764. Xining: Mtsho sngon mi rigs dpe skrun khang.

Dor zhi gdong drug snyems pa'i blo gros, ed. 1991. *Gsang chen thabs lam nyer mkho rnal 'byor snying nor*. Beijing: Mi rigs dpe skrun khang.

'Dul ba rnam par 'byed pa. 1934. In *Lha sa bka' 'gyur*, 100 vols., vol. 5, f. 30a–vol. 8, f.386a. Blockprint preserved at Tibet House, New Delhi. Lhasa: Zhol par khang. Available at TBRC W26071.

'Dul ba gzhi. 1934. In *Lha sa bka' 'gyur*, 100 vols., vols. 1–4. Blockprint preserved at Tibet House, New Delhi. Lhasa: Zhol par khang. Available at TBRC W26071.

Sde srid Sangs rgyas rgya mtsho. N.d.a. *Thams cad mkhyen pa drug pa blo bzang rin chen tshangs dbyangs rgya mtsho'i thun mong phyi'i rnam*. Blockprint. 'Bras spungs, Lhasa: Dga' ldan pho brang. Available at TBRC W2CZ7844.

——. N.d.b. *Ngag dbang blo bzang rgya mtsho'i rnam thar 'phro 'thus*. 3 vols. Blockprint. 'Bras spungs, Lhasa: Dga' ldan pho brang. Available at TBRC W8239.

——. 1973a. *Bai ḍūr sṅon po: Being the Text of Gso ba rig pa'i bstan bcos sman bla'i dgongs rgyan rgyud bzhi'i gsal byed bai dur sngon po'i ma lli ka*. 4 vols. From a Lcags po ri blockprint, carved 1888–92. Leh: T. Y. Tashigangpa.

——. 1973b. *Mchod sdoṅ 'dzam gliṅ rgyan gcig rten gtsug lag khaṅ daṅ bcas pa'i dkar chag thar gliṅ rgya mtshor bgrod pa'i gru rdziṅs byin rlabs kyi baṅ mdzod*. 2 vols. From a blockprint. Tespal Taikhang: New Delhi.

——. 1978. *Man ngag yon tan rgyud kyi lhan thabs zug rngu'i tsha gdung sel ba'i karpūra dus min 'chi zhags gcod pa'i ral gri*. From a 1733 Sde dge blockprint. Leh: T. S. Tashigangpa.

——. 1979. *Blaṅ dor gsal bar ston pa'i draṅ thig dwaṅs śel me loṅ*. From a Lhasa Zhol blockprint. Dolanji, India: Sonam Drakpa.

——. 1982. *Dpal ldan gso ba rig pa'i khog 'bugs legs bshad baiḍūrya'i me long drang srong dgyes pa'i dga' ston*. Lanzhou: Kansu'u mi rigs dpe skrun khang.

Nor brang o rgyan. 2006. "Sde srid sangs rgyas rgya mtsho'i skor gleng ba dad pa'i char 'bebs." In *Nor brang o rgyan gyi gsung rtsom phyogs bsdus*, 233–316. Beijing: Krung go'i bod rig pa dpe skrun khang.

Pa sangs yon tan. 1988. *Bod kyi gso ba rig pa'i lo rgyus kyi bang mdzod g.yu thog bla ma dran pa'i pho nya*. Leh: Yuthok Institute of Tibetan Medicine.

Dpa' bo gtsug lag phreng ba. 1986. *Chos 'byung mkhas pa'i dga' ston*. 2 vols. Beijing: Mi rigs dpe skrun khang.

Spyan snga Rin chen ldan. 1982. *Sbas bshad kyi zhal gdams sbas pa gdon kyi gter mdzod*. In *Collected Writings of Rgyal ba yang dgon pa rgyal mtshan dpal*, 3 vols., vol. 2, 1–11. From a manuscript at Rta mgo. Thimphu: Rta mgo Monastery.

Pha gol, Dpal ldan. 2006. *Yan lag brgyad pa'i snying bsdus kyi rang 'grel*. Beijing: Mi rigs dpe skrun khang.

Bi ci'i pu ti kha ser. 2005. Lhasa: Bod ljongs mi dmangs dpe skrun khang.

Bu ston Rin chen grub. 1988. *Bde bar gshegs pa'i bstan pa'i gsal byed chos kyi 'byung gnas gsung rab rin po che'i mdzod*. Beijing: Krung go'i bod kyi shes rig dpe skrun khang.

——. 2000. *Chos kyi dkon mchog mdzod 'dzin de'i phyag tu 'bul*. In *Gsung 'bum*, 26 vols., vol. 26, 149–352. Blockprint. Lhasa: Zhol par khang. Available at TBRC W1934.

Bod kyi thang ga. 1985. Ed. Bod rang skyong ljongs rig dngos do dam U yon lhan khang. Beijing: Rig dngos dpe skrun khang.

Bod mkhas pa Mi pham dge legs rnam rgyal. 1986. *Gso rig gzhi lam 'bras bu'i rnam gzhag la dpyad pa'i dris lan legs bshad rin po che'i snang pa'i gsar pa*. From a manuscript from the library of Tenzin Chodrak. Dharamsala: Library of Tibetan Works & Archives.

Bai ro tsa na? 2007. *'Khrungs dpe g.yu yi phreng ba zhes bya ba'i ma dpe*. In *Gso rig sman gyi ro nus ngos 'dzin gsal ston phyogs sgrig rin chen sgron me*, ed. Dpal brtsegs bod yig dpe rnying zhib 'jug khang, 414–50. Beijing: Krung go'i bod rig pa dpe skrun khang.

Byang pa Rnam rgyal grags bzang. 2001. *Bshad rgyud kyi 'grel chen bdud rtsi'i chu rgyun*. Chengdu: Si khron mi rigs pe skrun khang.

——. 2008. *Bdud rtsi snying po yan lag brgyad pa gsang ba man ngag gi rgyud bzhi'i rtsa ba'i 'grel pa legs pa bzhugs pa'i dbu phyogs*. In *Rgyud bzhi'i rtsa ba'i grel pa: Yongs gtad rgyud kyi 'grel chen*, 4–186. Beijing: Mi rigs dpe skrun khang.

Byams pa 'phrin las. 1996a. *Bod kyi gso rig rgyud bzhi'i nang don mtshon pa'i sman thang bris cha'i skor la rags tsam dpyad pa*. In *Byams pa 'phrin las kyi gsung rtsom phyogs bsgrigs*, 370–81. Beijing: Krung go'i bod kyi shes rig dpe skrun khang.

——. 1996b. *Sde srid sangs rgyas rgya mtsho'i 'khrungs rabs dang mdzad rjes dad brgya'i padma rnam par bzhad pa'i phreng ba*. In *Byams pa 'phrin las kyi gsung rtsom phyogs bsgrigs*, 402–42. Beijing: Krung go'i bod kyi shes rig dpe skrun khang.

——. 2000. *Gangs ljongs gso rig bstan pa'i nyin byed rim byon gyi rnam thar phyogs bsgrigs*. Beijing: Mi rigs dpe skrun khang.

Brang ti Dpal ldan 'tsho byed. N.d. *Bdud rtsi snying po yan lag brgyad pa gsang ba man ngag gis [sic] rgyud kyi spyi don shes bya rab gsal rgyas pa*. Incomplete manuscript in Library of the Cultural Palace of Minorities in Beijing. Photocopy made by Leonard van der Kuijp.

——. 1977. *Rtsa ba'i rgyud gyis* [sic] *'brel pa rgyud don rab gsal rgyas ba*. In *Commentaries on the Rtsa rgyud, Bśad rgyud, and Phyi ma rgyud*, by Gong sman Dkon mchog bde legs, 1–195. From a manuscript. Leh: T. Sonam Tashigang.

Dbang 'dus. 1983. *Bod gangs can pa'i so ba rig pa'i dpal ldan rgyud bzhi sogs kyi brda dang dka' gnad 'ga' zhig bkrol ba sngon byon mkhas pa'i gsung rgyun g.yu thog dgong rgyan*. Beijing: Mi rigs dpe skrun khang.

——. 2004. "Bod sman gyi gshags las dar rgyas rgud gsum gyi skor rags tsam gleng ba." In *Bod sman slob gso dang zhib 'jug* 9: 1–11.

Sman dpyad zla ba'i rgyal po. 1985. Beijing: Mi rigs dpe skrun khang.

Tshul khrims rgyal mtshan. n.d. "Sman thang las 'brel pa rtsa yi gnas lugs kyi dpe ris skor gsal bar bshad pa." In *Krung go'i mtho rim bod sman shib 'jug bgro gleng 'dzin grwa'i rtsom yig gces bsdus*, 83–91. Lhasa: Bod rang skyong ljongs sman rtsis khang.

Tshe'i rig byed mtha' dag gi snying po bsdus pa. 1982–1985. In *Sde dge bstan 'gyur*, 213 vols., vol. No (= TBRC vol. 211), 599–670. From the Sde dge blocks carved 1737–44. Delhi: Delhi Karmapae Choedhey.

Btsan lha ngag dbang tshul khrims. 1997. *Brda dkrol gser gyi me long*. Beijing: Mi rigs dpe skrun khang.

'Tsho byed dkon skyabs. 1976. *Rtsod bzlog gegs sel 'khor lo*. In *G.yu thog cha lag bco brgyad* 1976, vol. 1, 303–29.

Rdza dpal sprul O rgyan 'jigs med chos kyi dbang po. 2003. *Gzhon nu blo ldan gyi dris lan legs bshad blo gros snying po*. In *Dpal sprul o rgyan 'jigs med chos kyi dbang po'i gsung 'bum*, 8 vols., vol. 1, 31–55. Chengdu: Si khron mi rigs dpe skrun khang.

Wang le and Byams pa 'phrin las, eds. 2004. *Bod lugs gso rig rgyud bzhi'i nang don bris cha ngo mtshar mthong ba don ldan*. Lhasa: Bod ljongs mi dmangs dpe skrun khang.

Zur mkhar ba Blo gros rgyal po. 1980–85. *Mes po'i zhal lung*, 4 vols. From a Lhasa Zhol block-print. Leh: Tashigangpa.

——. 1986. *Rgyud bzhi bka' dang bstan bcos rnam par dbye ba mun sel sgron me*. In *Bod kyi sman rtsis ched rtsom phyogs bsdus*, ed. Bod rang skyong ljongs sman rtsis khang, 64–71. Lhasa: Bod ljongs mi dmangs dpe skrun khang.

——. 1989. *Rgyud bzhi'i 'grel pa mes po'i zhal lung*, vol.1. Beijing: Krung go'i bod kyi shes rig dpe skrun khang.

——. 2001. *Sman pa rnams kyis mi shes su mi rung ba'i shes bya spyi'i khog dbubs*. Chengdu: Si khron mi rigs dpe skrun khang.

——. 2003a. *Rdo ring mdzes byed kyi dris lan rgan po'i kha chems mtshan mo mun nag gi glog 'od lta bu*. In *Zur mkhar blo gros rgyal po'i gsung rtsom gces btus*, ed. 'Bum kho and Grags pa, 10–76. Kunming: Yu nan mi rigs dpe skrun khang.

——. 2003b. *Drang srong chen po mnyam nyid rdo rje'i rnam par thar pa 'gog pa med pa'i yi ge'i gtam chen po*. In *Zur mkhar blo gros rgyal po'i gsung rtsom gces btus*, ed. 'Bum kho and Grags pa, 77–132. Kunming: Yu nan mi rigs dpe skrun khang.

Yang dgon pa Rgyal mtshan dpal. 1976. *Rdo rje lus kyi sbas bshad*. In *The Collected Works (Gsuṅ 'bum) of Yaṅ-dgon-pa Rgyal-mtshan-dpal*, 3 vols., vol. 2, 421–97. From a manuscript at Pha jo ldings. Thimphu: Kunsang Topgey, 1976.

Ye shes gzungs, Sum ston. 1976. *'Grel ba 'bum chung gsal sgron nor bu'i 'phreng mdzes*. In *G.yu thog cha lag bco brgyad* 1976, vol. 1, 157–301.

——. 1981. *G.yu thog snying thig las byin rlabs bla ma sgrub pa'i chos skor sdug bsngal mun sel thugs rje'i nyi 'od ces pa'i thog mar lo rgyus dge ba'i lcags kyu*. In *G.yu thog sñiṅ thig* 1981, 5–41.

——. 1998. *Bshad rgyud 'grel pa 'bum nag gsal sgron*, Beijing: Mi rigs dpe skrun khang.

Ye shes gzungs and Gzhon nu ye shes (?). 1976. *Brgyud pa'i rnam thar med thabs med pa*. In *G.yu thog cha lag bco brgyad* 1976, vol. 2, 133–41.

G.yang dga'. 2002. *Lus kyi phung po'i gnas lugs rgyas par bstan pa ro bkra don gsal me long*. M.A. thesis, Gangs ljongs bod lugs gso rig slob gling (Tibetan Medical College), Lhasa.

G.yu thog cha lag bco brgyad. 1976. *A Corpus of Tibetan Medical Teachings Attributed to Gyu-thog the Physician*, 2 vols. From a Lhasa Zhol blockprint, seventeenth century. Delhi: Topden Tshering, 1976.

G.yu thog sñiṅ thig gi yig cha: The Collected Basic Texts and Ritual Works of the Medical Teachings Orally Passed from G'yu-thog Yon-tan-mgon-po. 1981. From a Lcags po ri blockprint, ed. Khams smyon Dharma seng ge, carved 1888. Leh: D. L. Tashigang.

G.yu thog sman pa dge bshes mañju (?). 1976. *Ṭīkka mun sel sgron me*. In *G.yu thog cha lag bco brgyad* 1976, vol. 1, 121–56.

[G.yu thog Yon tan mgon po]. 1975. *Rgyud bzhi*. From a Zung cu ze blockprint, eighteenth century. Leh: S. W. Tashigangpa.

[——.] 1976a. *Skor tshoms stong thun bcu gcig las gnyis pa chu'i stong thun*. In *G.yu thog cha lag bco brgyad* 1976, vol. 1, 343–50.

[——.] 1976b. *Rgyud chung bdud rtsi snying po*. In *G.yu thog cha lag bco brgyad* 1976, vol. 2, 203–449.

[——.] 1978. *Rgyud bzhi*. From a Lcags po ri blockprint, carved 1888. Leh: T. Sonam Tashigang.

[——.] 199?. *Rgyud bzhi*. From a Lho 'brug blockprint. Leh: T. Sonam Tashigang.

[——.] 1992. *Bdud rtsi snying po yan lag brgyad pa gsang ba man ngag gi rgyud*, Lhasa: Bod ljongs mi dmangs dpe skrun khang.

[——.] 1999?. *Rgyud bzhi*. Imprint from blocks carved in 1733. Sde dge: Sde dge par khang chen mo. Available at TBRC W00EGS1016257.

——. 2003. *Yan lag brgyad pa'i gzhung las bsdus pa nor bu'i 'phreng ba*. Chengdu: Si khron mi rigs dpe skrun khang.

[——.] 2005a. *Grwa thang rgyud bzhi*. Beijing: Mi rigs dpe skrun khang.

——. 2005b. *Bu don ma*. Beijing: Mi rigs dpe skrun khang.

——. 2006. *Sngo 'bum sman gyi gter mdzod*. Beijing: Mi rigs dpe skrun khang.

G.yu thog Yon tan mgon po sogs. 2005. *Sngo 'bum sman gyi gter mdzod*. Beijing: Mi rigs dpe skrun khang.

Ra se Dkon mchog rgya mtsho. 2003. *Gangs ljongs skyes ma'i lo rgyus spyi bshad*. Lhasa: Bon ljongs mi dmangs dpe skrun khang.

Rang byung rdo rje. 1970? *Zab mo nang don rtsa ba*. Blockprint. Gangtok, Sikkim: Rum btegs Chos sgar. Available at TBRC W679.

Śākya mchog ldan, Gser mdog Paṇ chen. 1975. *Byang pa bdag po rnam rgyal grags pa la bde gshegs snying po sman dpyad dus 'khor gsum gyi dri ba*. In *The Complete Works of Gser-mdog Paṇ-chen Śākya-mchog-ldan*, vol. 17, 325–29. From an eighteenth-century manuscript in Bhutan. Thimphu: Kunzang Tobgey.

Sa skya Paṇḍita Kun dga' rgyal mtshan. 1992–93. *Sdom gsum rab dbye*. In *Sa skya bka' 'bum*, by Kun dga' snying po et al., 15 vols., vol. 12, 1–96. From a Sde dge blockprint, carved 1736. Dehra Dun, India: Sakya Center.

Sa chen Kun dga' snying po. 1983. *Lam 'bras gzhung bshad sras don ma*. In *Lam 'bras slob bśad: The Sa-skya-pa Teachings of the Path and the Fruit, According to the Tshar-pa Transmission*, 21 vols., vol. 12, 1–446. From a Sde dge blockprint. Dehra Dun, India: Sakya Center.

Sa dpyad sgag mos rngam thabs. 1976. In *G.yu thog cha lag bco brgyad* 1976, vol. 1, 77–120.

Sangs rgyas gling pa. 1981–84. *Bla-ma dgongs-'dus: A Complete Cycle of Buddhist Practice*. 18 vols. From a manuscript at Mtshams brag dgon. Paro, Bhutan: Lama Ngodrup and Sherab Drimey.

Sog bzlog pa Blo gros rgyal mtshan. 1975. *Rgyud bzhi'i bka' bsgrub nges don snying po*. In *Collected Writings of Sog-bzlog-pa Blo-gros-rgyal-mtshan*, 2 vols., vol. 2, 213–42. From an incomplete manuscript from the library of Bdud 'joms Rin po che. New Delhi: Sanji Dorji.

Bsod nams tshe mo. 1992–93. *Chos la 'jug pa'i go*. In *Sa skya bka' 'bum* by Kun dga' snying po et al., 15 vols., vol. 4, 525–633. From a Sde dge blockprint, carved 1736. Dehra Dun, India: Sakya Center.

Bsod nams ye shes rgyal mtshan, Rin spungs pa'i bla sman. 2002. *Bshad rgyud kyi 'grel chen dri med kun gsal*. 2 vols. Chengdu: Si khron mi rigs dpe skrun khang.

Lha sdings Byams pa skal bzang, ed. 2000. *A Mirror of the Murals in the Potala (Pho brang po ta la' ldebs bris ri mo'i 'byung khungs lo rgyus gsal ba'i me long)*. Beijing: Jiu zhou tu shu chu ban she.

WESTERN-LANGUAGE WORKS AND MODERN EDITIONS

Abeysekara, Ananda. 2002. *Colors of the Robe: Religion, Identity, and Difference*. Columbia: University of South Carolina Press.

Adams, Vincanne. 2001. "The Sacred in the Scientific: Ambiguous Practices of Science in Tibetan Medicine." *Cultural Anthropology* 16 (4): 542–75.

——. 2007. "Integrating Abstraction: Modernising Medicine at Lhasa's Mentsikhang." In *Soundings in Tibetan Medicine: Anthropological and Historical Perspectives: PIATS 2003: Tibetan Studies: Proceedings of the Tenth Seminar of the International Association for Tibetan Studies, Oxford, 2003*, ed. Mona Schrempf, 29–45. Leiden: Brill.

Adams, Vincanne, and Fei Fei Li. 2008. "Integration or Erasure? Modernizing Medicine at Lhasa's Mentsikhang." In *Exploring Tibetan Medicine in Contemporary Context: Perspectives in Social Sciences*, ed. Laurent Pordie, 105–31. London: Routledge.

Adams, Vincanne, Mona Schrempf, and Sienna R. Craig, eds. 2011. *Medicine Between Science and Religion: Explorations on Tibetan Grounds*. New York: Berghahn Books.

Ali, Daud. 2004. *Courtly Culture and Political Life in Early Medieval India*. Cambridge: Cambridge University Press.

Almond, Philip C. 1988. *The British Discovery of Buddhism*. Cambridge: Cambridge University Press.

Aris, Michael. 1994. "India and the British According to a Tibetan Text of the Later Eighteenth Century." In *Tibetan Studies: Proceedings of the 6th Seminar of the International Association for Tibetan Studies, Fagernes*, ed. Per Kværne. Oslo: Institute for Comparative Research in Human Culture.

——. 1995. *Jigs-med-gling-pa's "Discourse on India" of 1789: a critical edition and annotated translation of the Lho-phyogs rgya-gar-gyi gtam brtag-pa brgyad-kyi me-long*. Occasional Paper Series, vol. 9. Tokyo: The International Institute for Buddhist Studies.

Ary, Elijah Sacvan. 2007. "Logic, Lives, and Lineage: Jetsun Chokyi Gyaltsen's Ascension and the 'Secret Biography of Khedrup Geleg Pelzang.'" Ph.D. diss., Harvard University.

Asaṅga. 1907. *Mahāyāna-Sūtrālaṃkāra*, vol. 1. Ed. Sylvain Lévi. Paris: Librarie Honoré Champion.

Bacot, Jacques. 1956. "Reconnaissance en Haute Asie Septentrionale par Cinq Envoyés Ouigours au VIIIe Siècle." *Journal Asiatique* 244: 137–53.

Baker, Ian A. 2000. *The Dalai Lama's Secret Temple: Tantric Wall Paintings from Tibet*. New York: Thames & Hudson.

Barthes, Roland. 1978. "The Rhetoric of the Image." In *Image-Music-Text*, 32–51. New York: Macmillan.

——. 1989. "The Reality Effect." In *The Rustle of Language*, trans. Richard Howard, 141–48. Berkeley: University of California Press.

Beal, Samuel, trans. 1973. *The Life of Hiuen-Tsiang by the Shaman Hwui Li*. 1911; reprint, Westport, CT: Hyperion Press.

Beckwith, Christopher I. 1979. "The Introduction of Greek Medicine Into Tibet in the Seventh and Eighth Centuries." *Journal of the American Oriental Society* 99 (2) (April 1): 297–313.

Bellah, Robert Neelly. 1957. *Tokugawa Religion: The Values of Pre-industrial Japan*. Glencoe, IL: Free Press.

Berger, Patricia Ann. 2003. *Empire of Emptiness: Buddhist Art and Political Authority in Qing China*. Honolulu: University of Hawaii Press.

Bernbaum, Edwin. 1980. *The Way to Shambhala*. Garden City, NY: Anchor Books.

Bhishagratna, Kaviraj Kunjalal, ed. 1998–99. *Suśruta Saṁhitā*, 3 vols. Varanasi: Chowkhamba Sanskrit Series Office.

Birnbaum, Raoul. 1979. *The Healing Buddha*. Boulder, CO: Shambhala.

Bjerken, Zeff. 2001. "The Mirror Work of Tibetan Religious Historians: A Comparison of Buddhist and Bon Historiography." Ph.D. diss., University of Michigan.

Blundell, Boyd. 2010. *Paul Ricoeur Between Theology and Philosophy: Detour and Return*. Bloomington: Indiana University Press.

Bodhi, Bhikkhu, trans. 2000. *The Connected Discourses of the Buddha: A Translation of Saṃyutta Nikāya*. Somerville, MA: Wisdom.

——. 2005. *In the Buddha's Words: An Anthology of Discourses from the Pāli Canon*. Teachings of the Buddha. Boston: Wisdom.

Bolsokhoyeva, Natalia. 2007. "Tibetan Medical Illustrations from the History Museum of Buryatia, Ulan Ude." *Asian Medicine* 3 (2): 347–67.

Braun, Erik Christopher. 2013. *The Birth of Insight: Ledi Sayadaw, Meditation, and Modern Buddhism in Burma*. Chicago: University of Chicago Press.

Bronkhorst, Johannes. 2006. "Commentaries and the History of Science in India." *Asiatische Studien/Études Asiatiques* 60 (4): 773–78.

Bronner, Yigal. 2002. "What Is New and What Is Navya: Sanskrit Poetics on the Eve of Colonialism." *Journal of Indian Philosophy* 30 (5): 441–62.

Brunnhölzl, Karl. 2007. *In Praise of Dharmadhātu: Nāgārjuna and the Third Karmapa, Rangjung Dorje*. Ithaca, NY: Snow Lion.

Buddhaghosa, Bhadantācariya. 1999. *The Path of Purification (Visuddhimagga)*. Trans. Bhikkhu Ñyāṇamoli. Seattle: BPS Pariyatti Editions.

Lo Bue, Erberto. 2003. "Scholars, Artists and Feasts." In *Lhasa in the Seventeenth Century: The Capital of the Dalai Lamas*, ed. Françoise Pommaret, 179–98. Leiden: Brill.

Butler, Judith. 1990. *Gender Trouble: Feminism and the Subversion of Identity*. New York: Routledge.

Capitanio, Joshua. 2010. "Indian and Chinese Conceptions of the Body in the Medical Writings of Tiantai Zhiyi." Paper delivered at the American Academy of Religion Annual Conference in Atlanta, Georgia.

Chakrabarty, Dipesh. 2000. *Provincializing Europe: Postcolonial Thought and Historical Difference*. Princeton, NJ: Princeton University Press.

Chang, Chen-chi, ed. 1983. *A Treasury of Mahāyāna Sūtras: Selections from the Mahāratnakūta Sūtra*. University Park: Pennsylvania State University Press.

Clark, Barry. 1995. *The Quintessence Tantras of Tibetan Medicine*. Ithaca, NY: Snow Lion.

Collins, Steven. 2010. *Nirvana: Concept, Imagery, Narrative*. Cambridge: Cambridge University Press.

Craig, Sienna R. 2012. *Healing Elements: Efficacy and the Social Ecologies of Tibetan Medicine*. Berkeley: University of California Press.

Cuevas, Bryan J. 2011. "Illustrations of Human Effigies in Tibetan Ritual Texts: With Remarks on Specific Anatomical Figures and Their Possible Iconographic Source." *Journal of the Royal Asiatic Society* (Third Series) 21 (1): 73–97.

Cullen, Christopher. 2001. "Yi'an: The Origins of a Genre of Chinese Medical Literature." In *Innovation in Chinese Medicine*, ed. Elisabeth Hsu, 297–323. Cambridge: Cambridge University Press.

Cüppers, Christoph, Leonard van der Kuijp, and Ulrich Pagel. 2012. *Handbook of Tibetan Iconometry: A Guide to the Arts of the Seventeenth Century*. Chinese introduction by Dobis Tsering Gyal. Leiden: Brill.

Czaja, Olaf. 2005a. "Zurkharwa Lodro Gyalpo (1509–1579) on the Controversy of the Indian Origin of the rGyud bzhi." *Tibet Journal* 30/31 (4/1): 133–54.

——. 2005b. "A Hitherto Unknown 'Medical History' of mTsho smad mkhan chen (b. 16th cent.)." *Tibet Journal* 30/31 (4/1): 155–74.

——. 2007. "The Making of the *Blue Beryl*—Some Remarks on the Textual Sources of the Famous Commentary of Sangye Gyatsho (1653–1705)." In *Soundings in Tibetan Medicine: Anthropological and Historical Perspectives: PIATS 2003: Tibetan Studies: Proceedings of the Tenth Seminar of the International Association for Tibetan Studies, Oxford, 2003*, ed. Mona Schrempf, 345–72. Leiden: Brill.

——. 2008. "Mi pham dge legs rnam rgyal (1618–1685) and His Reply to the Famous Public Letter of Zur mkhar ba Blo gros rgyal po (1509–1579?)." In *Written Treasures of Bhutan: Mirror of the Past and Bridge to the Future*," ed. John A. Ardussi and Sonam Tobgay, vol. 2, 75–142. Thimphu, Bhutan: National Library of Bhutan.

Dalton, Jacob Paul. 2011. *The Taming of the Demons: Violence and Liberation in Tibetan Buddhism*. New Haven: Yale University Press.

Das, Rahul Peter. 2003. *The Origin of the Life of a Human Being: Conception and the Female According to Ancient Indian Medical and Sexological Literature.* Delhi: Motilal Banarsidass.

Das, Rahul Peter, and Ronald E. Emmerick, eds. 1998. *Vāgbhaṭa's Aṣṭāṅgahṛdayasaṃhitā: the Romanised Text Accompanied by Line and Word Indexes.* Groningen: Forsten.

Das, Sarat Chandra. 1981. *Tibetan-English Dictionary.* 1902; reprint, Kyoto: Rinsen-Shoten Bookstore.

Das, Veena. 2012. "Ordinary Ethics." In *A Companion to Moral Anthropology*, ed. Didier Fassin, 133–49. Hoboken, NJ: Wiley-Blackwell.

Dasgupta, Surendranath. 1952. *A History of Indian Philosophy*, vol. 2. 1922; reprint, Delhi: Motilal Banarsidass.

Daston, Lorraine, and Peter Galison. 2007. *Objectivity.* Brooklyn, NY: Zone Books.

Davidson, Ronald M. 1990. "Appendix: An Introduction to the Standards of Scriptural Authenticity in Indian Buddhism." In *Chinese Buddhist Apocrypha*, ed. Robert E. Buswell, 291–326. Honolulu: University of Hawaii Press.

Demieville, Paul. 1985. *Buddhism and Healing: Demiéville's Article "Byô" from Hôbôgirin (1937).* Trans. Mark Tatz. Lanham: University Press of America.

Desideri, Ippolito. 1981. *Il T'o-raṅs = L'aurora.* Trans. Giuseppe M. Toscano. Roma: Istituto italiano per il Medio ed Estremo Oriente.

Dollfus, Pascale. 1991. "Peintures Tibétaines De La Vie De Mi-la-ras-pa." *Arts Asiatiques* XLVI (1): 50–71.

Dreyfus, Georges B. J. 1994. "Cherished Memories, Cherished Communities: Proto-nationalism in Tibet." In *Tibetan Studies: Proceedings of the 6th Seminar of the International Association for Tibetan Studies, Fagernes, 1992*, ed. Per Kværne, vol. 1, 205–18. Oslo: Institute for Comparative Research in Human Culture.

——. 2003. *The Sound of Two Hands Clapping: The Education of a Tibetan Buddhist Monk.* Berkeley: University of California Press.

Dudjom Rinpoche. 1991. *The Nyingma School of Tibetan Buddhism: Its Fundamentals and History.* 2 vols. Trans. Gyurme Dorje and Matthew Kapstein. Boston: Wisdom.

Dunhuang yan jiu yuan. 1997. *Anxi Yulin ku.* Beijing: Wen wu zhu ban she.

Dussel, Enrique. 2006. "World-System and 'Trans'-Modernity." In *Unbecoming Modern: Colonialism, Modernity, Colonial Modernities*, ed. Saurabh Dube and Ishita Banerjee-Dube, 165–89. New Delhi: Social Science Press.

Dutt, Sukumar. 1962. *Buddhist Monks and Monasteries of India: Their History and Their Contribution to Indian Culture.* London: George Allen and Unwin.

Eberhardt, Nancy. 2006. *Imagining the Course of Life: Self-Transformation in a Shan Buddhist Community.* Honolulu: University of Hawaii Press.

Egmond, Florike, Paul G. Hoftijzer, and Robert P. W. Visser, eds. 2007. *Carolus Clusius in a New Context: Towards a Cultural History of a Renaissance Naturalist.* Amsterdam: Koninklijke Nederlandse Akademie van Wetenschappen.

Eisenstadt, S. N. 2000. "Multiple Modernities." *Daedalus* 129 (1) (January 1): 1–29.

Eliade, Mircea. 1958. *Yoga: Immortality and Freedom.* Trans. William R. Trask. Bollingen Series 56. London: Routledge & Kegan Paul.

Elias, Norbert. 1994. "The Civilizing Process: The History of Manners and State Formation and Civilization." In *The Civilizing Process*, trans. Edmund Jephcott, 168–78. Oxford: Blackwell.

Elman, Benjamin A. 2005. *On Their Own Terms: Science in China, 1550–1900*. Cambridge, MA: Harvard University Press.

Emmerick, Ronald E. 1977. "Sources of the Four Tantras." In *Zeitschrift Der Deutschen Morgenländischen Gessellschaft*, ed. W. Voigt. Supplement 3.2, 135–142. Wiesbaden: Franz Steriner.

——. 1980–82. *The Siddhasāra of Ravigupta*. 2 vols. Wiesbaden: Steiner.

——. 1981. "Mi-chos." In *Ludwik Sternbach Felicitation*, ed. J. P. Sinha, 883–85. Lucknow: Akhila Bharatiya Sanskrit Parishad.

——. 1990. "Rgas Pa Gso Pa." In *Indo-Tibetan Studies: Papers in Honour and Appreciation of Professor David L. Snellgrove's Contribution to Indo-Tibetan Studies*, ed. Tadeusz Skorupski, 89–99. Tring, UK: Institute of Buddhist Studies.

——. 1991. "Some Remarks on Tibetan Sphygmology." In *Medical Literature from India, Sri Lanka and Tibet*, ed. G. Jan Meulenbeld, 66–72. Leiden: Brill.

——. 1993. "Some Tibetan Medical Tankas." *Bulletin of Tibetology, Special Volume: Aspects of Classical Tibetan Medicine* 29: 56–78.

——. 1995. [Untitled review of Parfionovitch et al.]. *Bulletin of the School of Oriental and African Studies, University of London* 58 (2): 403–406.

Emmerick, Ronald E., trans. 1996. *Sutra of Golden Light, Being a Translation of the* Suvarṇabhāsottamasūtra. 1970; reprint, Oxford: Pali Text Society.

Emmerick, Ronald E., Michael D. Willis, and Iravatham Mahadevan. 1988. "Notes and Communications." *Bulletin of the School of Oriental and African Studies* 51 (3): 537–41.

Erhard, Franz-Karl. 2007. "A Short History of the g.Yu thog snying thig." In *Indica et Tibetica 66: Festschrift für Michael Hahn*, ed. Konrad Klaus and Jens-Uwe Hartmann. Wien: Arbeitskreis für Tibetische und Buddhistische Studien, 151–70.

Farrow, G. W., and I. Menon, trans. 1992. *The Concealed Essence of the Hevajra Tantra: With the Commentary Yogaratnamālā*. Delhi: Motilal Banarsidass.

Fenner, Todd. 1996. "The Origin of the rGyud Bzhi: A Tibetan Medical Tantra." In *Tibetan Literature: Studies in Genre*, ed. Roger R. Jackson and José Ignacio Cabezón, 458–69. Ithaca, NY: Snow Lion.

Ferrari, Alfonsa, trans. 1958. *Mk'yen brtse's Guide to the Holy Places of Central Tibet*. Ed. Luciano Petech and Hugh Richardson. Roma: Istituto italiano per il Medio ed Estremo Oriente.

Filliozat, Jean. 1975. *La doctrine classique de la médecine indienne; ses origines et ses parallèles grecs*. 1949; reprint, Paris: Ecole Française D'Extréme-Orient.

Finnegan, Damchö. 2007. "From Lhasa to the Kingdom of Guge: Painting a Place for Us." In *From Lhasa to the Kingdom of Guge*. http://acrosstibet.blogspot.com/2007/08/painting-place-for-us.html.

Gaonkar, Dilip Parameshwar. 2001. "On Alternative Modernities." In *Alternative Modernities*, ed. Dilip Parameshwar Gaonkar, 1–23. Durham, NC: Duke University Press.

Garrett, Frances. 2006. "Buddhism and the Historicising of Medicine in Thirteenth-century Tibet." *Asian Medicine* 2 (2): 204–24.

——. 2007. "Critical Methods in Tibetan Medical Histories." *The Journal of Asian Studies* 66 (2): 363–87.

——. 2008. *Religion, Medicine and the Human Embryo in Tibet*. London: Routledge.

—. 2009. "The Alchemy of Accomplishing Medicine (*Sman Sgrub*): Situating the Yuthok Heart Essence (*G.yu Thog Snying Thig*) in Literature and History." *Journal of Indian Philosophy* 37 (3): 207–30.

Garrett, Frances, and Vincanne Adams. 2008. "The Three Channels in Tibetan Medicine. With a Translation of Tsultrim Gyaltsen's 'A Clear Explanation of the Principal Structure and Location of the Circulatory Channels as Illustrated in the Medical Paintings.'" *Traditional South Asian Medicine* 8: 86–114.

Gerke, Barbara. 1999. "On the History of the Two Tibetan Medical Schools Janglug and Zurlug." *AyurVijnana* 6 (Spring): 17–25.

—. 2001. "The Authorship of the Tibetan Medical Treatise *Cha Lag Bco Brgyad* (Twelfth Century A.D.) and a Description of Its Historical Background." *Traditional South Asian Medicine* 6: 27–50.

Gerke, Barbara, and Natalia Bolsokhoeva. 1999. "Namthar of Zurkha Lodo Gyalpo (1509–1579): A Brief Biography of a Tibetan Physician." *AyurVijnana* 6 (Spring): 26–38.

Goble, Andrew Edmund. 2011. *Confluences of Medicine in Medieval Japan: Buddhist Healing, Chinese Knowledge, Islamic Formulas, and Wounds of War*. Honolulu: University of Hawai'i Press.

Gold, Jonathan C. 2008. *The Dharma's Gatekeepers: Sakya Pandita on Buddhist Scholarship in Tibet*. Albany: State University of New York Press.

Gombrich, Richard F. 1988. *Theravāda Buddhism: A Social History from Ancient Benares to Modern Colombo*. London: Routledge & Kegan Paul.

Gray, David B. 2007. "Compassionate Violence? On the Ethical Implications of Tantric Buddhist Ritual." *Journal of Buddhist Ethics* 14: 240–71.

Guenther, Herbert V., trans. 1971. *Jewel Ornament of Liberation by Sgam Po Pa*. Berkeley, CA: Shambala.

Gyaltsen, Sakya Pandita Kunga. 2000. *Ordinary Wisdom: Sakya Pandita's Treasury of Good Advice*. Trans. John T. Davenport, Sallie D. Davenport, and Losang Thonden. Boston: Wisdom.

Gyaltshen, Sakya Pandita Kunga. 2002. *A Clear Differentiation of the Three Codes: Essential Distinctions Among the Individual Liberation, Great Vehicle, and Tantric System*. Trans. Jared Rhoton. Ed. Victoria R. M. Scott. Albany: State University of New York Press.

Gyatso, Desi Sangyé. 2010. *A Mirror of Beryl: A Historical Introduction to Tibetan Medicine*. Trans. Gavin Kilty. Boston: Wisdom.

Gyatso, Janet. 1981. "A Literary Transmission of the Traditions of Thang-Stong Rgyal-Po: A Study of Visionary Buddhism in Tibet." Ph.D. diss., University of California, Berkeley.

—. 1986. "Signs, Memory and History: A Tantric Buddhist Theory of Scriptual Transmission." *Journal of the International Association of Buddhist Studies* 9 (2): 7–35.

—. 1992. "Autobiography in Tibetan Religious Literature: Reflections on Its Modes of Self-Presentation." In *Tibetan Studies: Proceedings of the 5th International Association of Tibetan Studies Seminar*, ed. Shoren Ihara and Zuiho Yamaguchi, vol. 2, 465–78. Narita: Naritasan Institute for Buddhist Studies.

—. 1993. "The Logic of Legitimation in the Tibetan Treasure Tradition." *History of Religions* 33 (2): 97–134.

—. 1996. "Drawn from the Tibetan Treasury: The Gter-ma Literature." In *Tibetan Literature: Studies in Genre*, ed. José Ignacio Cabezón and Roger Jackson. Albany: State University of New York Press, 147–69.

——. 1998. *Apparitions of the Self: The Secret Autobiographies of a Tibetan Visionary: A Translation and Study of Jigme Lingpa's* Dancing Moon in the Water *and Ḍākki's* Grand Secret-talk. Princeton, NJ: Princeton University Press.
——. 1999. "Healing Burns with Fire: The Facilitations of Experience in Tibetan Buddhism." *Journal of the American Academy of Religion* 67 (1): 113–47.
——. 2003. "One Plus One Makes Three: Buddhist Gender, Monasticism, and the Law of the Non-excluded Middle." *History of Religions* 43 (2): 89–115.
——. 2004. "The Authority of Empiricism and the Empiricism of Authority: Medicine and Buddhism in Tibet on the Eve of Modernity." *Comparative Studies of South Asia, Africa and the Middle East* 24 (2): 83–96.
——. 2005. "Sex." In *Critical Terms for the Study of Buddhism*, ed. Donald S. Lopez, 271–90. Buddhism and Modernity. Chicago: University of Chicago Press.
——. 2009. "Spelling Mistakes, Philology, and Feminist Criticism: Women and Boys in Tibetan Medicine." In *Tibetan Studies in Honor of Samten Karmay*, ed. Françoise Pommaret and Jean-Luc Achard, 81–98. Dharamsala, India: Amnye Machen Institute.
——. 2011a. "Looking for Gender in the Medical Paintings of Desi Sangye Gyatso, Regent of the Tibetan Buddhist State." *Asian Medicine* 6 (2): 217–92.
——. 2011b. "Moments of Tibetan Modernity: Methods and Assumptions." In *Mapping the Modern in Tibet*, ed. Gray Tuttle, 1–44. Beiträge Zur Zentralasienforschung 24. Andiast, Switzerland: International Institute for Tibetan and Buddhist Studies GmbH.
——. 2014. "Buddhist Practices and Ideals in Desi Sangye Gyatso's Medical Paintings." In *Bodies in Balance: The Art of Tibetan Medicine*, ed. Theresia Hofer, 198–220. Seattle: University of Washington Press.
Gyatso, Lobsang. 1998. *Memoirs of a Tibetan Lama*. Trans. Gareth Sparham. Ithaca, NY: Snow Lion.
Gyatso, Tenzin (the XIVth Dalai Lama). 2005a. *The Universe in a Single Atom: The Convergence of Science and Spirituality*. New York: Morgan Road Books.
——. 2005b. "Our Faith in Science." *The New York Times*, November 12, Opinion section. http://www.nytimes.com/2005/11/12/opinion/12dalai.html.
Halbfass, Wilhelm. "Karma, *Apūrva*, and 'Natural' Causes: Observations on the Growth and Limits of the Theory of *Saṃsāra*." In *Karma and Rebirth in Classical Indian Traditions*, ed. Wendy Doniger O'Flaherty, 268–302. Berkeley: University of California Press, 1980.
Haldar, J. R. 1977. *Medical Science in Pali Literature*. Calcutta: Indian Museum Calcutta.
Hallisey, Charles. 2010. "Between Intuition and Judgment: Moral Creativity in Theravada Buddhist Ethics." In *Ethical Life in South Asia*, ed. Anand Pandian and Daud Ali, 141–53. Bloomington: Indiana University Press.
Hallisey, Charles, and Frank Reynolds. 1987. "Buddhism: An Overview." In *The Encyclopedia of Religion*, ed. Mircea Eliade, vol. 2, 334–51. New York: Macmillan.
Hansen, Anne. 2007. *How to Behave: Buddhism and Modernity in Colonial Cambodia, 1860–1930*. Honolulu: University of Hawai'i Press.
Hanson, Marta. 2003. "The 'Golden Mirror' in the Imperial Court of the Qianlong Emperor, 1739–1742." *Early Science and Medicine* 8 (2): 111–47.
Haskell, Francis. 1993. *History and Its Images: Art and the Interpretation of the Past*. New Haven: Yale University Press.

Havnevik, Hanna. 1999. "The Life of Jetsun Lochen Rinpoche (1865–1951) as Told in Her Autobiography." Ph.D. diss., University of Arts, Oslo.

Heller, Amy. 1985. "An Early Tibetan Ritual: Rkyal 'Bud." In *Soundings in Tibetan Civilization*, ed. Barbara Nimri Aziz and Matthew Kapstein, 257–67. New Delhi: Manohar.

——. 1999. *Tibetan Art: Tracing the Development of Spiritual Ideals and Art in Tibet, 600—2000* A.D. Milano: Jaca Book.

——. 2006. "Preliminary Remarks on the Donor Inscriptions and Iconography of an 11th-Century Mchod Rten at Tholing." In *Tibetan Art and Architecture in Context: Tibetan Studies: Proceedings of the Eleventh Seminar of the International Association for Tibetan Studies, Königswinter 2006*, ed. Erberto Lo Bue and Christian Luczanits, 43–74. Halle: International Institute for Tibetan Studies.

——. 2009. *Hidden Treasures of the Himalayas: Tibetan Manuscripts, Paintings and Sculptures of Dolpo*. London: Serindia.

——. In press. "Preliminary Remarks on Painted Coffin Panels from Tibetan Tombs." In *Scribes, Texts, and Rituals in Early Tibet and Dunhuang: Proceedings of the 12th Seminar of the International Association for Tibetan Studies*, ed. Brandon Dotson, K. Iwao, and T. Takeuchi. Wiesbaden: Reichert Verlag.

Heruka, Tsangnyön. 2010. *The Life of Milarepa*. Trans. Andrew Quintman. New York: Penguin.

Hoernle, A. F. Rudolf. 1907. *Studies in the Medicine of Ancient India. Part I. Osteology, or The Bones of the Human Body*. Oxford: Clarendon Press.

——, ed. 1983. *The Bower Manuscript: Facsimile Leaves, Nagari Transcript, Romanised Transliteration, and English Translation with Notes*. Trans. A. F. Rudolf Hoernle. 3 vols. 1893–1912; reprint, New Delhi: Sharada Rani.

Hofer, Theresia. 2007. "Preliminary Investigations Into New Oral and Textual Sources on Byang Lugs—The 'Northern School' of Tibetan Medicine." In *Soundings in Tibetan Medicine: Anthropological and Historical Perspectives: PIATS 2003: Tibetan Studies: Proceedings of the Tenth Seminar of the International Association for Tibetan Studies, Oxford, 2003*, ed. M. Schrempf, 373–410. Leiden: Brill.

——. 2012. *The Inheritance of Change: Transmission and Practice of Tibetan Medicine in Ngamring*. Wien: Arbeitskreis für Tibetische und Buddhistische Studien, Universität Wien.

Hopkins, Jeffrey. 2002. *Reflections on Reality: The Three Natures and Non-Natures in the Mind-Only School: Dynamic Responses to Dzong-ka-ba's* The Essence of Eloquence. Vol. 2. Berkeley: University of California Press.

Horner, Isaline Blew. 1997–2000. *The Book of the Discipline* (Vinaya-Piṭaka). 5 vols. Oxford: Pali Text Society.

Hsu, Elisabeth. 2008. "A Hybrid Body Technique: Does the Pulse Diagnostic *cun guan chi* Method Have Chinese-Tibetan Origins?" *Gesnerus* 65: 5–29.

Hsüan-tsang. 1884. *Si-yu-ki (Buddhist Records of the Western World)*. Trans. Samuel Beal. 2 vols. London: Trübner & Co.

Huang, Mingxin, and Jiujin Chen. 1987. *Zangli de yuanli yu shijian: ju Shangzhuote Sangre yu Mayang Suobajiacan zangwen yuan zhu fanyi he yanjiu*. Beijing: Minzu chubanshe.

Huber, Toni. 1990. "Where Exactly Are Caritra, Devikota and Himavat? A Sacred Geography Controversy and the Development of Tantric Buddhist Pilgrimage Sites in Tibet." *Kailash* 16 (3 and 4): 121–64.

Irigaray, Luce. 1981. "This Sex Which Is Not One." In *New French Feminisms: An Anthology*, ed. Elaine Marks and Isabelle De Courtivron, trans. Claudia Reeder, 99–106. New York: Schocken.

——. 1991. "Sexual Difference." In *The Irigaray Reader*, ed. Margaret Whitford, trans. Seán Hand. Cambridge, MA: Basil Blackwell.

Ishihama, Yumiko. 1993. "On the Dissemination of the Belief in the Dalai Lama as a Manifestation of the Bodhisattva Avalokiteśvara." *Acta Asiatica* 64: 38–56.

Israel, Jonathan I. 2001. *Radical Enlightenment: Philosophy and the Making of Modernity, 1650–1750*. Oxford: Oxford University Press.

Jackson, David. 1996. *A History of Tibetan Painting*. Wien: Austrian Academy of Sciences.

Jacoby, Sarah Hieatt. 2007. "Consorts and Revelation in Eastern Tibet: The Auto/biographical Writings of the Treasure Revealer Sera Khandro (1892–1940)." Ph.D. diss., University of Virginia.

Jameson, Fredric. 2002. *A Singular Modernity: Essay on the Ontology of the Present*. London: Verso.

Jordens, J. T. F. 1978. *Dayānanda Sarasvatī: His Life and Ideas*. Delhi: Oxford University Press.

Kālidāsa. 1928. *Le Raghuvamça*. Ed. Louis Renou. Paris: Paul Geuthner.

Kaneko, Eiichi. 1982. *Ko-Tantora Zenshū Haidai Mokuroku* [Catalogue of the *Rnying ma'i rgyud 'bum*]. Tokyo: Kokusho Kankōkai.

Kapstein, Matthew. 1989. "The Purificatory Gem and Its Cleansing: A Late Tibetan Polemical Discussion of Apocryphal Texts." *History of Religions* 28 (3): 217–44.

——. 1992. "Remarks on the *Mani Bka 'bum* and the Cult of Avalokitesvara in Tibet." In *Tibetan Buddhism: Reason and Revelation*, ed. Steven D. Goodman and Ronald M. Davidson, 79–94. Albany: State University of New York Press.

——. 1996. "*gDams Ngag*: Tibetan Technologies of the Self." In *Tibetan Literature: Studies in Genre*, ed. José Ignacio Cabezón and Roger R. Jackson, 275–89. Albany: State University of New York Press.

——. 2002. "What Is 'Tibetan Scholasticism'? Three Ways of Thought." In Matthew Kapstein, *The Tibetan Assimilation of Buddhism: Conversion, Contestation, and Memory*, 85–120. Oxford: Oxford University Press.

——. 2011. "Just Where on Jambudvīpa Are We? New Geographical Knowledge and Old Cosmological Schemes in Eighteenth-century Tibet." In *Forms of Knowledge in Early Modern Asia: Explorations in the Intellectual History of India and Tibet, 1500–1800*, ed. Sheldon Pollock, 336–64. Durham, NC: Duke University Press.

Karmay, Samten. 1988a. *The Great Perfection* (rDzogs Chen)*: A Philosophical and Meditative Teaching of Tibetan Buddhism*. Leiden: Brill.

——. 1988b. *Secret Visions of the Fifth Dalai Lama: The Gold Manuscript in the Fournier Collection Musée Guimet, Paris*. London: Serindia.

——. 1998a. "The Soul and the Turquoise: A Ritual for Recalling the *Bla*." In Samten Karmay, *The Arrow and the Spindle: Studies in History, Myths, Rituals and Beliefs in Tibet*, 310–38. Kathmandu: Mandala Book Point.

——. 1998b. "The Wind-horse and the Well-being of Man." In Samten Karmay, *The Arrow and the Spindle: Studies in History, Myths, Rituals and Beliefs in Tibet*, 413–23. Kathmandu: Mandala Book Point.

——. 1998c. "The Four Tibetan Medical Treatises and Their Critics." In Samten Karmay, *The Arrow and the Spindle: Studies in History, Myths, Rituals and Beliefs in Tibet*, 228–37. Kathmandu: Mandala Book Point.

——. 2003. "The Fifth Dalai Lama and His Reunification of Tibet." In *Lhasa in the Seventeenth Century: The Capital of the Dalai Lamas*, ed. Françoise Pommaret, 65–80. Leiden: Brill.

——, trans. 2014. *The Illusive Play: The Autobiography of the Fifth Dalai Lama*. Chicago: Serindia.

Kaviraj, Sudipta. 2005a. "The Sudden Death of Sanskrit Knowledge." *Journal of Indian Philosophy* 33 (1): 119–42.

——. 2005b. "An Outline of a Revisionist Theory of Modernity." *European Journal of Sociology* 46 (3): 497–526.

Klimburg-Salter, Deborah E. 1988. *The Life of the Buddha in Western Himalayan Monastic Art and Its Indian Origins: Act One*. Roma: Istituto italiano per il Medio ed Estremo Oriente.

Kollmar-Paulenz, K. 1992. "Utopian Thought in Tibetan Buddhism: A Survey of the Śambhala Concept and Its Sources." *Studies in Central and East Asian Religions: Journal of the Seminar for Buddhist Studies* 5/6: 78–96.

Kongtrul, Jamgön 1996. *Creation and Completion: Essential Points of Tantric Meditation*. Trans. Sarah Harding. Boston: Wisdom.

Köppl, Heidi I. 2008. *Establishing Appearances as Divine: Rongzom Chözang on Reasoning, Madhyamaka, and Purity*. Ithaca, NY: Snow Lion.

Kossak, Steven, Jane Casey Singer, and Robert Bruce-Gardner. 1998. *Sacred Visions: Early Paintings from Central Tibet*. New York: Metropolitan Museum of Art.

Kritzer, Robert. 1998a. "Garbhāvakrāntisūtra: A Comparison of the Contents of Two Versions." *Maranatha: Bulletin of the Christian Culture Research Institute* 6: 4–13.

——. 1998b. "Semen, Blood, and the Intermediate Existence." *Indogaku Bukkyōgaku Kenkyū* XLVI (2): 30–36.

——. 2009. "Life in the Womb: Conception and Gestation in Buddhist Scripture and Classical Indian Medical Literature." In *Imagining the Fetus: The Unborn in Myth, Religion, and Culture*, ed. Vanessa Sasson and Jane Marie Law, 73–89. New York: Oxford University Press.

van der Kuijp, Leonard. 2005. "The Dalai Lamas and the Origins of Reincarnate Lamas." In *The Dalai Lamas: A Visual History*, ed. Martin Brauen, 14–31. Chicago: Serindia Publications in association with Ethnographic Museum of University of Zurich.

——. 2010. "Za Hor and Its Contribution to Tibetan Medicine, Part One: Some Names, Places and Texts." *Bod Rig Pa'i Dus Deb* 6: 21–50.

Kuriyama, Shigehisa. 2002. *The Expressiveness of the Body and the Divergence of Greek and Chinese Medicine*. New York: Zone Books.

Kusukawa, Sachiko. 2009. "Image, Text and Observatio: The Codex Kentmanus." *Early Science and Medicine* 14 (4): 445–75.

La Vallée Poussin, Louis de, trans. 1923–31. *L'abhidharmakośa De Vasubandhu*. 5 vols. Paris: Paul Geuthner.

——. 1962. *Catalogue of the Tibetan Manuscripts from Tun-huang in the India Office Library*. London: Oxford University Press.

LaCapra, Dominick. 1983. *Rethinking Intellectual History: Texts, Contexts, Language*. Ithaca, NY: Cornell University Press.

Lamotte, Etienne. 1988. Karmasiddhiprakaraṇa: *The Treatise on Action by Vasubandhu*. Trans. Leo M. Pruden. Berkeley, CA: Asian Humanities Press.

Lange, Kristina. 1964. "Eine Anatomische Tafel Zur Lamaistischen Heilkunde." *Annals of the Naprstek Museum* 3: 65–85.

—. 1976. *Die Werke Des Regenten Sans Rgyas Rgya Mco, 1653–1705*. Berlin: Akademie-Verlag.

Laqueur, Thomas Walter. 1990. *Making Sex: Body and Gender from the Greeks to Freud*. Cambridge, MA: Harvard University Press.

Law, Narendra Nath, ed. 1949. Sphuṭārtha Abhidharmakośa-vyākhyā *of Yaśomitra*. London: Luzac and Co.

Lévi, Sylvain. 1886. "Notes sur les Indo-Scythes." *Journal Asiatique* 8: 444–84.

Lin, Nancy. 2011. "Adapting the Buddha's Biographies: A Cultural History of the 'Wish-Fulfilling Vine' in Tibet, Seventeenth to Eighteenth Centuries." Ph.D. diss., University of California, Berkeley.

Liu, Yanchi. 1988. *The Essential Book of Traditional Chinese Medicine*. Ed. Peter Eckman and Kathleen Vian. Trans. Fang Tingyu and Chen Laidi. New York: Columbia University Press.

Lopez, Donald S. 1996. "Polemical Literature (*dGag Lan*)." In *Tibetan Literature: Studies In Genre*, ed. Jose Ignacio Cabezon and Roger R. Jackson, 217– 28. Ithaca, NY: Snow Lion.

—. 2008. *Buddhism and Science: A Guide for the Perplexed*. Buddhism and Modernity. Chicago: University of Chicago Press.

Lyssenko, Viktoria. 2004. "The Human Body Composition in Statics and Dynamics: Āyurveda and the Philosophical Schools of Vaiśeṣika and Sāṃkhya." *Journal of Indian Philosophy* 32 (1): 31–56.

MacQueen, Graeme. 1981. "Inspired Speech in Early Mahāyāna Buddhism I." *Religion* 11 (4): 303–19.

—. 1982. "Inspired Speech in Early Mahāyāna Buddhism II." *Religion* 12 (1): 49–65.

Manṣūr ibn Muḥammad ibn Aḥmad ibn Yūsuf ibn Ilyās. 2004. "Islamic Medical Manuscripts in the NLM—MS P 18: Tashrīḥ-i badan-i insān (The Anatomy of the Human Body)." Digital Library Collections. *Islamic Medical Manuscripts at the National Library of Medicine*. January 15. http://www.nlm.nih.gov/hmd/arabic/p18.html.

Martin, Dan. 2007. "An Early Tibetan History of Indian Medicine." In *Soundings in Tibetan Medicine: Anthropological and Historical Perspectives: PIATS 2003: Tibetan Studies: Proceedings of the Tenth Seminar of the International Association for Tibetan Studies, Oxford, 2003*, ed. Mona Schrempf, 307–26. Leiden: Brill.

—. 2010. "Greek and Islamic Medicines' Historical Contact with Tibet: A Reassessment in View of Recently Available but Relatively Early Sources on Tibetan Medical Eclecticism." In *Islam and Tibet: Interactions Along the Musk Routes*, ed. Anna Akasoy, Charles S. F. Burnett, and Ronit Yoeli-Tlalim, 117–44. Surrey: Ashgate.

Massin, Christophe. 1982. *La Médecine Tibétaine*. Paris: Guy Trédaniel.

Masuzawa, Tomoko. 2005. *The Invention of World Religions, or, How European Universalism Was Preserved in the Language of Pluralism*. Chicago: University of Chicago Press.

McKeown, Arthur Philip. 2010. "From Bodhgaya to Lhasa to Beijing: The Life and Times of Sariputra (c. 1335–1426), Last Abbot of Bodhgaya." Ph.D. diss., Harvard University.

McMahan, David L. 2008. *The Making of Buddhist Modernism*. Oxford; New York: Oxford University Press.

Meeks, Lori Rachelle. 2010. *Hokkeji and the Reemergence of Female Monastic Orders in Premodern Japan*. Honolulu: University of Hawaii Press.

Meulenbeld, G. Jan. 1974. *The* Mādhavanidāna *and Its Chief Commentary, Chapters 1–10*. Leiden: Brill.

——. 1999–2002. *A History of Indian Medical Literature*. 5 vols. Groningen Oriental Studies 15. Groningen: E. Forsten.

Meyer, Fernand. 1981. *Gso-ba rig-pa: Le système médical tibétain*. Paris: Centre national de la recherche scientifique.

——. 1992. "Introduction: The Medical Paintings of Tibet." In *Tibetan Medical Paintings: Illustrations to the Blue Beryl Treatise of Sangyé Gyamtso (1653–1705)*, ed. Yuri Parfionovitch, Fernand Meyer, and Gyurme Dorje, 2–13. New York: Abrams.

——. 2003. "The Golden Century of Tibetan Medicine." In *Lhasa in the Seventeenth Century: The Capital of the Dalai Lamas*, ed. Françoise Pommaret, 99–117. Leiden: Brill.

Millard, Colin, and Geoffrey Samuel, eds. Forthcoming. *Perspectives on the History, Theory and Practice of Bon Medicine*. Kathmandu: Vajra Books.

Miller, Willa Blythe. 2013. *Secrets of the Vajra Body: Dngos po'i gnas lugs and the Apotheosis of the Body in the work of Rgyal ba Yang dgon pa*. Ph.D. diss., Harvard University.

Minow, Martha. 1990. *Making All the Difference: Inclusion, Exclusion, and American Law*. Ithaca, NY: Cornell University Press.

Mitra, Jyotir. 1985. *A Critical Appraisal of Āyurvedic Material in Buddhist Literature, with Special Reference to Tripiṭaka*. Varanasi: Jyotiralok Prakashan.

Murthy, K. R. Srikantha, trans. 1991. *Vāgbhaṭa's Aṣṭāṅga Hṛdayam*. 2 vols. Varanasi: Krishnadas Academy.

Nappi, Carla. 2009. *The Monkey and the Inkpot: Natural History and Its Transformations in Early Modern China*. Cambridge, MA: Harvard University Press.

Nārada, Mūla Paṭṭhāna Sayadaw, U., trans. 1969. *Conditional Relations (*Paṭṭhāna*)*. London: Luzac, for the Pali Text Society.

Nattier, Jan. 1991. *Once Upon a Future Time: Studies in a Buddhist Prophecy of Decline*. Berkeley, CA: Asian Humanities Press.

Newman, John. 1996. "Itineraries to Shambhala." In *Tibetan Literature: Studies in Genre*, ed. Roger R. Jackson and José Ignacio Cabezón, 485–99. Ithaca, NY: Snow Lion.

Norman, K. R. 1969. *The Elders' Verses. 1. Theragāthā*. Pali Text Society Translation Series 38. London: Luzac and Co., Ltd. for the Pali Text Society.

Oakley, Ann. 1972. *Sex, Gender, and Society*. London: Maurice Temple Smith.

Obermiller, Eugéne, trans. 1931. *History of Buddhism (*Chos-hbyung*)*. Leipzig: O. Harrassowitz.

Olivelle, Patrick. 2005. *Manu's Code of Law: A Critical Edition and Translation of the* Mānava-Dharmaśāstra. Oxford: Oxford University Press.

Overbey, Ryan Richard. 2010. "Memory, Rhetoric, and Education in the 'Great Lamp of the Dharma Dharaṇi Scripture.'" Ph.D. diss., Harvard University.

Pal, Pratapaditya. 2008. *Ocean of Wisdom, Embodiment of Compassion: Glimpses of Tibetan Buddhist Art: A Selection from the Pritzker Collection*. Chicago: Margot L. and Thomas J. Pritzker.

Parfionovitch, Yuri, Fernand Meyer, and Gyurme Dorje, ed. 1992. *Tibetan Medical Paintings: Illustrations to the Blue Beryl Treatise of Sangyé Gyamtso (1653–1705)*. 2 vols. London: Serindia.

Patrul Rinpoche. 1998. *The Words of My Perfect Teacher*. Trans. Padmakara Translation Group. Boston: Shambhala.

Petech, Luciano. 1944–46. *I Missionari Italiani Nel Tibet e Nel Nepal*. 7 vols. Il Nuovo Ramusio. Roma: Libreria dello Stato.

——. 1972. *China and Tibet in the Early Eighteenth Century: History of the Establishment of the Chinese Protectorate in Tibet*. 2nd rev. ed. Leiden: Brill.

Pingree, David. 1996. "Indian Reception of Muslim Versions of Ptolemaic Astronomy." In *Tradition, Transmission, Transformation: Proceedings of Two Conferences on Pre-modern Science Held at the University of Oklahoma*, ed. F. Jamil Ragep, Sally P. Ragep, and Steven John Livesey, 471–85. Leiden: Brill.

Pittman, Don Alvin. 2001. *Toward a Modern Chinese Buddhism: Taixu's Reforms*. Honolulu: University of Hawaii Press.

Pocock, J. G. A. 1999–2005. *Barbarism and Religion*. 4 vols. Cambridge: Cambridge University Press.

Pollock, Sheldon. 1998. "India in the Vernacular Millennium: Literary Culture and Polity, 1000–1500." *Daedalus* 127 (3): 41–74.

——. 2000. "Indian Knowledge Systems on the Eve of Colonialism." *Intellectual History Newsletter* 22: 1–16.

——. 2001a. "The Death of Sanskrit." *Comparative Studies in History and Society* 43 (2): 392–426.

——. 2001b. "New Intellectuals in Seventeenth-century India." *Indian Economic & Social History Review* 38 (1): 3–31.

——. 2005. *The Ends of Man at the End of Premodernity*. Amsterdam: Royal Netherlands Academy of Arts and Sciences.

——. 2006. *The Language of the Gods in the World of Men: Sanskrit, Culture, and Power in Premodern India*. Berkeley: University of California Press.

——. 2011. *Forms of Knowledge in Early Modern Asia: Explorations in the Intellectual History of India and Tibet, 1500–1800*. Durham, NC: Duke University Press.

Pommaret-Imaeda, Françoise. 2003. "Chapter Twelve: Scholars, Artists and Feasts." In *Lhasa in the Seventeenth Sentury: The Capital of the Dalai Lamas*, ed. Françoise Pommaret-Imaeda, 179–98. Leiden: Brill.

Pomplun, Trent. 2010. *Jesuit on the Roof of the World: Ippolito Desideri's Mission to Eighteenth-century Tibet*. Oxford: Oxford University Press.

Prets, Ernst. 2000. "Theories of Debate, Proof and Counter-Proof in the Early Indian Dialectical Tradition." In *On the Understanding of Other Cultures*, ed. Piotr Blcerowicz and Marek Mejor, vol. 7, 333–46. Warszawa: Instytut Orientalistyczny, Uniwersytet Warszawski.

Rabgay, Lobsang. 1981a. "Pulse Analysis in Tibetan Medicine." *Tibetan Medicine* (Gso rig) 3: 45–52.

——. 1981b. "Urine Analysis in Tibetan Medicine." *Tibetan Medicine* (Gso rig) 3: 53–60.

Rangdrol, Tsele Natsok. 1993. *Empowerment and the Path of Liberation*. Trans. Erik Pema Kunsang. Hong Kong: Rangjung Yeshe.

Rechung Rinpoche. 1976. *Tibetan Medicine: Illustrated in Original Texts*. Berkeley: University of California Press.

rGya-mTSHo, Saṅs-Rgyas. 1999. *Life of the Fifth Dalai Lama*, Volume IV, Part 2. Trans. Zahiruddin Ahmad. New Delhi: International Academy of Indian Culture and Aditya Prakashan.

Rhys Davids, T. W., trans. 1977. *Dialogues of the Buddha* (Dīgha-nikāya). Vol. 1. 1910; reprint, London: Pali Text Society.

Richardson, Hugh. 1977. "The Dharma That Came Down from Heaven: A Tun-huang Fragment." In *Buddhist Thought and Asian Civilization: Essays in Honor of Herbert V. Guenther on His Sixtieth Birthday*, ed. Leslie S. Kawamura and Keith Scott, 219–29. Emeryville, CA: Dharma Publishing.

——. 1987. "Early Tibetan Inscriptions: Some Recent Discoveries." *Bulletin of Tibetology* 23 (3): 5–18.

——. 1998. "The Decree Appointing Sangs-rgyas Rgya-mtsho as Regent." In *High Peaks, Pure Earth: Collected Writings on Tibetan History and Culture*, ed. Michael Aris, 440–61. 1980; reprint, London: Serindia.

Roerich, George, trans. 1976. *The Blue Annals*. 1949–50; reprint, Delhi: Motilal Banarsidass.

Ruch, Barbara. 1990. "The Other Side of Culture in Medieval Japan." In *The Cambridge History of Japan, Vol. 3: Medieval Japan*, ed. Kozo Yamamura, 500–43. Cambridge: Cambridge University Press.

——, ed. 2002. *Engendering Faith: Women and Buddhism in Premodern Japan*. Michigan Monograph Series in Japanese Studies 43. Ann Arbor: Center for Japanese Studies, University of Michigan.

Ruegg, David Seyfort. 1995. *Ordre spirituel et ordre temporel dans la pensée bouddhique de l'Inde et du Tibet: quatre conférences au Collège de France*. Paris: Collège de France.

Salguero, Pierce. 2010. "Buddhist Medicine in Medieval China: Disease, Healing, and the Body in Crosscultural Translation (Second to Eighth Centuries C.E.)." Ph.D. diss., Johns Hopkins University.

——. 2013. "'On Eliminating Disease': Translations of the Medical Chapter from the Chinese Versions of the *Sutra of Golden Light*." *eJournal of Indian Medicine* 6: 21–43, http://www.indianmedicine.nl/.

——. 2014. *Chinese Buddhist Medicine: Disease, Healers, and the Body in Crosscultural Translation*. Philadelphia: University of Pennsylvania Press.

Samuel, Geoffrey. 1993. *Civilized Shamans: Buddhism in Tibetan Societies*. Washington, DC: Smithsonian Institution Press.

Sangpo, Khetsun. 1974. *Gtam dpe sna tshogs daṅ Gźas tshig khaśas: A Collection of Proverbs and Various Poetical Pieces Written on Various Occasions*. Dharamsala: Library of Tibetan Works and Archives.

Śāntideva. 1995. *Bodhicaryāvatāra*. Trans. Kate Crosby and Andrew Skilton. Oxford: Oxford University Press.

Sasaki, R., ed. 1965. *Mahāvyutpatti*. 2 vols. Tokyo: Suzuki Gakujutsu Zaidan.

Savage-Smith, Emilie. 1997. "The Depiction of Human Anatomy in the Islamic World." In *Science, Tools and Magic: Body and Spirit, Mapping the Universe*, ed. Francis Maddison and Emilie Savage-Smith, vol. 12, 14–24. London: The Nour Foundation in association with Azimuth Editions and Oxford University Press.

Schaeffer, Kurtis R. 2003. "Textual Scholarship, Medical Tradition, and Mahāyāna Buddhist Ideals in Tibet." *Journal of Indian Philosophy* 31 (5–6): 621–41.

——. 2004. *Himalayan Hermitess: The Life of a Tibetan Buddhist Nun*. Oxford: Oxford University Press.

—. 2006. "Ritual, Festival and Authority Under the Fifth Dalai Lama." In *Power, Politics, and the Reinvention of Tradition: Tibet in the Seventeenth and Eighteenth Centuries (Proceedings of the Tenth Seminar of the IATS, 2003, 3)*, ed. Bryan J. Cuevas and Kurtis R. Schaeffer, 187–202. Leiden: Brill.

—. 2007. "Death, Prognosis, and the Physician's Reputation in Tibet." In *Heroes and Saints: The Moment of Death in Cross-cultural Perspectives*, ed. Phyllis Granoff and Koichi Shinohara, 159–72. Newcastle-upon-Tyne, U. K.: Cambridge Scholars.

—. 2009. *The Culture of the Book in Tibet*. New York: Columbia University Press.

—. 2011. "New Scholarship in Tibet, 1650–1700." In *Forms of Knowledge in Early Modern Asia: Explorations in the Intellectual History of India and Tibet, 1500–1800*, ed. Sheldon Pollock, 291–310. Durham, NC: Duke University Press.

Schaeffer, Kurtis R., Matthew T. Kapstein, and Gray Tuttle, eds. 2013. *Sources of Tibetan Tradition*. New York: Columbia University Press.

Schapiro, Meyer. 1996. *Words, Script, and Pictures: Semiotics of Visual Language*. New York: G. Braziller.

Scheier-Dolberg, Joseph. 2005. "Treasure House of Tibetan Culture: Canonization, Printing, and Power in the Derge Printing House." M.A. thesis, Harvard University.

Schopen, Gregory Robert. 1978. "The Bhaisajyaguru-Sutra and the Buddhism of Gilgit." Ph.D. diss., The Australian National University.

Schuh, Dieter. 1970. "Studien Zur Geschichte Der Mathematik Und Astronomie in Tibet. Teil 1: Elementare Arithmetik." *Zentralasiatische Studien Des Seminars Für Sprach- Und Kulturwissenschaft Zentralasiens Der Universität Bonn* 4: 81–181.

—. 1972. "Über die Möglichkeit der Identifizierung tibetischer Jahresangaben anhand der sMe-ba dgu." *Zentralasiatische Studien des Seminars für Sprach- und Kulturwissenschaft Zentralasiens der Universität Bonn* 6: 485–504.

—. 1973a. "Die Darlegungen des 5. Dalai Lama Ngag dbang blo bzang rgya mtsho zur Kalkulation der neun sMe-ba." *Zentralasiatische Studien des Seminars für Sprach- und Kulturwissenschaft Zentralasiens der Universität Bonn* 7: 285–99.

—. 1973b. "Der chinesische Steinkreis. Ein Beitrag zur Kenntnis der sino-tibetischen Divinationskalkulationen." *Zentralasiatische Studien des Seminars für Sprach- und Kulturwissenschaft Zentralasiens der Universität Bonn* 7: 353–423.

—. 1973c. *Untersuchungen Zur Geschichte Der Tibetischen Kalenderrechnung*. Vol. 16. Verzeichnis Der Orientalischen Handschriften in Deutschland. Supplementband. Wiesbaden: F. Steiner.

—. 1974. "Grundzüge der Entwicklung der tibetischen Kalenderrechnung." *Zeitschrift der Deutschen Morgenländischen Gesellschaft, Supplement* II (XVIII): 554–66.

—. 1978. "Ergebnisse und Aspekte tibetischer Urkundenforschung." In *Proceedings of the Csoma de Kőrös Memorial Symposium Held at Mátrafüred, Hungary 24–30 September 1976*, vol. 1, 411–25. Budapest.

—. 1981a. *Grundlagen Tibetischer Siegelkunde: Eine Untersuchung Über Tibetische Siegelaufschriften in ʾPhags-pa-Schrift*. Monumenta Tibetica Historica. Abteilung III, Diplomata Et Epistolae 5. Sankt Augustin: VGH Wissenschaftsverlag.

—. 1981b. "Zum Entstehungsprozess von Urkunden in den Tibetischen Herrscherkanzleien." In *Contributions on Tibetan Language, History and Culture, Proceedings of the Csoma de Kőrös Symposium held at Velm-Vienna, Austria, 13–19 September, 1981*, vol. 1, 303–28. Budapest.

——. 2004. "Politik und Wissenschaft in Tibet im 13. und 17. Jahrhundert (Politics and Science in Tibet in the 13th and the 17th Century)." *Zentralasiatische Studien des Seminars für Sprach- und Kulturwissenschaft Zentralasiens der Universität Bonn* 33: 1–23.

Scott, Joan W. 1993. "How Did the Male Become the Normative Standard for Clinical Drug Trials?" *Food and Drug Law Journal* 48: 187–93.

Selby, Martha Ann. 2005. "Narratives of Conception, Gestation, and Labour in Sanskrit Āyurvedic Texts." *Asian Medicine* 1 (2): 254–75.

Senart, Emile. 1875. *Essai sur la Légende du Buddha: Son Caractère et ses Origines*. Paris: Imprimerie nationale.

Shakabpa, Tsepon W. D. 1967. *Tibet, A Political History*. New Haven: Yale University Press.

Shapin, Steven. 1994. *A Social History of Truth: Civility and Science in Seventeenth-century England*. Science and Its Conceptual Foundations. Chicago: University of Chicago Press.

——. 1996. *The Scientific Revolution*. Chicago: University of Chicago Press.

——. 2010. "Cordelia's Love: Credibility and the Social Studies of Science." In *Never Pure: Historical Studies of Science as If It Was Produced by People with Bodies, Situated in Time, Space, Culture, and Society, and Struggling for Credibility and Authority*, 17–31. Baltimore: Johns Hopkins University Press.

Sharma, Priya Vrat, trans. 2000. *Suśruta-saṃhitā: With English Translation of Text and Ḍalhaṇa's Commentary Along with Critical Notes*. Varanasi: Chaukhambha Visvabharati.

Sharma, Ram Karan, and Vaidya Bhagwan Dash, trans. 1976. *Agniveśa's Caraka Saṃhitā*. 2 vols. Varanasi: Chowkhamba Sanskrit Series Office.

Shastri, Swami Dwarikadas. 1970–73. *Abhidharmakośam and Bhāsya of Acharya Vasubandhu with Sphutārtha Commentary of Ācārya Yaśomitra*. 4 vols. Varanasi: Bauddha Bharati.

Smith, E. Gene. 2001a. "A Tibetan Encyclopedia from the Fifteenth Century." In *Among Tibetan Texts: History and Literature of the Himalayan Plateau*, ed. Kurtis R. Schaeffer, 209–24. Boston: Wisdom.

——. 2001b. "Jam Mgon Kong Sprul and the Nonsectarian Movement." In *Among Tibetan Texts: History and Literature of the Himalayan Plateau*, ed. Kurtis R. Schaeffer, 235–72. Boston: Wisdom.

Sperling, Elliot. 2003. "Tibet's Foreign Relations During the Epoch of the Fifth Dalai Lama." In *Lhasa in the Seventeenth Century: The Capital of the Dalai Lamas*, ed. Françoise Pommaret, 119–32. Leiden: Brill.

Stede, William, and T. W. Rhys Davids, eds. 1997. *Pali-English Dictionary*. 2nd ed. Delhi: Motilal Banarsidass.

Stein, Rolf A. 1961. *Une chronique ancienne de bSam-yas, sBa-bžed*. Paris: Institut des hautes études chinoises.

——. 1972. *Tibetan Civilization*. Trans. J. E. Stapleton Driver. Stanford, CA: Stanford University Press.

——. 2010. *Rolf Stein's* Tibetica Antiqua*: With Additional Materials*. Trans. Arthur P. McKeown. Leiden: Brill.

Sterckx, Roel. 2002. *The Animal and the Daemon in Early China*. Albany: State University of New York Press.

——. 2008. "The Limits of Illustration: Animalia and Pharmacopeia from Guo Pu to Bencao Gangmu." *Asian Medicine* 4 (2): 357–94.

Subrahmanyam, Sanjay. 1997. "Connected Histories: Notes Towards a Reconfiguration of Early Modern Eurasia." *Modern Asian Studies* 31 (3) (July 1): 735–62.

——. 1998. "Hearing Voices: Vignettes of Early Modernity in South Asia, 1400–1750." *Daedalus* 127 (3): 75–104.

Takakusu, Junijrō, and Makoto Nagai, trans. 1999. *Samantapāsādikā: Buddhaghosa's Commentary on the Vinaya Piṭaka.* London: Oxford University Press for the Pali Text Society.

Taube, Manfred. 1980. "Tibetische Autoren Zur Geschichte Der rGyud-bzhi." *Acta Orientalia Academiae Scientiarum Hungaricae* 34: 297–304.

——. 1981. *Beiträge zur Geschichte der medizinischen Literatur Tibets.* Sankt Augustin: VGH Wissenschaftsverlag.

Teiser, Stephen F. 2006. *Reinventing the Wheel: Paintings of Rebirth in Medieval Buddhist Temples.* Seattle: University of Washington Press.

Thondup, Tulku, and Harold Talbott. 1986. *Hidden Teachings of Tibet: An Explanation of the Terma Tradition of the Nyingma School of Buddhism.* London: Wisdom.

Thon-mi Sambhoṭa. 1973. *Sum cu pa dan Rtags kyi jug pa bzugs so.* Ed. Zuiho Yamaguchi. Tokyo: Toyo Bunko.

Tominaga Nakamoto. 1990. *Emerging from Meditation.* Trans. Michael Pye. London: Duckworth.

Toulmin, Stephen Edelston. 1990. *Cosmopolis: The Hidden Agenda of Modernity.* New York: Free Press.

Townsend, Dominique. 2012. "Materials of Buddhist Culture: Aesthetics and Cosmopolitanism at Mindroling Monastery." Ph.D. diss., Columbia University.

Trésors du Tibet: Région autonome du Tibet/Chine. 1987. Paris: Muséum National d'Histoire Naturelle.

Tsering, Tashi. 2005. "Outstanding Women in Tibetan Medicine." In *Women in Tibet,* ed. Janet Gyatso and Hanna Havnevik, 169–94. New York: Columbia University Press.

Tsuda, Shiníchi, ed. 1974. *The Saṁvarodaya-tantra: Selected Chapters.* Tokyo: Hokuseido Press.

Tsyrempilov, Nicolay. 2006. "Dge Lugs Pa Divided: Some Aspects of the Political Role of Tibetan Buddhism in the Expansion of the Qing Dynasty." In *Power, Politics, and the Reinvention of Tradition: Tibet in the Seventeenth and Eighteenth Centuries (Proceedings of the Tenth Seminar of the IATS, 2003, 3),* ed. Bryan J. Cuevas and Kurtis R. Schaeffer, 47–66. Leiden: Brill.

Tucci, Giuseppe. 1936. *Indo-tibetica. III, parte II. Templi del Tibet occidentale e il loro simbolismo artistico.* Roma: Reale accademia d'Italia.

——. 1949. *Tibetan Painted Scrolls.* Roma: Libreria dello Stato.

——. 1971. *Minor Buddhist Texts: Part III, Third Bhāvanākrama.* Roma: Istituto Italiano per il medio ed estremo oriente.

——. 1980. *The Religions of Tibet.* Trans. Geoffrey Samuel. Berkeley: University of California Press.

——. 1986. *Minor Buddhist Texts: Parts One and Two.* 1956–58; reprint, Delhi: Motilal Banarsidass.

Tuttle, Gray. 2006. "A Tibetan Buddhist Mission to the East: The Fifth Dalai Lama's Journey to Beijing 1652–1653." In *Power, Politics, and the Reinvention of Tradition: Tibet in the*

Seventeenth and Eighteenth Centuries (Proceedings of the Tenth Seminar of the IATS, 2003, 3), ed. Bryan J. Cuevas and Kurtis R. Schaeffer, 65–87. Leiden: Brill.

——. 2011. "Challenging Central Tibet's Dominance of History: The Oceanic Book, a Nineteenth Century Politico-religious Geographic History." In *Mapping the Modern in Tibet*, ed. Gray Tuttle, 135–72. Beiträge Zur Zentralasienforschung 24. Andiast, Switzerland: International Institute for Tibetan and Buddhist Studies GmbH.

Ui, Hakuju et al. 1934. *A Complete Catalogue of the Tibetan Buddhist Canons.* Sendai, Japan: Tôhoku Imperial University.

Uray, Géza. 1972. "The Narrative of Legislation and Organization of the Mkhas-pa'i Dga'-ston." *Acta Orientalia Academiae Scientiarum Hungaricae* 26 (1): 11–68.

Veith, Ilza. 1960. *Medizin in Tibet.* Leverkusen: Bayer.

Vernant, Jean-Pierre. 1989. "Dim Body, Dazzling Body." In *Fragments for a History of the Human Body*, ed. Michel Feher, Ramona Naddaff, and Nadia Tazi, trans. Anne M. Wilson, 19–47. New York, Cambridge, MA: Zone Books.

Vitali, Roberto. 1990. *Early Temples of Central Tibet.* London: Serindia.

Vogel, Claus. 1963. "On Bu-ston's View of the Eight Parts of Indian Medicine." *Indo-Iranian Journal* 6 (3–4): 290–95.

Vovelle, Michel. 1990. *Ideologies and Mentalities.* Trans. Eamon O'Flaherty. Chicago: University of Chicago Press.

Wallace, B. Alan. 2003. *Buddhism and Science: Breaking New Ground.* New York: Columbia University Press.

——. 2007. *Contemplative Science: Where Buddhism and Neuroscience Converge.* New York: Columbia University Press.

Wallace, Vesna A., trans. 2010. *The Kālacakra Tantra: The Chapter on Sādhanā, Together with the Vimalaprabhā Commentary.* New York: American Institute of Buddhist Studies.

Walsh, E. H. C. 1910. "XXIX. The Tibetan Anatomical System." *Journal of the Royal Asiatic Society (New Series)* 42 (4): 1215–45.

Walshe, Maurice, trans. 1987. *Thus Have I Heard: The Long Discourses of the Buddha Dīgha Nikāya.* London: Wisdom.

Wedemeyer, Christian. 2011. "Locating Tantric Antinomianism: An Essay Toward an Intellectual History of the 'Practices/Practice Observance' (*Caryā/Caryāvrata*)." *Journal of the International Association of Buddhist Studies* 34 (1–2): 349–419.

——. 2013. *Making Sense of Tantric Buddhism: History, Semiology, and Transgression in the Indian Traditions.* New York: Columbia University Press.

Weintraub, Karl Joachim. 1978. *The Value of the Individual: Self and Circumstance in Autobiography.* Chicago: University of Chicago Press.

White, David Gordon. 1996. *The Alchemical Body: Siddha Traditions in Medieval India.* Chicago: University of Chicago Press.

——. 2003. *Kiss of the Yoginī: "Tantric Sex" in Its South Asian Contexts.* Chicago: University of Chicago Press.

Williamson, Laila, and Serinity Young, ed. 2009. *Body and Spirit: Tibetan Medical Paintings.* New York: American Museum of Natural History, in association with University of Washington Press.

Wood, James. 2005. "The Blue River of Truth." *The New Republic*, August 1, 23–27.

Wu, Wi-Li. 2008. "The Gendered Medical Iconography of the Golden Mirror (*Yuzuan Yizong Jinjian*, 1742)." *Asian Medicine* 4 (2): 452–91.

Wujastyk, Dagmar. 2012. *Well-Mannered Medicine: Medical Ethics and Etiquette in Classical Ayurveda*. New York: Oxford University Press.

Wujastyk, Dominik. 2005. "Change and Creativity in Early Modern Indian Medical Thought." *Journal of Indian Philosophy* 33 (1): 95–118.

Wylie, Turrell V. 1962. *The Geography of Tibet According to the* 'Dzam-gling-rGyas-bShad. Roma: Istituto italiano per il Medio ed Estremo Oriente.

——, ed. 1970. *A Tibetan Religious Geography of Nepal*. Roma: Instituto italiano per il Medio ed Estremo Oriente.

Xu, Xinguo. 2006. *Western China and the Eastern and Western Civilizations* (*Xi chui zhi di yu dong xi fang wen ming*). Beijing: Beijing yan shan chu ban she.

Yamada, Isshi. 1968. *Karuṇāpuṇḍarīka: The White Lotus of Compassion*. 2 vols. London: School of Oriental and African Studies.

Yamaguchi, Zuiho. 1999. "The Emergence of the Regent Sangs rgyas rgya mtsho and the Denouement of the Dalai Lama's First Administration." *Memoirs of the Research Department of the Tōyō Bunko* 57: 113–36.

Yang Ga. 2010. "The Sources for the Writing of the 'Rgyud bzhi,' Tibetan Medical Classic." Ph.D. diss., Harvard University.

——. 2014. "Sources for the Tibetan Medical Classic *Four Tantras*." In *Bodies in Balance: The Art of Tibetan Medicine*, ed. Theresia Hofer, 154–77. Seattle: University of Washington Press.

Yates, Frances Amelia. 1966. *The Art of Memory*. Chicago: University of Chicago Press.

Yoeli-Tlalim, Ronit. 2004. "Contemporary Oral Teachings of Kālacakra in Exile: a Dialogue Between Tradition and Change." Ph.D. diss., University of London, School of Oriental and African Studies.

——. 2010a. "Islam and Tibet: Cultural Interactions—an Introduction." In *Islam and Tibet: Interactions Along the Musk Routes*, ed. Anna Akasoy, Charles S. F. Burnett, and Ronit Yoeli-Tlalim. Surrey: Ashgate.

——. 2010b. "On Urine Analysis and Tibetan Medicine's Connections with the West." In *Studies of Medical Pluralism in Tibetan History and Society: PIATS 2006: Proceedings of the Eleventh Seminar of the International Association for Tibetan Studies, Königswinter 2006*, ed. Sienna R. Craig, M. Cuomo, Frances Garrett, and Mona Schrempf. Andiast, Switzerland: International Institute for Tibetan and Buddhist Studies GmbH.

——. 2012a. "Re-visiting 'Galen in Tibet.'" *Medical History* 56 (3): 355–65.

——. 2012b. "Central Asian Melange: Early Tibetan Medicine from Dunhuang." In *Scribes, Texts, and Rituals in Early Tibet and Dunhuang*, ed. Brandon Dotson, Kazushi Iwao, and Tsuguhito Takeuchi, 53–60. Wiesbaden: Reichert Verlag.

Yongdan, Lobsang. 2011. "Tibet Charts the World: Btsan Po No Mon Han's *The Detailed Description of the World*, an Early Major Scientific Work in Tibet." In *Mapping the Modern in Tibet*, ed. Gray Tuttle, 73–134. Beiträge Zur Zentralasienforschung 24. Andiast, Switzerland: International Institute for Tibetan and Buddhist Studies GmbH.

Yonten, Pasang, Arya. 2014. "External Therapies in Tibetan Medicine—The *Gyushi*, Contemporary Practice, and a Preliminary History of Tibetan Medical Surgery." In *Bodies in Balance: The Art of Tibetan Medicine,* ed. Theresia Hofer, 64–89. Seattle: University of Washington Press.

Ziegler, Vereena. 2010. "A Preliminary Report on the Life of Buddha Śākyamuni in the Murals of the Circumambulatory of the Prajñāpāramitā Chapel in Zha lu." In *The Art of Tibetan Painting: Recent Research on Manuscripts, Murals and Thangkas of Tibet, the Himalayas and Mongolia (11th–19th Century)*, ed. Amy Heller. Proceedings of the Twelfth Seminar of the International Association for Tibetan Studies in Vancouver, 2010. *Asianart.com.* http://www.asianart.com/articles/iats/index.html.

Zwilling, Leonard. 1992. "Homosexuality as Seen in Indian Buddhist Texts." In *Buddhism, Sexuality, and Gender*, ed. José Ignacio Cabezón, 203–14. Albany: State University of New York Press.

Zysk, Kenneth G. 1991. *Asceticism and Healing in Ancient India: Medicine in the Buddhist Monastery.* New York: Oxford University Press.

INDEX

Page numbers in *italics* indicate illustrations.